Progress in Probability

Volume 56

Stochastic Inequalities and Applications

Evariste Giné
Christian Houdré
David Nualart
Editors

Springer Basel AG

Editors' addresses:

Evariste Giné
Department of Mathematics, U-3009
University of Connecticut
Storrs, CT 06268
USA
gine@uconnvm.uconn.edu

David Nualart
Universitat de Barcelona
Facultat de Matemàtiques
Gran Via, 585
08007 Barcelona
Spain
nualart@mat.ub.es

Christian Houdré
Laboratoire d'Analyse et de
Mathématiques Appliquées
CNRS UMR 8050
Université de Paris XII
94010 Créteil Cedex, France

and

School of Mathematics
Georgia Institute of Technology
Atlanta, GA 30332
USA
houdre@math.gatech.edu

2000 Mathematics Subject Classification 28C10, 35B65, 37A50, 37H15, 44A60, 46B09, 49R50, 58C35, 58J65, 60B11, 60B99, 60E15, 60F05, 60F07, 60F10, 60F99, 60Gxx, 60Hxx, 62B10, 62E17, 62E20, 62F12, 62G30, 65M12, 65M15, 93D20, 93E11, 94A17

A CIP catalogue record for this book is available from the Library of Congress, Washington D.C., USA

Bibliographic information published by Die Deutsche Bibliothek
Die Deutsche Bibliothek lists this publication in the Deutsche Nationalbibliografie; detailed bibliographic data is available in the Internet at <http://dnb.ddb.de>.

ISBN 978-3-0348-9428-9 ISBN 978-3-0348-8069-5 (eBook)
DOI 10.1007/978-3-0348-8069-5

Originally published by Birkhäuser Verlag Basel in 2003
Softcover reprint of the hardcover 1st edition 2003
Printed on acid-free paper produced from chlorine-free pulp. TCF ∞
ISBN 978-3-0348-9428-9

9 8 7 6 5 4 3 2 1 www.birkhäuser-science.com

Table of Contents

Preface vii

I. Geometric Inequalities

Sergey G. Bobkov and Fedor L. Nazarov
Large Deviations of Typical Linear Functionals on a Convex Body with Unconditional Basis 3

Christian Houdré and Nicolas Privault
A Concentration Inequality on Riemannian Path Space 15

Ioannis Kontoyiannis and Ali Devin Sezer
A Remark on Unified Error Exponents: Hypothesis Testing, Data Compression and Measure Concentration 23

Paul-Marie Samson
Concentration Inequalities for Convex Functions on Product Spaces .. 33

II. Independent Random Vectors, Chaos, Martingales and Lévy Processes

Christian Houdré and Patricia Reynaud-Bouret
Exponential Inequalities, with Constants, for U-statistics of Order Two 55

Stanisław Kwapień and Vaja Tarieladze
On a.s. Unconditional Convergence of Random Series in Banach Spaces 71

Rafał Latała and Rafał Łochowski
Moment and Tail Estimates for Multidimensional Chaoses Generated by Positive Random Variables with Logarithmically Concave Tails ... 77

Malwina J. Luczak
A Quantitative Law of Large Numbers via Exponential Martingales... 93

Michael B. Marcus and Jan Rosiński
Sufficient Conditions for Boundedness of Moving Average Processes .. 113

Laurent Miclo
Notes on the Speed of Entropic Convergence in the Central Limit Theorem 129

Krzysztof Oleszkiewicz
On a Nonsymmetric Version of the Khinchine-Kahane Inequality 157

Iosif Pinelis
Dimensionality Reduction in Extremal Problems for Moments of Linear Combinations of Vectors with Random Coefficients 169

III. Empirical Processes

Miguel A. Arcones
Moderate Deviations of Empirical Processes ... 189

Olivier Bousquet
Concentration Inequalities for Sub-Additive Functions Using the Entropy Method ... 213

Evarist Giné, Vladimir Koltchinskii and Jon A. Wellner
Ratio Limit Theorems for Empirical Processes ... 249

Erich Haeusler and David M. Mason
Asymptotic Distributions of Trimmed Wasserstein Distances Between the True and the Empirical Distribution Functions ... 279

IV. Stochastic Differential Equations

István Gyöngy and Nicolai Krylov
On the Rate of Convergence of Splitting-up Approximations for SPDEs ... 301

Arturo Kohatsu-Higa
Lower Bounds for Densities of Uniformly Elliptic Non-homogeneous Diffusions ... 323

C. W. Li
Lyapunov Exponents of Nonlinear Stochastic Differential Equations with Jumps ... 339

David Nualart and Youssef Ouknine
Stochastic Differential Equations with Additive Fractional Noise and Locally Unbounded Drift ... 353

Preface

Stochastic inequalities have always been a fundamental building block of Probability Theory. One thinks, for instance, of the basic Kolmogorov and Lévy's inequality to deal with the asymptotic theory for sums of independent random variables. One also thinks of the fundamental Azuma-Hoeffding inequalities whose range of applications is impressive.

Striking advances on stochastic inequalities have taken place in recent years. Differential and geometric techniques have been borrowed from other areas and successfully incorporated in probability theory. This leads, for example, to concentration inequalities for various types of processes, somewhat analogous to the classical inequalities but with a much greater range of applicability. New techniques have been combined with refined classical ones and have given rise to sharp moment inequalities for sums of independent random vectors, as well as martingale and decoupling inequalities and inequalities in the Malliavin Calculus. These advances have had a tremendous impact in discrete mathematics, mathematical statistics, convex geometry and, of course, probability theory.

The collection of papers in the present volume reflects the breadth and depth of the subject, with many new results in several areas. The topics include, among others, entropic, large deviations and concentration inequalities for different types of processes, with applications; moment and exponential inequalities, in particular for processes of Rademacher type, chaos type, martingales, diffusion densities, moving averages; stochastic differential and partial differential equations; and random series, to name a few.

This volume is representative of the contributions presented at a Euroconference on *Stochastic Inequalities and their Applications* held in Barcelona, June 18–22, 2002. The participants should be congratulated on their fine contributions.

This conference was partly an offspring of a first conference of this type that was held at the Georgia Institute of Technology in Atlanta in 1997, as a AMS Special Session. The proceedings, *Advances in Stochastic Inequalities*, appeared in Contemporary Mathematics, 234, AMS, Providence, Rhode Island, 1999. Might we dream in thinking that the baton could be passed and that someone might organize a follow up in 2007, if not before!

The meeting was organized by the Centre de Recerca Matemàtica and received generous support from the European Union, the Generalitat de Catalunya and the Spanish Government, which we gratefully acknowledge. We owe special thanks to the Director of the CRM, Manuel Castellet, and to its staff, Maria Julià, Neus Portet and Consol Roca, for excellent planning and organization.

Evarist Giné
Christian Houdré
David Nualart
May, 2003

Part I

Geometric Inequalities

Progress in Probability, Vol. 56, 3–13

Large Deviations of Typical Linear Functionals on a Convex Body with Unconditional Basis

Sergey G. Bobkov and Fedor L. Nazarov

Abstract. We study large deviations of linear functionals on an isotropic convex set with unconditional basis. It is shown that suitably normalized ℓ_1-balls play the role of extremal bodies.

1. Introduction

Let K be a convex body in $\mathbf{R}^n$ with the properties:

1) $\mathrm{vol}_n(K) = 1$, where vol_n stands for the Lebesgue measure;
2) given $x \in K$, $y \in \mathbf{R}^n$ such that $|y_j| \leq |x_j|$, for all $j \leq n$, we have $y \in K$;
3) the integrals

$$\int_K x_j^2\, dx = L_K^2$$

do not depend on $j \leq n$.

By the assumption 2), the set K is centrally symmetric, and moreover, the canonical basis in $\mathbf{R}^n$ is unconditional for the norm associated to K. Under 2), the normalizing assumption 3) defines K as an isotropic body. This means, that linear functionals

$$f_\theta(x) = \theta_1 x_1 + \cdots + \theta_n x_n, \quad x \in \mathbf{R}^n,$$

parameterized by unit vectors $\theta = (\theta_1, \dots, \theta_n)$ have $L^2(K)$-norm equal to L_K.

Due to the hypotheses $1) - 3)$ on K, the quantity L_K satisfies $c_1 \leq L_K \leq c_2$, for some absolute constants $c_1, c_2 > 0$ (cf. [2]). Moreover, according to Borell's lemma ([4], Lemma 3.1), $L^p(K)$-norms of f_θ are at most Cp, for all $p \geq 1$ and some numerical constant C. This can be written in terms of the Young function $\psi_1(t) = e^{|t|} - 1$ and the corresponding Orlicz norm on K as one inequality

$$\|f_\theta\|_{\psi_1} \leq C_1.$$

Key words and phrases. Convex bodies, unconditional basis, Gaussian tails, comparison theorem.
Supported in part by NSF grants.

A natural general question in this direction (regarding of the unconditionality assumption) is how to determine whether or not, for some unit vector θ, or moreover, for most of them, we have a stronger inequality

$$\|f_\theta\|_{\psi_2} \leq C_2 \tag{1.1}$$

with respect to the Young function $\psi_2(t) = e^{|t|^2} - 1$. The inequality (1.1) is equivalent to the property that f_θ admits a gaussian bound on the distribution of tails,

$$\mathrm{vol}_n\{x \in K : |f_\theta(x)| \geq t\} \leq 2\, e^{-t^2/C}, \quad t \geq 0$$

(with C proportional to C_2). The study of this question was initiated by J. Bourgain [3] who related it to the slicing problem in Convex Geometry. While for this problem it is important to know how to control $\sup_\theta \|f_\theta\|_{\psi_2}$ as a quantity depending on K, it turns out non-trivial to see in general whether the inequality (1.1) holds true for at least one vector θ with a universal C_2 (a question posed and propagandized over the years by V. D. Milman). Recently, G. Paouris studied the problem for several families of isotropic bodies including zonoids and those that are contained in the Euclidean ball of radius of order $\sqrt{n}$. See [7] where one can also find further references and comments on the relationship to the slicing problem. In [1], it is shown that, under the hypotheses 1) – 3), the inequality (1.1) holds always true for the main direction, that is, for the functional

$$f(x) = \frac{x_1 + \cdots + x_n}{\sqrt{n}}, \quad x \in \mathbf{R}^n.$$

In this paper we suggest another approach to this result which allows one to involve into consideration arbitrary linear functionals f_θ and thus to study their possible behavior on average.

Theorem 1.1. *For every vector $\theta \in \mathbf{R}^n$,*

$$\|f_\theta\|_{\psi_2} \leq 4\, \|\theta\|_\infty \sqrt{3n}. \tag{1.2}$$

Here, $\|\theta\|_\infty = \max_{j \leq n} |\theta_j|$. The inequality (1.2) may be applied to f itself which yields (1.1) with a dimension free constant.

Up to an absolute factor, the right hand in (1.2) cannot be improved. This can be shown on the example of the normalized ℓ^1-balls, see Proposition 2.1 below. On the other hand, the average value of $\|\theta\|_\infty \sqrt{n}$ with respect to the uniform measure σ_{n-1} on the unit sphere S^{n-1} is about $\sqrt{\log n}$. Therefore, one cannot hope that (1.1) will hold for most of the unit vectors in the sense of σ_{n-1}, so other norms or rates for distribution tails have to be examined in order to describe the (worst) typical behavior of linear functionals on K.

Theorem 1.2. *There exist positive numerical constants c_1, c_2 and t_0 with the following property. For all $\theta \in S^{n-1}$ except possibly for a set of σ_{n-1}-measure at most n^{-c_1},*

$$\mathrm{vol}_n\{x \in K : |f_\theta(x)| \geq t\} \leq \exp\left\{-\frac{c_2\, t^2}{\log t}\right\}, \quad t \geq t_0. \tag{1.3}$$

Moreover, c_1 can be chosen arbitrarily large at the expense of suitable c_2 and t_0.

Thus, in the worst case, the tails of f_θ are "almost" Gaussian. In particular, for most unit vectors, we have a weakened version of (1.1),

$$\|f_\theta\|_{\psi_\alpha} \leq C_\alpha,$$

which is fulfilled for all $\alpha \in [1,2)$ with respect to the Young functions $\psi_\alpha(t) = e^{|t|^\alpha} - 1$ (with C_α depending on α, only).

Introduce the unit ball of the space ℓ_1^n,

$$B_1 = \{x \in \mathbf{R}^n : |x_1| + \cdots + |x_n| \leq 1\}.$$

It is known that the basic assumptions $1) - 3)$ imply a set inclusion $K \subset CnB_1$, for some numerical C. This fact itself may inspire an idea that a number of essential properties of K could be inherited from the dilated ℓ_1^n-ball. One comparison claim of this kind is discussed in Section 3, where we also complete the proof of Theorems 1.1 and 1.2. The case of ℓ_1^n-ball has to be treated separately and is considered in Section 2.

2. Linear functionals on ℓ_1^n ball

Given a probability space (Ω, μ) and a Young function ψ on the real line $\mathbf{R}$ (i.e., a convex, even function such that $\psi(0) = 0$, $\psi(t) > 0$ for $t \neq 0$), one defines the corresponding Orlicz norm by

$$\|f\|_\psi = \|f\|_{L_\psi(\mu)} = \inf\left\{\lambda > 0 : \int \psi(f/\lambda)\, d\mu \leq 1\right\},$$

where f is an arbitrary measurable function on Ω. If $\psi(t) = |t|^p$ $(p \geq 1)$, we arrive at the usual Lebesgue space norm $\|f\|_p = \|f\|_{L^p(\mu)}$. It is well-known and easy to see that for $\psi = \psi_2$, the Orlicz norm $\|f\|_{\psi_2}$ is equivalent to $\sup_{p \geq 1} \frac{\|f\|_p}{\sqrt{p}}$. So, in order to get information on large deviations of f and, in particular, to bound its ψ_2-norm, it suffices to study the rate of growth of L^p-norms of f.

We equip $\Omega = B_1$, the unit ball of the space ℓ_1^n, with the uniform distribution μ_n. This probability measure has density

$$\frac{d\mu_n(x)}{dx} = \frac{n!}{2^n}\, 1_{B_1}(x), \quad x \in \mathbf{R}^n.$$

For any positive real numbers $p_1, \dots, p_n$, one has a well-known identity

$$\int_{\Delta_n} x_1^{p_1 - 1} \dots x_n^{p_n - 1}\, dx_1 \dots dx_n = \frac{\Gamma(p_1) \dots \Gamma(p_1)}{\Gamma(p_1 + \cdots + p_n + 1)},$$

where the integration is performed over $\Delta_n = \{x \in \mathbf{R}_+^n : x_1 + \cdots + x_n \leq 1\}$, the part of B_1 in the positive octant $\mathbf{R}_+^n = [0, +\infty)^n$. Together with the polynomial formula, this identity implies that, for any positive even integer $p = 2q$,

$$\int |f_\theta(x)|^{2q}\, d\mu_n(x) = \frac{n!\,(2q)!}{(n+2q)!} \sum_{q_1 + \cdots + q_n = q} \theta_1^{2q_1} \dots \theta_n^{2q_n}, \tag{2.1}$$

where the summation is performed over all non-negative integers $q_1, \dots, q_n$ such that $q_1 + \cdots + q_n = q$. One easily derives from this:

Proposition 2.1. *For every $\theta \in \mathbf{R}^n$,*

$$\frac{c_1 \|\theta\|_\infty}{\sqrt{n}} \le \|f_\theta\|_{L_{\psi_2}(\mu_n)} \le \frac{c_2 \|\theta\|_\infty}{\sqrt{n}}, \tag{2.2}$$

where c_1 and c_2 are absolute positive constants. One can take $c_1 = \frac{1}{\sqrt{6}}$, $c_2 = 2\sqrt{2}$.

In the sequel, we use notation $C_m^k = \frac{m!}{k!(m-k)!}$ for usual binomial coefficients (where $k = 0, 1, \dots m$).

Proof. From (2.1), setting $\alpha = \sqrt{n}\,\|\theta\|_\infty$, and recalling that the sum therein contains C_{n+q-1}^{n-1} terms, we get

$$\begin{aligned} \int |f_\theta|^{2q}\, d\mu_n \quad &\le \quad \frac{n!\,(2q)!}{(n+2q)!}\ \frac{(n+q-1)!}{(n-1)!\,q!}\ \frac{\alpha^{2q}}{n^q} \\ &= \quad \frac{n}{(n+2q)\dots(n+q)}\ \frac{C_{2q}^q\, q!\,\alpha^{2q}}{n^q} \ \le\ \frac{4^q\, q!\,\alpha^{2q}}{n^{2q}}. \end{aligned} \tag{2.3}$$

Therefore, by Taylor's expansion, for all $|\lambda| < \frac{1}{2\alpha}$,

$$\begin{aligned} \int e^{(\lambda n f_\theta)^2}\, d\mu_n &= 1 + \sum_{q=1}^\infty \frac{\lambda^{2q}}{q!} \int |f_\theta|^{2q}\, d\mu_n \\ &\le 1 + \sum_{q=1}^\infty 4^q \lambda^{2q} \alpha^{2q} = \frac{1}{1 - 4\lambda^2\alpha^2}. \end{aligned}$$

The last expression is equal to 2 for $\lambda = \frac{1}{2\sqrt{2}\alpha}$, so $n\|f_\theta\|_{L_{\psi_2}(\mu_n)} \le 2\sqrt{2}\,\alpha$. This gives the upper estimate in (2.2) with $c_2 = 2\sqrt{2}$.

For the lower estimate, we may assume $\theta_j \ge 0$, for all $j \le n$. It follows from (2.1) that all L^{2q}-norms $\theta \to \|f_\theta\|_{2q}$ as functions of θ and therefore the function $\theta \to \|f_\theta\|_{\psi_2}$ are coordinatewise increasing on $\mathbf{R}_+^n$. Consequently, $\|f_\theta\|_{\psi_2} \ge \|\theta\|_\infty \|f_1\|_{\psi_2}$ where $f_1(x) = x_1$. To bound from below $\|f_1\|_{\psi_2}$, note that, given a random variable ξ with $\|\xi\|_{\psi_2} = 1$, we always have $2 = \mathbf{E}\, e^{\xi^2} \ge \frac{1}{q!}\,\mathbf{E}\,\xi^{2q}$, so $\mathbf{E}\,\xi^{2q} \le 2q!$ for all integers $q \ge 1$. Therefore, by homogeneity of the norm, $\mathbf{E}\,\xi^{2q} \le 2q!\,\|\xi\|_{\psi_2}^{2q}$ in general. Applying this to $\xi = f_1$ and $q = n$, and recalling (2.1), we find that

$$\begin{aligned} \|f_1\|_{\psi_2}^{2n} \ge \frac{1}{2n!} \int |f_1(x)|^{2n}\, d\mu_n(x) &= \frac{1}{2n!}\ \frac{n!\,(2n)!}{(3n)!} \\ &= \frac{1}{2\,(2n+1)\dots(3n)} \ge \frac{1}{2\,(3n)^n}. \end{aligned}$$

Hence, $\|f_1\|_{\psi_2} \ge \frac{1}{\sqrt{6n}}$. This yields the lower bound in (2.2) with $c_1 = 1/\sqrt{6}$. □

Now, we will try to sharpen the bound (2.3) on L^{2q}-norms. Given a vector $\theta \in S^{n-1}$, $n \geq 2$, put

$$C_n(\theta) = \|\theta\|_\infty \sqrt{\frac{n}{\log n}}.$$

Thus, $C_n(\theta) = \frac{1}{\sqrt{\log n}}$ for the main direction $\theta_j = \frac{1}{\sqrt{n}}$. However, this quantity is of order 1 for a typical θ, that is, when this vector is randomly selected with respect to the uniform distribution σ_{n-1} on the unit sphere S^{n-1}.

Proposition 2.2. *On (B_1, μ_n), for all $\theta \in S^{n-1}$ and every real $p \geq 1$,*

$$n\,\|f_\theta\|_p \leq C \max\left\{\sqrt{p},\, C_n(\theta)\sqrt{p\log p}\right\}, \tag{2.4}$$

where C is an absolute constant. One can take $C = 4.8$.

The inequality (2.4) implies the upper estimate of Proposition 2.1. Indeed, if $p \leq n$, we get

$$n\,\|f_\theta\|_p \leq C \max\left\{\sqrt{p},\, C_n(\theta)\sqrt{p\log n}\right\} = C\,\|\theta\|_\infty\sqrt{np},$$

where we also applied $1 \leq \|\theta\|_\infty \sqrt{n}$ (due to the normalization assumption $|\theta| = 1$). Thus,

$$\|f_\theta\|_p \leq \frac{C\,\|\theta\|_\infty}{\sqrt{n}}\,\sqrt{p}, \quad 1 \leq p \leq n.$$

The latter easily yields the right hand side of (2.2).

Introduce the full homogeneous polynomial of degree q,

$$P_q(a) \;=\; \sum_{q_1+\dots+q_n=q} a_1^{q_1}\dots a_n^{q_n}, \quad a \in \mathbf{R}^n, \tag{2.5}$$

where the summation is performed over all non-negative integers $q_1, \dots, q_n$ such that $q_1 + \dots + q_n = q$. According to the exact formula (2.1), we need a more accurate estimate on $P_q(a)$ (with $a_i = \theta_i^2$) in comparison with what was used for the proof of Proposition 2.1 on the basis of the trivial bound $a_i \leq \|a\|_\infty$.

Lemma 2.3. *For any $q \geq 1$ integer and every $a \in \mathbf{R}^n$ such that $\sum_{i=1}^n |a_i| = 1$,*

$$P_q(a) \leq \left(2e\max\left\{\frac{1}{q},\, \|a\|_\infty\right\}\right)^q.$$

Proof. We may assume $a_i \geq 0$. First we drop the condition $\|a\|_1 = \sum_{i=1}^n |a_i| = 1$ but assume $a_i \leq \frac{1}{2}$, for all $i \leq n$. Then $\frac{1}{1-a_i} \leq e^{2a_i}$ and, performing the summation in (2.5) over all integers $q_i \geq 0$, we get

$$P_q(a) \leq \sum_{q_i\geq 0} a_1^{q_1}\dots a_n^{q_n} = \frac{1}{(1-a_1)\dots(1-a_n)} \leq e^{2\|a\|_1}.$$

Applying this estimate to the vector ta instead of a and assuming $\|a\|_1 = 1$, we thus obtain that

$$P_q(a) \leq \frac{e^{2t}}{t^q}, \quad \text{whenever } 0 \leq t \leq \frac{1}{2\|a\|_\infty}.$$

Optimization over all admissible t leads to

$$P_q(a) \leq \begin{cases} (\frac{2e}{q})^q, & \text{if } \|a\|_\infty \leq \frac{1}{q}, \\ (2\|a\|_\infty)^q \exp\{\frac{1}{\|a\|_\infty}\}, & \text{if } \|a\|_\infty \geq \frac{1}{q}. \end{cases}$$

It remains to apply $\exp\{\frac{1}{\|a\|_\infty}\} \leq e^q$ in the second case $\|a\|_\infty \geq \frac{1}{q}$. □

Proof of Proposition 2.2. Applying Lemma 2.3 and (2.1) with $a_i = \theta_i^2$, we obtain that, for every integer $q \geq 1$,

$$\|f_\theta\|_{2q}^{2q} \leq \frac{n!\,(2q)!}{(n+2q)!} \left(2e \max\left\{\frac{1}{q}, \|a\|_\infty\right\}\right)^q.$$

Using $(n+2q)! \geq n!\,n^{2q}$ and $(2q)! \leq (2q)^{2q}$, we thus get

$$\begin{aligned} n\,\|f_\theta\|_{2q} &\leq 2q\sqrt{2e} \max\left\{\frac{1}{\sqrt{q}}, \sqrt{\|a\|_\infty}\right\} \\ &= 2\sqrt{2e} \max\{\sqrt{q}, q\,\|\theta\|_\infty\}. \end{aligned} \tag{2.6}$$

Now, starting from a real number $p \geq 2$, take the least integer q such that $p \leq 2q$. Then, $q \leq p$ and

$$\begin{aligned} n\,\|f_\theta\|_p \leq n\,\|f_\theta\|_{2q} &\leq 2\sqrt{2e} \max\{\sqrt{q}, q\,\|\theta\|_\infty\} \\ &\leq 2\sqrt{2e} \max\{\sqrt{p}, p\,\|\theta\|_\infty\} \\ &= 2\sqrt{2e} \max\left\{\sqrt{p}, C_n(\theta)\, p\sqrt{\frac{\log n}{n}}\right\}. \end{aligned}$$

Assume $p \leq n$. The function $\frac{\log x}{x}$ is decreasing in $x \geq e$, so $\frac{\log n}{n} \leq \frac{\log p}{p}$ in case $p \geq e$. If $2 \leq p \leq e$, we have anyway $\frac{\log n}{n} \leq \frac{\log 9}{\log 8} \frac{\log p}{p}$. Thus, in the range $2 \leq p \leq n$,

$$n\,\|f_\theta\|_p \leq 2\sqrt{2e} \max\left\{\sqrt{p}, 1.03\, C_n(\theta)\, \sqrt{p\log p}\right\}.$$

For $p \geq n$, this inequality is immediate since then

$$n\,\|f_\theta\|_p \leq n\,\|f_\theta\|_\infty = n\|\theta\|_\infty \leq C_n(\theta)\, \sqrt{p\log p}.$$

It remains to note that $2\sqrt{2e}\,1.03 < 4.8$ and that the case $1 \leq p \leq 2$ follows from the estimate $n\,\|f_\theta\|_p \leq n\,\|f_\theta\|_2 < \sqrt{2}$. Proposition 2.2 has been proved. □

Corollary 2.4. *For all $\theta \in S^{n-1}$ and every real $p \geq 2$,*

$$n\,\|f_\theta\|_p \leq 7\max\{1, C_n(\theta)\}\sqrt{p\log p}. \tag{2.7}$$

3. The general case

Now, let us return to the general case of a body K in $\mathbf{R}^n$ satisfying the basic properties 1) – 3). As we already mentioned, within the class of such bodies, a suitably normalized ℓ_1^n-ball is the largest set. More precisely, we have

$$K \subset V \equiv Cn\, B_1,$$

for some universal constant C. As shown in [1], one may always take $C = \sqrt{6}$, and moreover $C = 1$, if additionally K is symmetric under permutations of coordinates. Note that the sets K and V have similar volume radii of order $\sqrt{n}$. Our nearest goal will be to sharpen this property in terms of distributions of certain increasing functionals with respect to the uniform distributions on these bodies which we denote by μ_K and μ_V, respectively.

Denote by $\mathcal{F}_n$ the family of all functions F on $\mathbf{R}^n$ that are symmetric about coordinate axes and representable on $\mathbf{R}_+^n$ as

$$F(x) = \pi\left([0, x_1] \times \cdots \times [0, x_n]\right), \quad x \in \mathbf{R}_+^n, \tag{3.1}$$

for some positive Borel measure π (finite on all compact subsets). If π is absolutely continuous, the second assumption on F is equivalent to the representation

$$F(x) = \int_0^{x_1} \cdots \int_0^{x_n} q(t)\, dt, \quad x \in \mathbf{R}_+^n,$$

for some non-negative measurable function q. Of a special interest will be the functions of the form $F(x) = |x_1|^{p_1} \dots |x_n|^{p_n}$ with $p_1, \dots, p_n \geq 0$.

With these assumptions, we have:

Proposition 3.1. *If $F \in \mathcal{F}_n$, for all $t \geq 0$,*

$$\mu_K\{x \in \mathbf{R}^n : F(x) \geq t\} \leq \mu_V\{x \in \mathbf{R}^n : F(x) \geq t\}, \tag{3.2}$$

where the constant C defining V may be taken to be $\sqrt{6}$ (and one may take $C = 1$ in case K is symmetric under permutations of coordinates). In particular,

$$\int F(x)\, d\mu_K(x) \leq \int F(x)\, d\mu_V(x). \tag{3.3}$$

Proof. Consider

$$u(\alpha_1, \dots, \alpha_n) = \mu_K\{x \in K : |x_1| \geq \alpha_1, \dots, |x_n| \geq \alpha_n\}, \quad \alpha_j \geq 0.$$

By the Brunn–Minkowski inequality and since K is symmetric about coordinate axes, the function $u^{1/n}$ is concave on the convex set $K^+ = K \cap \mathbf{R}_+^n$. In addition, $u(0) = 1$. Without loss in generality, assume that u is continuously differentiable in K^+. Note that $\frac{\partial u(\alpha)}{\partial \alpha_j}\big|_{\alpha=0} = -\mathrm{vol}_{n-1}(K_j)$, where K_j are the sections of K by the coordinate hyperspaces $\{x \in \mathbf{R}^n : x_j = 0\}$. As known (cf. eg. [1]), for every hyperspace H in $\mathbf{R}^n$,

$$\mathrm{vol}_{n-1}(K \cap H) \geq c$$

with $c = \frac{1}{\sqrt{6}}$. Moreover, for all coordinate hyperspaces, one has a sharper estimate $\mathrm{vol}_{n-1}(K_j) \geq 1$, provided that K is invariant under permutations of coordinates. Consequently, $\frac{\partial u(\alpha)}{\partial \alpha_j}\big|_{\alpha=0} \leq -c$. By concavity, partial derivatives of $u^{1/n}$ are coordinatewise non-increasing, so, for all points α in K^+,

$$\frac{\partial u^{1/n}(\alpha)}{\partial \alpha_j} \leq \frac{\partial u^{1/n}(\alpha)}{\partial \alpha_j}\bigg|_{\alpha=0} = \frac{1}{n}\,\frac{\partial u(\alpha)}{\partial \alpha_j}\bigg|_{\alpha=0} \leq -\frac{c}{n}.$$

Thus, necessarily $u^{1/n}(\alpha) - u^{1/n}(0) \leq -\frac{c}{n}\,(\alpha_1 + \cdots + \alpha_n)$, that is,

$$\mu_K\{|x_1| \geq \alpha_1, \dots, |x_n| \geq \alpha_n\} \leq \left(1 - \frac{c\,(\alpha_1 + \cdots + \alpha_n)}{n}\right)^n. \tag{3.4}$$

Note that on the complement $\mathbf{R}^n_+ \setminus K^+$ the left hand side of (3.4) is zero, so this inequality extends automatically from K^+ to the whole octant $\mathbf{R}^n_+$.

Now, it is useful to observe that, whenever $\alpha_1 + \cdots + \alpha_n \leq Cn$, $\alpha_j \geq 0$,

$$\begin{aligned}\mu_V\{|x_1| \geq \alpha_1, \dots, |x_n| \geq \alpha_n\} &= \frac{\mathrm{vol}_n\{x \in V : |x_1| \geq \alpha_1, \dots, |x_n| \geq \alpha_n\}}{\mathrm{vol}_n(V)} \\ &= \frac{\mathrm{vol}_n\left((Cn - \alpha_1 - \dots \alpha_n)\,B_1\right)}{\mathrm{vol}_n(CnB_1)} \\ &= \left(1 - \frac{\alpha_1 + \cdots + \alpha_n}{Cn}\right)^n.\end{aligned}$$

Thus, (3.4) is exactly the desired inequality (3.2) with $C = \frac{1}{c}$ for the characteristic function $F_\alpha(x) = 1_{\{|x_1| \geq \alpha_1, \dots, |x_n| \geq \alpha_n\}}$. In the representation (3.1), this function corresponds to the measure π that assigns unit mass to the point α. To get (3.1) for all other $F's$, it remains to take into account an obvious fact that the functions F_α form a collection of all extremal "points" for the cone $\mathcal{F}_n$. Proposition 3.1 follows. □

As a basic example, we apply (3.3) to compare the absolute mixed moments of the measures μ_K and μ_V: for all $p_1, \dots, p_n \geq 0$,

$$\int |x_1|^{p_1} \dots |x_n|^{p_n}\, d\mu_K(x) \leq \int |x_1|^{p_1} \dots |x_n|^{p_n}\, d\mu_V(x). \tag{3.5}$$

As in case of the ℓ_1^n unit ball B_1, such moments are involved through the polynomial formula in the representation

$$\int |f_\theta|^{2q}\, d\mu = (2q)! \sum \frac{\theta_1^{2q_1} \dots \theta_n^{2q_n}}{(2q_1)! \dots (2q_1)!} \int x_1^{2q_1} \dots x_n^{2q_n}\, d\mu(x),$$

where q is a natural number, and where μ may be an arbitrary probability measure on $\mathbf{R}^n$ symmetric about the coordinate axes (as before, the summation performed over all non-negative integers $q_1, \dots, q_n$ such that $q_1 + \cdots + q_n = q$). In view of (3.5), we thus get:

Corollary 3.2. *For every $\theta \in \mathbf{R}^n$, and for all integers $q \geq 0$,*

$$\int |f_\theta|^{2q}\, d\mu_K \leq \int |f_\theta|^{2q}\, d\mu_V. \tag{3.6}$$

To complete the proof of Theorem 1.1, it remains to note that, with Taylor's expansion for $e^{(\lambda f_\theta)^2}$, (3.6) readily implies an analogous inequality for ψ_2-norm,

$$\|f_\theta\|_{L_{\psi_2}(\mu_K)} \leq \|f_\theta\|_{L_{\psi_2}(\mu_V)}.$$

Since $V = CnB_1$, the right hand side in terms of the uniform measure μ_n on B_1 is just $Cn\, \|f_\theta\|_{L_{\psi_2}(\mu_n)}$. Thus, by Proposition 2.1,

$$\|f_\theta\|_{L_{\psi_2}(\mu_K)} \leq 2\sqrt{2}\, C\, \|\theta\|_\infty \sqrt{n}.$$

The constant $2\sqrt{2}\, C$ does not exceed $4\sqrt{3}$, as it is claimed in (1.2), and does not exceed $2\sqrt{2}$ in case K is symmetric under permutations of the coordinates. This proves Theorem 1.1. □

Now, combining (3.6) with the moment estimate (2.6) on B_1, we obtain

$$\|f_\theta\|_{L^{2q}(\mu_K)} \leq 2C\sqrt{2e}\max\{\sqrt{q},\, q\,\|\theta\|_\infty\}.$$

Moreover, with the same argument leading from (2.6) to (2.7) in the proof of Proposition of 2.2, the above estimate implies a precise anologue of Corollary 2.4, i.e., the inequality

$$\|f_\theta\|_{L^p(\mu_K)} \leq 7C\max\{1, C_n(\theta)\}\sqrt{p\log p}, \tag{3.7}$$

which holds true for every real $p \geq 2$ and for all $\theta \in S^{n-1}$ with $C_n(\theta) = \|\theta\|_\infty \sqrt{\frac{n}{\log n}}$. Also recall that we may take $C = \sqrt{6}$. To reach the statement of Theorem 1.2, one needs to transform (3.7) into an appropriate deviation inequality.

Lemma 3.3. *Given a measurable function ξ on some probability space $(\Omega, \mathbf{P})$, assume that its L^p-norms satisfy, for all $p \geq 2$ and some constant $A \geq 1/2$,*

$$\|\xi\|_p \leq A\sqrt{p\log p}. \tag{3.8}$$

Then, for all $t \geq 2Ae$,

$$\mathbf{P}\{|\xi| \geq t\} \leq \exp\left\{-\frac{t^2}{8A^2 e\, \log t}\right\}. \tag{3.9}$$

Proof. Put $\eta = \xi^2$ and write the assumption (3.8) as $\int \eta^q\, d\mathbf{P} \leq (Bq\log(2q))^q$, $q \geq 1$, where $B = 2A^2$. Hence, by Chebyshev's inequality, for any $x > 0$,

$$\mathbf{P}\{\eta \geq x\} \leq x^{-q}\int \eta^q\, d\mathbf{P} \leq \left(\frac{B\, q\log(2q)}{x}\right)^q.$$

Apply it to q of the form $\frac{cx}{\log(cx)}$, $c > 0$, to get

$$\mathbf{P}\{\eta \geq x\} \leq \left(Bc \, \frac{\log \frac{2cx}{\log(cx)}}{\log(cx)} \right)^{\frac{cx}{\log(cx)}}$$

Assume $cx \geq e$. Since the function $\frac{z}{\log z}$ increases in $z \geq e$, the requirement $q \geq 1$ is fulfilled and, in addition, $\log \frac{2cx}{\log(cx)} \leq \log(2cx) \leq 2\log(cx)$. Thus, we may simplify the above estimate as $\mathbf{P}\{\eta \geq x\} \leq (2Bc)^{\frac{cx}{\log(cx)}}$. Choosing $c = \frac{1}{2Be}$, we obtain that

$$\mathbf{P}\{\eta \geq x\} \leq \exp\left\{ -\frac{cx}{\log(cx)} \right\}, \quad \text{provided that} \quad x \geq 2Be^2.$$

Equivalently, replacing x with t^2, $\mathbf{P}\{|\xi| \geq t\} \leq \exp\{-\frac{ct^2}{\log(ct^2)}\}$, for all $t \geq 2Ae$. Since $c = \frac{1}{4A^2e} \leq 1$, we have $\log(ct^2) \leq \log(t^2)$, so

$$\mathbf{P}\{|\xi| \geq t\} \leq \exp\left\{ -\frac{ct^2}{2\log t} \right\}, \quad t \geq 2Ae,$$

which is the desired inequality (3.9). □

Proof of Theorem 1.2. According to (3.7), for any $\theta \in S^{n-1}$, the linear functional $\xi = f_\theta$ on $(\Omega, \mathbf{P}) = (K, \mu_K)$ satisfies the assumption of Lemma 3.3 with constant $A(\theta) = 7\sqrt{6}\max\{1, C_n(\theta)\}$. As a function of θ, this constant has relatively small deviations with respect to the uniform measure σ_{n-1} on the sphere S^{n-1}. Indeed, consider the function $g(\theta) = \max_{j \leq n} |\theta_j|$. Since it has Lipschitz seminorm 1, by a concentration inequality on the sphere (cf. e.g. [6], [5]), for all $h > 0$,

$$\sigma_{n-1}\{g \geq m + h\} \leq e^{-nh^2/2}, \tag{3.10}$$

where m is σ_{n-1}-meadian for g. As is known, the median does not exceed $\alpha\sqrt{\frac{\log n}{n}}$, for some numerical $\alpha > 0$. Taking h proportional to m in (3.10), we obtain that $\sigma_{n-1}\{g \geq \beta\sqrt{\frac{\log n}{n}}\} \leq n^{-(\beta-\alpha)^2/2}$, for every $\beta > \alpha$. Equivalently, in terms of $C_n(\theta)$, $\sigma_{n-1}\{C_n(\theta) \geq \beta\} \leq n^{-(\beta-\alpha)^2/2}$, so,

$$\sigma_{n-1}\{A(\theta) \geq 7\sqrt{6}\,\beta\} \leq \frac{1}{n^{(\beta-\alpha)^2/2}}.$$

Thus, starting with a constant $c_1 > 0$, take $\beta > \alpha$ such that $(\beta - \alpha)^2/2 = c_1$. Then, with $A = 7\sqrt{6}\,\beta$, we get, by Lemma 3.3,

$$\mu_K\{|f_\theta| \geq t\} \leq \exp\left\{ -\frac{t^2}{8A^2e\,\log t} \right\}, \quad t \geq t_0 \equiv 2Ae.$$

This inequality holds true for all θ in S^{n-1} except for a set on the sphere of measure at most n^{-c_1}. □

References

[1] Bobkov, S. G., Nazarov, F. L. On convex bodies and log-concave probability measures with unconditional basis. *Geom. Aspects of Func. Anal., Lect. Notes in Math.*, to appear.

[2] Bourgain, J. On high-dimensional maximal functions associated to convex bodies. *Amer. J. Math.*, 108 (1986), No. **6**, 1467–1476.

[3] Bourgain, J. On the distribution of polynomials on high dimensional convex sets. *Geom. Aspects of Func. Anal., Lect. Notes in Math.*, **1469** (1991), 127–137.

[4] Borell, C. Convex measures on locally convex spaces. *Ark. Math.*, **12** (1974), 239–252.

[5] Ledoux, M. The concentration of measure phenomenon. *Math. Surveys and Monographs*, vol. **89**, 2001, AMS.

[6] Milman, V. D., Schechtman, G. Asymptotic theory of finite dimensional normed spaces. *Lecture Notes in Math.*, **1200** (1986), Springer-Verlag.

[7] Paouris, G. Ψ_2-estimates for linear functionals on zonoids. *Geom. Aspects of Func. Anal., Lect. Notes in Math.*, to appear.

School of Mathematics, University of Minnesota, Minneapolis, MN 55455
E-mail address: `bobkov@math.umn.edu`

Department of Mathematics, Michigan State University, East Lansing, MI 48824–1027
E-mail address: `fedja@math.msu.edu`

Progress in Probability, Vol. 56, 15–21

A Concentration Inequality on Riemannian Path Space

Christian Houdré and Nicolas Privault

Abstract. The covariance representations method to obtain concentration inequalities is applied to functionals of Riemannian Brownian motion. This recovers, in particular, tail estimates for Brownian motion on a manifold.

1. Introduction and Background

Concentration and deviation inequalities have been obtained on the Wiener and Poisson spaces and for discrete random walks in [7] using the method of covariance representations ([1], [6]). In this note we present an application of this method to concentration inequalities on path space. The concentration results for the laws of general random variables presented here on Riemannian path space are slightly weaker than the ones obtained from logarithmic Sobolev inequalities (see [8]). Nevertheless they allow us to recover some classical bounds such as the one for the law of the supremum of the distance of Riemannian Brownian motion to the origin. The covariance representation method relies on the Clark formula on path space, a short proof of which is obtained by showing that the damped and flat gradients have the same adapted projections, as a consequence of an intertwining formula using Skorohod integrals.

In [7] we showed that the use of semi-groups for covariance representations allows us to recover the concentration and deviation inequalities obtained from logarithmic Sobolev inequalities and the Herbst method [8]. In particular, it turned out that covariance representations written in terms of the Clark formula generally yield weaker results than covariance representations written in terms of semi-groups. In the path case, however, covariance representations using Ornstein-Uhlenbeck type semi-groups are unknown to the authors. If available, they would allow us to recover the concentration results that follow from the Herbst method and the logarithmic Sobolev inequalities of [2].

Received by the editors January 7, 2003.
2000 *Mathematics Subject Classification.* 60F99, 60H07, 58J65, 58C35.
Key words and phrases. Concentration inequalities, Riemannian Brownian motion, covariance identities, damped gradient, Clark formula.
Research supported in part by a NSF Grant.

2. Preliminaries and notation

Let $(x(t))_{t\in[0,T]}$ denote the $\mathbb{R}^d$-valued Brownian motion on the Wiener space W with Wiener measure μ, generating the filtration $(\mathcal{F}_t)_{t\in[0,T]}$. Let M be a Riemannian manifold of dimension d whose Ricci curvature is uniformly bounded from below, and let $O(M)$ denote the bundle of orthonormal frames over M. The Levi-Civita parallel transport defines d canonical horizontal vector fields $A_1, \ldots, A_d$ on $O(M)$, and the Stratonovich stochastic differential equation

$$\begin{cases} dr(t) = \sum_{i=1}^{i=d} A_i(r(t)) \circ dx^i(t), & t \in [0,T], \\ r(0) = (m_0, r_0) \in O(M), \end{cases}$$

defines an $O(M)$-valued process $(r(t))_{t\in[0,T]}$. Let $\pi : O(M) \longrightarrow M$ be the canonical projection, let $\gamma(t) = \pi(r(t))$, $t \in [0,T]$, be the Brownian motion on M and let the Itô parallel transport along $(\gamma(t))_{t\in[0,T]}$ be defined as

$$t_{t\leftarrow 0} = r(t)r_0^{-1} : T_{m_0}M \simeq \mathbb{R}^d \longrightarrow T_{\gamma(t)}M, \quad t \in [0,T].$$

Let $\mathcal{C}_0([0,T];\mathbb{R}^d)$ denote the space of continuous $\mathbb{R}^d$-valued functions on $[0,T]$ vanishing at the origin. Let also $\mathbf{P}(M)$ denote the set of continuous paths on M starting at m_0, let

$$\begin{aligned} I \ : \ & \mathcal{C}_0([0,T];\mathbb{R}^d) \longrightarrow \mathbf{P}(M) \\ & (\omega(t))_{t\in[0,T]} \mapsto I(\omega) = (\gamma(t))_{t\in[0,T]} \end{aligned}$$

be the Itô map, and let ν denote the image measure on $\mathbf{P}(M)$ of the Wiener measure μ by I. Let Ω_r denote the curvature tensor of M, and let $\mathrm{ric}_r : \mathbb{R}^d \longrightarrow \mathbb{R}^d$ be the Ricci tensor at the frame $r \in O(M)$. Given an adapted process $(z_t)_{t\in[0,T]}$ with absolutely continuous trajectories, we let $(\hat{z}(t))_{t\in[0,T]}$ be defined by

$$\dot{\hat{z}}(t) = \dot{z}(t) + \frac{1}{2}\mathrm{ric}_{r(t)} z(t), \quad t \in [0,T], \quad \hat{z}(0) = 0. \tag{2.1}$$

We recall that $z \mapsto \hat{z}$ can be inverted, i.e. there exists a process $(\tilde{z}_t)_{t\in[0,T]}$ such that $\hat{\tilde{z}} = z$, cf. Sect. 3.7 of [5]. Finally, let $Q_{t,s} : \mathbb{R}^d \longrightarrow \mathbb{R}^d$, be defined as

$$\frac{dQ_{t,s}}{dt} = -\frac{1}{2}\mathrm{ric}_{r(t)} Q_{t,s}, \quad Q_{s,s} = \mathrm{Id}_{T_{m_0}}, \quad 0 \le s \le t,$$

and let

$$q(t,z) = -\int_0^t \Omega_{r(s)}(\circ dx(s), z(s)), \quad t \in [0,T],$$

where $\circ dx(s)$ denotes the Stratonovich differential. Let $Q_{t,s}^*$ be the adjoint of $Q_{t,s}$, let $\mathbb{H} = L^2([0,T],\mathbb{R}^d)$, and let $\mathbf{H} = L^\infty(\mathbf{P}(M), \mathbb{H}; d\nu)$. Let finally $\mathcal{C}_c^\infty(M^n)$ denote the space of infinitely differentiable functions with compact support in M^n.

We now recall various notions of gradient developed for the analysis on path space. The following definitions are found in [5]. The notation ∇_i^M denotes the gradient on M applied to the i-th variable of f.

i) (flat gradient) Given $F = f(x(t_1), \dots, x(t_n))$, $f \in \mathcal{C}_c^\infty((\mathbb{R}^d)^n)$, let

$$D_t F = \sum_{i=1}^{i=n} 1_{[0,t_i]}(t) \nabla_i^{\mathbb{R}^d} f(x(t_1), , \dots, x(t_n)), \quad t \in [0,T].$$

ii) (intrinsic gradient) Given $F = f(\gamma(t_1), \dots, \gamma(t_n))$, $f \in \mathcal{C}_c^\infty(M^n)$, let:

$$\hat{D}_t F = \sum_{i=1}^{i=n} t_{0 \leftarrow t_i} \nabla_i^M f(\gamma(t_1), \dots, \gamma(t_n)) 1_{[0,t_i]}(t), \quad t \in [0,T].$$

iii) (damped gradient) Given $f \in \mathcal{C}_c^\infty(M^n)$, let:

$$\tilde{D}_t F = \sum_{i=1}^{i=n} 1_{[0,t_i]}(t) Q_{t_i,t}^* t_{0 \leftarrow t_i} \nabla_i^M f(\gamma(t_1), \dots, \gamma(t_n)), \quad t \in [0,T].$$

Let $\operatorname{Dom}(D)$, $\operatorname{Dom}(\hat{D})$ and $\operatorname{Dom}(\tilde{D})$ denote the respective domains of D, $\hat{D}$ and $\tilde{D}$ as closable operators.

Still developing our background, let us recall the intertwining formula between $\hat{D}$ and D (Th. 2.6 of [3]). This can be written as

$$\int_0^T \dot{z}(t) \cdot (\hat{D}_t F) \circ I \, dt = \int_0^T \dot{z}(t) \cdot D_t(F \circ I) \, dt - \int_0^T (\hat{D}_t F) \circ I \cdot q(t,z) \odot dx(t),$$

where $\odot dx(t)$ is the Stratonovich differential. We use a different intertwining formula which is stated in terms of the Itô–Skorohod differential $dx(t)$ instead of the Stratonovich integral $\odot dx(t)$, cf. Cor. 5.2.1 and Cor. 5.4.1 of [9].

Proposition 2.1. *([9]) Let $z : W \times [0,T] \to \mathbb{R}^d$ be adapted with absolutely continuous trajectories and $\dot{z} \in L^2(W \times [0,T])$. For $F = f(\gamma(t_1), \dots, \gamma(t_n))$, $f \in \mathcal{C}_c^\infty(M^n)$ we have:*

i) *(intrinsic gradient)*

$$\int_0^T \dot{z}(t) \cdot (\hat{D}_t F) \circ I \, dt = \int_0^T \dot{\hat{z}}(t) \cdot D_t(F \circ I) \, dt + \int_0^T q(t,z) D_t(F \circ I) \cdot dx(t), \quad (2.2)$$

ii) *(damped gradient):*

$$\int_0^T \dot{z}(t) \cdot (\tilde{D}_t F) \circ I \, dt = \int_0^T \dot{z}(t) \cdot D_t(F \circ I) dt + \int_0^T q(t,\tilde{z}) D_t(F \circ I) \cdot dx(t). \quad (2.3)$$

The intertwining formula will now lead us to a Clark type formula. Indeed, since $E_\mu \left[\int_0^T q(t,\tilde{z}) D_t(F \circ I) \cdot dx(t) \right] = 0$, it follows directly from (2.2) and (2.3) that the processes DF and $\tilde{D}F$ have the same adapted projections:

$$E_\mu[D_t(F \circ I) \mid \mathcal{F}_t] = E_\mu[(\tilde{D}_t F) \circ I \mid \mathcal{F}_t], \quad t \in [0,T], \quad (2.4)$$

$F = f(\gamma(t_1), \dots, \gamma(t_n))$. Using this relation and the classical Clark formula on Wiener space

$$F \circ I = E_\mu[F \circ I] + \int_0^T E_\mu[D_t(F \circ I) \mid \mathcal{F}_t] \cdot dx(t),$$

we obtain the expression of the Clark formula on path space:

Proposition 2.2. *([5]) Let $F \in \mathrm{Dom}(\tilde{D})$, then*

$$F \circ I \quad = \quad E_\nu[F] + \int_0^T E_\mu[(\tilde{D}_t F) \circ I \mid \mathcal{F}_t] \cdot dx(t).$$

3. Concentration inequalities on Riemannian path space

Let $\mathrm{Cov}_\nu(F,G) = E_\nu[FG] - E_\nu[F]E_\nu[G]$ for $F, G \in L^2(\mathbf{P}(M), \nu)$. The following covariance identity is an immediate consequence of the Clark formula.

Proposition 3.1. *Let $F, G \in \mathrm{Dom}(\tilde{D})$, then*

$$\mathrm{Cov}_\nu(F,G) = E_\mu\left[\int_0^T (\tilde{D}_t F) \circ I \cdot E_\mu[(\tilde{D}_t G) \circ I \mid \mathcal{F}_t] \, dt\right]. \tag{3.1}$$

Next, we apply this covariance representation to the proof of a concentration inequality on path space.

Lemma 3.2. *Let $F \in \mathrm{Dom}(\tilde{D})$. If $\|\tilde{D}F\|_{L^2([0,T], L^\infty(\mathrm{P}(M)))} \leq C$, for some $C > 0$, then*

$$\nu(F - E_\nu[F] \geq x) \leq \exp\left(-\frac{x^2}{2C\|\tilde{D}F\|_{\mathbf{H}}}\right), \quad x \geq 0. \tag{3.2}$$

In particular, $E[e^{\lambda F^2}] < \infty$, for $\lambda < (2C\|\tilde{D}F\|_{\mathbf{H}})^{-1}$.

Proof. We first consider a bounded random variable $F \in \mathrm{Dom}(\tilde{D})$. The general case follows by approximating $F \in \mathrm{Dom}(\tilde{D})$ by the sequence $\max(-n, \min(F, n))$, $n \geq 1$. Let

$$\eta_F(t) = E_\mu[(\tilde{D}_t F) \circ I \mid \mathcal{F}_t], \quad t \in [0,T].$$

Assuming first that $E_\nu[F] = 0$, we have

$$\begin{aligned}
E_\nu[Fe^{sF}] &= E_\mu\left[\int_0^T (\tilde{D}_u e^{sF}) \circ I \cdot \eta_F(u) du\right] \\
&= sE_\mu\left[e^{sF\circ I}\int_0^T (\tilde{D}_u F) \circ I \cdot \eta_F(u) du\right] \\
&\leq sE_\mu\left[e^{sF\circ I}\|\tilde{D}F\|_{\mathbb{H}} \circ I \ \|\eta_F\|_{\mathbb{H}}\right] \\
&\leq sE_\nu\left[e^{sF}\right] \|\tilde{D}F\|_{\mathbf{H}} \|\eta_F\|_{L^\infty(W,\mathbb{H})} \leq sCE_\nu\left[e^{sF}\right] \|\tilde{D}F\|_{\mathbf{H}}.
\end{aligned}$$

In the general case, letting $L(s) = E_\nu[e^{s(F - E_\nu[F])}]$ we obtain:

$$\begin{aligned}
\log(E_\nu[e^{t(F-E_\nu[F])}]) &= \int_0^t \frac{L'(s)}{L(s)} ds \leq \int_0^t \frac{E_\nu[(F - E_\nu[F])e^{s(F-E_\nu[F])}]}{E_\nu[e^{s(F-E_\nu[F])}]} ds \\
&= \frac{1}{2} t^2 C \|\tilde{D}F\|_{\mathbf{H}}, \quad 0 \leq t \leq T.
\end{aligned}$$

We now have for all $x \in \mathbb{R}_+$ and $t \in [0,T]$:

$$\nu(F - E_\nu[F] \geq x) \leq e^{-tx} E_\nu[e^{t(F-E_\nu[F])}] \leq \exp\left(\frac{1}{2}t^2 C\|\tilde{D}F\|_{\mathbf{H}} - tx\right),$$

which yields (3.2) after minimization in $t \in [0,T]$. □

Since $\|\tilde{D}F\|_{\mathbf{H}} \leq \|\tilde{D}F\|_{L^2([0,T],L^\infty(\mathrm{P}(M)))}$, the inequality (3.2) is weaker than:

$$\nu(F - E_\nu[F] \geq x) \leq \exp\left(-\frac{x^2}{2\|\tilde{D}F\|^2_{\mathbf{H}}}\right), \quad x \geq 0, \tag{3.3}$$

which follows (via the Herbst method [8]) from the logarithmic Sobolev inequalities of [2], whose proof relies on the Itô formula and the Clark formula (see also [4]). The method presented in this paper is, somehow, simpler, and moreover (3.2) is sufficient to recover some classical tail estimates for functionals of $(\gamma(t))_{t\in\mathbb{R}_+}$, see [8] and references therein. Below, let ρ denote the distance on M.

Corollary 3.3. *Let the Ricci tensor be bounded from below by a constant $-K \in \mathbb{R}$. Then,*

$$\nu\left(\sup_{0\leq t\leq T}\rho(\gamma(t),m_0) \geq E_\nu\left[\sup_{0\leq t\leq T}\rho(\gamma(t),m_0)\right] + x\right) \leq \exp\left(-\frac{x^2}{2Te^{KT}}\right), x \geq 0, \tag{3.4}$$

Proof. Let $f(m_1,\ldots,m_n) = \max_{1\leq i\leq n}\rho(m_i,m_0)$ and

$$F = f(\gamma(t_1),\ldots,\gamma(t_n)) = \max_{1\leq i\leq n}\rho(\gamma(t_i),m_0),$$

where $0 < t_1 < \cdots < t_n < T$ determines a partition of $[0,T]$. Following the argument of [8], p. 196, we obtain for a certain partition $(A_j)_{1\leq j\leq n}$ of $\mathbf{P}(M)$:

$$\begin{aligned}|\tilde{D}_t F| &\leq \sum_{i=1}^{i=n} 1_{(t_{i-1},t_i]}(t)\sum_{j=i}^{j=n} e^{K(t_j-t)/2}|\nabla_j^M f| \\ &\leq e^{KT/2}\sum_{i=1}^{i=n} 1_{(t_{i-1},t_i]}(t)\sum_{j=i}^{j=n} 1_{A_j} \leq e^{KT/2},\end{aligned}$$

hence $\|\tilde{D}F\|_{L^2([0,T],L^\infty(\mathrm{P}(M)))} \leq Te^{KT}$, which gives (3.4) by monotone convergence as the mesh of the partition goes to zero. □

The Herbst method applied to the logarithmic Sobolev inequalities of [2] would yield the same estimate as (3.4), since here the bound on $\|\tilde{D}_t F\|_\infty$ is uniform in t.

When $K = 0$ this leads to the known estimate (cf. [8] and references therein):

$$\nu\left(\sup_{0\leq t\leq T}\rho(\gamma(t),m_0) \geq E_\nu\left[\sup_{0\leq t\leq T}\rho(\gamma(t),m_0)\right] + x\right) \leq \exp\left(-\frac{x^2}{2T}\right), \quad x \geq 0.$$

In the following we assume that the Ricci tensor of M is bounded from above and below.

Theorem 3.4. *Assume that the Ricci tensor of M is bounded by $K \geq 0$, i.e. $\|\mathrm{ric}_u\| \leq K$, $u \in O(M)$, and let $F \in \mathrm{Dom}(\hat{D})$ be such that $\|\hat{D}F\|_{L^2([0,T],L^\infty(\mathrm{P}(M)))} \leq C$, for some $C > 0$. Then*

$$\nu(F - E_\nu[F] \geq x) \leq \exp\left(-\frac{x^2}{2e^{K/2}(C + (e^{K/2}-1)\|\hat{D}F\|_{\mathbf{H}})\|\hat{D}F\|_{\mathbf{H}}}\right), \quad x \geq 0.$$

Proof. Again, we may start by assuming that F is bounded. We have the bound

$$\begin{aligned} |\tilde{D}_t F|^2 &\leq \left(|\hat{D}_t F| + \int_t^T \left\|\frac{d}{ds}Q^*_{t,s}\right\| |\hat{D}_s F| ds\right)^2 \\ &\leq \left(|\hat{D}_t F| + \frac{1}{2}K \int_t^T e^{K(s-t)/2}|\hat{D}_s F| ds\right)^2 \\ &\leq \left(|\hat{D}_t F| + \frac{1}{2}\sqrt{K(e^{K(T-t)}-1)}\|\hat{D}F\|_{\mathbb{H}}\right)^2 \\ &\leq |\hat{D}_t F|^2 + \sqrt{K(e^{K(T-t)}-1)}|\hat{D}_t F|\|\hat{D}F\|_{\mathbb{H}} \\ &\quad + \frac{1}{4}K(e^{K(T-t)}-1)\|\hat{D}F\|^2_{\mathbb{H}} \end{aligned}$$

(see p. 75 of [2]), which implies

$$\|\tilde{D}F\|^2_{\mathbb{H}} \leq \left(1 + \frac{1}{2}\sqrt{e^{KT} - KT - 1}\right)^2 \|\hat{D}F\|^2_{\mathbb{H}} \leq e^{KT}\|\hat{D}F\|^2_{\mathbb{H}},$$

and

$$\begin{aligned} \|\tilde{D}F\|_{L^2([0,T],L^\infty(\mathrm{P}(M)))} &\leq \|\hat{D}F\|_{L^2([0,T],L^\infty(\mathrm{P}(M)))} \\ &\quad + \frac{1}{2}\sqrt{e^{KT} - KT - 1}\|\hat{D}F\|_{\mathbf{H}} \\ &\leq C + (e^{KT/2} - 1)\|\hat{D}F\|_{\mathbf{H}}. \end{aligned}$$

It remains to apply Lemma 3.2. □

As a corollary we obtain

$$\nu(F - E_\nu[F] \geq x) \leq \exp\left(-\frac{x^2}{2e^{KT}C\|\hat{D}F\|_{\mathbf{H}}}\right), \quad x \geq 0. \tag{3.5}$$

A damped gradient satisfying a Clark formula can also be defined for diffusion processes on Riemannian manifolds, and in this case similar bounds hold between $\tilde{D}_t F$ and $\hat{D}_t F$, cf. Sect. 3 of [2] and the references therein. Hence the argument of Th. 3.4 will also be valid on the space of a diffusion process on a manifold.

For the finite dimensional random variable $F = f(\gamma(t))$ we have $\hat{D}F = 1_{[0,t]} t_{0\leftarrow t} \nabla^M f(\gamma(t))$ and

$$\|\hat{D}F\|^2_{L^2([0,T],L^\infty(\mathrm{P}(M)))} = t|\nabla^M f|^2_\infty = \|\hat{D}F\|^2_{\mathbf{H}},$$

hence both norms coincide and (3.5) becomes:

$$\nu(f(\gamma(t)) - E_\nu[f(\gamma(t))] \geq x) \leq \exp\left(-\frac{x^2}{2e^{KT}t|\nabla^M f|_\infty^2}\right), \quad x \geq 0, \tag{3.6}$$

References

[1] S. Bobkov, F. Götze, and C. Houdré. On Gaussian and Bernoulli covariance representations. *Bernoulli*, **7**(3):439–451, 2001.

[2] M. Capitaine, E.P. Hsu, and M. Ledoux. Martingale representation and a simple proof of logarithmic Sobolev inequalities on path spaces. *Electron. Comm. Probab.*, **2**:71–81 (electronic), 1997.

[3] A.B. Cruzeiro and P. Malliavin. Renormalized differential geometry on path space: Structural equation, curvature. *J. Funct. Anal.*, **139**:119–181, 1996.

[4] S. Fang. Integration by parts formula and logarithmic Sobolev inequality on the path space over loop groups. *Ann. Probab.*, **27**(2):664–683, 1999.

[5] S. Fang and P. Malliavin. Stochastic analysis on the path space of a Riemannian manifold. I: Markovian stochastic calculus. *J. Funct. Anal.*, **118**(1):249–274, 1993.

[6] C. Houdré. Remarks on deviation inequalities for functions of infinitely divisible random vectors. *Ann. Probab.*, **30**(3):1223–1237, 2002.

[7] C. Houdré and N. Privault. Concentration and deviation inequalities in infinite dimensions via covariance representations. *Bernoulli*, **8**(6):697–720, 2002.

[8] M. Ledoux. Concentration of measure and logarithmic Sobolev inequalities. In *Séminaire de Probabilités XXXIII, Lecture Notes in Math.* **1709**, pages 120–216. Springer, 1999.

[9] J.J. Prat and N. Privault. Explicit stochastic analysis of Brownian motion and point measures on Riemannian manifolds. *J. Funct. Anal.*, **167**:201–242, 1999.

Laboratoire d'Analyse et de Mathématiques Appliquées, CNRS UMR 8050, Université Paris XII, 94010 Créteil Cedex, France
and
School of Mathematics, Georgia Institute of Technology, Atlanta, Ga 30332 USA.
E-mail address: `houdre@math.gatech.edu`

Département de Mathématiques, Université de La Rochelle, 17042 La Rochelle, France.
E-mail address: `nprivaul@univ-lr.fr`

Progress in Probability, Vol. 56, 23–32

A Remark on Unified Error Exponents: Hypothesis Testing, Data Compression and Measure Concentration

Ioannis Kontoyiannis and Ali Devin Sezer

Abstract. Let A be finite set equipped with a probability distribution P, and let M be a "mass" function on A. A characterization is given for the most efficient way in which A^n can be covered using spheres of a fixed radius. A covering is a subset C_n of A^n with the property that most of the elements of A^n are within some fixed distance from at least one element of C_n, and "most of the elements" means a set whose probability is exponentially close to one (with respect to the product distribution P^n). An efficient covering is one with small mass $M^n(C_n)$. With different choices for the geometry on A, this characterization gives various corollaries as special cases, including Marton's error-exponents theorem in lossy data compression, Hoeffding's optimal hypothesis testing exponents, and a new sharp converse to some measure concentration inequalities on discrete spaces.

1. Introduction

Let A be a finite set and P a probability distribution on A. Suppose that the distance (or "distortion") $\rho(x,y)$ between any two points $x,y \in A$ is measured by a given nonnegative function $\rho : A\times A \to [0,\infty)$, and for strings $x_1^n = (x_1, x_2, \ldots, x_n)$ and $y_1^n = (y_1, y_2, \ldots, y_n)$ in A^n let $\rho_n(x_1^n, y_1^n)$ be the corresponding coordinate-wise distance (or single-letter distortion measure) on $A^n \times A^n$:

$$\rho_n(x_1^n, y_1^n) = \frac{1}{n}\sum_{i=1}^{n} \rho(x_i, y_i).$$

Since A is a finite set, the function ρ is bounded above by

$$D_{\max} \triangleq \max_{x,y\in A} \rho(x,y) = \max_{x_1^n, y_1^n \in A^n} \rho_n(x_1^n, y_1^n).$$

Received by the editors September 22, 2003.
2000 *Mathematics Subject Classification.* Primary 62B10; Secondary 60F10.
Key words and phrases. Sphere-covering, hypothesis testing, measure concentration, data compression, large deviations, error-exponents.
This research was partially supported by NSF grants #0073378-CCR and DMS-9615444, and by USDA-IFAFS grant #00-52100-9615.

Without loss of generality we assume throughout that $P(a) > 0$ for all $a \in A$, and that for each $a \in A$ there exists a $b \in A$ with $\rho(a,b) = 0$ (otherwise we may consider $\rho'(x,y) = [\rho(x,y) - \min_{z\in A}\rho(x,z)]$ instead of $\rho(x,y)$).

Given a $D \geq 0$, we want to cover "most" of A^n using balls $B(y_1^n, D)$, where

$$B(y_1^n, D) = \{x_1^n \in A^n \ : \ \rho_n(x_1^n, y_1^n) \leq D\}$$

is the closed ball of radius D centered at $y_1^n \in A^n$. To be precise, given a set $C_n \subset A^n$, we write $[C_n]_D$ for the *D-blowup of* C_n,

$$[C_n]_D \triangleq \bigcup_{y_1^n \in C_n} B(y_1^n, D).$$

A *D-covering of* A^n is a sequence of subsets C_n of A^n, $n \geq 1$, such that the P^n-probability of the part of A^n which is not covered by C_n within distance D has exponentially small probability,

$$\Pr\{\text{"error"}\} \triangleq 1 - P^n([C_n]_D) \quad \approx \quad 2^{-nE}, \tag{1.1}$$

for some $E > 0$. We are interested in "efficient" coverings of A^n, that is, given a "mass function" $M : A \to (0, \infty)$, we want to find D-coverings $\{C_n\}$ that satisfy (1.1) and also have small mass

$$M^n(C_n) \triangleq \sum_{y_1^n \in C_n} M^n(y_1^n) = \sum_{y_1^n \in C_n} \prod_{i=1}^{n} M(y_i).$$

Clearly there is a trade-off between finding coverings $\{C_n\}$ with small mass, and coverings with a good (i.e., large) error-exponent E as in (1.1). Typically, the better the error-exponent, the larger the C_n, and the bigger their mass would tend to be.

Motivated, in part, by the following example and by the applications illustrated in the examples of the following section, in our main result we give a precise characterization of this trade-off.

Example: Measure Concentration on the Binary Cube.

Consider the n-dimensional binary cube $A^n = \{0,1\}^n$. We measure distance on A^n by the proportion of mismatches between two binary strings x_1^n and y_1^n, i.e., we take $\rho_n(x_1^n, y_1^n)$ to be the Hamming distance,

$$\rho_n(x_1^n, y_1^n) = \frac{1}{n}\sum_{i=1}^{n} \mathbb{I}_{\{x_i \neq y_i\}}, \quad x_1^n, y_1^n \in A^n, \tag{1.2}$$

which also coincides with the normalized graph distance when A^n is equipped with the nearest-neighbor graph structure. For simplicity, in this example we consider natural logarithms and exponentials.

A well-known measure concentration inequality [10, Prop. 2.1.1][9, Thm. 3.5] gives a precise lower bound on the sphere-covering error probability of an arbitrary

C_n: For any $D \geq 0$, any product distribution P^n on A^n, and any $C_n \subset A^n$,

$$\Pr\{\text{"error"}\} = 1 - P^n([C_n]_D) \leq \frac{e^{-nD^2/2}}{P^n(C_n)}.$$

Therefore, if $\{C_n\}$ is *any* D-covering consisting of sets with $P^n(C_n) \approx e^{-nr}$ for some $r > 0$, then the union of the balls $B(y_1^n, D)$ centered at the points $y_1^n \in C_n$ covers all of A^n except for a set of probability no greater than

$$\approx e^{-n\left(\frac{D^2}{2} - r\right)}. \tag{1.3}$$

It is then natural to ask, what is the *best* achievable error exponent among all D-coverings $\{C_n\}$ with probability no greater that $\approx e^{-nr}$? In other words, we are asking for small sets with the largest possible "boundary," sets C_n with "volume" $P^n(C_n)$ no greater than e^{-nr} but whose D-blowups $[C_n]_D$ cover as much of A^n as possible. As pointed in [6], this question can be thought of as the opposite of the usual isoperimetric problem.

Taking $M = P$ in the general setting described above, we obtain the answer to this question as a corollary to our general result in the following section; see Corollary 3.

2. Results

Given any $D \geq 0$ and any $R \in \mathbb{R}$, let $E(R, D)$ denote the best achievable error-exponent among all D-coverings with mass asymptotically bounded by 2^{nR}. Letting $\mathcal{C}(R)$ denote the collection of all sequences of subsets C_n of A^n with $\limsup_n \frac{1}{n} \log M^n(C_n) \leq R$, define,

$$E(R, D) \triangleq \sup_{\{C_n\} \in \mathcal{C}(R)} \liminf_{n \to \infty} -\frac{1}{n} \log \left[1 - P^n([C_n]_D)\right],$$

where 'log' denotes the logarithm taken to base 2.

A weaker version of this problem was recently considered in [6], where it was shown that the probability of error can only decrease to zero if R is greater than $R(D; P, M)$,

$$R(D; P, M) \triangleq \inf_{X \sim P,\ E\rho(X,Y) \leq D} \left\{H(P_{X,Y} \| P \times P_Y) + E[\log M(Y)]\right\}, \tag{2.1}$$

where the infimum is taken over all jointly distributed random variables (X, Y) such that X has distribution P and $E\rho(X, Y) \leq D$, and $P_{X,Y}$ denotes the joint distribution of X, Y, P_Y denotes the marginal distribution of Y, and $H(\mu \| \nu)$ denotes the relative entropy between two probability measures μ and ν on the same finite set S,

$$H(\mu \| \nu) \triangleq \sum_{s \in S} \mu(s) \log \frac{\mu(s)}{\nu(s)}.$$

Therefore, the error-exponent $E(R,D)$ can only be nontrivial (i.e., nonzero) for $R > R(D;P,M)$. Also note that any $C_n \subset A^n$ has

$$\frac{1}{n}\log M^n(C_n) \leq \frac{1}{n}\log M^n(A^n) = \log M(A).$$

Hence, from now on we restrict attention to the range of interesting values for R between $R(D;P,M)$ and $R_{\max} \triangleq \log M(A)$.

Theorem. *For all $D \in [0, D_{\max})$ and all $R(D;P,M) < R < R_{\max}$, the best achievable exponent of the error probability, among all D-coverings $\{C_n\}$ with mass asymptotically bounded by 2^{nR}, is*

$$E(R,D) = E^*(R,D) \triangleq \inf_{Q\,:\,R(D;Q,M)>R} H(Q\|P),$$

where $R(D;P,M)$ is defined in (2.1) and $H(Q\|P)$ denotes the relative entropy (or Kullback–Leibler divergence) between two distributions P and Q.

Remarks.

1. A slightly different error-exponent. Alternatively, we can define a version of the optimal error-exponent by considering only D-coverings $\{C_n\}$ with mass bounded by 2^{nR} for *all* n:

$$E'(R,D) \triangleq \liminf_{n\to\infty} -\frac{1}{n}\log\left\{\min_{C_n\,:\,M^n(C_n)\leq 2^{nR}} \Big[1 - P^n([C_n]_D)\Big]\right\}.$$

From the theorem it easily follows that $E'(R,D)$ is also equal to $E^*(R,D)$ at all points R where $E^*(R,D)$ is continuous and, since it is nondecreasing in R, $E^*(R,D)$ is indeed continuous at all except countably many values of R. But in general it may fail to be continuous everywhere, as illustrated in the discussions by Marton [7] and Ahlswede [1] for the special case of lossy data compression (which corresponds to taking $M(x) \equiv 1$; see Example 2 below).

2. Proof. The proof of the theorem is a modification of Marton's [7] original argument for the case of error-exponents in lossy data compression. The optimal sets $\{C_n\}$ achieving $E^*(R,D)$ are randomly generated, and they are *universal* in that their construction only depends on R, D, and M. Therefore, they achieve the optimal error-exponent simultaneously for all distributions P. □

Example 1. Hypothesis Testing.

Let P_0 and P_1 be two probability distributions A with all positive probabilities. Suppose that the null hypothesis that a sample $X_1^n = (X_1, X_2, \ldots, X_n)$ of n independent observations comes from P_0 is to be tested against the simple alternative that X_1^n comes from P_1. Any test between these two hypotheses is simply a decision region $C_n \subset A^n$: If $X_1^n \in C_n$ we declare that $X_1^n \sim P_1^n$, otherwise we declare $X_1^n \sim P_0^n$. The set C_n is called the *critical region*, and the type-I and type-II probabilities of error associated with the test are, respectively,

$$\alpha_n = P_0^n(C_n) \quad \text{and} \quad \beta_n = P_1^n(C_n^c).$$

Clearly we wish to have α_n and β_n both decrease to zero as fast as possible. In particular, we may ask how quickly β_n can decay to zero if we require that α_n decays exponentially at some rate $r > 0$, i.e., $\alpha_n \approx 2^{-nr}$. In statistical terminology, we are asking for the fastest rate of decay of the type-II error probability among all tests with significance level $\alpha_n \leq 2^{-nr}$.

Formally, we want to identify the best exponent of the error probability $\beta_n = 1 - P_1^n(C_n)$ among all C_n with $P_0^n(C_n) \leq 2^{-nr}$. Taking $P = P_1$, $M = P_0$, $R = -r$, and allowing *no* distortion, this question reduces exactly to the our earlier sphere-covering problem. [To be precise, allowing no distortion means we take $D = 0$ with $\rho(x, y)$ being Hamming distortion as in (1.2).] Accordingly, $R(D; P, M) = R(0; P_1, P_0)$ turns out to be equal to $-H(P_1\|P_0)$, and from the theorem we immediately obtain the following classical result of Hoeffding. Also see [2, Thms. 9, 10] and [3, Ex.12, p.43] for versions of this result in the information theory literature.

Corollary 1. *(Hypothesis Testing)* [5] *Let $\{C_n\}$ be an arbitrary sequence of tests with associated error probabilities α_n and β_n as above. Among all tests with*

$$\limsup_{n\to\infty} \frac{1}{n} \log \alpha_n \leq -r$$

for some $r \in (0, H(P_1\|P_0))$, the fastest achievable asymptotic rate of decay of β_n is

$$\lim_{n\to\infty} -\frac{1}{n} \log \beta_n = \inf_{Q\,:\,H(Q\|P_0)<r} H(Q\|P_1).$$

As mentioned earlier, the decision regions C_n in the Corollary are randomly generated. Therefore, although they do achieve asymptotically optimal performance, they are not optimal for finite n in the Neyman-Pearson sense.

Example 2. Lossy Data Compression.

Suppose data $X_1^n = (X_1, X_2, \ldots, X_n)$ is generated by a stationary, memoryless source, i.e., X_1^n are i.i.d. (independent and identically distributed) random variables, with distribution P on the finite alphabet A. The objective of lossy data compression is to find efficient representations $y_1^n \in A^n$ for all source strings $x_1^n \in A^n$. In particular, suppose that the maximum amount of distortion $\rho_n(x_1^n, y_1^n)$ that we are willing to tolerate between a source string x_1^n and its representation y_1^n is some $D \geq 0$, where $\{\rho_n\}$ is a family of single-letter distortion measures as in (1.1). Then the problem is to find an efficient codebook $C_n \subset A^n$ such that for most of the source strings x_1^n there is a $y_1^n \in C_n$ with $\rho_n(x_1^n, y_1^n) \leq D$.

Here, an efficient codebook C_n is one that leads to good compression, i.e., one whose size is as small as possible. And, on the other hand, we also want to make sure that the probability that a source string cannot be represented by any element of C_n with distortion D or less, is small. Taking M to be counting measure ($M(x) = 1$ for all $x \in A$), the mass $M^n(C_n)$ of the codebook becomes its size $|C_n|$, and the problem of finding a good codebook reduces to the earlier sphere-covering question. Accordingly, the rate-function $R(D; P; M)$ reduces to Shannon's rate-distortion function $R(D;P)$, and the theorem yields Marton's error-exponents result.

Corollary 2. *(Lossy Data Compression)* [7] *Let $D \geq 0$ be a given distortion level, and $R(D;P) < R < \log|A|$. Among all sequences of codebooks $\{C_n\}$ with asymptotic rate no greater than R bits/symbol,*

$$\limsup_{n\to\infty} \frac{1}{n} \log |C_n| \leq R,$$

the fastest achievable asymptotic rate of decay of the probability of error is

$$\lim_{n\to\infty} -\frac{1}{n} \log \Big[1 - P^n([C_n]_D)\Big] = \inf_{Q\,:\,R(D;Q)>R} H(Q\|P).$$

Example 3. Measure Concentration on the Binary Cube.

Consider again the setting of the example described in the introduction. There we asked for the *best* achievable error exponent among all D-coverings $\{C_n\}$ with probability no greater that $\approx e^{-nr}$. Taking $M = P$ in the theorem, we obtain the answer to this question in the following Corollary. Let $H_e(P\|Q)$ denote the relative entropy expressed in nats rather than bits, $H_e(P\|Q) = (\log_e 2)H(P\|Q)$, and similarly write $R_e(D;P,M) = (\log_e 2)R(D;P,M)$.

Corollary 3. *(Converse Measure Concentration) Let $D \geq 0$ and $0 < r < -R_e(D;P,P)$. Among all D-coverings $\{C_n\}$ with*

$$\limsup_{n\to\infty} \frac{1}{n} \log_e P^n(C_n) \leq -r,$$

the fastest achievable asymptotic rate of decay of the probability of error is

$$\lim_{n\to\infty} -\frac{1}{n} \log_e \Big[1 - P^n([C_n]_D)\Big] = \mathcal{E}^*(r,D),$$

where

$$\mathcal{E}^*(r,D) \stackrel{\triangle}{=} \inf_{Q\,:\,R_e(D;Q,P)>-r} H_e(Q\|P).$$

Although the exponent $\mathcal{E}^*(r,D)$ above is not as explicit as $(\frac{D^2}{2} - r)$ in (1.3), it is easy to evaluate numerically and it contains much more useful information. For example, Figure 1 shows the graph of $\mathcal{E}^*(r,D)$ as a function of r, for $D = 0.3$, P being the Bernoulli(0.4) distribution, and r running over the range $r \in (0.6109, 0.6393)$ where $\mathcal{E}^*(r,D)$ is nontrivial (i.e., finite and nonzero). In this case, (1.3) is only useful when $(\frac{D^2}{2} - r)$ is positive, i.e., for $r \in (0, 0.045)$: There (1.3) says that, whenever $P^n(C_n) \approx e^{-nr}$ for some $r \in (0, 0.045)$, the probability of error decays exponentially fast. But in that range, and in fact for all r up to ≈ 0.61, we have $\mathcal{E}^*(r,D) = \infty$ so there are sets C_n with $P^n(C_n) \approx e^{-nr}$ and probability of error decaying *super*-exponentially fast. Moreover, in the range $r \in (0.6109, 0.6393)$ where $\mathcal{E}^*(r,D)$ is nontrivial, we can choose C_n with $P^n(C_n) \approx e^{-nr}$ and Pr{"error"} $\approx e^{-n\mathcal{E}^*(r,D)}$.

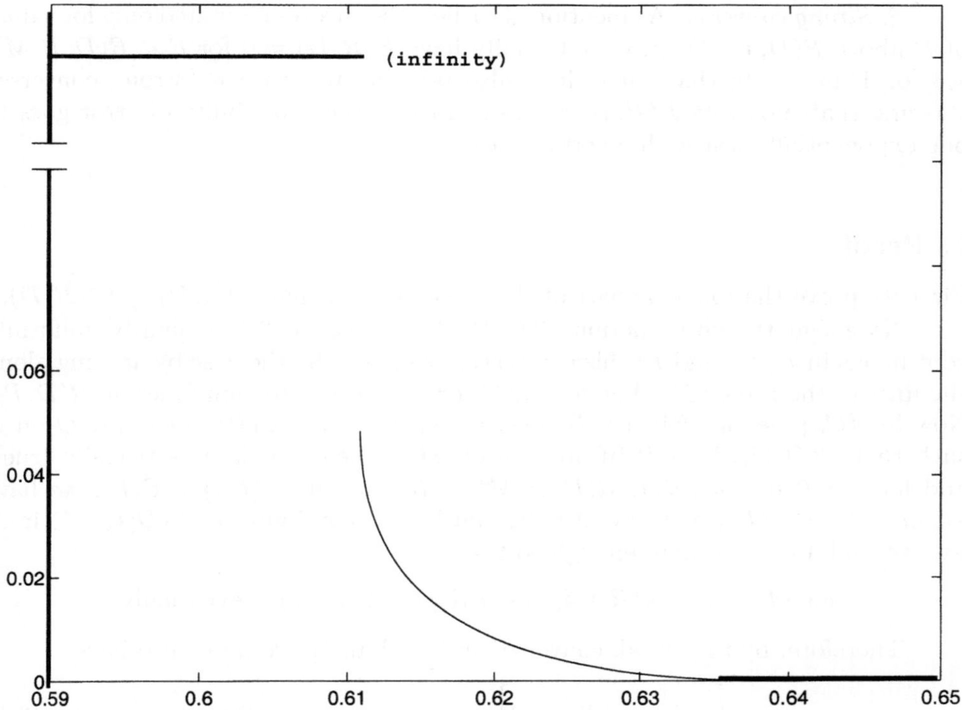

FIGURE 1. Graph of the error-exponent function $\mathcal{E}^*(r, D)$ in Corollary 3 as a function of r, for $D = 0.3$ and $P(1) = 0.4$. Note that $\mathcal{E}^*(r, D)$ is infinite for all $r \in (0, 0.6109)$, and that it is zero for $r > 0.6393$.

Finally we remark that the "extremal" sets in the classical isoperimetric problem, namely, those C_n that achieve equality in (1.3), are very different from the extremal sets in Corollary 3. The former are well-known to be Hamming balls B_n centered at $0^n = (0, 0, \ldots, 0) \in A^n$, $B_n = \{x_1^n : \rho_n(x_1^n, 0^n) \leq \delta\}$ (see [4][8, p. 174][10, Sec. 2.3]), while the latter are collections of strings y_1^n randomly selected from a collection of suitable strings.

Extensions.

1. Different alphabets. Although we assumed from the start that $\rho(x, y)$ is a distortion measure on $A \times A$, it is straightforward to generalize the main result as well as the subsequent discussion above to the case when $\rho(x, y)$ is a distortion measure between the "source" alphabet A and a different ("reproduction") alphabet $\hat{A}$, as long as it is still the case that for each $a \in A$ there exists a $b \in \hat{A}$ with $\rho(a, b) = 0$. The necessary modifications to the statements and proofs follow exactly as in the case of Marton's result; see [3, Sec. 2.4].

2. Strong converse. As mentioned earlier, the theorem is stated only for values of R above $R(D;P,M)$ since we trivially have $E(R,D)=0$ for $R<R(D;P,M)$; see [6, Thm. 1]. In that range it is also possible to prove a "strong converse" showing that, not only $E(R,D)=0$, but in fact the probability of error goes to one exponentially fast with a certain rate.

3. Proof

First we prove the *converse* part of the theorem, i.e., that $E(R,D)\le E^*(R,D)$.

Note that the rate-function $R(D;P,M)$ defined in (2.1) is jointly uniformly continuous in $D\ge 0$ and P; this can be easily seen to be the case by arguing along the lines of the proof of [3, Lemma 2.2.2] for the rate-distortion function $R(D;P)$. Now let $\{C_n\}$ be an arbitrary D-covering with $\{C_n\}\in\mathcal{C}(R)$. Take any Q on A such that $R(D;Q,M)>R$ (if no such Q exists then the claim is trivially true), and let $\delta>0$ be such that $R(D;Q,M)>R+\delta$. Since $\{C_n\}\in\mathcal{C}(R)$, we have $\log M^n(C_n)<n(R+\delta/2)$, eventually, and by the continuity of $R(D;Q,M)$ in D we can find an $\eta>0$ small enough so that

$$\log M^n(C_n)<n(R+\delta/2)<nR(D+\eta;Q,M),\quad \text{eventually.}$$

Therefore, by the "weak converse" in [6, Thm. 1], we must also have

$$E_{Q^n}\left[\min_{y_1^n\in C_n}\rho_n(X_1^n,y_1^n)\right]>D+\eta,\quad \text{eventually,} \tag{3.1}$$

where X_1^n denote n i.i.d. random variables with distribution Q^n. Writing

$$Z_n\triangleq\min_{y_1^n\in C_n}\rho_n(X_1^n,y_1^n),$$

the bound in equation (3.1) implies that

$$D+\eta<E[Z_n]\le D\,Q^n(Z_n\le D)+D_{\max}\,Q^n(Z_n>D),$$

i.e.,

$$Q^n(Z_n>D)>\frac{\eta}{D_{\max}-D}.$$

From Stein's lemma [3, Cor. 1.1.2] we also know that, for any P and any $\epsilon>0$,

$$\lim_{n\to\infty}\frac{1}{n}\log\left[\min_{B_n\subset A^n\,:\,Q^n(B_n)>\epsilon}P^n(B_n)\right]=-D(Q\|P).$$

Taking $\epsilon=\eta/(D_{\max}-D)>0$ and applying this to the events

$$B_n\triangleq\{Z_n>D\}=[C_n]_D^c,$$

yields

$$\liminf_{n\to\infty}\frac{1}{n}\log\left[1-P^n([C_n]_D)\right]\ge -D(Q\|P),$$

and since this holds for all Q with $R(D;Q,M) > R$, we obtain

$$\limsup_{n\to\infty} -\frac{1}{n}\log\left[1 - P^n([C_n]_D)\right] \leq E^*(R,D).$$

Finally, since the sequence $\{C_n\} \in \mathcal{C}(R)$ was arbitrary, this establishes that $E(R,D) \leq E^*(R,D)$, as required.

To prove the *direct* part of the theorem, asserting the existence of a D-covering $\{C_n\} \in \mathcal{C}(R)$ such that

$$\liminf_{n\to\infty} -\frac{1}{n}\log\left[1 - P^n([C_n]_D)\right] \geq E^*(R,D),$$

we follow the same outline as in the proof of the direct part of [3, Thm. 2.4.5].

Using the joint uniform continuity of $R(D;P,M)$ in $D \geq 0$ and P, the proof of the type-covering lemma [3, Lemma 2.4.1] can be generalized to the corresponding statement with $R(D;P,M)$ in place of $R(D;P)$. The main new observation here is that, since all the elements y_1^n of the covering set B are drawn from the set $T^n_{[Y^*]}$ of Y^*-typical strings, where (X^*,Y^*) achieve the infimum in the definition (2.1) of $R(D;P,M)$, their mass $M^n(y_1^n)$ satisfies

$$\frac{1}{n}\log M^n(y_1^n) \leq E[\log M(Y^*)] + \delta_n\left[\sum_y \log M(y)\right],$$

where the sequence $\delta_n \to 0$ as $n \to \infty$.

Finally, following the same steps as in the proof of the direct part of [3, Thm. 2.4.5] and replacing $R(D;P)$ by $R(D;P,M)$, we obtain the existence of a D-covering $\{C_n\} \in \mathcal{C}(R)$ with error exponent no worse than $E^*(R,D) - \delta$, where $\delta > 0$ is an arbitrary constant. This proves that $E(R,D) \geq E^*(R,D)$, and completes the proof. □

Acknowledgments

We wish to thank Amir Dembo and Neri Merhav for asking us (independently) whether the results of [6] could be extended to the case of error-exponents.

References

[1] R. Ahlswede. Extremal properties of rate-distortion functions. *IEEE Trans. Inform. Theory*, **36**(1):166–171, 1990.

[2] R.E. Blahut. Hypothesis testing and information theory. *IEEE Trans. Inform. Theory*, **20**(4):405–417, 1974.

[3] I. Csiszár and J. Körner. *Information Theory: Coding Theorems for Discrete Memoryless Systems.* Academic Press, New York, 1981.

[4] L.H. Harper. Optimal numberings and isoperimetric problems on graphs. *J. Combinatorial Theory*, **1**:385–393, 1966.

[5] W. Hoeffding. Asymptotically optimal tests for multinomial distributions. *Ann. Math. Statist.*, **36**:369–408, 1965.

[6] I. Kontoyiannis. Sphere-covering, measure concentration, and source coding. *IEEE Trans. Inform. Theory*, **47**:1544–1552, May 2001.

[7] K. Marton. Error exponent for source coding with a fidelity criterion. *IEEE Trans. Inform. Theory*, **20**:197–199, 1974.

[8] C. McDiarmid. On the method of bounded differences. In *Surveys in combinatorics (Norwich, 1989)*, pages 148–188. London Math. Soc. Lecture Note Ser., **141**, Cambridge Univ. Press, Cambridge, 1989.

[9] C. McDiarmid. Concentration. In *Probabilistic methods for algorithmic discrete mathematics*, pages 195–248. Algorithms Combin., **16**, Springer, Berlin, 1998.

[10] M. Talagrand. Concentration of measure and isoperimetric inequalities in product spaces. *Inst. Hautes Études Sci. Publ. Math.*, No. **81**:73–205, 1995.

Division of Applied Mathematics, and Department of Computer Science, Brown University, 182 George Street, Providence, RI 02912, USA
E-mail address: `yiannis@dam.brown.edu`

Division of Applied Mathematics, Brown University, 182 George Street, Providence, RI 02912, USA
E-mail address: `ali_sezer@brown.edu`

Progress in Probability, Vol. 56, 33–52

Concentration Inequalities for Convex Functions on Product Spaces

Paul-Marie Samson

Abstract. Let $\mu = \mu_1 \otimes \cdots \otimes \mu_n$ denote a product probability measure on $\mathbb{R}^n$ with compact support. We present a simple proof to get concentration results for convex functions on $\mathbb{R}^n$ under μ. We use the infimum convex convolution description of the concentration phenomenon introduced by Maurey [14]. The main result of this paper is a transportation inequality for the measure μ comparable to the classical transportation inequality for the canonical Gaussian measure on $\mathbb{R}^n$. In application, we present Khinchine Kahane inequalities for norms of random series with non-symmetric Bernoulli coefficients.

1. Introduction

Following the work by Talagrand [9], [10], Maurey [14], Bobkov, Ledoux and Gentil [2], [3], Marton [12], [13], and [15], the measure concentration phenomenon may be stated by proving transportation inequalities. To clarify, let us recall the classical transportation inequality satisfied by the canonical Gaussian measure γ_n on $\mathbb{R}^n$. Given two probability measures μ and ν, let us define the relative entropy of a measure ν with respect to a measure μ, or the so-called informational divergence or Kullback distance of ν with respect to μ: if ν is absolutely continuous with respect to μ with density $d\nu/d\mu$,

$$\mathrm{Ent}_\mu\left(\frac{d\nu}{d\mu}\right) = \int \frac{d\nu}{d\mu} \log \frac{d\nu}{d\mu} d\mu.$$

The canonical Gaussian measure satisfies the following transportation inequality (see [11]), for every probability measure ν absolutely continuous with respect to γ_n,

$$W_2^2(\gamma_n, \nu) \leq \mathrm{Ent}_{\gamma_n}\left(\frac{d\nu}{d\gamma_n}\right).$$

Received by the editors September 22, 2003.
Key words and phrases. measure concentration, transportation inequalities, infimum convolution.

W_2 is the L^2-Wasserstein distance for the quadratic cost function $C(h) = |h|^2/2$, $h \in \mathbb{R}^n$, where $|h|$ denotes the Euclidean norm of h. One has

$$W_2^2(\gamma_n, \nu) = \inf_\pi \iint C(x-y)d\pi(x,y),$$

where the infimum is taken over all probability measures π on $\mathbb{R}^n \times \mathbb{R}^n$ with marginals ν and γ_n. The Monge–Kantorovitch Theorem provides a dual representation of W_2

$$W_2^2(\gamma_n, \nu) = \sup_f \left[\int g d\nu - \int f d\gamma_n \right],$$

where the supremum is running over all bounded measurable functions f and g on $\mathbb{R}^n$ such that

$$g(x) \leq f(y) + C(x-y)$$

for every $x, y \in \mathbb{R}^n$. For a given function f, the infimum convolution $Q_C f$ with quadratic cost C,

$$Q_C f(x) = \inf_{y \in \mathbb{R}^n} [f(y) + C(x-y)]$$

achieves the optimal choice for g. As is well known (see e. g. [1],[4]), if f is bounded then $Q_C f$ is bounded and Lipschitz. Consequently, we may state the Gaussian transportation inequality as follows: for every bounded measurable function f, for every probability measure ν absolutely continuous with respect to γ_n,

$$\int Q_C f d\nu - \int f d\gamma_n \leq \mathrm{Ent}_{\gamma_n}\left(\frac{d\nu}{d\gamma_n}\right).$$

As we know from [2] this inequality is equivalent to

$$\int e^{Q_C f} d\gamma_n \leq e^{\int f d\gamma_n}.$$

The L^2-Wasserstein distance with the quadratic cost is well suited to describe the concentration phenomena for the Gaussian measure (see [2]).

The main purpose of this note is to present similar transportation inequalities for any product measure $\mu = \mu_1 \otimes \cdots \otimes \mu_n$ on $\mathbb{R}^n$ with compact support. The basic idea, which may be found in Maurey's paper is the following: by a restriction of the class of functions to convex functions on $\mathbb{R}^n$, we may tensorise the transportation inequality in dimension one to get a transportation inequality in dimension n. The main interest of this paper is to provide a simple systematic proof in dimension one to reach a powerful cost function associated to any kind of probability measure with compact support on $\mathbb{R}$. The result in dimension n is the following (see Theorem 2.1): for every probability measures ν_1 and ν_2 on $\mathbb{R}^n$ absolutely continuous with respect to the product measure μ and for every convex measurable function f on $\mathbb{R}^n$, one has

$$\int Q_{C_\alpha} f d\nu_1 - \int f d\nu_2 \leq \frac{1}{\alpha} \mathrm{Ent}_\mu\left(\frac{d\nu_1}{d\mu}\right) + \frac{1}{\beta} \mathrm{Ent}_\mu\left(\frac{d\nu_2}{d\mu}\right),$$

where α and β are some non-negative parameters with $\alpha+\beta=1$. The cost function C_α which defines the infimum convolution $Q_{C_\alpha}f$, depends on the diameter of the support of each probability measure μ_i. If these diameters are less than one, C_α may be compared to the quadratic cost function with $C_\alpha(h) \geq |h|^2/2$, $h \in \mathbb{R}^n$. The above transportation inequality is symmetric with respect to ν_1, ν_2 and μ and is equivalent to the next exponential inequality,

$$\left(\int e^{\alpha Q_{C_\alpha} f} d\mu\right)^{\frac{1}{\alpha}} \left(\int e^{-\beta f} d\mu\right)^{\frac{1}{\beta}} \leq 1. \tag{1.1}$$

If we adopt the convention $+\infty.0 \leq 1$, we see easily that the above inequality extends to all convex measurable functions taking values in $\mathbb{R} \cup \{+\infty\}$.

In section 2.4, this method is applied to non-symmetric Bernoulli measures. Let p be a parameter in $]0,1[$, if for every $1 \leq i \leq n$, $\mu_i = p\delta_{(1-p)} + (1-p)\delta_{-p}$, then the cost function C_α may be replaced by a better one (see Theorem 2.5).

From the inequality (1.1), we deduce that for every convex Lipschitz function f with Lipschitz norm $\|f\|_{\mathrm{Lip}}$,

$$\left(\int e^{\alpha\lambda f} d\mu\right)^{\frac{1}{\alpha}} \left(\int e^{-\beta\lambda f} d\mu\right)^{\frac{1}{\beta}} \leq e^{\lambda^2\|f\|^2_{\mathrm{Lip}}/2}. \tag{1.2}$$

By Jensen, this implies the classical bound for the Laplace transform of convex Lipschitz functions,

$$\int e^{\lambda(f-\int f d\mu)} d\mu \leq e^{\lambda^2\|f\|^2_{\mathrm{Lip}}/2}.$$

This bound follows from the log-Sobolev method introduced by Ledoux [6]. Nevertheless, up to now, the log-Sobolev method does not provide inequality (1.2).

Let us now describe the concentration phenomenon in terms of sets. For every convex set $A \subset \mathbb{R}^n$, let us consider the convex function ϕ_A defined as $\phi_A(x) = 0$ if $x \in A$ and $\phi_A(x) = +\infty$ otherwise. ϕ_A satisfies (1.1) which provides

$$\int e^{\alpha d_{C_\alpha}(x,A)} d\mu(x) \leq \frac{1}{\mu(A)^{\frac{\alpha}{1-\alpha}}},$$

where $d_{C_\alpha}(x,A) = \inf_{y\in A} C_\alpha(x-y)$. This result has to be compared with Theorem 4.2.4 by Talagrand [9]. Actually, this result is quite equivalent to Theorem 4.2.4 restricted to convex sets. With the notations of Talagrand, setting $\alpha = \frac{\gamma}{\gamma+1}$, $\gamma \in \mathbb{R}^+$, we exactly have using the definition (4.2.1) [9] of ξ and f_γ,

$$\alpha C_\alpha(h) = \xi(\gamma, h)$$

and, for every convex set A,

$$\alpha d_{C_\alpha}(x,A) \leq f_\gamma(A,x).$$

Theorem 4.2.4 mentions that

$$\int e^{f_\gamma(A,x)} d\mu(x) \leq \frac{1}{\mu(A)^\gamma}.$$

From the functional form (1.1) of this "convex" concentration phenomenon, we easily deduce Khintchine–Kahane inequalities with explicit constants for the norms of sum of vectors with independent random coefficients (see section 2.3). Let $e_1, \ldots, e_n$ be some vectors on a Banach space $(E, \|\cdot\|)$. In the case of non-symmetric Bernoulli sequences (see section 2.4), when $\varepsilon_1, \ldots, \varepsilon_n$ are independent random variables with centered law $p\delta_{(1-p)} + (1-p)\delta_{-p}$, we get for every $q \geq 2$,

$$\|Z\|_q \leq 2\sqrt{\frac{q}{p}}\|Z\|_2,$$

where $\|Z\|_q$ denotes the L_q norm of the random variable

$$Z = \left\| \sum_{i=1}^{n} \varepsilon_i e_i \right\|.$$

In this Khintchine–Kahane inequality, the constant 2 is not optimal. Let us note that in the real case, when e_i are some real number and $Z = |\sum_{i=1}^n \varepsilon_i e_i|$, Oleszkiewicz recently obtained the optimal constant in the Khinchine–Kahane Inequality [8] (in the real case for symetric Bernoulli sequences see also [5]).

2. Statement of the results

2.1. A transportation inequality and an exponential inequality

Let us first present some cost functions associated to the convex concentration phenomena for any probability measure μ on $\mathbb{R}$ with compact support. Let w and v be some non-negative convex functions defined as follows, for every $h \leq 1$,

$$w(h) = (1-h)\log(1-h) + h,$$

and $w(h) = +\infty$ otherwise; and for every $h < 1$,

$$v(h) = -\log(1-h) - h,$$

and $v(h) = +\infty$ otherwise. These two functions are directly linked to the informational divergence by the following identities: if μ and ν are two probability measures and θ is the density of ν with respect to μ,

$$\int w(1-\theta)d\mu = \mathrm{Ent}_\mu\left(\frac{d\nu}{d\mu}\right), \tag{2.1}$$

and

$$\int v\left(1 - \frac{1}{\theta}\right) d\nu = \mathrm{Ent}_\mu\left(\frac{d\nu}{d\mu}\right).$$

For every $0 < \alpha, \beta < 1$, $\alpha + \beta = 1$, let us define the non-negative function c_α on $\mathbb{R}$ by

$$c_\alpha(h) = \frac{1}{\alpha\beta}(\alpha w(h) - w(\alpha h)), \quad h \leq 1,$$

and $c_\alpha(h) = +\infty$ if $h > 1$. By differentiating, we check that c_α is convex. We may compare the functions c_α, v, w. If $h \geq 0$ one has

$$v(h) \geq c_\alpha(h) \geq w(h) \geq \frac{h^2}{2}. \tag{2.2}$$

Moreover, since $\dfrac{w(\alpha h)}{\alpha} \leq \alpha \dfrac{h^2}{2}$, one has

$$c_\alpha(h) \underset{\alpha \to 0}{\longrightarrow} w(h).$$

In a same way, if α goes to 1, since

$$c_\alpha(h) = \frac{h}{\alpha} \frac{w(h) - w(\alpha h)}{(1-\alpha)h} - \frac{w(h)}{\alpha},$$

one has

$$c_\alpha(h) \underset{\alpha \to 1}{\longrightarrow} hw'(h) - w(h) = v(h).$$

We may say that c_α is an interpolation function between w and v.

One of the main results of this note is the following transportation inequality valid for any probability measure with compact support. The proof is given in Section 3.

Theorem 2.1. *Let $\mu = \mu_1 \otimes \cdots \otimes \mu_n$ be a product probability measure with for every $1 \leq i \leq n$, $0 < \operatorname{diam}(\operatorname{supp} \mu_i) \leq D_i$, where $\operatorname{diam}(\operatorname{supp} \mu_i)$ is the diameter of the support of μ_i. Let us consider some parameters $\alpha > 0$, $\beta > 0$ with $\alpha + \beta = 1$. Let us define the convex cost function C_α on $\mathbb{R}^n$ with*

$$C_\alpha(h) = \sum_{i=1}^n c_\alpha\left(\frac{|h_i|}{D_i}\right), \quad h = (h_1, \ldots, h_n) \in \mathbb{R}^n.$$

For every probability measures ν_1 and ν_2 on $\mathbb{R}^n$ absolutely continuous with respect to μ and for every convex measurable function f on $\mathbb{R}^n$, one has

$$\int Q_{C_\alpha} f \, d\nu_1 - \int f \, d\nu_2 \leq \frac{1}{\alpha} \operatorname{Ent}_\mu\left(\frac{d\nu_1}{d\mu}\right) + \frac{1}{\beta} \operatorname{Ent}_\mu\left(\frac{d\nu_2}{d\mu}\right); \tag{2.3}$$

$Q_{C_\alpha} f$ is the infimum convolution of f associated to the cost function C_α, for every $x \in \mathbb{R}^n$,

$$Q_{C_\alpha} f(x) = \inf_{y \in \mathbb{R}^n} \left[f(y) - C_\alpha(x-y)\right].$$

Equivalently for every convex measurable function f on $\mathbb{R}^n$, one has

$$\left(\int e^{\alpha Q_{C_\alpha} f} d\mu\right)^{\frac{1}{\alpha}} \left(\int e^{-\beta f} d\mu\right)^{\frac{1}{\beta}} \leq 1. \tag{2.4}$$

The equivalence between the transportation inequality (2.3) and the exponential inequality (2.4) is easy to prove using the same argument as in [2]. Note that if $\frac{d\nu}{d\mu} = \frac{e^g}{\int e^g d\mu}$, then

$$\mathrm{Ent}_\mu\left(\frac{d\nu}{d\mu}\right) = \int g d\nu - \log \int e^g d\mu.$$

Applying the transportation inequality (2.3) to the measures ν_1 and ν_2 with respective densities $\frac{d\nu_1}{d\mu} = \frac{e^{\alpha Q_{C_\alpha} f}}{\int e^{\alpha Q_{C_\alpha} f} d\mu}$ and $\frac{d\nu_2}{d\mu} = \frac{e^{-\beta f}}{\int e^{-\beta f} d\mu}$, we get

$$\frac{1}{\alpha} \log \int e^{\alpha Q_{C_\alpha} f} + \frac{1}{\beta} \log \int e^{-\beta f} d\mu \leq 0.$$

Then we deduce (2.4) by taking the exponential in this inequality. Conversely, the transportation inequality (2.3) follows from the exponential inequality (2.4) combined with the identity

$$\log \int e^g d\mu = \sup_\nu \left[\int g d\nu - \mathrm{Ent}_\mu \left(\frac{d\nu}{d\mu}\right) \right],$$

where the supremum runs over all probability measures ν on $\mathbb{R}^n$. Let V and W be the cost functions on $\mathbb{R}^n$ associated to the convex functions v and w on $\mathbb{R}$, for every $h = (h_1, \ldots, h_n) \in \mathbb{R}^n$,

$$W(h) = \sum_{i=1}^n w(|h_i|), \qquad V(h) = \sum_{i=1}^n v(|h_i|).$$

When α goes to zero, the exponential inequality (2.4) provides

$$\int e^{Q_V f} d\mu \leq e^{\int f d\mu},$$

where $Q_V f$ is the infimum convolution associated to the cost V. Similarly, when α goes to one, from (2.4) we get

$$\int e^{-f} d\mu \leq e^{-\int Q_W f d\mu}.$$

These two inequalities provide concentration results for regular convex functions as we will show in the next section.

2.2. Exponential inequality and concentration

In this section, we present the method mentioned in [3] to derive exponential inequalities from Theorem 2.1. Let f be a smooth convex function, for every x, y in $\mathbb{R}^n$

$$f(y) \geq f(x) - \sum_{i=1}^n \partial_i f(x)(x_i - y_i) = f(x) - \langle \nabla f(x), x - y \rangle,$$

where $\partial_i f$ is the i-th partial derivative of f, ∇f is the gradient of f and $\langle \ , \ \rangle$ denotes the usual scalar product on $\mathbb{R}^n$. By convenience we assumed that f is smooth, nevertheless the last inequality is still true replacing the gradient vector $\nabla f(x)$ by any vector of the subdifferential of f at x. By this way, we may extend the next results to any convex function f.

From the definition of $Q_{C_\alpha} f$, we get that for every $x \in \mathbb{R}^n$, one has

$$f(x) \geq Q_{C_\alpha} f(x) \geq f(x) - \sup_{y \in \mathbb{R}^n} \sum_{i=1}^{n} \left[|\partial_i f(x)||x_i - y_i| - c_\alpha(|x_i - y_i|) \right].$$

Let c_α^* denote the Legendre transform of c_α, $c_\alpha^*(\ell) = \sup_{h \in \mathbb{R}}[h\ell - c_\alpha(\ell)], \ell \in \mathbb{R}$. Using (2.2), we may compare c_α^* with v^* and w^*, for every $\ell \geq 0$,

$$v^*(\ell) \leq c_\alpha^*(\ell) \leq w^*(\ell) \leq \frac{\ell^2}{2}.$$

Consequently if $C_\alpha^*(h) = \sum_{i=1}^{n} c_\alpha^*(|h_i|)$, $h = (h_1, \dots, h_n) \in \mathbb{R}^n$, one has

$$0 \leq f(x) - Q_{C_\alpha} f(x) \leq C_\alpha^*(\nabla f(x)) \leq \frac{|\nabla f(x)|^2}{2}.$$

In other words, the difference $f - Q_{C_\alpha} f$ is controlled by the smoothness of f. Using this estimate, from inequality (2.4) of Theorem 2.1 we deduce the next Corollary.

Corollary 2.2. *Let $\mu = \mu_1 \otimes \cdots \otimes \mu_n$ be a probability measure on $\mathbb{R}^n$, with for every $1 \leq i \leq n$, $0 < \operatorname{diam}(\operatorname{supp} \mu_i) \leq D_i$. Let $\alpha > 0$, $\beta > 0$, $\alpha + \beta = 1$. For every smooth convex function f on $\mathbb{R}^n$ one has*

$$\left(\int e^{\alpha[f - C_\alpha^*(\nabla f)]} d\mu \right)^{\frac{1}{\alpha}} \left(\int e^{-\beta f} d\mu \right)^{\frac{1}{\beta}} \leq 1. \tag{2.5}$$

If α goes to zero we get

$$\int e^{[f - V^*(\nabla f)]} d\mu \leq e^{\int f d\mu},$$

and if α goes to one

$$\int e^{-f} d\mu \leq e^{-\int [f - W^*(\nabla f)] d\mu}.$$

To illustrate Corollary 2.2, assume f is a convex Lipschitz function, for every $x \in \mathbb{R}^n$ $|\nabla f(x)| \leq \|f\|_{\mathrm{Lip}}$.

Applying Corollary 2.2 to λf, $\lambda \geq 0$, since

$$C_\alpha^*(\lambda \nabla f) \leq \frac{\lambda^2 |\nabla f|^2}{2} \leq \frac{\lambda^2 \|f\|_{\mathrm{Lip}}^2}{2},$$

we obtain

$$\left(\int e^{\alpha \lambda f} d\mu \right)^{\frac{1}{\alpha}} \left(\int e^{-\beta \lambda f} d\mu \right)^{\frac{1}{\beta}} \leq e^{\lambda^2 \|f\|_{\mathrm{Lip}}^2 / 2}. \tag{2.6}$$

This inequality generalises the classical Gaussian type inequalities for Lipschitz convex function

$$\int e^{\lambda(f - \int f d\mu)} d\mu \leq e^{\lambda^2 \|f\|_{\mathrm{Lip}}^2 / 2}, \qquad \int e^{-\lambda(f - \int f d\mu)} d\mu \leq e^{\lambda^2 \|f\|_{\mathrm{Lip}}^2 / 2},$$

which follows when α goes to zero or one. The Tchebichev inequality then provides the Gaussian deviation inequalities of f around its mean under μ.

Now, assume moreover f is a non-negative function. Let us apply Corollary 2.2 to λf^2. One has $|\nabla(\lambda f^2)|^2 = 4\lambda^2 f^2|\nabla f|^2$. Consequently, if f is Lipschitz then $C^*_\alpha(\nabla(\lambda f^2)) \leq 2\lambda^2\|f\|^2_{\mathrm{Lip}} f^2$. From Corollary 2.2 we get

$$\left(\int e^{\alpha\lambda f^2(1-2\lambda\|f\|^2_{\mathrm{Lip}})}d\mu\right)^{\frac{1}{\alpha}}\left(\int e^{-\beta\lambda f^2}d\mu\right)^{\frac{1}{\beta}} \leq 1,$$

for every $\lambda \geq 0$. This inequality is interesting for small λ, if $\lambda \leq 1/(4\|f\|^2_{\mathrm{Lip}})$, then one has

$$\left(\int e^{\alpha\lambda f^2/2}d\mu\right)^{\frac{1}{\alpha}}\left(\int e^{-\beta\lambda f^2}d\mu\right)^{\frac{1}{\beta}} \leq 1. \tag{2.7}$$

If α goes to zero (or by Jensen inequality), we found the classical inequality: for every $\lambda \leq 1/(4\|f\|^2_{\mathrm{Lip}})$,

$$\int e^{\lambda f^2/2}d\mu \leq e^{\lambda\int f^2 d\mu}. \tag{2.8}$$

Clearly, using the result of Bobkov, Götze [2], inequality (2.8) is a consequence of the logarithmic Sobolev inequality for smooth convex function obtained by Ledoux [6], nevertheless the stronger form (2.7) can not be deduced from the same logarithmic Sobolev inequality. As we will see in the next section, (2.8) gives a simple proof of Khintchine–Kahane inequalities.

2.3. Application to Orlicz norm and Khintchine-Kahane inequalities

The inequality (2.8) may be applied to the norms of sum of vectors with random coefficients. Then we get an estimate of the usual Orlicz norm ψ_2 for these norms of random series. We also deduce Khintchine–Kahane inequalities from this ψ_2 estimate with explicit constants. The proofs are short. We will not use the Gaussian estimate of the distribution tails as in [7]. Let us consider the Young function $\Psi_2 : \mathbb{R}^+ \longrightarrow \mathbb{R}^+$ (i. e., an even, non-negative convex function with $\psi_2(0) = 0$ and $\lim_{x\to\infty}\psi_2(x) = \infty$), for every $x \in \mathbb{R}^+$,

$$\psi_2(x) = e^{x^2} - 1.$$

Let $L_{\psi_2} = L_{\psi_2}(\Omega, A, \mathbb{P})$ denote the Orlicz space of all random variables on a probability space $(\Omega, A, \mathbb{P})$ such that $\mathbb{E}\psi_2(|X|/c)) < \infty$, for some $c > 0$. The space L_{ψ_2} equipped with the norm $\|X\|_{\psi_2} = \inf\{c > 0, \mathbb{E}\psi_2(|X|/c)) \leq 1\}$ is a Banach space. Let $X_1, \ldots, X_n$ be independent random variables on $(\Omega, A, \mathbb{P})$. We assume that for every $1 \leq i \leq n$, one has $a_i \leq X_i \leq b_i$ with $|b_i - a_i| \leq 1$. Let $e_1, \ldots, e_n$ be vectors in a Banach space E with norm $\|\cdot\|$. We consider the random variable

$$Z = \left\|\sum_{i=1}^n X_i e_i\right\|.$$

Let f denote the convex function,

$$f(x) = \left\| \sum_{i=1}^{n} x_i e_i \right\| = \sup_{\|e^*\|_* \leq 1} \sum_{i=1}^{n} x_i e^*(e_i), \qquad x = (x_1, \ldots, x_n) \in \mathbb{R}^n,$$

where $\| \cdot \|_*$ is the dual norm of $\| \cdot \|$ and e^* is any element of the dual space of E. Clearly $Z = f(X_1, \ldots, X_n)$. The quantity

$$\sigma^2 = \sup_{\|e^*\|_* \leq 1} \sum_{i=1}^{n} e^*(e_i)^2,$$

is a kind of weak variance, σ is the Lipschitz norm of f. Replacing $\|f\|_{\mathrm{Lip}}$ by σ, f satisfies the inequality (2.8). If μ is the law of $(X_1, \ldots X_n)$, we get for every $0 \leq \lambda \leq 1/(4\sigma^2)$,

$$\mathbb{E}\left(e^{\lambda Z^2/2}\right) \leq e^{\lambda \mathbb{E}(Z^2)}. \tag{2.9}$$

Let us assume that there exist a constant K independent of n such that $\sigma^2 \leq K\mathbb{E}(Z^2)$. This assumption follows the Bernoulli model when $X_i = \varepsilon_i, 1 \leq i \leq n$ are symmetric random variables on $\{-1, 1\}$, by orthogonality one has

$$\sigma^2 = \sup_{\|e^*\|_* \leq 1} \mathbb{E}\left[\left(\sum_{i=1}^{n} \varepsilon_i e^*(e_i)\right)^2\right] \leq \mathbb{E}(Z^2).$$

Let us note that this inequality is still satisfied when the ε_i's are independent standard Gaussian random variables. In the same way, if the X_i's are centered, then by orthogonality one has

$$\sigma^2 \leq \frac{1}{\inf_i \mathbb{E}(X_i^2)} \sup_{\|e^*\|_* \leq 1} \mathbb{E}\left[\left(\sum_{i=1}^{n} X_i e^*(e_i)\right)^2\right] \leq \frac{1}{\inf_i \mathbb{E}(X_i^2)} \mathbb{E}(Z^2),$$

and therefore $\sigma^2 \leq K\mathbb{E}(Z^2)$, with $K = 1/\inf_i \mathbb{E}(X_i^2)$.

Now, choosing $\lambda = \dfrac{1}{4K\mathbb{E}(Z^2)}$, we deduce from (2.9) the following ψ_2 estimate,

$$\mathbb{E}\left(\exp \frac{Z^2}{8K\mathbb{E}(Z^2)}\right) \leq e^{1/(4K)},$$

and if $K \geq \dfrac{1}{4\log 2}$ then

$$\|Z\|_{\psi_2} \leq 8K\|Z\|_2^2,$$

where $\|Z\|_q, q \geq 1$ denotes the L^q-norm of Z, $\|Z\|_q^q = \mathbb{E}(Z^q)$. We turn now to the proof of the Khintchine–Kahane inequalities. By the Cauchy–Schwarz inequality, one has

$$\|Z\|_q^q \leq \mathbb{E}\left(Z^{2q} \exp - \frac{Z^2}{8K\mathbb{E}(Z^2)}\right)^{\frac{1}{2}} \mathbb{E}\left(\exp \frac{Z^2}{8K\mathbb{E}(Z^2)}\right)^{\frac{1}{2}}.$$

Then, using the previous ψ_2 estimate, we get for every $q \geq 1$,

$$\|Z\|_q^q \leq e^{1/(8K)} \sup_{t\geq 0}\left(t^q \exp -\frac{t^2}{16K\mathbb{E}(Z^2)}\right),$$

that is

$$\|Z\|_q \leq 2\sqrt{\frac{2K}{e}} e^{1/(8Kq)}\sqrt{q}\|Z\|_2.$$

2.4. Exponential inequalities for Bernoulli sequences

We hope that the method used to get Theorem 2.1 may be applied to reach concentration results for some specific probability measures. In this section, we consider the case of non symmetric Bernoulli sequences $(\varepsilon_i)_i$. The goal is to improve the constants in the latter exponential inequalities. For $1 \leq i \leq n$, let $p_i \in]0,1[$ and ε_i be a random variable with distribution

$$\mu_i = (1-p_i)\delta_{-p_i} + p_i\delta_{(1-p_i)}.$$

The mean of each ε_i is zero. Let $\mu = \mu_1 \otimes \cdots \otimes \mu_n$ be the law of $(\varepsilon_1, \ldots, \varepsilon_n)$.

As in the first part, we define a cost function associated to the measure μ. For any parameter $p \in]0,1[$, let us first introduce a useful convex function u_p, for every $-\frac{1-p}{p} \leq h \leq 1$,

$$u_p(h) = \frac{1-p}{p}\left(1+\frac{p}{1-p}h\right)\log\left(1+\frac{p}{1-p}h\right) + (1-h)\log(1-h),$$

and $u_p(h) = +\infty$ otherwise. Notice that

$$u_p(h) = \frac{1-p}{p} w\left(-\frac{p}{1-p}h\right) + w(h), \quad h \in \mathbb{R}. \tag{2.10}$$

This function is connected to the informational divergence as follows. If ν is a probability measure with density θ with respect to $\mu = (1-p)\delta_{-p} + p\delta_{(1-p)}$, we have

$$(1-p)\theta(-p) + p\theta(1-p) = 1.$$

Then, the identity

$$\mathrm{Ent}_\mu\left(\frac{d\nu}{d\mu}\right) = \int w(1-\theta)d\mu$$

may be rewritten as

$$\mathrm{Ent}_\mu\left(\frac{d\nu}{d\mu}\right) = pu_p(1-\theta(1-p)) = (1-p)u_{1-p}(1-\theta(-p)). \tag{2.11}$$

From the proof of Theorem 2.1 in dimension one (see section 3), we know that the expression of the cost function c_α, $0 \leq \alpha \leq 1$ directly follows from the identity (2.1). Similarly, using the identity (2.11) and adapting the proof to Bernoulli measures, we get the following expression for an analogue of the function c_α, $0 \leq \alpha \leq 1$. Let d_α be the following quantity, for every $0 \leq h \leq 1$,

$$d_\alpha(p,h) = \inf_{\tau_1,\tau_1 \geq 1} \left[\frac{1}{\alpha\tau_1} u_p(1-\tau_1) + \frac{1}{\beta\tau_1} u_p(1-\tau_1(1-h)) \right], \tag{2.12}$$

and for every $h > 1$, $d_\alpha(p,h) = +\infty$. The cost function on $\mathbb{R}$ associated to the Bernoulli measure $\mu = (1-p)\delta_{-p} + p\delta_{(1-p)}$ is defined as

$$d_{p,\alpha}(h) = d_\alpha(p,h), \qquad \text{for } h \geq 0,$$

and

$$d_{p,\alpha}(h) = d_\alpha(1-p,-h), \qquad \text{for } h < 0.$$

From the definition (2.10) of u_p and the identity (3.2), one has

$$d_{p,\alpha}(h) \geq c_\alpha(|h|),$$

for every $h \in \mathbb{R}$. Therefore, for the non symmetric Bernoulli measure, the following lemma improve Theorem 2.1 in dimension one.

Lemma 2.3. *Let μ denote the non symmetric Bernoulli measure with parameter $p \in]0,1[$. Let $\alpha > 0$, $\beta > 0$ denote some parameters with $\alpha + \beta = 1$. For every convex functions f on $\mathbb{R}$, one has*

$$\left(\int e^{\alpha Q_{p,\alpha} f} d\mu \right)^{\frac{1}{\alpha}} \left(\int e^{-\beta f} d\mu \right)^{\frac{1}{\beta}} \leq 1, \tag{2.13}$$

where $Q_{p,\alpha} f$ is the infimum convolution of f associated to the cost function $d_{p,\alpha}$.

The proof of this Lemma is presented in Section 3. Obviously the definition of $d_{p,\alpha}$ is not simple to use. As in the proof of Theorem 2.1, we need to prove that $d_{p,\alpha}$ is a convex cost function to tensorise (2.13) by applying Lemma 3.2. We will not prove the convexity of $d_{p,\alpha}$. However, in the next Proposition we compare $d_{p,\alpha}$ to some powerful convex functions.

Proposition 2.4.

(i) *For every $0 < p < 1$, $h \geq 0$,*

$$d_\alpha(p,h) \underset{\alpha\to 0}{\longrightarrow} u_p(h), \qquad \textit{and} \qquad d_\alpha(p,h) \underset{\alpha\to 1}{\longrightarrow} (1-p) c_{1-p}\left(\frac{h}{1-p}\right).$$

(ii) *Let $p = 1/2$, for every $h \geq 0$,*

$$h^2 \leq u(h) \leq d_\alpha(1/2,h),$$

where $u(h) = u_{1/2}(h) = w(h) + w(-h)$.

(iii) *For every $0 < p < 1$, $h \geq 0$,*

$$d_\alpha(p,h) \geq \max(c_\alpha(h), 2ph^2).$$

Proof. In order to prove (i) when α tends to zero, we first note that for $h \geq 0$,

$$\liminf_{\alpha\to 0} d_\alpha(p,h) \leq \limsup_{\alpha\to 0} d_\alpha(p,h) \leq u_p(h). \tag{2.14}$$

For any $0 \leq h \leq 1$ fixed, let ψ denotes the function

$$\psi(\tau_1) = \frac{1}{\alpha\tau_1} u_p(1-\tau_1) + \frac{1}{\beta\tau_1} u_p(1-\tau_1(1-h)), \ \ \tau_1 \geq 0.$$

Differentiating in τ_1, we get

$$\psi'(\tau_1) = -\frac{1}{\beta\alpha\tau_1^2}\left[\beta\log\left(1-\frac{p}{1-p}(\tau_1-1)\right)+\alpha\log\left(1-\frac{p}{1-p}(\tau_1(1-h)-1)\right)\right].$$

Consequently we see that the infimum of ψ for $0 \leq h \leq 1$ is reached for the unique $\hat{\tau}_1 \geq 1$ satisfying

$$\left(1-\frac{p}{1-p}(\hat{\tau}_1-1)\right)^\beta\left(1-\frac{p}{1-p}(\hat{\tau}_1(1-h)-1)\right)^\alpha = 1.$$

Therefore, since $\hat{\tau}_1 \geq 1$, one has $(1-h)\hat{\tau}_1 \leq 1$ and $\inf_{\tau_1\geq 0}\psi(\tau_1) = d_\alpha(p,h)$, that is

$$d_\alpha(p,h) = \left[\frac{1}{\alpha\hat{\tau}_1} u_p(1-\hat{\tau}_1) + \frac{1}{\beta\hat{\tau}_1} u_p(1-\hat{\tau}_1(1-h))\right],$$

for every $0 \leq h \leq 1$. Let us note that the point $\hat{\tau}_1$ depends on α. When α goes to zero, using (2.14), we conclude that $\hat{\tau}_1$ tends to 1. Finally, we get

$$\liminf_{\alpha\to 0} d_\alpha(p,h) = \limsup_{\alpha\to 0} d_\alpha(p,h) = u_p(h).$$

This ends the proof of the first part of (i). We use identical arguments for the second part of (i).

No we turn to the proof of (ii). One easily checks that $h^2 \leq u(h)$. The second inequality of (ii) requires more works. Minimizing over all $0 \leq \alpha \leq 1$, the variational identity,

$$\inf_\alpha\left(\frac{A}{\alpha}+\frac{B}{1-\alpha}\right) = \left(\sqrt{A}+\sqrt{B}\right)^2, \ A, B \geq 0, \tag{2.15}$$

provides

$$d_\alpha(1/2,h) \geq \inf_{\tau_1} H^2(\tau_1),$$

for every $0 \leq h \leq 1$, where the infimum is taken over all $\tau_1 \geq 1$ satisfying $\tau_1(1-h) \leq 1$, and

$$H^2(\tau_1) = \frac{1}{\tau_1}\left(\sqrt{u(1-\tau_1)}+\sqrt{u(1-\tau_1(1-h))}\right)^2.$$

Noting that $H^2(1) = u(h)$, the proof of (ii) is finished provided H is an increasing function. One has

$$H'(\tau_1) = -\frac{1}{\tau_1^{3/2}}\left(\frac{\log(2-\tau_1)}{\sqrt{u(1-\tau_1)}}+\frac{\log(2-\tau_1(1-h))}{\sqrt{u(1-\tau_1(1-h))}}\right).$$

Since $1-\tau_1 \le 0$ and $1-\tau_1(1-h) \ge 0$, and using

$$\log(1+h) \le h \le \sqrt{u(h)} \le -\log(1-h),$$

the result follows. We get $\sqrt{u(h)} \le -\log(1-h)$ by rewriting $(\log(1-h))^2 - u(h)$ as the product of two non-negative factors for $h \ge 0$,

$$(\log(1+h) - \log(1-h))\,(-h - \log(1-h))\,.$$

This ends the proof of (ii).

In order to prove (iii), we first check that $u_p''(h) \ge 4p$ for every $h \in \mathbb{R}$. This implies that $u_p(h) \ge 2ph^2$, for every $h \in \mathbb{R}$. Then from the definition (2.12) of d_α and using the variational identity (2.15), we obtain

$$d_\alpha(p,h) \ge 2ph^2.$$

The proof of (iii) is complete. □

The functions u_p, c_{1-p}, u which appear in (i),(ii) of Proposition 2.4 are convex. Therefore we may tensorise Lemma 2.3 to get an analogue of Theorem 2.1 for the product measure $\mu = \mu_1 \otimes \cdots \otimes \mu_n$ of Bernoulli measures μ_i with parameter p_i.

Theorem 2.5. *Let μ be the product measure $\mu_1 \otimes \cdots \otimes \mu_n$ where μ_i is the Bernoulli measure $\mu_i = (1-p_i)\delta_{-p_i} + p_i\delta_{(1-p_i)}$, $p_i \in]0,1[$. For $0 < \alpha < 1$, let D_α be the convex cost function on $\mathbb{R}^n$ defined by*

$$D_\alpha(h) = \sum_{i=1}^n \max\left[c_\alpha(|h_i|), 2p_i h_i^2 1\!\!1_{h_i \ge 0} + 2(1-p_i)h_i^2 1\!\!1_{h_i<0}\right]$$

for all $h = (h_1, \ldots, h_n) \in \mathbb{R}^n$. For every convex function f on $\mathbb{R}^n$,

$$\left(\int e^{\alpha Q_{D_\alpha} f} d\mu\right)^{\frac{1}{\alpha}} \left(\int e^{-\beta f} d\mu\right)^{\frac{1}{\beta}} \le 1, \tag{2.16}$$

where $\alpha+\beta = 1$ and Q_{D_α} is the infimum convolution associated to the cost function D_α.

When α goes to one or zero, we may improve the last inequality, one has

$$\int e^{Q_{V_p} f} d\mu \le e^{\int f d\mu},$$

and

$$\int e^{-f} d\mu \le e^{-\int Q_{U_p} f d\mu},$$

for every convex function. U_p and V_p denote the cost functions defined as

$$U_p(h) = \sum_{i=1}^n \left[u_{p_i}(h_i) 1\!\!1_{h_i \ge 0} + u_{1-p_i}(-h_i) 1\!\!1_{h_i<0}\right], \qquad \textit{and}$$

$$V_p(h) = \sum_{i=1}^n \left[(1-p_i)c_{1-p_i}\left(\frac{h_i}{1-p_i}\right) 1\!\!1_{h_i \ge 0} + p_i c_{p_i}\left(\frac{-h_i}{p_i}\right) 1\!\!1_{h_i<0}\right],$$

for every $h = (h_1, \ldots, h_n) \in \mathbb{R}^n$.

If $p_i = 1/2$ for every i, we may use the result (ii) of Proposition 2.4. Therefore, the cost function D_α may be replaced in (2.16) by

$$U(h) = \sum_{i=1}^{n} u(h_i), \quad h = (h_1, \ldots, h_n) \in \mathbb{R}^n.$$

To illustrate, let us present the Khintchine–Kahane inequality which derive from Theorem 2.5. Assume that $p_i = p$ for every $1 \leq i \leq n$ which means that the ε_i's are identically distributed. From inequality (2.2), the cost function V_p satisfies, for every $h \in \mathbb{R}^n$

$$V_p(h) \geq \frac{h^2}{2\max(p, 1-p)},$$

and therefore

$$V_p^*(h) \leq \frac{\max(p, 1-p)h^2}{2}.$$

For convenience, assume that $0 \leq p \leq 1/2$. From Theorem 2.5 we get that for every non-negative Lipschitz convex function f, for every $\lambda \geq 0$,

$$\int e^{\lambda f^2 (1 - 2\lambda(1-p)\|f\|_{Lip}^2)} d\mu \leq e^{\lambda \int f^2 d\mu}.$$

For $Z = \left\| \sum_{i=1}^{n} \varepsilon_i e_i \right\|$, one has $\sigma^2 \leq \frac{1}{p(1-p)} E(Z^2)$. Consequently, we get the following ψ_2 estimate:

$$\mathbb{E}\left(\exp \frac{pZ^2}{8\mathbb{E}(Z^2)}\right) \leq e^{\frac{p}{4}}.$$

As previously, this provides the next Khintchine–Kahane inequality: for every $q \geq 1$,

$$\|Z\|_q \leq 2\sqrt{\frac{2}{e}} e^{p/(8q)} \sqrt{\frac{q}{p}} \|Z\|_2,$$

and since $2\sqrt{\frac{2}{e}} e^{1/16} \approx 1,826... \leq 2$, for every $q \geq 2$,

$$\|Z\|_q \leq 2\sqrt{\frac{q}{p}} \|Z\|_2.$$

3. Proofs of Theorem 2.1 and Theorem 2.5

The structure of the proof is the same for Theorem 2.1 and Theorem 2.5. We first prove the result in dimension one, then this inequality may be tensorised using Lemma 3.2.

3.1. Proofs in dimension one

Proof. We first present the proof of Theorem 2.1 in dimension one. Let μ be a probability measure on $\mathbb{R}$. with $0 < \operatorname{diam}(\operatorname{supp}\mu) \leq D$. Let ν_1 and ν_2 be two probability measures with respective densities θ_1 and θ_2 with respect to μ. We want to show that for every convex function f on $\mathbb{R}$,

$$\int Q_{c_\alpha} f d\nu_1 - \int f d\nu_2 \leq \frac{1}{\alpha}\mathrm{Ent}_\mu\left(\frac{d\nu_1}{d\mu}\right) + \frac{1}{\beta}\mathrm{Ent}_\mu\left(\frac{d\nu_2}{d\mu}\right). \tag{3.1}$$

By a simple scaling argument, we may assume that $D = 1$ and by convenience, we assume that μ is a probability measure on $[0,1]$. Actually, since

$$\int w(1-\theta_1)d\mu = \mathrm{Ent}_\mu\left(\frac{d\nu_1}{d\mu}\right),$$

it suffices that for every $\theta_1 \geq 0$, $\theta_2 \geq 0$ and for every $x \in [0,1]$

$$Q_{c_\alpha} f(x)\theta_1 - f(x)\theta_2 \leq \frac{1}{\alpha} w(1-\theta_1) + \frac{1}{\beta} w(1-\theta_2),$$

the result then follows by integrating this inequality with respect to μ. Setting $\theta = \theta_2/\theta_1$, we just need that for every $\theta \geq 0$

$$Q_{c_\alpha} f(x) \leq \theta f(x) + \inf_{\theta_1 \geq 0}\left[\frac{1}{\alpha\theta_1} w(1-\theta_1) + \frac{1}{\beta\theta_1} w(1-\theta\theta_1)\right].$$

We exactly have that

$$\inf_{\theta_1 \geq 0}\left[\frac{1}{\alpha\theta_1} w(1-\theta_1) + \frac{1}{\beta\theta_1} w(1-\theta\theta_1)\right] = c_\alpha(1-\theta), \tag{3.2}$$

for every $\theta \geq 0$. Now the problem is reduced to prove that for every $\theta \geq 0$ and for every $x \in [0,1]$,

$$Q_{c_\alpha} f(x) \leq \theta f(x) + c_\alpha(1-\theta).$$

The inequality (3.1) is unchanged if one adds a constant to f. Therefore, we may assume that $\inf_{z\in[0,1]} f(z) = 0$. Recall that

$$Q_{c_\alpha} f(x) = \inf_{y\in\mathbb{R}}\left[f(y) + c_\alpha(|x-y|)\right].$$

Let us consider $\varepsilon > 0$ and $a \in [0,1]$ such that $f(a) \leq \varepsilon$. Choosing $y = (1-\theta)a + \theta x$, $0 \leq \theta \leq 1$, by convexity of the function f, one has for every $0 \leq \theta \leq 1$,

$$Q_{c_\alpha} f(x) \leq (1-\theta)f(a) + \theta f(x) + c_\alpha((1-\theta)|x-a|).$$

c_α is increasing on $\mathbb{R}^+$ and if $x \in [0,1]$, then $|x-a| \leq 1$; consequently, when ε goes to zero, we get for every $x \in [0,1]$,

$$Q_{c_\alpha} f(x) \leq \inf_{0\leq\theta\leq 1}[\theta f(x) + c_\alpha(1-\theta)].$$

Let us now observe that the infimum over all $\theta \geq 0$ of $[\theta f(x) + c_\alpha(1-\theta)]$ is reached for $0 \leq \theta \leq 1$ since $f(x) \geq 0$ for $x \in [0,1]$. This ends the proof of Theorem 2.1 in dimension one. □

Proof. We turn now to the proof of Lemma 2.3. The approach is exactly the same as for the proof of Theorem 2.1. If ν_1 and ν_2 are two probability measures with respective densities θ_1 and θ_2 with respect to $\mu = (1-p)\delta_{-p} + p\delta_{(1-p)}$, $p \in]0,1[$, one has $(1-p)\theta_1(-p)+p\theta_1(1-p) = 1$ and $(1-p)\theta_2(-p)+p\theta_2(1-p) = 1$. Inequality (3.1) may be rewritten as

$$\begin{aligned}[Q_{p,\alpha}f(-p)(1-p)\theta_1(-p) + Q_{p,\alpha}f(1-p)p\theta_1(1-p)] \\ - [f(-p)(1-p)\theta_2(-p) + f(1-p)p\theta_2(1-p)] \\ \leq \frac{1}{\alpha}[(1-p)w(1-\theta_1(-p)) + pw(1-\theta_1(1-p))] \\ + \frac{1}{\beta}[(1-p)w(1-\theta_2(-p)) + pw(1-\theta_2(1-p))].\end{aligned}$$

As previously by translation, we may assume that $\min(f(-p), f(1-p)) = 0$. Now, we have to consider two different cases, either $f(-p) = 0$ either $f(1-p) = 0$.

First assume that $f(-p) = 0$ and $f(1-p) \geq 0$, then we have $Q_{p,\alpha}f(-p) \leq 0$. Since for $i = 1, 2$,

$$1 - \theta_i(-p) = \frac{p}{1-p}(1 - \theta_i(1-p)),$$

setting $\tau_i = \theta_i(1-p)$ for $i = 1, 2$, it suffices to prove that for every $\tau_1, \tau_2 \geq 0$

$$\begin{aligned}Q_{p,\alpha}f(1-p)\tau_1 \leq \tau_2 f(1-p) + \frac{1}{\alpha}\left[\frac{1-p}{p}w\left(\frac{p}{1-p}(\tau_1 - 1)\right) + w(1-\tau_1)\right] \\ + \frac{1}{\beta}\left[\frac{1-p}{p}w\left(\frac{p}{1-p}(\tau_2 - 1)\right) + w(1-\tau_2)\right].\end{aligned}$$

In other words, using the definition (2.10) of u_p and setting $\theta = \tau_2/\tau_1$, we have to check that for every $\theta \geq 0$ and $\tau_1 \geq 0$,

$$Q_{p,\alpha}f(1-p) \leq \theta f(1-p) + \frac{1}{\alpha\tau_1}u_p(1-\tau_1) + \frac{1}{\beta\tau_1}u_p(1-\theta\tau_1). \tag{3.3}$$

For every $\tau_1 \geq 0$, the infimum

$$\inf_{\theta, \theta \geq 0}\left[\theta f(1-p) + \frac{1}{\beta\tau_1}u_p(1-\theta\tau_1)\right]$$

is reached for $\tilde{\theta}(\tau_1)$ satisfying $u_p'(1-\tilde{\theta}(\tau_1)\tau_1) = \beta f(1-p)$, that is

$$\tilde{\theta}(\tau_1)\tau_1 = \frac{1}{(1-p)e^{\beta f} + p}.$$

Using $f(1-p) \geq 0$, we easily see that $\tilde{\theta}(\tau_1)\tau_1 \leq 1$ for every $\tau_1 \geq 0$. The minimum value of the function

$$\tau_1 \longrightarrow \tilde{\theta}(\tau_1)f(1-p) + \frac{1}{\alpha\tau_1}u_p(2(1-\tau_1)) + \frac{1}{\beta\tau_1}u_p(2(1-\tilde{\theta}(\tau_1)\tau_1)),$$

is reached for $\hat{\tau}_1 \in]0, 1/(1-p)]$. Let us note $\tilde{\theta}(\hat{\tau}_1) = \hat{\theta}$. One has $0 \le \hat{\theta}\hat{\tau}_1 \le 1$ and the point $(\hat{\theta}, \hat{\tau}_1)$ minimizes the right hand side of (3.3) over all $\tau_1 \ge 0$ and $\theta \ge 0$. Now let us consider

$$\inf_{\tau_1, \tau_1 \ge 0} \left[\frac{1}{\alpha\tau_1} u_p(1-\tau_1) + \frac{1}{\beta\tau_1} u_p(1-\hat{\theta}\tau_1) \right].$$

This infimum is achieved at the point $\hat{\tau}_1$. Consequently, by differentiating the above expression in τ_1, we get that the point $(\hat{\theta}, \hat{\tau}_1)$ satisfies

$$\left(1 - \frac{1}{1-p}(\hat{\tau}_1 - 1)\right)^{\beta} \left(1 - \frac{1}{1-p}(\hat{\tau}_1\hat{\theta} - 1)\right)^{\alpha} = 1.$$

From this inequality and since $\hat{\theta}\hat{\tau}_1 \le 1$, we get that $\hat{\tau}_1 \ge 1$. Finally, the right hand side of (3.3) is minimum at the point $(\hat{\theta}, \hat{\tau}_1)$, with $0 \le \hat{\theta} \le \hat{\theta}\hat{\tau}_1 \le 1$. By the definition (2.12) of $d_{p,\alpha}$, one has

$$d_\alpha(p, 1-\hat{\theta}) = \inf_{\tau_1, \tau_1 \ge 1} \left[\frac{1}{\alpha\tau_1} u(2(1-\tau_1)) + \frac{1}{\beta\tau_1} u(2(1-\theta\tau_1)) \right].$$

Using $0 \le \hat{\theta} \le 1$, this infimum may be restricted to $\tau_1 \ge 1$ and $\hat{\theta}\tau_1 \le 1$. Now the problem is reduced to get that for every $0 \le \theta \le 1$,

$$Q_{p,\alpha} f(1-p) \le \theta f(1-p) + d_\alpha(p, 1-\theta).$$

Let us recall the definition of $Q_{p,\alpha} f(1-p)$,

$$Q_{p,\alpha} f(1-p) = \inf_{y \in \mathbb{R}} \left[f(y) + d_{p,\alpha}((1-p) - y) \right].$$

Choosing $y = \theta(1-p) + (1-\theta)(-p)$, with $0 \le \theta \le 1$, by the convexity of f and since $f(-p) = 0$ we obtain

$$Q_{p,\alpha} f(1-p) \le \theta f(1-p) + d_{p,\alpha}(1-\theta).$$

This ends the proof when $f(-p) = 0$ since $d_{p,\alpha}(1-\theta) = d_\alpha(p, 1-\theta)$ for $0 \le \theta \le 1$.

In the second case, when $f(1-p) = 0$ and $f(-p) \ge 0$, as previously, we first reduce the problem to show that for every $0 \le \theta \le 1$,

$$Q_{p,\alpha} f(-p) \le \theta f(-p) + d_\alpha(1-p, 1-\theta).$$

Then, choosing $y = \theta(-p) + (1-\theta)(1-p)$, $0 \le \theta \le 1$, in the definition of $Q_{p,\alpha} f(-p)$ we obtain that, for every $0 \le \theta \le 1$,

$$Q_{p,\alpha} f(-p) \le \theta f(-p) + d_{p,\alpha}(-(1-\theta)).$$

The proof is finished since $d_{p,\alpha}(-(1-\theta)) = d_\alpha(1-p, 1-\theta)$ for $0 \le \theta \le 1$. □

3.2. Tensorisation

In this section, we present the tensorisation argument of Lemma 3.2 that may be found in Maurey's paper [14]. Let us consider two probability measures μ_1 and μ_2 on some Banach spaces E_1 and E_2. Let c_1 and c_2 be two convex cost functions on E_1 and E_2. For $i = 1, 2$, we define the infimum convolution of a real function g on E_i associated to the cost function c_i, for every $x_i \in E_i$,

$$Q_i g(x_i) = \inf_{y_i \in E_i} [g(y_i) + c_i(x_i - y_i)].$$

Definition 3.1. Let $\alpha, \beta > 0$, with $\alpha + \beta = 1$. A probability measure ν on E satisfies the convex concentration property associated to the cost function c if for every convex measurable function g on E,

$$\left(\int e^{\alpha Q g} d\nu\right)^{\frac{1}{\alpha}} \left(\int e^{-\beta g} d\nu\right)^{\frac{1}{\beta}} \leq 1,$$

where Qg is the infimum convolution associated to the cost function c.

Lemma 3.2. *If the probability measure μ_1 on E_1, respectively μ_2 on E_2, satisfies the convex concentration property associated to the cost function c_1, respectively c_2, then $\mu_1 \otimes \mu_2$ satisfies the convex concentration property associated to the cost function c on $E_1 \times E_2$, defined by*

$$c(h_1, h_2) = c_1(h_1) + c_2(h_2), \quad h_1 \in E_1,\, h_2 \in E_2.$$

Proof. Let f denote a convex function on $E_1 \times E_2$, and let Qf be the infimum convolution of f associated to the cost function c. For every $x_2 \in E_2$, f^{x_2} denotes the marginal function

$$f^{x_2} : x_1 \longrightarrow f(x_1, x_2).$$

For every $x_1 \in E_1$, the function $x_2 \longrightarrow Q_1 f^{x_2}(x_1)$ is convex. Indeed, by the convex property of the functions f and c_1 on E_1, for every $0 \leq \theta \leq 1$, for every $x_2, x_2' \in E_2$,

$$\begin{aligned} f(\theta y_1 + (1-\theta) y_1', \theta x_2 + (1-\theta) x_2') + c_1(x_1 - (\theta y_1 + (1-\theta) y_1')) \\ \leq \theta f(y_1, x_2) + (1-\theta) f(y_1', x_2') + \theta c_1(x_1 - y_1) + (1-\theta) c_1(x_1' - y_1'). \end{aligned}$$

For $\varepsilon > 0$, let us choose y_1 and y_1' in E_1 such that

$$f^{x_2}(y_1) + c(x_1 - y_1) \leq Q_1 f^{x_2}(x_1) + \varepsilon,$$

and

$$f^{x_2'}(y_1') + c(x_1' - y_1') \leq Q_1 f^{x_2'}(x_1) + \varepsilon.$$

Taking ε going to 0, we obtain for every x_1,

$$Q_1 f^{\theta x_2 + (1-\theta) x_2'}(x_1) \leq \theta Q_1 f^{x_2}(x_1) + (1-\theta) Q_1 f^{x_2'}(x_1).$$

Let ϕ be the function on E_2 defined by

$$\phi(x_2) = \frac{1}{\alpha} \log \int e^{\alpha Q_1 f^{x_2}} d\mu_1, \; x_2 \in E_2.$$

By Hölder inequality, ϕ is a convex function on E_2. Consequently, since μ_2 satisfies the convex concentration property associated to the cost function c_2, we have

$$\left(\int e^{\alpha Q_2 \phi} d\mu_2\right)^{\frac{1}{\alpha}} \left(\int e^{-\beta\phi} d\mu_2\right)^{\frac{1}{\beta}} \leq 1. \tag{3.4}$$

For every x_2 in E_2, f^{x_2} is a convex function on E_1. The measure μ_1 satisfies the convex concentration property associated to the cost function c_1, therefore

$$\int e^{-\beta\phi(x_2)} d\mu_2(x_2) = \int \left(\int e^{\alpha Q_1 f^{x_2}} d\mu_1\right)^{-\frac{\beta}{\alpha}} d\mu_2(x_2) \geq \iint e^{-\beta f^{x_2}} d\mu_1 d\mu_2(x_2).$$

We also have,

$$\begin{aligned}
&\iint e^{\alpha Q f}\, d\mu_1\, d\mu_2 \\
&= \iint \exp\left(\alpha \inf_{y_1,y_2} [f(y_1,y_2) + c_1(x_1-y_1) + c_2(x_2-y_2)]\right) d\mu_1(x_1)\, d\mu_2(x_2) \\
&= \iint \inf_{y_2} \left[e^{\alpha Q_1 f^{y_2}(x_1) + \alpha c_2(x_2-y_2)}\right] d\mu_1(x_1)\, d\mu_2(x_2) \\
&\leq \int \inf_{y_2} \left[\int e^{\alpha Q_1 f^{y_2}(x_1)} d\mu_1(x_1)\; e^{\alpha c_2(x_2-y_2)}\right] d\mu_2(x_2) \\
&= \int e^{\alpha Q_2 \phi(x_2)} d\mu_2(x_2).
\end{aligned}$$

Consequently, from (3.4), we obtain that $\mu_1 \otimes \mu_2$ satisfies the convex concentration property associated to the cost function c. □

References

[1] Barles G. (1994). Solutions de viscosité des équations de Hamilton-Jacobi. Springer.

[2] Bobkov S. , Götze F. , (1998). Exponential integrability and transportation cost related to logarithmic Sobolev inequalities, *J. Funct. Anal.*, **163** 1,1-28.

[3] Bobkov S. , Gentil I. , Ledoux M. , (2000). Hypercontractivity of Hamilton-Jacobi equations. *Geom. Funct. Anal.* **10** 1028-1052.

[4] Evans L. C. (1997). Partial differential equations. *Graduate Studies in Math. 19. Amer. Math. Soc.*

[5] Haagerup Uffe, (1982).The best constants in the Khintchine inequality. *Studia Mathematica,* T. LXX.

[6] Ledoux M. , (1996). Talagrand deviation inequalities for product measures. *ESAIM: Probab. Statist.* **1** 63-87.

[7] Ledoux M. , Talagrand M. , (1991). *Probability in Banach spaces.* Springer-Verlag.

[8] Oleszkiewicz, K. (2002). On a Non-symmetric Version of the Khinchine-Kahane Inequality. In this Volume, 161-172.

[9] Talagrand, M. (1995). Concentration of measure and isoperimetric inequalities in product spaces. *Publications Mathématiques de l'I.H.E.S.* **81** 73-205.

[10] Talagrand, M. (1996). New concentration inequalities in product spaces. *Invent. Math.* **126** 505-563.

[11] Talagrand, M. (1996). Transportation cost for Gaussian and other product measures. *Geom. and Func. Anal.* **6**, 587-600

[12] Marton, K. (1996). Bounding $\overline{d}$-distance by information divergence: a method to prove measure concentration. *Ann. Proba.* **24** 927-939.

[13] Marton, K. (1997). A measure concentration inequality for contracting Markov chains. *Geom. Funct. Anal.* **6** 556-571.

[14] Maurey, B. (1991). Some deviation inequalities, *Geom. Func. Anal.* **1**, 188-197.

[15] Samson, P.-M. (2000). Concentration of measure inequalities for Markov chains and ϕ-mixing processes. *Ann. Proba.* **28**, 416-461.

UMR 8050 CNRS,

Laboratoire d'analyse et de mathématiques appliquées, UMR 8050 CNRS, Université Marne-La-Vallée, Champs sur Marne, 77454 Marne-La-Vallée Cedex 2, France
E-mail address: samson@math.univ-mlv.fr

Part II

Independent Random Vectors, Chaos, Martingales and Lévy Processes

Progress in Probability, Vol. 56, 55–69

Exponential Inequalities, with Constants, for U-statistics of Order Two

Christian Houdré and Patricia Reynaud-Bouret

Abstract. A martingale proof of a sharp exponential inequality (with constants) is given for U-statistics of order two as well as for double integrals of Poisson processes.

1. Introduction

We wish in these notes to further advance our knowledge of exponential inequalities for U-statistics of order two. These types of inequalities are already present in Hoeffding seminal papers [6], [7] and have seen further development since then. For example, exponential bounds were obtained (in the (sub)Gaussian case) by Hanson and Wright [5], by Bretagnolle [1], and most recently by Giné, Latała, and Zinn [4] (and the many references therein). As indicated in [4], the exponential bound there is optimal since it involves a mixture of exponents corresponding to a Gaussian chaos of order two behavior, and (up to logarithmic factors) to the product of a normal and of a Poisson random variable and to the product of two independent Poisson random variables. These various behaviors can be obtained as limits in law of triangular arrays of canonical U-statistics of degree two (with possibly varying kernels).

The methods of proof of [4] rely on precise moment inequalities of Rosenthal type which are of independent interest (and which are valid for U–statistics of arbitrary order). In case of order two, these moment inequalities together with Talagrand inequality for empirical processes provided exponential bounds. Here, we present a different proof of their result which also provide information about the constants which is often needed in statistical applications [9]. Our approach still rely on Talagrand inequality but replaces the moment estimates by martingales types inequalities. As also indicated [4] the moment estimates and the exponential inequality are equivalent to one another and so our approach also provides sharp moment estimates. The methods presented here are robust enough that they can

2000 *Mathematics Subject Classification.* AMS numbers 60E15, 60G42, 62E17.
Key words and phrases. U–statistics, exponential inequalities, Poisson integrals.
Research supported in part by a NSF Grant.

be adapted to provide exponential inequalities for double integrals with respect to Poisson processes.

2. Background

Let us recall some known facts about U-statistics of order two. Throughout these notes, let $T_1, \dots, T_n$, be independent real random variables defined on a probability space $(\Omega, \mathcal{F}, \mathbb{P})$.

A canonical U-statistics of order two is generally defined for all positive integer n as

$$\sum_{i=1}^{n}\sum_{j=1}^{n} f_{i,j}(T_i, T_j), \tag{2.1}$$

where the $f_{i,j} : \mathbb{R} \times \mathbb{R} \to \mathbb{R}$ are Borel measurable functions.

We will not be concerned in this work with the diagonal part

$$\sum_{i=1}^{n} f_{i,i}(T_i, T_i),$$

nor with the part of (2.1) made of sums of independent random variables. Indeed for these parts, exponential tail inequalities are well known and a "$u/2$ argument", combined with our results, provides exponential bounds for canonical U–statistics (of order two). Hence we will deal with degenerate U-statistics of order two, defined for all integer $n \geq 2$, by

$$\begin{aligned} \mathcal{U}_n \quad = \quad & \sum_{i=1}^{n}\sum_{j\neq i} \Big[f_{i,j}(T_i, T_j) - \mathbb{E}(f_{i,j}(T_i, T_j)|T_j) - \\ & \qquad -\mathbb{E}(f_{i,j}(T_i, T_j)|T_i) + \mathbb{E}(f_{i,j}(T_i, T_j)) \Big]. \end{aligned} \tag{2.2}$$

This is equivalent to considering for all integer $n \geq 2$,

$$U_n = \sum_{i=2}^{n}\sum_{j=1}^{i-1} g_{i,j}(T_i, T_j), \tag{2.3}$$

where the $g_{i,j} : \mathbb{R} \times \mathbb{R} \to \mathbb{R}$ are Borel measurable functions verifying

$$\mathbb{E}(g_{i,j}(T_i, T_j)|T_i) = 0 \text{ and } \mathbb{E}(g_{i,j}(T_i, T_j)|T_j) = 0, \tag{2.4}$$

and where $\mathbb{E}$ is the expectation with respect to $\mathbb{P}$. Indeed it is sufficient to take

$$\begin{aligned} g_{i,j}(T_i, T_j) = {} & f_{i,j}(T_i, T_j) + f_{j,i}(T_j, T_i) - \mathbb{E}(f_{i,j}(T_i, T_j) + f_{j,i}(T_j, T_i)|T_i) - \\ & \mathbb{E}(f_{i,j}(T_i, T_j) + f_{j,i}(T_j, T_i)|T_j) + \mathbb{E}(f_{i,j}(T_i, T_j) + f_{j,i}(T_j, T_i)). \end{aligned}$$

Throughout these notes, U_n is now given by (2.3) and satisfies (2.4).

For any $n \geq 1$, let $\mathcal{F}_n$ be the σ-field generated by $\{T_1, \ldots, T_n\}$, $\mathcal{F}_0 = \{\Omega, \emptyset\}$ and for any $n \geq 2$, let

$$X_n = \sum_{j=1}^{n-1} g_{n,j}(T_n, T_j).$$

As in (2.3), U_n is only defined for $n \geq 2$, we set $U_1 = 0$ and also $X_1 = 0$. The following is an easy, known, but important lemma:

Lemma 2.1. *$(U_n, n \in \mathbb{N})$ is a discrete time martingale with respect to the filtration $(\mathcal{F}_n, n \in \mathbb{N})$ and for all n, $\mathbb{E}(X_n|\mathcal{F}_{n-1}) = 0$.*

Proof. Let $n \geq 2$. Then clearly, X_n is $\mathcal{F}_n$-measurable. Moreover

$$\mathbb{E}(X_n|\mathcal{F}_{n-1}) = \sum_{j=1}^{n-1} \mathbb{E}(g_{n,j}(T_n, T_j)|\mathcal{F}_{n-1}) = \sum_{j=1}^{n-1} \mathbb{E}(g_{n,j}(T_n, T_j)|T_j) = 0,$$

since the T_i's are independent random variables and by (2.4). Finally, since $U_n = \sum_{i=1}^{n} X_i$, $\mathbb{E}(U_n|\mathcal{F}_{n-1}) = U_{n-1} + \mathbb{E}(X_n|\mathcal{F}_{n-1}) = U_{n-1}$. □

Throughout the sequel, and for all i and j, we use the notation

$$\mathbb{E}_{(i)}(g_{i,j}(T_i, T_j)) = \mathbb{E}(g_{i,j}(T_i, T_j)|T_j)$$

and

$$\mathbb{E}^{(j)}(g_{i,j}(T_i, T_j)) = \mathbb{E}(g_{i,j}(T_i, T_j)|T_i).$$

3. Exponential Inequalities

Let V_n^2 be the angle bracket [12, p. 148] of U_n, i.e. let $V_n^2 = \sum_{i=1}^{n} \mathbb{E}(X_i^2|\mathcal{F}_{i-1})$ and let also $B_n = \sup_{i \leq n} |X_i|$. Let us present a first result which is not quite the one obtained in [4] (because of the extra term F present below) but which already provides some knowledge of the constants.

Theorem 3.1. *Let $u > 0, \varepsilon > 0$ and let $|g_{i,j}| \leq A$, for all i, j. Then*

$$\begin{aligned}\mathbb{P}\Big[U_n \geq (1+\varepsilon)C\sqrt{2u} + \left(2\sqrt{\kappa}D + \frac{1+\varepsilon}{3}F\right)u \\ + \left(\sqrt{2}\kappa(\varepsilon) + \frac{2\sqrt{\kappa}}{3}\right)Bu^{3/2} + \frac{\kappa(\varepsilon)}{3}Au^2\Big] \\ \leq 3e^{-u} \wedge 1. \end{aligned} \tag{3.1}$$

Above,

$$C^2 = \sum_{i=2}^{n} \sum_{j=1}^{i-1} \mathbb{E}(g_{i,j}(T_i, T_j)^2), \tag{3.2}$$

$$D = \sup\left\{\mathbb{E}\left(\sum_{i=2}^{n}\sum_{j=1}^{i-1} g_{i,j}(T_i,T_j)a_i(T_i)b_j(T_j)\right) : \right. \tag{3.3}$$

$$\left. \mathbb{E}\left(\sum_{i=2}^{n} a_i(T_i)^2\right) \le 1, \mathbb{E}\left(\sum_{j=1}^{n-1} b_j(T_j)^2\right) \le 1\right\},$$

$$F = \mathbb{E}\left(\sup_{i,t}\left|\sum_{j=1}^{i-1} g_{i,j}(t,T_j)\right|\right), \tag{3.4}$$

$$B^2 = \max\left\{\sup_{t,i}\left(\sum_{j=1}^{i-1}\mathbb{E}^{(j)}(g_{i,j}(t,T_j)^2)\right), \sup_{t,j}\left(\sum_{i=j+1}^{n}\mathbb{E}_{(i)}(g_{i,j}(T_i,t)^2)\right)\right\}, \tag{3.5}$$

while κ and $\kappa(\varepsilon)$ can be chosen respectively equal to 4 *and* $(2.5 + 32\varepsilon^{-1})$.

As a preparation for the proof, we first obtain bounds on V_n^2 and B_n.

Lemma 3.2. *Let $u > 0$ and let $\varepsilon > 0$. With probability larger than $1 - 2e^{-u}$,*

$$\sqrt{V_n^2} \le (1+\varepsilon)C + D\sqrt{2\kappa u} + \kappa(\varepsilon)Bu$$

and

$$B_n \le (1+\varepsilon)F + B\sqrt{2\kappa u} + \kappa(\varepsilon)Au,$$

where κ and $\kappa(\varepsilon)$ can be chosen respectively equal to 4 *and* $(2.5 + 32\varepsilon^{-1})$.

To prove this lemma, we use Talagrand's inequality [16] for empirical processes, and more precisely the version with constants obtained by Massart [11] (see also Ledoux [10]).

(Talagrand's inequality) Let $X_1 = (X_1^1, \dots, X_1^N), \dots, X_n = (X_n^1, \dots, X_n^N)$ be independent random variables with values in $[-b, b]^N$, for some positive real b. Let

$$Z = \sup_{1\le t\le N}\left|\sum_{i=1}^{n}(X_i^t - \mathbb{E}(X_i^t))\right|, \tag{3.6}$$

and let

$$v = \sup_{1\le t\le N}\sum_{i=1}^{n}\mathrm{Var}(X_i^t). \tag{3.7}$$

Then for all $\varepsilon > 0, z > 0$

$$\mathbb{P}[Z \ge (1+\varepsilon)\mathbb{E}(Z) + \sqrt{2\kappa vz} + \kappa(\varepsilon)bz] \le e^{-z}, \tag{3.8}$$

where κ and $\kappa(\varepsilon)$ can be respectively chosen equal to 4 and $2.5 + 32/\varepsilon$.

Proof of **Lemma 3.2.** It is easy to see by the independence property of the variables that

$$V_n^2 = \sum_{i=2}^{n}\mathbb{E}_{(i)}\left(\left[\sum_{j=1}^{i-1} g_{i,j}(T_i,T_j)\right]^2\right).$$

Therefore, by duality, we have that:

$$\sqrt{V_n^2} = \sup_{\sum_{i=2}^n \mathbb{E}(a_i(T_i)^2)=1} \left| \sum_{i=2}^n \mathbb{E}_{(i)} \left(a_i(T_i) \sum_{j=1}^{i-1} g_{i,j}(T_i, T_j) \right) \right|$$
$$= \sup_{\sum_{i=2}^n \mathbb{E}(a_i(T_i)^2)=1} \left| \sum_{j=1}^{n-1} \sum_{i=j+1}^{n} \mathbb{E}_{(i)}(a_i(T_i) g_{i,j}(T_i, T_j)) \right|,$$

and

$$B_n = \sup_{i \le n} |X_i| \le \sup_{i \le n} \sup_t \left| \sum_{j=1}^{i-1} g_{i,j}(t, T_j) \right| := \tilde{B}_n.$$

By density, we can restrict the previous suprema to a countable deterministic dense subset of parameters : for V_n^2, the set of $\mathbb{L}^2$ functions is separable and for $\tilde{B}_n$, the set of t is $\mathbb{R}$ which is also separable. By monotone limit, we can restrict ourselves to take a finite subset of parameters and then pass to the limit. These suprema can then be interpreted as suprema of the form $\sup_{u \in \mathcal{T}} \sum_{j=1}^{n-1} X_j^u$, where $\mathcal{T}$ is finite and the $(X_j^u, u \in \mathcal{T})$'s are centered, independent and bounded. Therefore, applying Talagrand's inequality, and passing to the limit give the following results:

Let $u > 0$ and let $\varepsilon > 0$. With probability larger than $1 - e^{-u}$,

$$\sqrt{V_n^2} \le (1+\varepsilon)\mathbb{E}(\sqrt{V_n^2}) + \sqrt{2\kappa v_1 u} + \kappa(\varepsilon) b_1 u, \tag{3.9}$$

where

$$v_1 = \sup_{\sum_{i=2}^n \mathbb{E}(a_i(T_i)^2)=1} \sum_{j=1}^{n-1} \mathrm{Var}^{(j)} \left(\sum_{i=j+1}^{n} \mathbb{E}_{(i)}(a_i(T_i) g_{i,j}(T_i, T_j)) \right)$$

and

$$b_1 = \sup_{t,j,\sum_{i=2}^n \mathbb{E}(a_i(T_i)^2)=1} \left| \sum_{i=j+1}^{n} \mathbb{E}_{(i)}(a_i(T_i) g_{i,j}(T_i, t)) \right|.$$

For $\tilde{B}_n$ we have with probability larger than $1 - e^{-u}$,

$$\tilde{B}_n \le (1+\varepsilon)\mathbb{E}(\tilde{B}_n) + \sqrt{2\kappa v_2 u} + \kappa(\varepsilon) b_2 u, \tag{3.10}$$

where

$$v_2 = \sup_{i,t} \sum_{j=1}^{i-1} \mathrm{Var}^{(j)} \left(g_{i,j}(t, T_j) \right)$$

and

$$b_2 = \sup_{t,j,x,i} \left| g_{i,j}(x,t) \right|.$$

So (3.9) and (3.10) hold true together on an event of probability larger than $1 - 2e^{-u}$. Using (2.4), we have $\mathbb{E}(\sqrt{V_n^2}) \le \sqrt{\mathbb{E}(V_n^2)} = C$, $v_1 = D^2$, $b_1 \le B$, $\mathbb{E}(\tilde{B}_n) = F$, $v_2 \le B^2$ and $b_2 = A$. The result follows. □

Proof of **Theorem 3.1.** First, define b and v by

$$\sqrt{v} = (1+\varepsilon)C + D\sqrt{2\kappa u} + \kappa(\varepsilon)Bu$$

and

$$b = (1+\varepsilon)F + B\sqrt{2\kappa u} + \kappa(\varepsilon)Au.$$

Next, let us now return to U_n. More precisely, let us define the stopping time T by $T+1 = \inf\{k \in \mathbb{N}, V_k > v \text{ or } \tilde{B}_k > b\}$. Then U_n^T, the martingale U_n stopped in T, is also a martingale with respect to the same filtration. As V_k and $\tilde{B}_k$ are nondecreasing, the angle bracket and the jumps of this new martingale are respectively bounded by v and b. Therefore, (see [12, Lemma VII-2-8, p. 154]), for all $\lambda > 0$,

$$\left(e^{\lambda U_n^T - \phi_b(\lambda)v}, n \in \mathbb{N}\right) \tag{3.11}$$

is a super-martingale where $\phi_b(\lambda) = (e^{\lambda b} - \lambda b - 1)/b^2$. Finally, performing some classical computation on the Laplace transform of U_n^T, we get via Chebyshev's inequality

$$\mathbb{P}\left(U_n^T \geq \sqrt{2vu} + \frac{b}{3}u\right) \leq e^{-u}.$$

Hence

$$\begin{aligned}\mathbb{P}\left(U_n \geq \sqrt{2vu} + \frac{b}{3}u\right) &\leq \mathbb{P}\left(U_n^T \geq \sqrt{2vu} + \frac{b}{3}u\right) + \mathbb{P}(T+1 \leq n)\\ &\leq 3e^{-u}\end{aligned}$$

by Lemma 3.2. □

As already indicated, Theorem 3.1 does not quite recover the exponential bound of [4] because of the extra term F. With a little more work, F can be removed. At first, we need the following simple lemma.

Lemma 3.3. *Let $(Y_n, n \in \mathbb{N})$ be a martingale. For all $k \geq 2$, let*

$$A_n^k = \sum_{i=1}^n \mathbb{E}\left((Y_i - Y_{i-1})^k | \mathcal{F}_{i-1}\right).$$

Then for all integers $n \geq 1$ and for all λ such that for all $i \leq n$, $\mathbb{E}[\exp(|\lambda(Y_i - Y_{i-1})|)] < +\infty$,

$$\mathcal{E}_n = \exp\left(\lambda Y_n - \sum_{k\geq 2} \frac{\lambda^k}{k!} A_n^k\right) \tag{3.12}$$

is a super-martingale.

Proof. For all integers $n \geq 1$,

$$\mathbb{E}(\mathcal{E}_n|\mathcal{F}_{n-1}) = \mathcal{E}_{n-1}\mathbb{E}(e^{\lambda(Y_n - Y_{n-1})}|\mathcal{F}_{n-1})$$

$$\exp\left(-\sum_{k\geq 2}\frac{\lambda^k}{k!}\mathbb{E}\left((Y_n - Y_{n-1})^k|\mathcal{F}_{n-1}\right)\right),$$

But

$$\mathbb{E}(e^{\lambda(Y_n - Y_{n-1})}|\mathcal{F}_{n-1}) = 1 + \mathbb{E}\left(\sum_{k\geq 2}\frac{\lambda^k}{k!}(Y_n - Y_{n-1})^k|\mathcal{F}_{n-1}\right).$$

The partial sums are dominated by $\exp(|\lambda(Y_n - Y_{n-1})|)$ which is integrable by assumption. Therefore, by dominated convergence for conditional expectations, we can exchange sum and expectation to obtain:

$$\begin{aligned}\mathbb{E}\left(e^{\lambda(Y_n - Y_{n-1})}|\mathcal{F}_{n-1}\right) &= 1 + \sum_{k\geq 2}\frac{\lambda^k}{k!}\mathbb{E}\left((Y_n - Y_{n-1})^k|\mathcal{F}_{n-1}\right)\\ &\leq \exp\left(\sum_{k\geq 2}\frac{\lambda^k}{k!}\mathbb{E}\left((Y_n - Y_{n-1})^k|\mathcal{F}_{n-1}\right)\right),\end{aligned}$$

giving the result. □

A_n^2 is the classical angle bracket. Assume $Y_0 = 0$. If the A_n^k are bounded by $w_n^k \geq 0$, we have for all $\lambda > 0$,

$$\mathbb{E}(e^{\lambda Y_n}) \leq \exp\left(\sum_{k\geq 2}\frac{\lambda^k}{k!}w_n^k\right), \tag{3.13}$$

since $\mathbb{E}(\mathcal{E}_n) \leq \mathbb{E}(\mathcal{E}_0) = 1$. This result is due to Pinelis [13, Theorem 8.5].

We now state our main result which recovers the exponential bound of [4] with estimates on the constants.

Theorem 3.4. *Let A, B, C, D be as in* Theorem 3.1. *For all $\varepsilon, u > 0$,*

$$\mathbb{P}(U_n \geq 2(1+\varepsilon)^{3/2}C\sqrt{u} + \eta(\varepsilon)Du + \beta(\varepsilon)Bu^{3/2} + \gamma(\varepsilon)Au^2) \leq 2.77e^{-u} \tag{3.14}$$

where

- $\eta(\varepsilon) = \sqrt{2\kappa}(2 + \varepsilon + \varepsilon^{-1})$,
- $\beta(\varepsilon) = e(1+\varepsilon^{-1})^2\kappa(\varepsilon) + \left[(\sqrt{2\kappa}(2+\varepsilon+\varepsilon^{-1})) \vee \frac{(1+\varepsilon)^2}{\sqrt{2}}\right]$,
- $\gamma(\varepsilon) = (e(1+\varepsilon^{-1})^2\kappa(\varepsilon)) \vee \frac{(1+\varepsilon)^2}{3}$,
- $\kappa = 4$,
- $\kappa(\varepsilon) = 2.5 + 32\varepsilon^{-1}$.

Proof. The A_n^k corresponding to the martingale U_n are

$$\sum_{i=2}^{n} \mathbb{E}_{(i)} \left[\left(\sum_{j=1}^{i-1} g_{i,j}(T_i,T_j) \right)^k \right] \leq V_n^k = \sum_{i=2}^{n} \mathbb{E}_{(i)} \left[\left| \sum_{j=1}^{i-1} g_{i,j}(T_i,T_j) \right|^k \right].$$

We now wish to estimate the V_n^k and this is the purpose of:

Lemma 3.5. *Let $\varepsilon > 0$ and $u > 0$. One has with probability larger than $1 - 1.77e^{-u}$, for all $k \geq 2$*

$$(V_n^k)^{1/k} \leq (1+\varepsilon)(\mathbb{E}(V_n^k))^{1/k} + \sigma_k\sqrt{2\kappa k u} + \kappa(\varepsilon) b_k k u,$$

where

$$\sigma_k^2 = \sup_{\sum_{i=2}^n \mathbb{E}(|a_i(T_i)|^{k/(k-1)})=1} \left\{ \sum_{j=1}^{n-1} \mathbb{E} \left(\left[\sum_{i=j+1}^{n} \mathbb{E}_{(i)}(a_i(T_i) g_{i,j}(T_i,T_j) \right]^2 \right) \right\},$$

$$b_k = \sup_{\sum_{i=2}^n \mathbb{E}(|a_i(T_i)|^{k/(k-1)})=1, j\leq n-1} \ \sup_t \left| \sum_{i=j+1}^{n} \mathbb{E}_{(i)} \left[g_{i,j}(T_i,t) a_i(T_i) \right] \right|$$

and where κ and $\kappa(\varepsilon)$ can be chosen respectively equal to 4 and $2.5 + 32/\varepsilon$.

Proof of Lemma 3.5. By Hölder's inequality, we have:

$$(V_n^k)^{1/k} = \sup_{\sum_{i=1}^n \mathbb{E}(|a_i(T_i)|^{k/(k-1)})=1} \left\{ \sum_{j=1}^{n-1} \sum_{i=j+1}^{n} \mathbb{E}_{(i)} \left(g_{i,j}(T_i,T_j) a_i(T_i) \right) \right\}.$$

Using the same method as before, we can view the V_n^k's as a limit of suprema of the form

$$\sup_{u\in\mathcal{T}} \sum_{j=1}^{n-1} X_j^u$$

where $\mathcal{T}$ is finite and where the $(X_j^u, u \in \mathcal{T})$'s are independent centered and bounded real random variables. Therefore we can again apply Talagrand's inequality (3.8): for all $k \geq 2$, all $z > 0$ and all $\varepsilon > 0$

$$\mathbb{P}\left((V_n^k)^{1/k} \geq (1+\varepsilon)\mathbb{E}((V_n^k)^{1/k}) + \sigma_k\sqrt{2\kappa z} + \kappa(\varepsilon) b_k z \right) \leq e^{-z}. \tag{3.15}$$

Applying (3.15) to $z = ku$ and summing over k, it follows that:

$$\mathbb{P}\left(\forall k \geq 2,\ (V_n^k)^{1/k} \geq (1+\varepsilon)\mathbb{E}((V_n^k)^{1/k}) + \sigma_k\sqrt{2\kappa k u} + \kappa(\varepsilon) b_k k u \right) \leq \sum_{k\geq 2} e^{-ku}.$$

In fact the above left hand side is more precisely dominated by

$$1 \wedge \sum_{k\geq 2} e^{-ku} \leq 1 \wedge \frac{1}{e^u(e^u-1)} \leq 1 \wedge \frac{1}{ue^u} \leq 1.77e^{-u}.$$

Finally, $\mathbb{E}((V_n^k)^{1/k}) \le (\mathbb{E}(V_n^k))^{1/k}$ and the result follows. □

We now bound the σ_k's and the b_k's. The easiest to bound are the b_k's: by Hölder's inequality,

$$b_k \le \sup_{j,t} \left(\sum_{i=j+1}^{n} \mathbb{E}_{(i)}(|g_{i,j}(T_i,t)|^k) \right)^{1/k} \le (B^2 A^{k-2})^{1/k},$$

where again B is given by (3.5) and since the $g_{i,j}$'s are bounded by A. The variance term is a bit more intricate.

$$\begin{aligned}
\sigma_k &= \sup_{\substack{\sum_{i=2}^n \mathbb{E}(|a_i(T_i)|^{k/(k-1)}) = 1 \\ \sum_{j=1}^{n-1} \mathbb{E}(|b_j(T_j)|^2) = 1}} \sum_{j=1}^{n-1} \mathbb{E}^{(j)} \left[\sum_{i=j+1}^{n} \mathbb{E}_{(i)}(g_{i,j}(T_i,T_j) a_i(T_i) b_j(T_j)) \right] \\
&= \sup_{\substack{\sum_{i=2}^n \mathbb{E}(|a_i(T_i)|^{k/(k-1)}) = 1 \\ \sum_{j=1}^{n-1} \mathbb{E}(|b_j(T_j)|^2) = 1}} \sum_{i=2}^{n} \mathbb{E}_{(i)} \left[\sum_{j=1}^{i-1} \mathbb{E}^{(j)}(g_{i,j}(T_i,T_j) b_j(T_j)) a_i(T_i) \right] \\
&= \sup_{\sum_{j=1}^{n-1} \mathbb{E}(|b_j(T_j)|^2)=1} \left[\sum_{i=2}^{n} \mathbb{E}_{(i)} \left[\sum_{j=1}^{i-1} \mathbb{E}^{(j)} \left(g_{i,j}(T_i,T_j) b_j(T_j) \right) \right]^k \right]^{1/k} \\
&\le (B^{k-2} D^2)^{1/k},
\end{aligned}$$

with D given by (3.3).

Next, since x^k is a convex function of x, applying the convexity property to $\left(\frac{\theta_1+\theta_2}{1+\varepsilon}\right)^k = \left(\frac{\theta_1}{1+\varepsilon} + \frac{\varepsilon\theta_2}{1+\varepsilon}\right)^k$, it easily follows that:

$$\forall k > 1, \theta_1, \theta_2, \varepsilon > 0, (\theta_1+\theta_2)^k \le (1+\varepsilon)^{k-1}\theta_1^k + (1+\varepsilon^{-1})^{k-1}\theta_2^k, \tag{3.16}$$

Using this previous inequality several times, with probability larger than $1 - 1.77e^{-u}$, for all $k \ge 2$, V_n^k is bounded by w_n^k, where w_n^k is given by

$$\begin{aligned}
w_n^k = (1+\varepsilon)^{2k-1}\mathbb{E}(V_n^k) + (2+\varepsilon+\varepsilon^{-1})^{k-1} D^2 B^{k-2} (\sqrt{2\kappa k u})^k \\
+ (1+\varepsilon^{-1})^{2k-2} B^2 A^{k-2} \kappa(\varepsilon)^k (ku)^k.
\end{aligned}$$

As in the proof of Theorem 3.1, let $T+1 = \inf\{p \in \mathbb{N}, \exists k, V_p^k \ge w_n^k\}$ and note that since the V_n^k are nondecreasing, by Lemma 3.5 $\mathbb{P}(T < n) \le 1.77e^{-u}$. Then stopping U_n at T, gives by Equation (3.13)

$$\mathbb{E}(e^{\lambda U_n^T}) \le \exp\left(\sum_{k \ge 2} \frac{\lambda^k}{k!} w_n^k \right).$$

It remains to simplify this last bound and to use Chebyshev's inequality.

$$\begin{aligned} q_n &= \sum_{k\geq 2} \frac{\lambda^k}{k!} w_n^k \\ &\leq \sum_{k\geq 2} \frac{\lambda^k}{k!}(1+\varepsilon)^{2k-1}\mathbb{E}(V_n^k) \\ &\quad +\sum_{k\geq 2} \frac{\lambda^k}{k!}(2+\varepsilon+\varepsilon^{-1})^{k-1}D^2B^{k-2}(\sqrt{2\kappa k u})^k \\ &\quad +\sum_{k\geq 2} \frac{\lambda^k}{k!}(1+\varepsilon^{-1})^{2k-2}B^2A^{k-2}\kappa(\varepsilon)^k(ku)^k. \end{aligned}$$

Let us respectively denote by α, β and γ, each one of the three previous sums. For the last sum, since for all k, $k! \geq (k/e)^k$ (see Stirling's formula with correction [3, p. 54]), setting $\delta(\varepsilon) = e(1+\varepsilon^{-1})^2\kappa(\varepsilon)$, we get

$$\gamma \leq \sum_{k\geq 2}(\delta(\varepsilon))^k B^2 A^{k-2}(\lambda u)^k = \frac{\lambda^2(B\delta(\varepsilon)u)^2}{1-(A\delta(\varepsilon)u)\lambda},$$

for $\lambda < (A\delta(\varepsilon)u)^{-1}$.

For the middle sum, since for all $k \geq 2$, $k! \geq k^{k/2}$ (again, see [3, p. 54]) and since moreover $2+\varepsilon+\varepsilon^{-1} \geq 4$, setting $\eta(\varepsilon) = \sqrt{2\kappa}(2+\varepsilon+\varepsilon^{-1})$, we similarly get

$$\beta \leq \frac{\lambda^2(D\eta(\varepsilon)\sqrt{u}/2)^2}{1-(B\eta(\varepsilon)\sqrt{u})\lambda},$$

for $\lambda < (B\eta(\varepsilon)\sqrt{u})^{-1}$.

The estimation of the first sum is more intricate:

$$\alpha = \frac{1}{1+\varepsilon}\sum_{i=1}^{n} \mathbb{E}_{(i)}\left(\mathbb{E}(\exp(\mu|C_i|)|T_i) - \mu\mathbb{E}(|C_i||T_i) - 1\right), \tag{3.17}$$

where $C_i = \sum_{j=1}^{i-1} g_{i,j}(T_i,T_j)$ and $\mu = \lambda(1+\varepsilon)^2$. As $e^\theta - \theta - 1 \geq 0$, for all θ, adding $\mathbb{E}(\exp(-\mu|C_i|)|T_i) + \mu\mathbb{E}(|C_i||T_i) - 1$ to (3.17), we get

$$\alpha \leq \frac{1}{1+\varepsilon}\sum_{i=1}^{n} \mathbb{E}_{(i)}\left(\mathbb{E}(\exp(\mu C_i)|T_i) - 1 + \mathbb{E}(\exp(-\mu C_i)|T_i) - 1\right).$$

Let us recall:

(Bernstein's inequality) Let $X_1, \ldots, X_n$ be n independent centered variables with values in $[-A, A]$. Let $S_n = X_1 + \cdots + X_n$ and let $v = \mathrm{Var}(S_n)$. Then for all $\mu > 0$,

$$\mathbb{E}(e^{\mu S_n}) \leq e^{\frac{\mu^2 v}{2-2\mu\frac{A}{3}}}.$$

Given T_i, C_i and $-C_i$ are sums of centered bounded i.i.d. quantities, it follows from Bernstein's inequality that

$$\alpha \leq \frac{2}{1+\varepsilon} \sum_{i=1}^{n} \mathbb{E}_{(i)} \left(e^{\frac{\mu^2 v_i(T_i)}{2-2\mu \frac{A}{3}}} - 1 \right), \tag{3.18}$$

where $v_i(T_i) = \sum_{j=1}^{i-1} \mathbb{E}^{(j)}(g_{i,j}(T_i, T_j)^2)$. But $v_i(T_i) \leq B^2$, thus

$$\sum_{i=1}^{n} \mathbb{E}_{(i)}(v_i(T_i)^k) \leq C^2 B^{2(k-1)},$$

where C is given by (3.2). Using these facts in (3.18) leads to

$$\alpha \leq \frac{(1+\varepsilon)^3 C^2 \lambda^2}{1 - \lambda(1+\varepsilon)^2 A/3 - \lambda^2 (1+\varepsilon)^4 B^2/2}.$$

The last expression can be upper bounded by:

$$\alpha \leq \frac{(1+\varepsilon)^3 C^2 \lambda^2}{1 - (1+\varepsilon)^2 \lambda (A/3 + B/\sqrt{2})},$$

for $\lambda \leq [(1+\varepsilon)^2 (A/3 + B/\sqrt{2})]^{-1}$. Finally one has,

$$\mathbb{E}(e^{\lambda U_n^T}) \leq \exp\left(\frac{\lambda^2 W^2}{1 - \lambda c} \right), \tag{3.19}$$

where

$$W = (1+\varepsilon)^{3/2} C + \eta(\varepsilon) D\sqrt{u}/2 + \delta(\varepsilon) B u,$$

and

$$c = \max\left((1+\varepsilon)^2 (A/3 + B/\sqrt{2}), \eta(\varepsilon) B\sqrt{u}, \delta(\varepsilon) A u \right).$$

Next, Chebyshev's inequality $\mathbb{P}(U_n^T \geq s) \leq e^{-\lambda s} \mathbb{E}(e^{\lambda U_n^T})$, in conjunction with (3.19) and for $\lambda = \frac{\sqrt{u}}{W + c\sqrt{u}}$, give

$$\mathbb{P}(U_n^T \geq 2W\sqrt{u} + cu) \leq e^{-u}.$$

Proceeding as in the end of the proof of Theorem 3.1, one then gets the bound

$$\mathbb{P}(U_n \geq 2W\sqrt{u} + cu) \leq 2.77 e^{-u}.$$

This inequality implies the result for $u > 1$, but if $u \leq 1$, $2.77 \exp(-u) > 1$. This finishes the proof of the theorem. □

Both Theorem 3.1 and Theorem 3.4 present some interest. The quadratic term in the first one is, as ε tends to 0, of the form $C\sqrt{2u}$ which is the optimal rate for the Central Limit Theorem since the variance term C^2 represents the true variance of the process.

The quadratic term in the second theorem is larger: it is of the form $2C\sqrt{u}$, the extra factor $\sqrt{2}$ coming from the use of symmetrization in the proof. This theorem gives precise constants which are unspecified in the result of [4]. Moreover

Theorem 3.4 has better order of magnitude than Theorem 3.1, as can be seen in the following example originating in statistics (see [9]).

Let $T_1, \dots, T_n$, be uniformly distributed on $[0,1)$. Let m be a regular partition of $[0,1)$, i.e. $[0,1) = \cup_{i=1}^{d}[\frac{i-1}{d}, \frac{i}{d})$.

We set

$$\forall (x,y) \in [0,1)^2, g(x,y) = d\sum_{I\in m}(\mathbb{1}_I(x) - 1/d)(\mathbb{1}_I(y) - 1/d).$$

Let U_n be the corresponding U-statistics (see the appendix of [9]). One has

$$A \leq 4d,\ B^2 \leq 2nd,\ C^2 \leq \frac{n(n-1)}{2}d,\ D \leq \frac{(n-1)}{2}.$$

F can also be computed (using Laplace transform) and is of the order of $d\ln n + n$.

For all ε and u positive, the following concentration inequalities hold true

- by applying Theorem 3.1: with probability smaller than $3e^{-u}$ one has

$$\frac{1}{n(n-1)}\sum_{i\neq j} g(T_i,T_j) = \frac{2U_n}{n(n-1)} \leq$$

$$2(1+\varepsilon)\sqrt{\frac{d}{n(n-1)}u} + \square\left(\frac{1}{n} + \frac{d\ln n}{n^2}\right)u +$$

$$+\square\frac{\sqrt{d/n}}{n-1}u^{3/2} + \square\frac{d}{n(n-1)}u^2.$$

- by applying Theorem 3.4: with probability smaller than $2.77e^{-u}$ one has

$$\frac{2U_n}{n(n-1)} \leq 2(1+\varepsilon)^3\sqrt{\frac{2d}{n(n-1)}u} + \square\frac{1}{n}u + \square\frac{\sqrt{d/n}}{n-1}u^{3/2} + \square\frac{d}{n(n-1)}u^2.$$

(The squares represent known but intricate constants.) The second inequality is sharper in the second term. In particular if d is of order n^2, the second one remains bounded while the first one tends to infinity with n.

4. The Poisson framework

The methodology of the previous sections can be easily adapted to obtain similar results for double integrals of Poisson processes. Let N be a time Poisson process with compensator Λ, and let $(M_t = N_t - \Lambda_t, t \geq 0)$ be the corresponding martingale.

The U-statistic or the double integral for the Poisson process is defined by

$$Z_t = \int_0^t \int_0^{y^-} f(x,y)dM_x dM_y$$

for $f : \mathbb{R}\times\mathbb{R} \to \mathbb{R}$ a Borel function. We do not need degeneracy assumptions, since we integrate with respect to a martingale and this implies that the expectations are zero.

Then we can easily obtain the corresponding version of Theorem 3.1.

Theorem 4.1. *Let $u, \varepsilon > 0$. If f is bounded by A, then*

$$\mathbb{P}\left[Z_t \geq (1+\varepsilon)C\sqrt{2u} + \left(2\sqrt{\kappa}D + \frac{1+\varepsilon}{3}F\right)u + \right.$$

$$\left. + \left(\sqrt{2}\kappa(\varepsilon) + \frac{2\sqrt{\kappa}}{3}\right)Bu^{3/2} + \frac{\kappa(\varepsilon)}{3}Au^2\right] \leq 3e^{-u},$$

where

$$C^2 = \int_0^t \int_0^y f(x,y)^2 d\Lambda_x \, d\Lambda_y,$$

$$D = \sup_{\int_0^t a_x^2 d\Lambda_x = 1, \int_0^t b_y^2 d\Lambda_y = 1} \int_0^t a_x \int_x^t b_y f(x,y) d\Lambda_y d\Lambda_x,$$

$$F = \mathbb{E}\left(\sup_{y \leq t}\left|\int_0^t \mathbb{1}_{x<y} f(x,y) dM_x\right|\right),$$

and

$$B^2 = \max\left\{\sup_{y \leq t}\int_0^y f(x,y)^2 d\Lambda_x, \sup_{x \leq t}\int_x^t f(x,y)^2 d\Lambda_y\right\}.$$

where $\kappa = 6$ and $\kappa(\varepsilon) = 1.25 + 32/\varepsilon$ are given by [15, Corollary 2].

Proof. Perform similar computations in continuous time, replacing Talagrand's inequality by [15, Corollary 2] and (3.11) by the corresponding Lemma derived by van de Geer in [17] or in [8, Theorem 23.17]. □

To conclude, we also state the Poisson version of Theorem 3.4.

Theorem 4.2. *For all $\varepsilon, u > 0$,*

$$\mathbb{P}(Z_t \geq 2(1+\varepsilon)^{3/2}C\sqrt{u} + 2\eta(\varepsilon)Du + \beta(\varepsilon)Bu^{3/2} + \gamma(\varepsilon)Au^2) \leq 2.77e^{-u}$$

where

- $\eta(\varepsilon) = \sqrt{2\kappa}(2 + \varepsilon + \varepsilon^{-1})$,
- $\beta(\varepsilon) = e(1+\varepsilon^{-1})^2\kappa(\varepsilon) + (\sqrt{2\kappa}(2+\varepsilon+\varepsilon^{-1})) \vee \frac{(1+\varepsilon)^2}{\sqrt{2}}$,
- $\gamma(\varepsilon) = (e(1+\varepsilon^{-1})^2\kappa(\varepsilon)) \vee \frac{(1+\varepsilon)^2}{3}$,
- $\kappa = 6$,
- $\kappa(\varepsilon) = 1.25 + 32/\varepsilon$.

Proof. Perform similar computations in continuous time, replacing Talagrand's inequality by [15, Corollary 2] and replacing Lemma 3.3 by its corresponding continuous time version [14, Proposition 4]. □

Potential statistical applications of the two previous theorems would be to construct tests for the Poisson intensity.

5. Concluding Remarks

In [4], the exponential bound is obtained for decoupled U-statistics, i.e. of the form

$$\sum_{i=1}^{n}\sum_{j=1}^{n} f_{i,j}(T_i, T'_j), \tag{5.1}$$

where $T_1, \dots, T_n, T'_1, \dots, T'_n$ are independent random variables. The decoupling inequality of de la Peña and Montgomery-Smith [2] states that, whenever $f_{i,i} = 0$ and $f_{i,j} = f_{j,i}$ for all i, j,

$$\mathbb{P}\left(\left|\sum_{i=1}^{n}\sum_{j=1}^{n} f_{i,j}(T_i, T_j)\right| \geq z\right) \leq C_2\mathbb{P}\left(C_2\left|\sum_{i=1}^{n}\sum_{j=1}^{n} f_{i,j}(T_i, T'_j)\right| \geq z\right), \tag{5.2}$$

for all $z > 0$ and for some unspecified constant $C_2 > 0$.

Our methods provide an exponential upper bound for the left hand side of (5.2) while [4] provides an exponential upper bound for its right hand side. However, Theorem 3.1 and Theorem 3.4 immediately imply their versions for decoupled U–statistics. Indeed, it is enough to take $n' = 2n$, $g'_{i,n+i} = g_{i,j}$ and $g'_{i,j} = 0$ if $1 \leq i, j \leq n$ or $n + 1 \leq i, j \leq 2n$.

The martingale part of the approach presented in these notes adapts easily to higher order U-statistics. However, we are lacking the corresponding version of (3.8). Even for suprema of U-statistics of order two, which will then imply results on U-statistics of order three, (3.8) is unknown. This problem deserves a closer attention.

References

[1] J. Bretagnolle. A new large deviation inequality for U-statistics of order 2. *ESAIM: Probability and Statistics*, 1999.

[2] V. De la Peña and S. Montgomery-Smith. Decoupling inequalities for the tail probabilities of multivariate u-statistics. *Ann. Probab.*, **23**:806–816, 1995.

[3] W. Feller. *An introduction to probability theory and its applications.*, volume 1. Wiley, 3rd edition, 1968.

[4] E. Giné, R. Latała, and J. Zinn. Exponential and moment inequalities for U-statistics. *High Dimensional Prob. II*, 2000.

[5] D.L. Hanson and F.T. Wright. A bound on tail probabilities for quadratic forms in independent random variables. *Ann. Math. Stat.*, **42**(3):1079–1083, 1971.

[6] W. Hoeffding. A class of statistics with asymptotically normal distribution. *Ann. Math. Stat.*, **19**(3), 1948.

[7] W. Hoeffding. Probability inequalities for sums of bounded random variables. *Journal of the American Statistical Association*, **58**, 1963.

[8] O. Kallenberg. *Foundations of modern probability.* Springer, 1997.

[9] B. Laurent. Adaptive estimation of a quadratic functional of a density by model selection. Prépublication, Département de Mathématiques, Université Paris XI, 2003.

[10] M. Ledoux. On Talagrand deviation inequalities for product measures. In *ESAIM: Probability and statistics 1*, 1996.

[11] P. Massart. About the constants in Talagrand's concentration inequalities for empirical processes. *Ann. Proba.*, **2**(28):863–884, 2000.

[12] J. Neveu. *Discrete-Parameter Martingales. Translated by T. P. Speed.* American Elsevier Publishing Company, Inc., 1975.

[13] I. Pinelis. Optimum bounds for the distributions of martingales in Banach spaces. *Ann. Prob.*, **22**:1679–1706, 1994.

[14] P. Reynaud-Bouret. Exponential inequalities for counting processes. Technical Report 1102-001, Georgia Institute of Technology, School of Mathematics, ftp://ftp.math.gatech.edu/pub/preprints/ available_preprints_som.html, 2002.

[15] P. Reynaud-Bouret. Adaptive estimation of the intensity of inhomogeneous Poisson processes via concentration inequalities. *P.T.R.F.*, 2003. D.O.I: 10.1007/s00440-003-0259-1.

[16] M. Talagrand. New concentration inequalities in product spaces. *Invent. Math.*, **126**(3):505–563, 1996.

[17] S. van de Geer. Exponential inequalities for martingales, with application to maximum likelihood estimation for counting processes. *Ann. Stat.*, **23**(5):1779–1801, 1995.

Laboratoire d' Analyse et de Mathématiques Appliquées, CNRS UMR 8050, Université Paris XII, 94010 Créteil Cedex, France
and
School of Mathematics, Georgia Institute of Technology, Atlanta, GA 30332, USA.
E-mail address: houdre@math.gatech.edu

School of Mathematics, Georgia Institute of Technology, Atlanta, GA 30332, USA
and
DMA, École Normale Supérieure, 45 rue d'Ulm, 75230 Paris Cedex 05, France.
E-mail address: Patricia.Reynaud-Bouret@ens.fr

Progress in Probability, Vol. 56, 71–75

On a.s. Unconditional Convergence of Random Series in Banach Spaces

Stanisław Kwapień and Vaja Tarieladze

Abstract. We prove that a random series $\sum x_i\gamma_i$, where (x_i) is a sequence of vectors in a Banach space and (γ_i) is a sequence of *i.i.d* Gaussian random variables is a.s. unconditionally convergent if and only if it is a.s. convergent and the series $\sum x_i$ is unconditionally convergent. A similar statement in which (γ_i) is replaced by a sequence of independent random variables is proved also. We give some applications to Karhunen–Loève representations of Gaussian processes.

1. Introduction

Let us remind that a series $\sum_{i=1}^{\infty} x_i$ in a Banach space F is said to be unconditionally convergent if the series $\sum_{i=1}^{\infty} x_{\sigma(i)}$ is convergent for each permutation σ of positive integers. Sequences of independent, real random variables will be denoted by (ξ_i), (η_i),.. The notation (γ_i) is reserved for a special case of such sequences in which all the random variables are distributed by the normal law $N(0,1)$ and such sequences will be called *Gaussian*. By a *Bernoulli sequence*, denoted by (ϵ_i), we will mean any sequence of independent random variables which are distributed by the common law $\mu(\pm 1) = \frac{1}{2}$. In the note we study a.s. unconditional convergence of random series of the form $\sum_i \xi_i x_i$ where (x_i) is a sequence of vectors in a Banach space and (ξ_i) a sequence of independent random variables. The main result of the paper is the following

Theorem 1. *Let (x_i) be a sequence of vectors in a Banach space and (ξ_i) a sequence of independent random variables.*

(i) If $\inf_i P(|\xi_i| > c) > 0$ for some $c > 0$ and the series $\sum_i \xi_i x_i$ is a.s. unconditionally convergent then the series $\sum_i x_i$ is unconditionally convergent.

(ii) If $sup_i E|\xi_i| < \infty$, the series $\sum_i \xi_i x_i$ is a.s. convergent and $\sum_i x_i$ is unconditionally convergent then $\sum_i \xi_i x_i$ is a.s. unconditionally convergent.

2000 *Mathematics Subject Classification.* Primary 60B11; Secondary 60G15.

Key words and phrases. unconditional convergence, random series, Banach spaces.
Research of the first author was partially supported by KBN grant #2 PO3A 027 22 The work was in a part conducted during his visit at the University of Tennessee, Knoxville.

For previous works on the subject we refer to [4],[5]. In particular the first part of the above theorem, was proved in [5], Theorem 2.1. For the sake of completeness we included the result and we give a proof. The second part of the theorem is a relatively easy consequence to a theorem proved by M. Talagrand, see [8], Theorem 4.12. An obvious consequence to the above theorem is

Corollary 1. *Let (x_i) be a sequence of vectors in a Banach space and (ξ_i) be an i.i.d. sequence of random variables with finite first moments and such that the series $\sum_i \xi_i x_i$ is a.s. convergent. Then the series is a.s unconditionally convergent if and only if the series $\sum_i x_i$ is unconditionally convergent.*

Corollary 1 applied to Gaussian sequences is of special interest because of applications to Karhunen–Loève representation of Gaussian processes which we give in the last Section. For this case we indicate an alternative proof of the corollary which is based on Slepian Lemma.

For convenience let us introduce the following definition and notations: A sequence (x_i) of vectors in a Banach space such that only a finite number of them are different from 0 will be called *finite*. For a finite sequence (x_i) let us put

$$|||(x_i)||| = \sup_{\varepsilon_i=\pm 1} ||\sum_i \varepsilon_i x_i|| = \sup_{x^* \in B(F^*)} \sum_i |x^*(x_i)|,$$

where F^* is the dual Banach space to F and $B(F^*)$ is its unite ball.

If (x_i) is a sequence of vectors and $n < m$ then $(x_i)_n^m$ will denote the sequence which is equal to x_i for $n \le i \le m$ and equal to 0 for the remaining i.

It is easy and a well known fact that a series $\sum_i x_i$ is unconditionally convergent if and only if $\lim_{n,m\to\infty} |||(x_i)_n^m||| = 0$.

2. Basic Inequalities and Proofs

The proof of Theorem 1 is based on the following inequality

Lemma 1. *If (x_i) is a finite sequence of vectors in a Banach space and (ξ_i) a sequence of independent random variables then*

$$\max\{E||\sum_i \xi_i x_i||, (\inf_i E|\xi_i|)|||(x_i)|||\} \le E|||(\xi_i x_i)|||$$

$$\le 4E||\sum_i \xi_i x_i|| + 3(\sup_i E|\xi_i|)|||(x_i)|||.$$

Proof. The inequality $E||\sum_i \xi_i x_i|| \le E|||(\xi_i x_i)|||$ is obvious. Since for each sequence of numbers (t_i) with $|t_i| \le 1$ for all i it is $|||(t_i x_i)||| \le |||(x_i)|||$ we get

$$(\inf_i E|\xi_i|)|||(x_i)||| \le |||(E|\xi_i|x_i)||| \le E|||(|\xi_i|x_i)||| = E|||(\xi_i x_i)|||.$$

These prove the left inequality of the lemma. To prove the right inequality let us assume at first that all the random variables ξ_i are symmetric. Consider the processes $(X_{x^*})_{x^*\in B(F^*)}$, $(Y_{x^*})_{x^*\in B(F^*)}$ which are defined by $X_{x^*} = \sum_i x^*(x_i)\xi_i$

and $Y_{x^*} = \sum_i |x^*(x_i)|\xi_i$. If $(\xi_i) = (\epsilon_i)$ is the Bernoulli sequence then it was proved by Talagrand, see [8],Theorem 4.12, that

$$E \sup_{x^* \in B(F^*)} Y_{x^*} \leq E \sup_{x^* \in B(F^*)} X_{x^*} \tag{2.1}$$

(in fact the statement of Theorem 4.12, [8] gives the above inequality with the constant 2, however its proof allows to reduce the constant to 1). In the case of symmetric random variables the sequence (ξ_i) is equidistributed with the sequence $(|\xi_i|\epsilon_i)$, where (ϵ_i) is a Bernoulli sequence which is independent of (ξ_i). Therefore, by the Fubini Theorem the inequality (2.1) holds true for all independent, symmetric sequences (ξ_i). By the Hahn-Banach Theorem $E \sup_{x^* \in B(F^*)} X_{x^*} = E||\sum_i \xi_i x_i||$. We will show that

$$\begin{aligned}
2E \sup_{x^* \in B(F^*)} Y_{x^*} &= E \sup_{x^* \in B(F^*)} \sum_i |x^*(x_i)| 2\xi_i \\
&\geq E(\sup_{x^* \in B(F^*)} \sum_i |x^*(x_i)|(|\xi_i| - E|\xi_i|)) \\
&\geq E(\sup_{x^* \in B(F^*)} \sum_i |x^*(x_i)||\xi_i|) - (\sup_i E|\xi_i|) \sup_{x^* \in B(F^*)} \sum_i |x^*(x_i)| \\
&= E|||(\xi_i x_i)||| - (\sup_i E|\xi_i|)|||(x_i)|||.
\end{aligned}$$

The last of these inequalities is obvious. The first one is proved by the following arguments. For random vectors ξ, η in R^d we write $\xi \prec_C \eta$ if for each convex function $\phi : R^d \to R^+$ it is $E\phi(\xi) \leq E\phi(\eta)$ (for this definition and the facts which we use below we refer to [7], Chapter 4). For each symmetric random variable ξ we have $|\xi| - E|\xi| \prec_C 2\xi$ therefore $(|\xi_1| - E|\xi_1|, \ldots, |\xi_d| - E|\xi_d|) \prec_C 2(\xi_1, \ldots, \xi_d)$. And this gives the first inequality. Combining these inequalities we obtain

$$E|||(\xi_i x_i)||| \leq 2E||\sum_i \xi_i x_i|| + (\sup_i E|\xi_i|)|||(x_i)||| \tag{2.2}$$

The general case is obtained by the symmetrization. Indeed, let $(\bar{\xi}_i)$ be an independent copy of (ξ_i) and let $\eta_i = \bar{\xi}_i - \xi_i$. The following inequalities are obvious

$$E|||(\eta_i x_i)||| \geq E|||(\xi_i x_i - E\xi_i x_i)||| \geq E|||(\xi_i x_i)||| - (\sup_i E|\xi_i|)|||(x_i)|||,$$

$$E||\sum_i \eta_i x_i|| \leq 2E||\sum_i \xi_i x_i|| \text{ and } \sup_i E|\eta_i| \leq 2 \sup_i E|\xi_i|.$$

Combining these inequalities with inequality (2.2) applied to the random variables (η_i) we conclude the proof of Lemma. □

Proof of Theorem 1. Let $\xi_i' = \xi_i I_{\{||\xi_i x_i|| < s\}}$ and $\xi_i'' = \xi_i I_{\{||\xi_i x_i|| \geq s\}}$ where s is any fixed constant. If the series $\sum_i \xi_i x_i$ is a.s. convergent then $\sum_i ||\xi_i'' x_i|| < \infty$ a.s and $\lim_{n\to\infty} E \sup_{n<m} ||\sum_{i=n}^m \xi_i' x_i|| = 0$. Thus the series $\sum_i \xi_i'' x_i$ is a.s. unconditionally convergent and therefore the series $\sum_i \xi_i x_i$ is a.s. unconditionally convergent if and only if $\sum_i \xi_i' x_i$ is also unconditionally convergent which in turn is equivalent to $\lim_{n\to\infty} E \sup_{n<m} |||(\xi_i' x_i)_n^m||| = 0$. To prove the first statement of Theorem 1 let

us choose $s > c\max_i ||x_i||$. Then $\inf_i E|\xi_i'| > 0$ and the left inequality of Lemma 1 gives that $\limsup_{n,m\to\infty} |||(x_i)_n^m||| = 0$ which means that the series $\sum_i x_i$ is unconditionally convergent. Conversely if the last series is unconditionally convergent then by the right inequality of Lemma 1 $\limsup_{n,m\to\infty} E|||(\xi_i' x_i)_n^m||| = 0$. This implies that $\limsup_{n,m\to\infty} |||(\xi_i' x_i)_n^m||| = 0$ a.s and this concludes the proof. □

The proof of Theorem 1 in the case when (ξ_i) is a Gaussian sequence (γ_i) can be made more direct. In this case the processes $(X_{x^*}), (Y_{x^*})$ are Gaussian and

$$\mathrm{Cov}(X_{x^*}, X_{y^*}) = \sum_i x^*(x_i)y^*(x_i) \le \sum_i |x^*(x_i)||y^*(x_i)| = \mathrm{Cov}(Y_{x^*}, Y_{y^*}),$$

$$\mathrm{D}^2(X_{x^*}) = \sum_i (x^*(x_i))^2 = \mathrm{D}^2(Y_{x^*}) \text{ for each } x^*, y^* \in B(F^*).$$

Thus in the Gaussian case the inequality (2.1) is a consequence of Slepian Lemma. Also, in this case we can avoid truncating of random variables because the a.s. convergence of Gaussian series is the same as the convergence in the mean.

By a result in [3], Section 2 we can prove that $|\gamma_i| - E|\gamma_i| \prec_C \gamma_i$ and therefore in the Gaussian case inequality (2.2) improves to

$$\max\{E||\sum_i \gamma_i x_i||, \sqrt{\frac{2}{\pi}}|||(x_i)|||\} \le E|||(x_i\gamma_i)||| \le E||\sum_i \gamma_i x_i|| + \sqrt{\frac{2}{\pi}}|||(x_i)|||.$$

Remark 1. W.Bednorz, [1], showed that in the case of Gaussian sequences Corollary 1 can be obtained by Fernique–Talagrand Majorizing Measure Theorem.

3. Applications and Comments

If X is a Gaussian, centered random vector in Banach space F then according to Karhunen–Loève Theorem there exists a.s. convergent Gaussian series such that $X = \sum_{i=1}^{\infty} x_i\gamma_i$ a.s. which we will call Karhunen–Loève representation. The choice of the sequence (x_i) is not unique. Each such sequence will be called representing sequence for X. Theorem 1, gives an answer to a question when the convergence in Karhunen–Loève representation is a.s. unconditional. For example if X is a Wiener process on $[0,1]$, which we treat as Gaussian random vector in $C[0,1]$, then representing sequences (x_i) are of the form $x_i(t) = \int_0^t e_i(s)ds$ where (e_i) is a complete orthonormal sequence in $L_2[0,1]$. By Theorem 1 the convergence in Karhunen–Loève representation is a.s. unconditional if and only if the series $\sum_i x_i$ is unconditionally convergent in $C[0,1]$ which is the same as the convergence of the series $\sum_i |x_i|$ in $C[0,1]$, (here $|x_i|$ are functions defined by $|x_i|(t) = |x_i(t)|$). Thus if we choose for (e_i) the trigonometrical polynomials $e_i(t) = \sqrt{2}\sin(\pi i t)$ (the Wiener construction) then because $\sum_i \int_0^1 |x_i|(t)dt = \infty$ the series $\sum x_i$ is not unconditionally convergent in $C([0,1])$ and hence the corresponding Karhunen–Loève representation is not a.s. unconditionally convergent. On the other hand side if we choose for (e_i) the Haar system (the Ciesielski construction) then the series $\sum x_i$ is unconditionally convergent and hence the corresponding Gaussian series is a.s.

unconditionally convergent. However this is obvious from the construction.

It was proved in [2] that each Gaussian vector in a Banach space F of cotype 2 is an image of a Gaussian vector in a Hilbert space by a continuous operator. Hence we deduce easily that it has a representing sequence which is unconditionally summable and the corresponding Karhunen–Loève representation is a.s. unconditionally convergent. On the other hand it was shown in [6] that there are Gaussian vectors in Banach spaces without unconditionally summable representing sequence, thus each of its Karhunen–Loève representation is not a.s. unconditionally convergent.

Remark 2. If (ξ_i) is an *i.i.d.* sequence of 1-stable, symmetric random variables and (x_i) is a sequence of numbers such that $\sum_i |x_i| < \infty$ and $\sum_i |x_i| \ln(1/|x_i|) = \infty$ then $\sum_i \xi_i x_i$ is a.s. convergent and it is not a.s. unconditionally convergent. This shows that we can not replace in Theorem 1, (ii) (resp. Corollary 1) the assumption $\sup_i E|\xi_i| < \infty$ by the condition that (ξ_i) is *i.i.d.*

References

[1] W. Bednorz. (2002) *Majorizing measures and Pietsch's measures* MD Thesis at Warsaw University (in polish).

[2] S. Chobanjan, V. Tarieladze. (1977)*Gaussian characterizations of certain Banach spaces* J. Multivariate Anal., **7**, 183–203.

[3] H. Koenig and S. Kwapień. (2001) *Best Khintchine type inequalities for sums of independent, rotationally invariant random vectors* Positivity, **5** , 115–152.

[4] V. Kvaratskhelia. (1980) *On unconditional convergence of random series in Banach spaces* Lecture Notes in Math. **828**, 162–166.

[5] V. Kvaratskhelia. (2000) *Unconditional convergence of random series and geometry of Banach spaces* Georgian Math. J.,**7**, no. **1**, 85-96.

[6] S. Kwapień and B. Szymański. (1980)*Some remarks on Gaussian measures in Banach space* Probab. Math. Statist. **1**, no. **1**, 59–65.

[7] S. Kwapień and W. Woyczyński. *Random Series and Stochastic Integrals; Single and Multiple* Birhausser 1992.

[8] M. Ledoux and M. Talagrand. *Probability in Banach Spaces*, Springer-Verlag, 1991.

E-mail address: kwapstan@mimuw.edu.pl

Institute of Mathematics, Warsaw University, Banacha 2, 02-097 Warszawa, POLAND

Muskhelishvili Institute of Computational Mathematics, Georgian Academy of Sciences, Tibilisi-93, GEORGI

E-mail address: tar@gw.acnet.ge

Progress in Probability, Vol. 56, 77–92

Moment and Tail Estimates for Multidimensional Chaos Generated by Positive Random Variables with Logarithmically Concave Tails

Rafał Latała and Rafał Łochowski

Abstract. In this paper we give estimates of tails and moments of multidimensional chaos $\sum a_{i_1,\ldots,i_d}X_{i_1}^{(1)}\cdots X_{i_d}^{(d)}$ $(a_{i_1,\ldots,i_d}\geq 0)$ generated by positive random variables $X_{i_1}^{(1)},\ldots,X_{i_d}^{(d)}$ with logarithmically concave tails. The estimates are exact up to constants depending only on the dimension d.

1. Introduction

Let $S=\sum a_{i_1,\ldots,i_d}X_{i_1}^{(1)}\cdots X_{i_d}^{(d)}$ be a random chaos of order d generated by random variables $X_i^{(r)}$. In this paper we present two sided estimates for moments and tails of S in the case when all coefficients are nonnegative and $X_i^{(r)}$ are positive r.v.'s with log-concave tails (precise definitions are given below). Many important inequalities for moments and tails of S are known, even in the more general case of U-statistics (see [2], [4], [7] and [8] for a review of some results). However bounds obtained here are much more precise with constants depending only on the order d.

The estimates for moments of linear combinations (case $d=1$) of symmetric log-concave random variables were obtained by Gluskin and Kwapień [5]. In the paper [9] similar inequalities were given for chaos of order 2 generated by such random variables (Gaussian chaos of order 2 were considered much earlier in [6]). Methods of both papers can be adapted to the nonnegative case (see [10]).

However there seems to be a significant difference between chaos of order $d\geq 3$ and $d=2$. As far as we know two sided estimates with deterministic quantities and universal constants are not known for moments and tails of Gaussian chaos of order 3 (Arcones and Giné [1] obtained some bounds but involving suprema

---

2000 *Mathematics Subject Classification.* Primary 60E15.
Key words and phrases. random chaos, moments.
Research partially supported by KBN grant #2 PO3A 027 22.

of empirical processes). We hope that our paper is just a first step towards more general results.

Let us comment on the organization of the paper. In the next section we present the necessary notation and definitions and formulate main results. Then, in Section 3, we prove moment estimates for the one dimensional case and in Section 4, by induction, for the multidimensional case.

2. Notation and Main Results

Assumptions and notation. Let d and n be positive integers. By $\mathbf{i}$ we denote d-dimensional multiindex $(i_1, \ldots, i_d)$, where $1 \leq i_1, \ldots, i_d \leq n$ and by $\mathbf{I}$ the set of all such multiindices.

For $I \subset \{1, 2, \ldots, d\}$ and $\mathbf{i} \in \mathbf{J} \subset \mathbf{I}$, let $\mathbf{i}_I$ denote $(i_k)_{k \in I}$ and $\mathbf{J}_I$ be the set of all such $\mathbf{i}_I$'s. For $I \subset \{1, 2, \ldots, d\}$ let I' be the set $\{1, 2, \ldots, n\} \setminus I$. With this notation we have for example $\mathbf{i}_{\{1\}'} = (i_2, \ldots, i_d)$. The symbol $\#I$ denotes the cardinality of the set I.

Henceforth $X_{i_1}^{(1)}, \ldots, X_{i_d}^{(d)}$, $1 \leq i_1, \ldots, i_d \leq n$ be independent positive random variables with logarithmically concave tails so that

$$\mathbf{P}\left(X_i^{(r)} \geq t\right) = e^{-N_i^{(r)}(t)}, \text{ for } 1 \leq r \leq d, 1 \leq i \leq n, t \geq 0$$

where $N_i^{(r)} : \mathbb{R}_+ \to \mathbb{R}_+$ is a convex, strictly increasing function normalized in such a way that

$$N_i^{(r)}(1) = 1. \tag{2.1}$$

Let us define a family of subsets of $\mathbb{R}_+^n$, $\{B_{\mathcal{N},p}^r\}_{1 \leq r \leq d}$, by the formula

$$B_{\mathcal{N},p}^r = \left\{ x \in \mathbb{R}_+^n : \sum_{i=1}^n N_i^{(r)}(x_i) \leq p \text{ and } (x_i = 0 \text{ or } x_i \geq 1) \text{ for } i = 1, 2, \ldots, n \right\}.$$

For the multidimensional table of reals, $(a_{\mathbf{i}})_{\mathbf{i} \in \mathbf{I}}$, we set

$$\left\|(a_{\mathbf{i}})_{\mathbf{i} \in \mathbf{I}}\right\|_{\mathcal{N},u} = \sup\left\{ \sum_{\mathbf{i} \in \mathbf{I}} a_{\mathbf{i}} \prod_{r=1}^d \left(1 + x_{i_r}^{(r)}\right) : x^{(1)} \in B_{\mathcal{N},u}^1, \ldots, x^{(d)} \in B_{\mathcal{N},u}^d \right\}.$$

By S we denote a random variable $S = \sum a_{\mathbf{i}} X_{i_1}^{(1)} \cdots X_{i_d}^{(d)}$. For $p \geq 1$ we put $\|S\|_p = (\mathbf{E} S^p)^{1/p}$.

We use letters c and C to denote universal positive constants, that may change from line to line and $c(d)$, $C(d)$ to denote positive constants, depending only on d ($c(d)$, $C(d)$ may also differ at each occurrence). Finally relation $A \sim_d B$ means that $c(d)\, A \leq B \leq C(d)\, A$.

Now we are ready to formulate the two main theorems.

Theorem 2.1. *There exist constants $0 < c_1(d), C_1(d) < \infty$ depending only on d such that for any $p \geq 1$ we have*

$$c_1(d) \left\|(a_{\mathbf{i}})_{\mathbf{i}\in\mathbf{I}}\right\|_{\mathcal{N},p} \leq \|S\|_p \leq C_1(d) \left\|(a_{\mathbf{i}})_{\mathbf{i}\in\mathbf{I}}\right\|_{\mathcal{N},p}.$$

We postpone the proof of Theorem 2.1 till the end of the paper and now present some applications and remarks.

Theorem 2.2. *There exist constants $0 < c_2(d), C_2(d) < \infty$ depending only on d such that for any $t \geq 0$ we have*

$$\mathbf{P}(S \geq C_2(d) \left\|(a_{\mathbf{i}})_{\mathbf{i}\in\mathbf{I}}\right\|_{\mathcal{N},t}) \leq e^{-t}$$

and

$$\mathbf{P}(S \geq c_2(d) \left\|(a_{\mathbf{i}})_{\mathbf{i}\in\mathbf{I}}\right\|_{\mathcal{N},t}) \geq \min(c_2(d), e^{-t}).$$

Proof. Notice that $\left\|(a_{\mathbf{i}})_{\mathbf{i}\in\mathbf{I}}\right\|_{\mathcal{N},t} = \left\|(a_{\mathbf{i}})_{\mathbf{i}\in\mathbf{I}}\right\|_{\mathcal{N},1}$ for $0 \leq t \leq 1$, so it is enough to consider only $t \geq 1$. First inequality then immediately follows by Theorem 2.1 and Chebyshev's inequality. To see the second inequality notice that for $p \geq 1$ by Theorem 2.1 we get

$$\|S\|_{2p} \leq C_1(d) \left\|(a_{\mathbf{i}})_{\mathbf{i}\in\mathbf{I}}\right\|_{\mathcal{N},2p} \leq 4^d C_1(d) \left\|(a_{\mathbf{i}})_{\mathbf{i}\in\mathbf{I}}\right\|_{\mathcal{N},p} \leq C(d)\|S\|_p$$

Now it is enough to notice that $\left\|(a_{\mathbf{i}})_{\mathbf{i}\in\mathbf{I}}\right\|_{\mathcal{N},\lambda p} \leq (2\lambda)^d \left\|(a_{\mathbf{i}})_{\mathbf{i}\in\mathbf{I}}\right\|_{\mathcal{N},p}$ for $\lambda \geq 1$ and use the Paley-Zygmund inequality as in [5]. □

Remark 2.3. We may define the norms $\left\|(a_{\mathbf{i}})_{\mathbf{i}\in\mathbf{I}}\right\|_{\mathcal{N},u}$ in a slightly different, a little bit more natural way. To do this it is enough to observe that

$$\begin{aligned}\left\|(a_{\mathbf{i}})_{\mathbf{i}\in\mathbf{I}}\right\|_{\mathcal{N},u} &\leq \sup\left\{\sum_{\mathbf{i}\in\mathbf{I}} a_{\mathbf{i}} \prod_{r=1}^{d}\left(1+x_{i_r}^{(r)}\right) : \sum_i N_i^{(r)}(x_i^{(r)}) \leq u, r = 1,2,\ldots,d\right\}\\ &\leq 2^d \left\|(a_{\mathbf{i}})_{\mathbf{i}\in\mathbf{I}}\right\|_{\mathcal{N},u}.\end{aligned}$$

Remark 2.4. All estimates obtained in Theorems 2.1 and 2.2 may be extended by simple approximation argument to the situation where functions $N_i^{(r)}$ are still convex but not necessarily continuous or strictly increasing on $\mathbb{R}$, i.e. they may take value 0 on some subinterval of $\mathbb{R}_+$ as well value $+\infty$ beginning from some number t_0 (it means that variable $X_i^{(r)}$ does not admit values greater than t_0). In this case condition (2.1) should be substituted by the following one

$$\inf\left\{t : N_i^{(r)}(t) \geq 1\right\} = 1.$$

Remark 2.5. Theorems 2.1 and 2.2 are also valid (with worse constants) for the undecoupled chaos of order d, that is for random variables of the form

$$S = \sum_{1\leq i_1<\ldots<i_d\leq n} a_{i_1,\ldots,i_d} X_{i_1}\cdots X_{i_d},$$

where coefficients $a_{i_1,\dots,i_d}$ are nonnegative and positive r.v.'s X_i are independent with log-concave tails. It is an immediate consequence of the result of de la Peña and Montgomery-Smith [3] that moments and tails of S are comparable (with constants depending only on d) with moments and tails of the decoupled chaos

$$\tilde{S} = \sum_{\pi} \sum_{1\le i_1\le\ldots\le i_d\le n} a_{i_1,\dots,i_d} X^{(1)}_{i_{\pi(1)}} \cdots X^{(d)}_{i_{\pi(d)}},$$

where the first sum is taken over all permutations π of the set $\{1,\dots,d\}$ and r.v.'s $X_i^{(r)}$ are independent copies of X_i.

Example. Let us consider a special case, when all variables $X^{(1)}_{i_1},\dots,X^{(d)}_{i_d}$, $1\le i_1,\dots,i_d\le n$ have the exponential distribution, i.e.

$$\mathbf{P}\left(X_i^{(r)}\ge t\right) = e^{-t} \text{ for } 1\le r\le d, 1\le i\le n.$$

In this case we may easily find a simpler expression for the norm $\left\|(a_{\mathbf{i}})_{\mathbf{i}\in\mathbf{I}}\right\|_{\mathcal{N},u}$. We have

$$\begin{aligned}\left\|(a_{\mathbf{i}})_{\mathbf{i}\in\mathbf{I}}\right\|_{\mathcal{N},u} &\le \sup\left\{\sum a_{\mathbf{i}}\prod_{r=1}^{d}\left(1+b^{(r)}_{i_r}\right) : \sum b^{(1)}_{i_1}\le u,\dots,\sum b^{(d)}_{i_d}\le u\right\}\\ &= \sup\left\{\sum_{I\subset\{1,2,\dots,d\}}\sum a_{\mathbf{i}}\prod_{r\in I} b^{(r)}_{i_r} : \sum b^{(1)}_{i_1}\le u,\dots,\sum b^{(d)}_{i_d}\le u\right\}.\end{aligned}$$

So

$$\begin{aligned}\left\|(a_{\mathbf{i}})_{\mathbf{i}\in\mathbf{I}}\right\|_{\mathcal{N},u} &\le \sum_{I\subset\{1,2,\dots,d\}} \sup\left\{\sum a_{\mathbf{i}}\prod_{r\in I} b^{(r)}_{i_r} : \sum b^{(r)}_{i_r}\le u \text{ for } r\in I\right\}\\ &= \sum_{I\subset\{1,2,\dots,d\}} u^{\#I}\max_{\mathbf{j}\in\mathbf{I}_I}\sum_{\mathbf{i}:\mathbf{i}_I=\mathbf{j}} a_{\mathbf{i}}. \qquad (2.2)\end{aligned}$$

On the other hand

$$\begin{aligned}\left\|(a_{\mathbf{i}})_{\mathbf{i}\in\mathbf{I}}\right\|_{\mathcal{N},u} &\ge \max_{I\subset\{1,2,\dots,d\}} \sup\left\{\sum a_{\mathbf{i}}\prod_{r\in I} b^{(r)}_{i_r} : \sum b^{(r)}_{i_r}\le u \text{ for } r\in I\right\}\\ &\ge 2^{-d}\sum_{I\subset\{1,2,\dots,d\}} u^{\#I}\max_{\mathbf{j}\in\mathbf{I}_I}\sum_{\mathbf{i}:\mathbf{i}_I=\mathbf{j}} a_{\mathbf{i}}. \qquad (2.3)\end{aligned}$$

From (2.2) and (2.3) and Theorem 2.1 we get that for all $p\ge 1$

$$\begin{aligned}\left\|\sum a_{\mathbf{i}} X^{(1)}_{i_1}\cdots X^{(d)}_{i_d}\right\|_p &\sim_d \sum_{I\subset\{1,2,\dots,d\}} p^{\#I}\max_{\mathbf{j}\in\mathbf{I}_I}\sum_{\mathbf{i}:\mathbf{i}_I=\mathbf{j}} a_{\mathbf{i}}\\ &= \sum_{i_1,\dots,i_d} a_{i_1,\dots,i_d} + p\max_{i_1}\sum_{i_2,\dots,i_d} a_{i_1,\dots,i_d} + p\max_{i_2}\sum_{i_1,i_3,\dots,i_d} a_{i_1,\dots,i_d} + \ldots +\\ &\quad + \ldots + p^d \max_{i_1,\dots,i_d} a_{i_1,\dots,i_d}.\end{aligned}$$

Remark 2.6. In the paper we do not put much emphasis on the rate of growth of constants $c_i^{-1}(d)$ and $C_i(d)$ that appear in Theorems 2.1 and 2.2. Following the presented proofs one can get $c_1^{-1}(d) \leq C^d$ and $c_2^{-1}(d), C_i(d) \leq (Cd)^d$. However there is no indication that this is their optimal rate of growth.

3. One-dimensional case

In this section we will prove Theorem 2.1 for $d = 1$. It will be the starting point of an induction proof of the general case. Let us start with the following simple Lemma.

Lemma 3.1. *We have*

$$e^{-1} \leq \mathbf{E}X_i^{(r)} \leq 1 + e^{-1}.$$

Proof. Let us observe that by (2.1), $\mathbf{E}X_i^{(r)} \geq \mathbf{P}(X_i^{(r)} \geq 1) \geq e^{-1}$ and on the other hand by the convexity of $N_i^{(r)}$ we have

$$\mathbf{E}X_i^{(r)} \leq 1 + \int_1^\infty \exp(-N_i^{(r)}(t))dt \leq 1 + \int_1^\infty e^{-t}dt = 1 + e^{-1}. \qquad \square$$

In order to ease the notation in the rest of this section, instead of writing $X_1^{(1)}, \ldots, X_n^{(1)}$ we will write $X_1, \ldots, X_n$ and instead of writing $N_1^{(1)}, \ldots, N_n^{(1)}$ we will write $N_1, \ldots, N_n$. So in this case for a sequence (a_i) of nonnegative real numbers and $u > 0$ we have $S = \sum a_i X_i$ and

$$\|(a_i)\|_{\mathcal{N},u} = \sup\left\{\sum a_i (1 + b_i) : (b_i) \in B^1_{\mathcal{N},u}\right\}.$$

We will show that for $p \geq 1$

$$\frac{1}{2e}\|(a_i)\|_{\mathcal{N},p} \leq \|S\|_p \leq 26\|(a_i)\|_{\mathcal{N},p}. \tag{3.1}$$

Proof. First we will prove the estimate from below, by the same method as in [5]. Since $\|S\|_1 \leq \|S\|_p$ by the definition of $\|(a_i)\|_{\mathcal{N},u}$ and Lemma 3.1 it is enough to show that

$$\sum a_i b_i \leq e\|S\|_p \tag{3.2}$$

for every sequence (b_i) of nonnegative reals such that $\sum N_i(b_i) \leq p$. We have

$$\begin{aligned}\|S\|_p = \left\|\sum a_i X_i\right\|_p &\geq \left(\sum a_i b_i\right) (\mathbf{P}(X_i \geq b_i : i = 1, 2, \ldots, n))^{1/p} \\ &= \left(\sum a_i b_i\right) \exp\left(-\frac{1}{p}\sum N_i(b_i)\right) \geq \frac{1}{e}\sum a_i b_i.\end{aligned}$$

So (3.2) holds.

Now we will prove the estimate from above. Let $M_i : \mathbb{R}_+ \to \mathbb{R}_+$ be the inverse of N_i restricted to $\mathbb{R}_+$. The variable X_i has distribution $M_i(\mu)$, where μ is the measure with density $e^{-x} I_{\{x \geq 0\}}$. This means in particular that

$$\mathbf{P}\left(\sum a_i X_i > t\right) = \mu^n \left(x \in \mathbb{R}_+^n : \sum a_i M_i(x_i) > t\right),$$

where μ^n is the product measure $\mu \otimes \mu \otimes \ldots \otimes \mu$ on $\mathbb{R}^n$. Let us define

$$A = \left\{x \in \mathbb{R}_+^n : \sum a_i M_i(x_i) \leq 3 \sum a_i\right\}.$$

By the Chebyshev inequality (since by Lemma 3.1, $\mathbf{E}S \leq \frac{3}{2}\sum a_i$) $\mu^n(A) \geq 1/2$ and therefore by a result of Talagrand (cf. [12] and [11] for a simpler proof) we get

$$\mu^n(A + V_s) \geq 1 - 2e^{-s},$$

where

$$V_s = \{x \in \mathbb{R}^n : \sum (|x_i| - 1)_+ = \sum \max(|x_i| - 1, 0) \leq 5s\}.$$

The last inequality is just an application of Lemma 4 from [11] for the couple $(\mu^n, w^{(n)})$ where $w^{(n)}(x) = \sum w(x_i)$ and $w(t) = \frac{1}{5}\max(|t| - 1, 0)$. It is easy to check that the couple $(\mu^n, w^{(n)})$ satisfies assumptions of the Lemma 4. Indeed, applying for a moment some notation from [11], it means that the couple $(\mu^n, w^{(n)})$ has property (τ), which follows from the Proposition from page 191, the inequality $w \leq W$ and Lemma 1 from page 190 of the cited paper.

Let now $\mathbb{R}_+^n \ni x = y + z$ with $y \in A$ and $z \in V_s$. By the convexity of N_i we have $|M_i(x_i) - M_i(y_i)| \leq M_i(|x_i - y_i|)$, so we obtain

$$\begin{aligned}
\sum a_i M_i(x_i) &\leq \sum a_i M_i(y_i) + \sum a_i |M_i(x_i) - M_i(y_i)| \\
&\leq 3\sum a_i + \sum a_i M_i(|z_i|) \\
&\leq 4\sum a_i + \|(a_i)\|_{\mathcal{N},5s},
\end{aligned}$$

since $\sum N_i((M_i(|z_i|) - 1)_+) I_{\{M_i(|z_i|) \geq 1\}} \leq \sum (N_i(M_i(|z_i|)) - 1)_+ = \sum (|z_i| - 1)_+ \leq 5s$. So

$$\mathbf{P}\left(S > 4\sum a_i + \|(a_i)\|_{\mathcal{N},5s}\right) \leq 2e^{-s} \tag{3.3}$$

and since $\|(a_i)\|_{\mathcal{N},\lambda u} \leq 2\lambda \|(a_i)\|_{\mathcal{N},u}$ for $\lambda \geq 1$ we have for $t \geq 1$

$$\mathbf{P}\left(S > 4\sum a_i + 2t\|(a_i)\|_{\mathcal{N},u}\right) \leq 2e^{-tu/5}.$$

Therefore, integrating by parts, we get

$$\begin{aligned}
\|S\|_p &\le 4\sum a_i + 2\|(a_i)\|_{\mathcal{N},p} + 2\|(a_i)\|_{\mathcal{N},p} \\
&\quad\times \left(\int_0^\infty pt^{p-1}\mathbf{P}\left(S > 4\sum a_i + 2(1+t)\|(a_i)\|_{\mathcal{N},p}\right)dt\right)^{1/p} \\
&\le 4\sum a_i + 2\|(a_i)\|_{\mathcal{N},p}\left(1 + \left(\int_0^\infty 2pt^{p-1}e^{-tp/5}dt\right)^{1/p}\right) \\
&\le 4\sum a_i + 2\|(a_i)\|_{\mathcal{N},p}\left(1 + 5\left(2\Gamma(p)/p^{p-1}\right)^{1/p}\right) \\
&\le 4\sum a_i + 22\|(a_i)\|_{\mathcal{N},p} \le 26\|(a_i)\|_{\mathcal{N},p}.
\end{aligned}$$

and the proof of (3.1) is now completed. □

Now, using (3.3), we will prove the following lemma, which will be used in the next section.

Lemma 3.2. *Let A be a finite subset of $\mathbb{R}_+^n$ and for $\mathbf{a} = (a_1, \ldots, a_n) \in A$ let us define $S_{\mathbf{a}} = \sum a_i X_i$, then the following estimate holds for any $p \ge 1$*

$$\left\|\max_{\mathbf{a}\in A}\left(S_{\mathbf{a}} - C\sum a_i\right)_+\right\|_p \le C(p + \ln \#A)\max_{\mathbf{a}\in A}\max_i a_i \tag{3.4}$$

Proof. By the convexity of N_i and by the condition (2.1) we have $N_i(x) \ge x$ for $x \ge 1$, thus we get for $s \ge 0$

$$\begin{aligned}
\|(a_i)\|_{\mathcal{N},s} &\le \sum a_i + \sup\left\{\sum a_i b_i : \sum b_i \le s\right\} \\
&= \sum a_i + s\max_i a_i.
\end{aligned}$$

So, applying (3.3), we obtain

$$\mathbf{P}\left(S_{\mathbf{a}} > 4\sum a_i + 5s\max_i a_i\right) \le 2e^{-s},$$

$$\mathbf{P}\left(\left(S_{\mathbf{a}} - C\sum a_i\right)_+ > Cs\max_{\mathbf{a}\in A}\max_i a_i\right) \le 2e^{-s}.$$

Thus

$$\begin{aligned}
&\mathbf{P}\left(\max_{\mathbf{a}\in A}\left(S_{\mathbf{a}} - C\sum a_i\right)_+ \ge Cs\max_{\mathbf{a}\in A}\max_i a_i\right) \\
&\quad\le \sum_{\mathbf{a}\in A}\mathbf{P}\left(\left(S_{\mathbf{a}} - C\sum a_i\right)_+ \ge Cs\max_{\mathbf{a}\in A}\max_i a_i\right) \le 2\#Ae^{-s}.
\end{aligned}$$

and

$$\begin{aligned}
&\mathbf{P}\left(\max_{\mathbf{a}\in A}\left(S_{\mathbf{a}} - C\sum a_i\right)_+ \ge C(1+s)(p + \ln \#A)\max_{\mathbf{a}\in A}\max_i a_i\right) \\
&\qquad\qquad\le 2\#Ae^{-(1+s)(p+\ln \#A)} \le 2e^{-sp}.
\end{aligned}$$

Integrating by parts we obtain

$$\left\|\max_{\mathbf{a}\in A}\left(S_{\mathbf{a}}-C\sum a_i\right)_+\right\|_p \leq C\,(p+\ln\#A)\max_{\mathbf{a}\in A}\max_i a_i$$

$$+\,C\,(p+\ln\#A)\max_{\mathbf{a}\in A}\max_i a_i\left(\int_0^\infty 2pt^{p-1}e^{-pt}dt\right)^{1/p}$$

$$\leq C\,(p+\ln\#A)\max_{\mathbf{a}\in A}\max_i a_i.$$

□

4. Multidimensional case

In this section we will prove Theorem 2.1 proceeding by induction on d. The first step was already established in the previous section. So in the sequel we will assume that Theorem 2.1 holds for $d-1$ and we will show it for d. First we will show the estimate from below that is

$$\|S\|_p \geq c_1\,(d)\,\left\|(a_{\mathbf{i}})_{\mathbf{i}\in\mathbf{I}}\right\|_{\mathcal{N},p}\,. \tag{4.1}$$

Proof. We will proceed in the similar way as in the proof of the lower estimate in the previous section. By the induction assumption and Lemma 3.1

$$\begin{aligned}
\|S\|_p &\geq c_1(d-1)\left\|\left\|\left(\sum_{\mathbf{i}:\mathbf{i}_{\{1\}'}=\mathbf{j}} a_{\mathbf{i}}X_{i_1}^{(1)}\right)_{\mathbf{j}\in\mathbf{I}_{\{1\}'}}\right\|_{\mathcal{N},p}\right\|_p \\
&\geq c_1\,(d-1)\,\mathbf{E}\left\|\left(\sum_{\mathbf{i}:\mathbf{i}_{\{1\}'}=\mathbf{j}} a_{\mathbf{i}}X_{i_1}^{(1)}\right)_{\mathbf{j}\in\mathbf{I}_{\{1\}'}}\right\|_{\mathcal{N},p} \\
&\geq c_1(d-1)\sup\left\{\mathbf{E}\sum_{\mathbf{i}} a_{\mathbf{i}}X_{i_1}^{(1)}\prod_{r>1}\left(1+x_{i_r}^{(r)}\right) : x^{(r)}\in B_{\mathcal{N},u}^r \text{ for } r>1\right\} \\
&\geq e^{-1}c_1(d-1)\sup\left\{\sum_{\mathbf{i}} a_{\mathbf{i}}\prod_{r>1}\left(1+x_{i_r}^{(r)}\right) : x^{(r)}\in B_{\mathcal{N},u}^r \text{ for } r>1\right\} \quad (4.2)
\end{aligned}$$

Let us now assume that (b_i) is a sequence of nonnegative reals with $\sum N_i^{(1)}(b_i)\leq p$, then again by the induction assumption

$$
\begin{aligned}
\|S\|_p &\geq c_1(d-1)\left\|\left\|\left(\sum_{\mathbf{i}:\mathbf{i}_{\{1\}'}=\mathbf{j}} a_{\mathbf{i}} X_{i_1}^{(1)}\right)_{\mathbf{j}\in\mathbf{I}_{\{1\}'}}\right\|_{\mathcal{N},p}\right\|_p \\
&\geq c_1(d-1)\left(\left\|\left(\sum_{\mathbf{i}:\mathbf{i}_{\{1\}'}=\mathbf{j}} a_{\mathbf{i}} b_{i_1}\right)_{\mathbf{j}\in\mathbf{I}_{\{1\}'}}\right\|_{\mathcal{N},p}^p \mathbf{P}\left(X_i^{(1)} \geq b_i,\ i=1,2,\ldots,n\right)\right)^{1/p} \\
&\geq c_1(d-1)\left\|\left(\sum_{\mathbf{i}:\mathbf{i}_{\{1\}'}=\mathbf{j}} a_{\mathbf{i}} b_{i_1}\right)_{\mathbf{j}\in\mathbf{I}_{\{1\}'}}\right\|_{\mathcal{N},p} \exp\left(-\frac{1}{p}\sum_{i=1}^n N_i^{(1)}(b_i)\right) \\
&\geq e^{-1}c_1(d-1)\sup\left\{\sum_{\mathbf{i}\in\mathbf{I}} a_{\mathbf{i}} b_{i_1} \prod_{r=2}^n \left(1+x_{i_r}^{(r)}\right) : x^{(2)} \in B_{\mathcal{N},u}^1, \ldots, x^{(d)} \in B_{\mathcal{N},u}^d\right\}.
\end{aligned}
$$

Combining the above with (4.2) we get the lower estimate (4.1) with $c_1(d) = (2e)^{-1}c_1(d-1)$. □

In order to prove the estimate from above that is

$$
\|S\|_p \leq C_1(d)\left\|(a_{\mathbf{i}})_{\mathbf{i}\in\mathbf{I}}\right\|_{\mathcal{N},p} \tag{4.3}
$$

we will need following two observations.

1. We may rearrange indexes $i_1, \ldots, i_d$ in such a way that for any $1 \leq r \leq d$

$$
i < j \Rightarrow \sum_{\mathbf{i}:i_r=i} a_{\mathbf{i}} \geq \sum_{\mathbf{i}:i_r=j} a_{\mathbf{i}}. \tag{4.4}
$$

2. We may split $\mathbf{I}$ into a sum of disjoint sets $\mathbf{I}_r, 1 \leq r \leq d$, in the following way.

   $\mathbf{I}_r = \{\mathbf{i} \in \mathbf{I} : i_r > \max(i_1, \ldots, i_{r-1}) \text{ and } i_r \geq \max(i_{r+1}, \ldots, i_d)\}.$

   Assuming that (4.3) holds for $d-1$, we may write

$$
\begin{aligned}
\|S\|_p = \left\|\sum_{r=1}^d \sum_{\mathbf{i}\in\mathbf{I}_r} a_{\mathbf{i}} X_{i_1}^{(1)} \cdots X_{i_d}^{(d)}\right\|_p &\leq \sum_{r=1}^d \left\|\sum_{\mathbf{i}\in\mathbf{I}_r} a_{\mathbf{i}} X_{i_1}^{(1)} \cdots X_{i_d}^{(d)}\right\|_p \\
&\leq C_1(d-1)\sum_{r=1}^d \left\|\left\|\left(\sum_{\mathbf{i}:\mathbf{i}_{\{r\}'}=\mathbf{j}} a_{\mathbf{i}} X_{i_r}^{(r)}\right)_{\mathbf{j}\in(\mathbf{I}_r)_{\{r\}'}}\right\|_{\mathcal{N},p}\right\|_p .
\end{aligned}
$$

So to show (4.3) it is sufficient to get for $r = 1, \ldots, d$

$$\left\| \left\| \left(\sum_{\mathbf{i}:\mathbf{i}_{\{r\}'}=\mathbf{j}} a_{\mathbf{i}} X_{i_r}^{(r)} \right)_{\mathbf{j}\in(\mathbf{I}_r)_{\{r\}'}} \right\|_{\mathcal{N},p} \right\|_p \leq C(d) \left\| (a_{\mathbf{i}})_{\mathbf{i}\in\mathbf{I}} \right\|_{\mathcal{N},p}. \tag{4.5}$$

Proof. We will consider only the summand corresponding to $r = 1$, since the proof for other values of r is entirely similar. Recall that $C(d)$ denotes the constant that may differ at each occurrence and that depends only on d (in particular it may depend on values of $c_1(d-1)$ and $C_1(d-1)$).

To ease the notation let us set $\mathbf{J}_1 = (\mathbf{I}_1)_{\{1\}'}$ and additionally define for $I \subset \{1, \ldots, d\}$

$$\left\| (a_{\mathbf{j}})_{\mathbf{j}\in\mathbf{I}_I} \right\|_{\mathcal{N},u} = \sup \left\{ \sum_{\mathbf{j}\in\mathbf{I}_I} a_{\mathbf{j}} \prod_{r\in I} \left(1 + x_{j_r}^{(r)}\right) : x^{(r)} \in B_{\mathcal{N},u}^r \text{ for } r \in I \right\}.$$

We have

$$\left\| \left(\sum_{\mathbf{i}:\mathbf{i}_{\{1\}'}=\mathbf{j}} a_{\mathbf{i}} X_{i_1}^{(1)} \right)_{\mathbf{j}\in\mathbf{J}_1} \right\|_{\mathcal{N},p}$$
$$= \sup_{x^{(r)}\in B_{\mathcal{N},p}^r, 2\leq r\leq d} \sum_{\mathbf{j}\in\mathbf{J}_1} \left(\sum_{\mathbf{i}\in\mathbf{I}_1:\mathbf{i}_{\{1\}'}=\mathbf{j}} a_{\mathbf{i}} X_{i_1}^{(1)} \right) \left(1 + x_{j_2}^{(2)}\right) \cdots \left(1 + x_{j_d}^{(d)}\right)$$

and for $x^{(r)} \in \mathbb{R}_+^n$, $2 \leq r \leq d$,

$$\left(1 + x_{j_2}^{(2)}\right) \cdots \left(1 + x_{j_d}^{(d)}\right) \leq \sum_{2\leq q\leq d} \prod_{\substack{r=2 \\ r\neq q}}^{d} \left(1 + x_{j_r}^{(r)}\right) + x_{j_2}^{(2)} \cdots x_{j_d}^{(d)}.$$

Therefore

$$\left\| \left(\sum_{\mathbf{i}:\mathbf{i}_{\{1\}'}=\mathbf{j}} a_{\mathbf{i}} X_{i_1}^{(1)} \right)_{\mathbf{j}\in\mathbf{J}_1} \right\|_{\mathcal{N},p}$$
$$\leq \sum_{2\leq q\leq d} \sup_{x^{(r)}\in B_{\mathcal{N},p}^r, 2\leq r\leq d} \sum_{\mathbf{j}\in\mathbf{J}_1} \left(\sum_{\mathbf{i}\in\mathbf{I}_1:\mathbf{i}_{\{1\}'}=\mathbf{j}} a_{\mathbf{i}} X_{i_1}^{(1)} \right) \prod_{\substack{r=2, \\ r\neq q}}^{n} \left(1 + x_{j_r}^{(r)}\right)$$
$$+ \sup_{x^{(r)}\in B_{\mathcal{N},p}^r, 2\leq r\leq d} \sum_{\mathbf{j}\in\mathbf{J}_1} \left(\sum_{\mathbf{i}\in\mathbf{I}_1:\mathbf{i}_{\{1\}'}=\mathbf{j}} a_{\mathbf{i}} X_{i_1}^{(1)} \right) x_{j_2}^{(2)} \cdots x_{j_d}^{(d)}.$$

By the inductive hypothesis, for any $2 \leq q \leq d$,

$$\left\| \sup_{x^{(r)} \in B^r_{\mathcal{N},p}, 2 \leq r \leq d} \sum_{\mathbf{j} \in \mathbf{J}_1} \left(\sum_{\mathbf{i} \in \mathbf{I}_1 : \mathbf{i}_{\{1\}'} = \mathbf{j}} a_{\mathbf{i}} X^{(1)}_{i_1} \right) \prod_{\substack{r=2 \\ r \neq q}}^{n} \left(1 + x^{(r)}_{j_r}\right) \right\|_p$$

$$= \left\| \sup_{x^{(r)} \in B^r_{\mathcal{N},p}, 2 \leq r \leq d} \sum_{\mathbf{j} \in (\mathbf{I}_1)_{\{q\}'}} \left(\sum_{\mathbf{i} \in \mathbf{I}_1 : \mathbf{i}_{\{q\}'} = \mathbf{j}} a_{\mathbf{i}} \right) X^{(1)}_{j_1} \prod_{\substack{r=2 \\ r \neq q}}^{n} \left(1 + x^{(r)}_{j_r}\right) \right\|_p$$

$$\leq c_1^{-1}(d-1) \left\| \sum_{\mathbf{j} \in (\mathbf{I}_1)_{\{q\}'}} \left(\sum_{\mathbf{i} \in \mathbf{I}_1 : \mathbf{i}_{\{q\}'} = \mathbf{j}} a_{\mathbf{i}} \right) \prod_{\substack{r=1 \\ r \neq q}}^{n} X^{(r)}_{j_r} \right\|_p$$

$$\leq C(d) \left\| \left(\sum_{\mathbf{i} \in \mathbf{I}_1 : \mathbf{i}_{\{q\}'} = \mathbf{j}} a_{\mathbf{i}} \right)_{\mathbf{j} \in (\mathbf{I}_1)_{\{q\}'}} \right\|_{\mathcal{N},p} \leq C(d) \left\| (a_{\mathbf{i}})_{\mathbf{i} \in \mathbf{I}} \right\|_{\mathcal{N},p}.$$

So it suffices to show that

$$\| \sup_{x^{(r)} \in B^r_{\mathcal{N},p}, 2 \leq r \leq d} \sum_{\mathbf{j} \in \mathbf{J}_1} (\sum_{\mathbf{i} \in \mathbf{I}_1 : \mathbf{i}_{\{1\}'} = \mathbf{j}} a_{\mathbf{i}} X^{(1)}_{i_1}) x^{(2)}_{j_2} \cdots x^{(d)}_{j_d} \|_p \leq C(d) \left\| (a_{\mathbf{i}})_{\mathbf{i} \in \mathbf{I}} \right\|_{\mathcal{N},p}. \tag{4.6}$$

Let us define for $2 \leq r \leq d$ the following sets

$$U^{(r)} = \left\{ x^{(r)} \in B^r_{\mathcal{N},p} : \ N^{(r)}_i \left(x^{(r)}_i \right) \leq 1 \text{ if } i < p \right\}$$
$$\cap \bigcap_{k=1}^{\infty} \left\{ x^{(r)} \in B^{(r)}_{\mathcal{N},p} : N^{(r)}_i \left(x^{(r)}_i \right) \leq k^3 \text{ if } i \in \left[2^{k-1}p; 2^k p\right) \right\},$$

$$V^{(r)} = \left\{ x^{(r)} \in B^r_{\mathcal{N},p} : x^{(r)}_i = 0 \text{ or } N^{(r)}_i \left(x^{(r)}_i \right) > 1 \text{ if } i < p \right\}$$
$$\cap \bigcap_{k=1}^{\infty} \left\{ x^{(r)} \in B^r_{\mathcal{N},p} : x^{(r)}_i = 0 \text{ or } N^{(r)}_i \left(x^{(r)}_i \right) > k^3 \text{ if } i \in \left[2^{k-1}p; 2^k p\right) \right\}.$$

For every $x^{(r)} \in B^r_{\mathcal{N},p}$ there exist $y^{(r)} \in U^{(r)}, z^{(r)} \in V^{(r)}$ such that $x^{(r)} = y^{(r)} + z^{(r)}$. Let $\mathcal{J}$ denote a family of subsets of indices defined in the following way:

$$\mathcal{J} = \left\{ I \subseteq \{1, 2, \ldots\} : \# I \leq p, \# I \cap \left[2^{k-1}p; 2^k p\right) \leq \frac{p}{k^3} \text{ for } k = 1, 2, \ldots \right\}.$$

Applying the inequality $\sum_{k \leq m} \binom{n}{k} \leq \left(\frac{Cn}{m}\right)^m$, we get an estimate of the cardinality of $\mathcal{J}$.

$$\# \mathcal{J} \leq 2^p \prod_{k=1}^{\infty} \left(\frac{C 2^{k-1} p}{p/k^3} \right)^{p/k^3} \leq C^p \prod_{k=1}^{\infty} \left(2^{k-1} k^3\right)^{p/k^3} \leq C^p. \tag{4.7}$$

Let us notice that for I such that $\#I \leq p$ and $2 \leq q \leq d$

$$\begin{aligned}
&\left\| \sup_{x^{(r)} \in B^r_{\mathcal{N},u}, 2 \leq r \leq d} \sum_{\mathbf{i} \in \mathbf{I}_1 : i_q \in I} a_{\mathbf{i}} X^{(1)}_{i_1} x^{(2)}_{i_2} \cdots x^{(d)}_{i_d} \right\|_p \\
&\leq c_1^{-1}(d-1) \left\| \sum_{\mathbf{i} \in \mathbf{I}_1 : i_q \in I} a_{\mathbf{i}} X^{(1)}_{i_1} X^{(2)}_{i_2} \cdots X^{(d)}_{i_d} \right\|_p \\
&\leq c_1^{-1}(d-1) \left\| \sup_{x^{(r)} \in B^r_{\mathcal{N},u}, r \in \{q\}'} \sum_{\mathbf{i} \in \mathbf{I}_1 : i_q \in I} a_{\mathbf{i}} \left(1 + x^{(1)}_{i_1}\right) \cdots X^{(q)}_{i_q} \cdots \left(1 + x^{(d)}_{i_d}\right) \right\|_p \\
&\leq C(d) \left\| (a_{\mathbf{i}})_{\mathbf{i} \in \mathbf{I}_1} \right\|_{\mathcal{N},p}. \qquad (4.8)
\end{aligned}$$

The last estimate follows by integration by parts from the following two remarks.

- The function
$$x^{(q)} \longmapsto \left\| \sup_{x^{(r)} \in B^r_{\mathcal{N},u}, r \in \{q\}'} \sum_{\mathbf{i} \in \mathbf{I}_1 : i_q \in I} a_{\mathbf{i}} (1 + x^{(1)}_{i_1}) \cdots x^{(q)}_{i_q} \cdots (1 + x^{(d)}_{i_d}) \right\|_p$$
is positive homogeneous on $\mathbb{R}^n_+$.
- The random variable $\sum_{i_q \in I} N^{(q)}_{i_q}\left(X^{(q)}_{i_q}\right)$ has the distribution $\Gamma(\#I, 1)$ and therefore (recall that $\#I \leq p$)

$$\mathbf{P}\left(\sum_{i_q \in I} N^{(q)}_{i_q}\left(\frac{1}{1+t} X^{(q)}_{i_q}\right) > p\right) \leq \mathbf{P}\left(\sum_{i_q \in I} N^{(q)}_{i_q}\left(X^{(q)}_{i_q}\right) > (1+t)\,p\right) \leq e^{-Ctp}.$$

From (4.8) and (4.7) it follows that

$$\begin{aligned}
&\left\| \sup_{x^{(r)} \in B^r_{\mathcal{N},u}, 2 \leq r \leq d, x^{(q)} \in V^{(q)}} \sum_{\mathbf{i} \in \mathbf{I}_1} a_{\mathbf{i}} X^{(1)}_{i_1} x^{(2)}_{i_2} \cdots x^{(d)}_{i_d} \right\|_p \\
&\leq \left\| \sup_{I \in \mathcal{J}} \sup_{x^{(r)} \in B^r_{\mathcal{N},u}, 2 \leq r \leq d} \sum_{\mathbf{i} \in \mathbf{I}_1 : i_q \in I} a_{\mathbf{i}} X^{(1)}_{i_1} x^{(2)}_{i_2} \cdots x^{(d)}_{i_d} \right\|_p \\
&\leq \left(\sum_{I \in \mathcal{J}} \left\| \sup_{x^{(r)} \in B^r_{\mathcal{N},u}, 2 \leq r \leq d} \sum_{\mathbf{i} \in \mathbf{I}_1 : i_q \in I} a_{\mathbf{i}} X^{(1)}_{i_1} x^{(2)}_{i_2} \cdots x^{(d)}_{i_d} \right\|_p^p \right)^{1/p} \\
&\leq C(d) \left((\#\mathcal{J}) \left\| (a_{\mathbf{i}})_{\mathbf{i} \in \mathbf{I}_1} \right\|^p_{\mathcal{N},p} \right)^{1/p} \leq C(d) \left\| (a_{\mathbf{i}})_{\mathbf{i} \in \mathbf{I}_1} \right\|_{\mathcal{N},p}. \qquad (4.9)
\end{aligned}$$

Let C_0 be the constant given by Lemma 3.2. Since

$$\sup_{x^{(r)} \in U^{(r)}, 2 \leq r \leq d} \sum_{\mathbf{i} \in \mathbf{I}_1} a_{\mathbf{i}} C_0 x^{(2)}_{i_2} \cdots x^{(d)}_{i_d} \leq C_0 \left\| (a_{\mathbf{i}})_{\mathbf{i} \in \mathbf{I}} \right\|_{\mathcal{N},p}$$

it follows by (4.9) that in order to prove (4.6) it is enough to show

$$\left\| \sup_{x^{(r)} \in U^{(r)}, 2 \le r \le d} \sum_{\mathbf{i} \in \mathbf{I}_1} a_{\mathbf{i}} (X_{i_1}^{(1)} - C_0) x_{i_2}^{(2)} \cdots x_{i_d}^{(d)} \right\|_p \le C(d) \left\| (a_{\mathbf{i}})_{\mathbf{i} \in \mathbf{I}} \right\|_{\mathcal{N},p}. \tag{4.10}$$

Now notice that since $N_i^{(r)}(x) \ge x$ for $x \ge 1, 1 \le r \le d, 1 \le i \le n$, then for $a \ge 1$ and arbitrary numbers $t_1, t_2, \ldots, t_n$

$$\sup \left\{ \sum_{\mathbf{i} \in \mathbf{I}_1} a_{\mathbf{i}} t_{i_1} x_{i_2}^{(2)} \cdots x_{i_d}^{(d)} : x^{(r)} \in B_{\mathcal{N},u}^r, \left\| \left(N_{i_r}^{(r)} \left(x_{i_r}^{(r)} \right) \right) \right\|_\infty \le a, 2 \le r \le d \right\}$$
$$\le \sup \left\{ \sum_{\mathbf{i} \in \mathbf{I}_1} a_{\mathbf{i}} t_{i_1} x_{i_2}^{(2)} \cdots x_{i_d}^{(d)} : \sum x_{i_r}^{(r)} \le p, \left\| x^{(r)} \right\|_\infty \le a, 2 \le r \le d \right\}$$
$$\le a^{d-1} \max_{\#I^r \le \lceil p/a \rceil, 2 \le r \le d} \sum_{\mathbf{i} \in \mathbf{I}_1 : i_r \in I^r, 2 \le r \le d} a_{\mathbf{i}} t_{i_1}$$

(above and below we use a standard convention $\sum_{\mathbf{i} \in \emptyset} b_{\mathbf{i}} = 0$). Hence by the definition of $\mathbf{I}_1$ and $U^{(r)}$ we get

$$\left\| \sup_{x^{(r)} \in U^{(r)}, 2 \le r \le d} \sum_{\mathbf{i} \in \mathbf{I}_1} a_{\mathbf{i}} (X_{i_1}^{(1)} - C_0) x_{i_2}^{(2)} \cdots x_{i_d}^{(d)} \right\|_p$$
$$\le \sum_{l \ge 1} \left\| \sup_{x^{(r)} \in U^{(r)}, 2 \le r \le d} \sum_{\mathbf{i} \in \mathbf{I}_1 : 2^{l-1} p \le i_1 < 2^l p} a_{\mathbf{i}} (X_{i_1}^{(1)} - C_0) x_{i_2}^{(2)} \cdots x_{i_d}^{(d)} \right\|_p$$
$$\le \sum_{1 \le l \le p^{1/3}} l^{3(d-1)} \left\| \max_{\substack{I^r \subset \{1, \ldots, 2^l p - 1\} \\ \#I^r \le \lceil p/l^3 \rceil, 2 \le r \le d}} \sum_{\substack{\mathbf{i} \in \mathbf{I}_1 : 2^{l-1} p \le i_1 < 2^l p, \\ i_r \in I^r, 2 \le r \le d}} a_{\mathbf{i}} (X_{i_1}^{(1)} - C_0) \right\|_p \tag{4.11}$$
$$+ p^{d-1} \sum_{l > p^{1/3}} \left\| \max_{\mathbf{j} \in \mathbf{I}_1 : 2^{l-1} p \le j_1 < 2^l p} \left(\sum_{\substack{\mathbf{i} \in \mathbf{I}_1 : \mathbf{i}_{\{1\}'} = \mathbf{j}_{\{1\}'} \\ 2^{l-1} p \le i_1 < 2^l p}} a_{\mathbf{i}} X_{i_1}^{(1)} - C_0 \right)_+ \right\|_p. \tag{4.12}$$

Applying Lemma 3.2 we will estimate (4.11) and (4.12). For $l \le p^{1/3}$ we have

$$\# \left\{ (I^r)_{1 \le r \le d} : I^r \subset \{1, \ldots, 2^l p - 1\}, \#I^r \le \lceil p/l^3 \rceil \text{ for } 1 \le r \le d \right\}$$
$$\le \left(\left(\frac{C 2^l p}{p/l^3} \right)^{2p/l^3} \right)^d \le C^{dp/l^2} \le C^{dp}. \tag{4.13}$$

Applying the ordering (4.4) we obtain

$$\max_{\substack{I^r\subset\{1,\ldots,2^lp-1\}\\ \#I^r\le\lceil p/l^3\rceil,2\le r\le d}}\ \max_{2^{l-1}p\le i<2^lp}\ \sum_{\substack{\mathbf{i}\in\mathbf{I}_1:i_1=i,\\ i^{(r)}\in I^r,2\le r\le d}} a_{\mathbf{i}}$$
$$\le \max_{i\ge 2^{l-1}p}\sum_{\mathbf{i}\in\mathbf{I}:i_1=i} a_{\mathbf{i}} \le \frac{1}{2^{l-1}p}\sum_{\mathbf{i}\in\mathbf{I}} a_{\mathbf{i}}. \tag{4.14}$$

From (3.4), (4.13) and (4.14) we get

$$\left\| \max_{I^r\subset\{1,\ldots,2^lp-1\}\#I^r\le\lceil p/l^3\rceil,2\le r\le d}\ \sum_{\substack{\mathbf{i}\in\mathbf{I}_1:2^{l-1}p\le i_1<2^lp,\\ i_r\in I^r,2\le r\le d}} a_{\mathbf{i}}\left(X^{(1)}_{i_1}-C_0\right)\right\|_p$$
$$\le C\left(p+\ln C^{dp}\right)\max_{\substack{I^r\subset\{1,\ldots,2^lp-1\}\\ \#I^r\le\lceil p/l^3\rceil,2\le r\le d}}\ \max_{2^{l-1}p\le i<2^lp}\ \sum_{\substack{\mathbf{i}\in\mathbf{I}_1:i_1=i,\\ i^{(r)}\in I^r,2\le r\le d}} a_{\mathbf{i}}$$
$$\le Cpd\frac{1}{2^{l-1}p}\sum_{\mathbf{i}\in\mathbf{I}} a_{\mathbf{i}} \le \frac{Cd}{2^{l-1}}\sum_{\mathbf{i}\in\mathbf{I}} a_{\mathbf{i}}.$$

So

$$\sum_{1\le l\le p^{1/3}} l^{3(d-1)}\left\| \max_{\substack{I^r\subset\{1,\ldots,2^lp-1\},\\ \#I^r\le\lceil p/l^3\rceil,2\le r\le d}}\ \sum_{\substack{\mathbf{i}\in\mathbf{I}_1:i_1=i,\\ i^{(r)}\in I^r,2\le r\le d}} a_{\mathbf{i}}(X^{(1)}_{i_1}-C_0)\right\|_p$$
$$\le \sum_{1\le l\le p^{1/3}} l^{3(d-1)}\frac{Cd}{2^{l-1}}\sum_{\mathbf{i}\in\mathbf{I}} a_{\mathbf{i}} \le C(d)\sum_{\mathbf{i}\in\mathbf{I}} a_{\mathbf{i}}. \tag{4.15}$$

For $l>p^{1/3}$ we similarly get

$$\max_{\mathbf{i}\in\mathbf{I}_1:2^{l-1}p\le i_1<2^lp} a_{\mathbf{i}} \le \frac{1}{2^{l-1}p}\sum_{\mathbf{i}\in\mathbf{I}} a_{\mathbf{i}}$$

and (since $\#\left\{\mathbf{j}\in\mathbf{I}_1:2^{l-1}p\le j_1<2^lp\right\}\le(2^lp)^d$)

$$\left\| \max_{\mathbf{j}\in\mathbf{I}_1:2^{l-1}p\le j_1<2^lp}\left(\sum_{\substack{\mathbf{i}\in\mathbf{I}_1:\mathbf{i}_{\{1\}'}=\mathbf{j}_{\{1\}'}\\ 2^{l-1}p\le i_1<2^lp}} a_{\mathbf{i}}(X^{(1)}_{i_1}-C_0)\right)_+\right\|_p \le \frac{Cld}{2^{l-1}}\sum_{\mathbf{i}\in\mathbf{I}} a_{\mathbf{i}}.$$

So we obtain

$$p^{d-1}\sum_{l>p^{1/3}}\left\|\max_{\mathbf{j}\in\mathbf{I}_1:2^{l-1}p\le j_1<2^lp}\left(\sum_{\substack{\mathbf{i}\in\mathbf{I}_1:\mathbf{i}_{\{1\}'}=\mathbf{j}_{\{1\}'}\\2^{l-1}p\le i_1<2^lp}}a_{\mathbf{i}}(X^{(1)}_{i_1}-C_0)\right)_+\right\|_p \tag{4.16}$$

$$\le p^{d-1}\sum_{l>p^{1/3}}\frac{Cld}{2^{l-1}}\sum_{\mathbf{i}\in\mathbf{I}}a_{\mathbf{i}}\le C(d)\sum_{\mathbf{i}\in\mathbf{I}}a_{\mathbf{i}}. \tag{4.17}$$

Finally, collecting (4.11), (4.12), (4.15) and (4.17) we get the desired estimate (4.10). □

References

[1] M. Arcones and E. Giné, *On decoupling, series expansions, and tail behavior of chaos processes*, J. Theoret. Probab. **6** (1993), 101–122.

[2] V. H. de la Peña and E. Giné, *Decoupling: From Dependence to Independence*, Springer Verlag, New York 1999.

[3] V. H. de la Peña and S. Montgomery-Smith, *Bounds for the tail probabilities of U-statistics and quadratic forms*, Bull. Amer. Math. Soc. **31** (1994), 223—227.

[4] E. Giné, R. Latała and J. Zinn, *Exponential and Moment Inequalities for U-statistics*, High Dimensional Probability II, 13–38, Progr. Probab. **47**, Birkhauser, Boston 2000.

[5] E. D. Gluskin and S. Kwapień, *Tail and moment estimates for sums of independent random variables with logarithmically concave tails*, Studia Math. **114** (1995), 303–309.

[6] D. L. Hanson and F. T. Wright, *A bound on tail probabilities for quadratic forms in independent random variables*, Ann. Math. Statist. **42** (1971), 1079–1083.

[7] R. Ibragimov and Sh. Sharakhmetov, *Analogues of Khintchin, Marcinkiewicz-Zygmund and Rosenthal inequalities for symmetric statistics*, Scand. J. Statist. **26** (1999), 621–623.

[8] M. Klass and K. Nowicki, *A symmetrization-desymmetrization procedure for uniformly good approximations of expectations involving arbitrary sums of generalized U-statistics*, Ann. Probab. **28** (2000), 1884–1907.

[9] R. Latała, *Tail and moment estimates for some type of chaos*, Studia Math. **135**, 39–53.

[10] R. Łochowski, *Estimation of moments of sums of independent random variables and of two dimensional chaos*, M. Sc. thesis, Warsaw University 2000 (in Polish.)

[11] B. Maurey, *Some deviation inequalities*, Geometric and Functional Analysis **1** (1991), 188–197.

[12] M. Talagrand *A new isoperimetric inequality and the concentration of measure phenomenon*, Israel Seminar (GAFA), Springer Verlag Lecture Notes in Math. **1469** (1991), 94–124.

Institute of Mathematics, Warsaw University, Banacha 2, 02-097 Warszawa, Poland
E-mail address: rlatala@mimuw.edu.pl, lochowsk@mimuw.edu.pl

Progress in Probability, Vol. 56, 93–111

A Quantitative Law of Large Numbers via Exponential Martingales

Malwina J. Luczak

Abstract. We propose a direct, quantitative technique for studying the asymptotic distribution of sequences of Markov processes in infinite dimensions. The "fluid limit" of such a sequence is constructed from the limiting generator, and convergence relies on the properties of an associated sequence of exponential martingales. Through a suitable localisation of the processes we obtain tight bounds on the deviation probabilities uniformly in all co-ordinates. Hence we are able to overcome the difficulties arising from an infinite-dimensional state space.

1. Introduction

In recent years there has been a lot of interest in the asymptotic behaviour of certain sequences of Markov jump processes arising in computer science, random graph models, queueing theory and communication theory. These processes often model load sharing among servers corresponding to communication links or computers, or the emergence of graph properties — see for instance [1, 7, 9, 11, 12, 21, 22, 23, 27, 28].

As an illustration consider the following well-known scenario with N servers. Customers arrive in a Poisson process at rate λ, where $0 < \lambda < 1$. Upon arrival each customer selects d servers uniformly at random with replacement, and joins the queue at a least-loaded server among those chosen. Customers are served according to the first-come first-served discipline. All service times are independent exponentially distributed random variables with mean 1. This model first made its appearance in the literature over 15 years ago, and since then has received a lot of attention [5, 7, 14, 18, 21, 23, 26, 30]. There is a related scenario where λ is allowed to be an arbitrary positive constant, and customers are served by infinite-capacity servers, so that there is no queueing [13, 25]. In both cases it is interesting to study the fraction $V_t^{N,k}$ of servers with load at least k at time t as N tends to infinity. It is not difficult to verify that $(V_t^{N,k} : k \in \mathbb{N})$ is a Markov process with an infinite-dimensional state space.

Received by the editors April 14, 2003.
Key words and phrases. law of large numbers, infinite dimensions, martingales.

Another model, with applications to telecommunications, is the "star" network defined as follows [9, 10, 27]. There are $N+1$ nodes numbered $v_0, v_1, \ldots, v_N$ and N undirected links joining node v_0 to each node v_i for $i = 1, \ldots, N$. Each link can simultaneously support up to c calls through its c communication channels, where $c \in \mathbb{N} \cup \{\infty\}$ is fixed (possibly infinite), or $c = c(N)$ varies with N. In the latter case, both from theoretical and practical perspectives, it is perhaps the most interesting to let $c(N) \to \infty$ as $N \to \infty$. Every call occupies a single channel on each one of d distinct links. There are $\binom{N}{d}$ call classes, each arriving in a Poisson stream at rate $\lambda N/\binom{N}{d}$, where λ is a positive constant; these call classes correspond to all the possible subsets of d distinct links. Call durations are independent exponentials with unit mean. Let $U_t^{N,k}$ denote the fraction of links with exactly k calls in progress at time t. The process $(U_t^{N,k} : k = 0, 1, \ldots, c)$ is non-Markov; however, it is easy to define an appropriate Markov process on a larger state space driving its evolution and including the processes $U_t^{N,k}$ among its components.

Under certain conditions a sequence of Markov processes (or its projection onto a smaller subspace of the state space) can converge to a solution of an ordinary differential equation; this can be interpreted as a functional law of large numbers. The Markov processes of interest to us will often be infinite-dimensional. Proving convergence becomes cumbersome in infinite dimensions, since it is hard to define a useful topology. The product topology is frequently used in applications. It yields convergence for each fixed coordinate, but gives no information about the asymptotic behaviour of coordinates with index $f(N)$ for a function $f(N)$ that is unbounded as $N \to \infty$.

Here we propose a new technique that will handle an infinite-dimensional process directly and with precision. The limit function solves an ordinary differential equation driven by the operator corresponding to the limit generator of the process. We establish convergence using concentration of measure inequalities derived from the associated sequence of exponential martingales. Thus our proof is constructive rather than existential. We do not require the technical machinery of the Skorohod space tied with non-constructive proofs of convergence. In contrast with such methods, our approach provides estimates for the rate of convergence, allowing true insight into the structure of the sequence and its limit. It handles a fine spectrum of deviations, and can be used to establish a central limit theorem.

Our approach naturally extends to a more general setting where one has a sequence of non-Markov but asymptotically Markov processes, and the evolution of each member of the sequence is determined by a Markov process defined on a (much) larger state space. The telephone network model described above can be placed in this framework. Many more such examples occur in random graph models. Extensions and refinements of the results presented in this paper will appear in [16, 17].

2. Exponential martingale

In this section we establish an exponential martingale inequality that applies in fairly general settings. It will form the basis for our main result, the law of large numbers in Section 3. The inequality was inspired by a finite-dimensional technique presented in [3].

Let $(X_t)_{t\geq 0}$ be a pure jump, right-continuous Markov process taking values in a countable subset E of $\mathbb{R}^{\mathbb{N}}$. If the process is in a state $x \in E$, then it jumps to a different state y at rate $g(x,y)$. We assume that each transition of $(X_t)_{t\geq 0}$ involves only a finite number of co-ordinates. The generator G of $(X_t)_{t\geq 0}$ is the matrix $G = [g(x,y)]_{x,y\in E}$; and we have, for each test function $f : E \to \mathbb{R}$,

$$Gf(x) = \sum_{y\in E} g(x,y)(f(y) - f(x)).$$

If T is the time of the first jump of $(X_t)_{t\geq 0}$, then

$$\mathbf{Pr}(T \in dt, X_T = y | T > t, X_0 = x) = g(x,y)dt.$$

It further follows from standard theory - see for instance [6, 24] - that

$$f(X_t) - \int_0^t Gf(X_s)ds \tag{2.1}$$

and

$$\frac{f(X_t)}{f(X_0)} \exp\left\{-\int_0^t \frac{g(X_s)}{f(X_s)} ds\right\} \tag{2.2}$$

are both local martingales.

For θ in the dual space $(\mathbb{R}^{\mathbb{N}})^*$, consider the Laplace transform

$$m(x,\theta) = \sum_{y\in E} e^{\langle\theta, y-x\rangle} g(x,y).$$

Note that each term in this summation is well-defined, since $y - x$ has only finitely many non-zero co-ordinates for each pair $x, y \in E$. Let $\|\cdot\|$ denote the supremum norm. We assume that for some $\eta_0 > 0$

$$\sup_{x\in E} \sup_{\|\theta\|\leq\eta_0} m(x,\theta) \leq c < \infty. \tag{2.3}$$

Fix $\eta \in (0, \eta_0)$. For $k \in \mathbb{N}$, let e_k be the unit vector in $\mathbb{R}^{\mathbb{N}}$ that has a 1 in the k-th coordinate and 0's in all the other co-ordinates. Let $m''(x,\theta)$ denote the Hessian matrix of second partial derivatives with respect to θ. Then there exists a sequence $(A_k : k \in \mathbb{N})$ of positive numbers such that for each $k \in \mathbb{N}$

$$\sup_{x\in E, \|\theta\|\leq\eta} \langle e_k, m''(x,\theta)e_k\rangle \leq A_k.$$

For $\theta \in (\mathbb{R}^{\mathbb{N}})^*$ we define

$$\phi(x,\theta) = \sum_{y\in E} \left\{e^{\langle\theta,y-x\rangle} - 1 - \langle\theta, y-x\rangle\right\} g(x,y).$$

Thus $\phi \geq 0$; and if $\| \theta \| \leq \eta$, then

$$\phi(x,\theta) = \int_0^1 \langle \theta, m''(x, r\theta)\theta \rangle (1-r)dr.$$

By (2.1),

$$M_t = X_t - X_0 - \int_0^t GX_s ds$$

is a local martingale, and (2.3) implies that it is in fact a martingale. Furthermore if we take

$$f(X_t) = \exp\{\langle \theta, X_t - X_0 \rangle\}$$

in formula (2.2) and use condition (2.3) once more, we find that for any $\theta \in (\mathbb{R}^{\mathbb{N}})^*$, $\| \theta \| \leq \eta$,

$$Z_t^\theta = \exp\left\{ \langle \theta, M_t \rangle - \int_0^t \phi(X_s, \theta) ds \right\}$$

is a non-negative martingale. Thus $\mathbf{E}[Z_t^\theta] = 1$ for all $t \geq 0$.

Lemma 2.1. *Let $k \in \mathbb{N}$. Then for all $\delta \in [0, A_k \eta t]$*

$$\mathbf{Pr}(\sup_{0 \leq s \leq t} |M_s^k| \geq \delta) \leq 2e^{-\delta^2/(2A_k t)}.$$

Proof. Take $\theta = e_k$ and consider the stopping time

$$T = \inf\{s \geq 0 : \langle \theta, M_s \rangle \geq \delta\}.$$

For $\epsilon \leq \eta$, we know that $Z_s^{\epsilon\theta}$ is a martingale. On the set $\{T \leq t\}$

$$Z_T^{\epsilon\theta} \geq e^{\delta\epsilon - A_k t \epsilon^2/2}.$$

By the optional stopping theorem,

$$\mathbf{E}[Z_{T\wedge t}^{\epsilon\theta}] = E[Z_0^{\epsilon\theta}] = 1.$$

Hence, using Markov's inequality,

$$\mathbf{Pr}(\sup_{0 \leq s \leq t} \langle \theta, M_s \rangle \geq \delta) = \mathbf{Pr}(T \leq t) \leq e^{-\delta\epsilon + A_k t \epsilon^2/2}.$$

When $\delta \leq A_k t \eta$ we can take $\epsilon = \delta / A_k t$ to obtain

$$\mathbf{Pr}(\sup_{0 \leq s \leq t} M_s^k \geq \delta) \leq e^{-\delta^2/2A_k t},$$

and hence

$$\mathbf{Pr}(\sup_{0 \leq s \leq t} |M_s^k| \geq \delta) \leq 2e^{-\delta^2/2A_k t}.$$

□

Now suppose there exists a sequence $(B_k : k \in \mathbb{N})$ of positive numbers, $B_k \leq A_k$ for each $k \in \mathbb{N}$, such that although it might not be true that

$$\sup_{x \in E, \|\theta\| \leq \eta} \langle e_k, m''(x,\theta) e_k \rangle \leq B_k,$$

the bounds B_k are "very likely" to hold throughout the interval $[0,t]$ for "reasonable" starting conditions. Specifically let $x_0 \in E$ and assume that for each $k \in \mathbb{N}$,

$$\mathbf{Pr}\left(\{X_0 = x_0\} \wedge \{\sup_{0\le s\le t, \|\theta\|\le\eta} \langle e_k, m''(X_s,\theta)e_k\rangle > B_k\}\right) \le p_k, \tag{2.4}$$

where $(p_k : k \in \mathbb{N})$ is a sequence of "small" positive numbers.

Lemma 2.2. *Let $k \in \mathbb{N}$. Then for all $\delta \in [0, B_k\eta t]$*

$$\mathbf{Pr}\left(\{\sup_{0\le s\le t, \|\theta\|\le\eta} \langle e_k, m''(X_s,\theta)e_k\rangle \le B_k\} \wedge \{\sup_{0\le s\le t} |M_s^k| \ge \delta\}\right) \le 2e^{-\delta^2/2B_k t}.$$

If (2.4) holds, then for all $\delta \in (0, B_k\eta t]$

$$\mathbf{Pr}\left(\{X_0 = x_0\} \wedge \{\sup_{0\le s\le t} |M_s^k| \ge \delta\}\right) \le 2e^{-\delta^2/2B_k t} + p_k.$$

Proof. The proof follows the lines of the proof of Lemma 2.1, using the fact that

$$\{T \le t\} \cap \{\sup_{0\le s\le t, \|\theta\|\le\eta} \langle e_k, m''(X_s,\theta)e_k\rangle \le B_k\} \subseteq \{Z_T^{\epsilon\theta} \ge e^{\delta\epsilon - B_k t\epsilon^2/2}\}. \qquad \square$$

3. Law of large numbers

For each $N \in \mathbb{N}$ let $(X_t^N)_{t\ge0}$ be a pure jump Markov process taking values in a countable space $E^N \subseteq \mathbb{R}^{\mathbb{N}}$. We study the asymptotic distribution of $(X_t^N)_{t\ge0}$ as $N \to \infty$. Given states $x, y \in E^N$ let $g^N(x,y)$ denote the jump rate of $(X_t^N)_{t\ge0}$ from x to y. Consider the Laplace transform

$$m^N(x,\theta) = \sum_{y\in E^N} e^{\langle\theta, y-x\rangle} g^N(x,y),$$

for $x \in E^N$ and $\theta \in (\mathbb{R}^{\mathbb{N}})^*$. We assume the existence of a constant κ such that uniformly in N and $x, y \in E^N$ with $g^N(x,y) \ne 0$, $|y-x|$ has at most κ non-zero co-ordinates, though this condition could be relaxed to some extent.

Let $E \subseteq \mathbb{R}^{\mathbb{N}}$ be a countable set such that $E^N \subseteq E$ for each N, and suppose that there exist a limit generator $G = [g(x,y)]_{x,y\in E}$ and a constant $\eta_0 > 0$ such that

$$\sup_{x\in\mathcal{A}^N} \sup_{\|\theta\|\le\eta_0} \left|\frac{m^N(x,N\theta)}{N} - m(x,\theta)\right| < \delta_N \quad \text{as } N \to \infty, \tag{3.1}$$

where δ_N is a sequence of positive numbers tending to 0 as $N \to \infty$, and the set $\mathcal{A}^N$ is defined below. Here the moment generating function

$$m(x,\theta) = \sum_{y\in E} e^{\langle\theta, y-x\rangle} g(x,y)$$

is assumed to satisfy the boundedness condition (2.3).

Note that (3.1) implies that as N tends to infinity transitions for $(X_t^N)_{t\geq 0}$ happen at a rate of order N, and the usual jump size is of order $1/N$. In typical applications considered here and in [16, 17], the rate of convergence in (3.1) will in fact be $O(1/N)$. The convergence of moment generating functions is not essential for the sequence $(X_t^N)_{t\geq 0}$ to obey a law of large numbers; it would suffice to place appropriate conditions on the θ-derivatives of these moment generating functions, that is the drift sequence. However, we choose to impose more restrictive assumptions for the sake of simplicity.

Let $b(x) = m'(x,0)$, that is $b(x)$ is the first derivative of $m(x,\theta)$ in θ evaluated at 0, assumed to be Lipschitz on E with respect to the supremum norm. Fix a point $x_0 \in E$ and denote by $(x_t)_{t\geq 0}$ the unique solution to the initial value problem $dx_t/dt = b(x_t)$ starting from x_0.

Let $(A_{N,k} : N, k \in \mathbb{N})$ be a doubly-indexed sequence of numbers such that $\sup_{k,N\in\mathbb{N}} NA_{N,k} \leq C$ for some constant $C > 0$. Define

$$\mathcal{A}_{N,k} = \{x \in E : \sup_{\|\theta\|\leq N\eta} \langle e_k, m''^N(x,\theta)e_k\rangle \leq A_{N,k}\}$$

and

$$\mathcal{A}_N = \cap_{k\in\mathbb{N}} \mathcal{A}_{N,k}.$$

Let $(\delta_{N,k} : N, k \in \mathbb{N})$ be a sequence of positive numbers, and assume that there exists a (small) subset of co-ordinates $\hat{S}^N \subseteq \mathbb{N}$ such that $\delta_{N,k}/NA_{N,k}$ is bounded uniformly in $k \in \hat{S}^N$, $N \in \mathbb{N}$. Without loss of generality we may assume that

$$\sup_{N\in\mathbb{N},k\in\hat{S}^N} \delta_{N,k}/NA_{N,k} \leq C. \tag{3.2}$$

Additionally, we require that for each $N \in \mathbb{N}$

$$\delta_{N,k'} \leq \delta_{N,k}, \qquad \forall \quad k' \in \mathbb{N}\setminus\hat{S}^N, k \in \hat{S}^N; \tag{3.3}$$

and that

$$\sup_{x\in\mathcal{A}^N} \sup_{k\in\hat{S}^N} |b^{N,k}(x) - b^k(x)|/\delta_{N,k} \to 0 \qquad \text{as } N \to \infty. \tag{3.4}$$

The latter assumption will follow as a consequence of (3.1) provided that $\limsup_{N\to\infty} \sup_{k\in\hat{S}^N} \delta_N/\delta_{N,k} = 0$.

Roughly speaking, $A_{N,k}$ will serve as an upper bound on the k-th diagonal entry of the matrix of the second derivatives of $m^N(X_t^N, N\theta)$, and $\delta_{N,k}$ will be the size of the deviation in the k-th co-ordinate of X_t^N. Also $\mathbb{N}\setminus\hat{S}^N$ will be the set of co-ordinates whose values, under appropriate initial conditions, are unlikely to experience big changes during the interval $[0,t]$ for any time $t > 0$. To ensure that the deviation probabilities of martingale components indexed by $k \in \hat{S}^N$ go to 0 we need

$$\sum_{k\in\hat{S}^N} e^{-\alpha\delta_{N,k}^2/A_{N,k}} \to 0 \quad \text{as } N \to \infty, \quad \forall \alpha > 0. \tag{3.5}$$

Initially, the components indexed by $k \in \hat{S}^N$ should be closely concentrated around the starting values of their deterministic limits, that is

$$\limsup_{N\to\infty} \sup_{k\in\hat{S}^N} \left[\frac{A_{N,k}}{\delta_{N,k}^2} \log \mathbf{Pr}(|X_0^{N,k} - x_0^k| \geq \alpha\delta_{N,k}) \right] < 0, \quad \forall \alpha > 0. \tag{3.6}$$

Note that if (3.6) holds, then it follows from (3.5) that

$$\limsup_{N\to\infty} \mathbf{Pr}(\sup_{k\in\hat{S}^N} \frac{|X_0^{N,k} - x_0^k|}{\delta_{N,k}} \geq \alpha) \to 0 \quad \text{as } N \to \infty, \quad \forall \alpha > 0.$$

Given a constant $\alpha > 0$ define the event

$$\mathcal{B}^N(\alpha) = \{ \sup_{k\in\mathbb{N}\setminus\hat{S}^N} \frac{|X_0^{N,k} - x_0^k|}{\delta_{N,k}} < \alpha \}.$$

We introduce a function $g(N)$ (where $g(N) \to \infty$ as $N \to \infty$) that will control the initial deviation probabilities of the (small) co-ordinates in $\mathbb{N} \setminus \hat{S}^N$. That is, for each $\alpha > 0$

$$\limsup_{N\to\infty} [g(N)]^{-1} \log \mathbf{Pr}(\overline{\mathcal{B}^N(\alpha)}) < 0. \tag{3.7}$$

Let $t_0 > 0$ and define

$$\mathcal{A}_{t_0}^{N,k} = \{X_t^N \in \mathcal{A}_{N,k} : \forall t \in [0, t_0]\}, \qquad \mathcal{A}_{t_0}^N = \wedge_{k\in\hat{S}^N} \mathcal{A}_{t_0}^{N,k}.$$

We assume that $\mathbf{Pr}(\overline{\mathcal{A}}_{t_0}^N) \to 0$ as $N \to \infty$. Assume also that for each pair of constants $\alpha_1, \alpha_2 > 0$,

$$\limsup_{N\to\infty} [g(N)]^{-1} \log \mathbf{Pr}(\{ \sup_{k\in\mathbb{N}\setminus\hat{S}^N} \sup_{0\leq t\leq t_0} \frac{|X_t^{N,k} - X_0^{N,k}|}{\delta_{N,k}} \geq \alpha_1 \} \wedge \mathcal{A}_{t_0}^N \wedge \mathcal{B}^N(\alpha_2)) < 0. \tag{3.8}$$

Thus if the path of the process over the interval $[0, t_0]$ lies in $\mathcal{A}_{t_0}^N$, and if the co-ordinates $X_0^{N,k}$ for $k \in \mathbb{N} \setminus \hat{S}^N$ are close to the corresponding co-ordinates of the deterministic vector x_0, then the $X_t^{N,k}$ are not likely to change much during $[0, t_0]$. We also need to ensure slow changes in x_t^k for $k \in \mathbb{N} \setminus \hat{S}^N$, that is

$$\limsup_{N\to\infty} \sup_{k\in\mathbb{N}\setminus\hat{S}^N} \sup_{0\leq t\leq t_0} \frac{|x_t^k - x_0^k|}{\delta_{N,k}} < \alpha, \quad \forall \alpha > 0. \tag{3.9}$$

The idea is to control the components $X_t^{N,k}$ for $k \in \hat{S}^N$ very precisely with the exponential martingale inequality of Lemma 2.2. For the remaining components, it may not be possible to obtain the exact asymptotics, but it is possible to upper bound them. Consequently, we obtain a quantitative law of large numbers that holds uniformly in all co-ordinates. In many applications of Theorem 3.1 below, such as the queueing example in Section 4, the interpretation is as follows. The event $\mathcal{A}_{t_0}^N$ can be regarded as a "big tube", and the event

$\mathcal{E}_{t_0}^N = \{\sup_{k\in\mathbb{N}} \sup_{0\le t\le t_0} \frac{|X_t^{N,k}-x_t^k|}{\delta_{N,k}} < \alpha\}$ as a "small tube" whose closure is contained in the big one, so that $\{\sup_{k\in\mathbb{N}} \sup_{0\le t\le t_0} \frac{|X_t^{N,k}-x_t^k|}{\delta_{N,k}} \le \alpha\} \subseteq \mathcal{A}_{t_0}^N$. While the process stays in the big tube, its transition rates are such that in fact it cannot get outside the small tube with very high probability; that is $\mathbf{Pr}(\mathcal{A}_{t_0}^N \cap \overline{\mathcal{E}}_{t_0}^N)$ is very small. But if its path lies in the small tube, then it is also contained in the big tube; thus $\mathbf{Pr}(\overline{\mathcal{E}}_{t_0}^N)$ is very small. More precisely, let $T_1 = \inf\{t \ge 0 : \overline{\mathcal{A}}_t^N \text{ occurs }\}$, let $T_2 = \inf\{t \ge 0 : \overline{\mathcal{E}}_t^N \text{ occurs }\}$, and set $T = T_1 \wedge T_2 \wedge t_0$. Then $\mathbf{Pr}(T = T_1)$ and $\mathbf{Pr}(T = T_2)$ both tend to 0 as N tends to infinity.

Theorem 3.1. *Let $\alpha > 0$ and let $t_0 > 0$. Under the assumptions (3.1) - (3.9), there exists a constant $\beta = \beta(t_0, \alpha)$ such that for all N sufficiently large*

$$\mathbf{Pr}(\overline{\mathcal{E}}_{t_0}^N) \le \sum_{k\in\hat{S}^N} e^{-\beta\delta_{N,k}^2/A_{N,k}} + e^{-\beta g(N)} + \mathbf{Pr}(\overline{\mathcal{A}_{t_0}^N}), \tag{3.10}$$

and thus tends to 0 as $N \to \infty$. Furthermore if

$$\left\{\sup_{k\in\mathbb{N}} \sup_{0\le t\le t_0} \frac{|X_t^{N,k} - x_t^k|}{\delta_{N,k}} \le \alpha\right\} \subseteq \mathcal{A}_{t_0}^N$$

then for N sufficiently large

$$\mathbf{Pr}(\overline{\mathcal{E}}_{t_0}^N) \le \sum_{k\in\hat{S}^N} e^{-\beta\delta_{N,k}^2/A_{N,k}} + e^{-\beta g(N)} \to 0 \quad \text{as } N \to \infty. \tag{3.11}$$

Proof. We prove (3.10); the proof of (3.11) is analogous. Set $b^N(x) = m'^N(x,0)$, and define $(M_t^N)_{t\ge0}$ by

$$X_t^N = X_0^N + M_t^N + \int_0^t b^N(X_s^N)ds.$$

Given a positive constant α set $\epsilon = \min\{e^{-Kt_0}\alpha/6, \epsilon_0\}$, where K is the Lipschitz constant of the operator b. Let $\alpha' = \min\{\alpha/3, \epsilon/Kt_0\}$ and let

$$\Omega_{t_0}^N = \{\sup_{k\in\hat{S}^N} \frac{|X_0^{N,k} - x_0^k|}{\delta_{N,k}} < \epsilon\} \wedge \mathcal{B}^N(\alpha') \wedge \mathcal{A}_{t_0}^N$$

$$\wedge\{\sup_{k\in\hat{S}^N} \sup_{t\le t_0} \frac{|M_t^{N,k}|}{\delta_{N,k}} < \epsilon\} \wedge \{\sup_{k\in\mathbb{N}\setminus\hat{S}^N} \sup_{t\le t_0} \frac{|X_t^{N,k} - X_0^{N,k}|}{\delta_{N,k}} < \alpha'\}.$$

By (3.4) there exists an integer N_0 such that whenever $N \ge N_0$

$$\sup_{x\in\mathcal{A}^N} \sup_{k\in\hat{S}^N} \frac{|b^{N,k}(x) - b^k(x)|}{\delta_{N,k}} < \epsilon/t_0;$$

and by (3.9), N_0 can be chosen such that for all $N \ge N_0$

$$\sup_{k\in\mathbb{N}\setminus\hat{S}^N} \sup_{0\le t\le t_0} \frac{|x_t^k - x_0^k|}{\delta_{N,k}} < \min\{\alpha/3, \epsilon/Kt_0\}.$$

For each $k \in \mathbb{N}$ we can write

$$X_t^{N,k} - x_t^k = (X_0^{N,k} - x_0^k) + M_t^{N,k}$$
$$+ \int_0^t (b^{N,k}(X_s^N) - b^k(X_s^N))ds + \int_0^t (b^k(X_s^N) - b^k(x_s))ds.$$

Note that

$$\sup_{k\in\mathbb{N}\setminus\hat{S}^N} \sup_{0\le s\le t} |X_s^{N,k} - x_s^k| \le \sup_{k\in\mathbb{N}\setminus\hat{S}^N} \sup_{0\le s\le t} |X_s^{N,k} - X_0^{N,k}|$$
$$+ \sup_{k\in\mathbb{N}\setminus\hat{S}^N} |X_0^{N,k} - x_0^k| + \sup_{k\in\mathbb{N}\setminus\hat{S}^N} \sup_{0\le s\le t} |x_0^k - x_s^k|.$$

Hence for $N \ge N_0$,

$$\sup_{k\in\mathbb{N}\setminus\hat{S}^N} \sup_{0\le t\le t_0} |X_t^{N,k} - x_t^k|/\delta_{N,k} < \alpha \quad \text{on } \Omega_{t_0}^N.$$

For $N \ge N_0$, $k \in \hat{S}^N$, and $t \le t_0$, we have on $\Omega_{t_0}^N$

$$|X_t^{N,k} - x_t^k| < 3\epsilon\delta_{N,k} + K\int_0^t (\sup_{k\in\hat{S}^N} |X_s^{N,k} - x_s^k| + \sup_{k\in\mathbb{N}\setminus\hat{S}^N} |X_s^{N,k} - x_s^k|)ds.$$

Also, using (3.3), for $N \ge N_0$

$$K\int_0^t \sup_{k\in\mathbb{N}\setminus\hat{S}^N} |X_s^{N,k} - x_s^k| ds < 3\epsilon\delta_{N,k} \quad \text{on } \Omega_{t_0}^N.$$

Hence, using Gronwall's Lemma,

$$\sup_{k\in\hat{S}^N} \sup_{0\le t\le t_0} |X_t^{N,k} - \tilde{x}_t^k|/\delta_{N,k} < \alpha \quad \text{on } \Omega_{t_0}^N$$

holds for all $N \ge N_0$; and hence $\Omega_{t_0}^N \subseteq \mathcal{E}_{t_0}^N$.

But now by Lemma 2.2, using the assumption (3.2), there exists a constant $\epsilon_0 > 0$ such that whenever $\epsilon \le \epsilon_0$ then

$$\mathbf{Pr}(\{\sup_{k\in\hat{S}^N} \sup_{t\le t_0} |M_t^{N,k}|/\delta_{N,k} \ge \epsilon\} \wedge \mathcal{A}_{t_0}^N) \le \sum_{k\in\hat{S}^N} e^{-\epsilon^2\delta_{N,k}^2/2A_{N,k}t_0}$$

for all $N \ge N_0$, and tends to 0 as $N \to \infty$ by (3.5). Combined with assumptions (3.6) - (3.8) this implies that N_0 can be chosen such that for some constant $\beta = \beta(t_0, \alpha)$

$$\mathbf{Pr}(\overline{\Omega}_{t_0}^N) \le \sum_{k\in\hat{S}^N} e^{-\beta\delta_{N,k}^2/A_{N,k}} + e^{-\beta g(N)} + \mathbf{Pr}(\overline{\mathcal{A}}_{t_0}^N)$$

holds for all $N \ge N_0$. Furthermore, the right hand side tends to 0 as $N \to \infty$, as required. □

Our localised estimates for each co-ordinate enable an extension of Theorem 3.1 to the case where the state space E^N of $(X_t^N)_{t\geq 0}$ can be expressed as a product $E^N = I^N \times J^N \subseteq \mathbb{R}^{d(N)} \times \mathbb{R}^{\mathbb{N}}$, with $d(N) \in \mathbb{N} \cup \{\infty\}$, and a law of large numbers holds for the projection $(\tilde{X}_t^N)_{t\geq 0}$ of $(X_t^N)_{t\geq 0}$ onto J^N. The extended version can be applied to the star network with N links described in the introduction. Let $(V_t^N)_{t\geq 0}$ be the random process with $\binom{N}{d}$ components where the r-th component is the number of ongoing calls of class r at time t. Thus $d(N) = \binom{N}{d}$ and $I^N = (\mathbb{N} \cup \{0\})^{d(N)}$. Let $J^N = \{0, 1/N, \ldots, 1\}^{\mathbb{N}}$, and for $k = 0, 1, \ldots, c$ let $U_t^{N,k}$ be the fraction of links with exactly k calls in progress at time t. Then the process $(U_t^N)_{t\geq 0}$, with state space J^N, obeys a law of large numbers. We refer the reader to [16] for details.

4. Queueing application

4.1. Preliminaries

We begin by recalling a pair of standard discrete concentration inequalities, which will be used in this section. If X is a binomial or Poisson random variable with mean μ, then for $0 \leq \epsilon \leq 1$

$$\mathbf{Pr}(X - \mu \leq -\epsilon\mu) \leq e^{-\frac{1}{2}\epsilon^2\mu} \tag{4.1}$$

and

$$\mathbf{Pr}(X - \mu \geq \epsilon\mu) \leq e^{-\frac{1}{3}\epsilon^2\mu}, \tag{4.2}$$

see for example Theorem 2.3 (c) and inequality (2.8) in [20].

Let $(Y_t^N)_{t\geq 0} = (Y_t^{N,i} : i = 1, \ldots, N)_{t\geq 0}$ represent the following system of N single-server queues. Customers arrive as a Poisson process of rate $N\lambda$, where $0 < \lambda < 1$, and all service times are independent exponentials with mean 1. Upon arrival each customer examines d queues chosen uniformly at random with replacement, and joins a queue with the smallest load among those examined.

It is known [7, 26] that, for each fixed positive integer k, as $N \to \infty$ the proportion of queues with length at least k converges weakly to a function v_t^k, where $v_t^0 = 1$ for all t and $(v_t^k : k \in \mathbb{N})$ is the unique [4, 26] solution of the ordinary differential equation

$$\frac{dv_t^k}{dt} = \lambda((v_t^{k-1})^d - (v_t^k)^d) - (v_t^k - v_t^{k+1}), \qquad k \in \mathbb{N}. \tag{4.3}$$

Furthermore, the system is *chaotic* starting close to a suitable deterministic initial state, that is the paths of members of a fixed subset of finitely many queues are asymptotically independent of one another uniformly on bounded time intervals. Chaoticity in fact implies a law of large numbers for the time evolution of the proportion of queues of different lengths, that is for the empirical measure on path space [7], which is stronger than the weak convergence result mentioned above. Yet even here convergence occurs with respect to a topology derived from the product topology on the infinite-dimensional state space, and so is only informative for co-ordinates with a fixed index as $N \to \infty$. Also, until now, no good bounds on the rate of convergence were known.

A careful study of this model was recently carried out in [14]. The analysis relies on a coupling and the (discrete) bounded differences inequality. It turns out that the proportion of queues with length at least k stays very near its mean over long periods of time, if we start close to a suitable deterministic state. Furthermore, for most such "nice" initial states the process converges rapidly to its stationary distribution, and hence the concentration result is also valid in equilibrium. The equilibrium mean fraction of queues with length at least k is very close to $\pi^k = \lambda^{\frac{d^k-1}{d-1}}$. The estimates hold uniformly in all co-ordinate indices $k \in \mathbb{N}$, and in particular show that for $d \geq 2$ the equilibrium maximum load of a queue is concentrated on two values $m_d(N)$ and $m_d(N)-1$, where $m_d(N) = \frac{\ln \ln N}{\ln d} + O(1)$. However, differential equations and convergence to a fluid limit are not explicitly considered in [14]. Our techniques, though akin to those of [14] in that they make use of a concentration inequality, allow as to handle a finer spectrum of deviations, and also yield a central limit theorem [17]. On the other hand, our results hold over much shorter time intervals and do not extend to the stationary distribution. This is to do with Gronwall's lemma, which enters in our proofs causing our bounds to grow exponentially with time. Gronwall's lemma is not needed for the discrete bounded differences approach in [14].

Denote by $V_t^{N,k}$ the number of queues of length at least k at time t; thus $V_t^{N,0} = N$. We shall focus on the rescaled quantities $\tilde{V}_t^{N,k} = N^{-1} V_t^{N,k}$. Set $\tilde{V}_t^N = (\tilde{V}_t^{N,k} : k \in \mathbb{Z}^+)_{t \geq 0}$. Then $(\tilde{V}_t^N)_{t\geq 0}$ is a pure jump Markov process with state space E^N contained in the space E defined by

$$E = \{v \in \mathbb{R}^{\mathbb{N}} : 1 = v^0 \geq v^1 \geq \ldots \geq 0\}.$$

The process makes transitions

$$\Delta \tilde{V}_t^{N,k} = 1/N \quad \text{at rate} \quad \lambda N \left(\left(\tilde{V}_t^{N,k-1}\right)^d - \left(\tilde{V}_t^{N,k}\right)^d \right)$$

$$\Delta \tilde{V}_t^{N,k} = -1/N \quad \text{at rate} \quad N(\tilde{V}_t^{N,k} - \tilde{V}_t^{N,k+1}).$$

Each transition involves exactly one co-ordinate $k \in \mathbb{N}$. Hence the Laplace transform of $(\tilde{V}_t^N)_{t\geq 0}$ is, for each $N \in \mathbb{N}$,

$$m^N(v, N\theta)/N = m(v,\theta) = \lambda \sum_{k\geq 1} \left((v^{k-1})^d - (v^k)^d\right) e^{\theta_k} + \sum_{k\geq 1} (v^k - v^{k+1}) e^{-\theta_k}.$$

Set $b(v) = (b^k(v) : k \in \mathbb{Z}^+)$, where $b^k(v) = \partial m(v,\theta)/\partial\theta_k$ evaluated at $\theta = 0$, so that

$$b^k(v) = \lambda \left((v^{k-1})^d - (v^k)^d\right) - (v^k - v^{k+1}), \quad k \geq 1.$$

We begin with a lemma which compares the queue lengths in our system during a time interval $[0, t_0]$ to the queue lengths in a system $(Z_t^N)_{t\geq 0} = (Z_t^{N,i} : i = 1, \ldots, N)_{t\geq 0}$ with identical arrival and choices processes, but without customer departures. Let us assume that the choices process always generates an ordered list of $d \geq 1$ queues at an arrival time, and the new customer joins a shortest queue among those chosen, breaking ties if necessary by choosing the first of the shortest queues in the list.

Lemma 4.1. *There exists a coupling of Y_t^N and Z_t^N such that if $Y_0^N \leq Z_0^N$ then $Y_t^N \leq Z_t^N$ for all $t \in [0, t_0]$.*

Proof. Arrivals appear at the same times in both processes, generated by a Poisson process at rate λN, and we choose the same list of d queues for the two processes. Monotonicity is straightforward: for if at time t a new customer joins queue i in the Y-process and $Y_t^{N,i} = Z_t^{N,i}$, then the customer also joins queue i in the Z-process. □

Let $t_0 > 0$. Lemma 4.1 will be used to upper bound the maximum size of high-indexed components of $\tilde{V}_t^N$ during the interval $[0, t_0]$.

Lemma 4.2. *Let $d \geq 1$. Let $f(N)$ be a function defined on the set of positive integers such that $f(N) \to \infty$ and $f(N) \ln f(N) - \eta N \to -\infty$ as $N \to \infty$ for each constant $\eta > 0$. Then for any constant $0 < \mu < 1$ there exists a constant $\alpha > 0$ such that the following holds. If*

$$\mathbf{Pr}(\sup_{k \leq f(N)/2} \tilde{V}_0^{N,k}/\mu^{2k} \geq \alpha) \leq p_N,$$

then

$$\mathbf{Pr}(\sup_{k \leq f(N)} \sup_{0 \leq t \leq t_0} \tilde{V}_t^{N,k}/\mu^{k-1} \geq 2\alpha) \leq p_N + f(N) e^{-N\mu^{f(N)}/\alpha}.$$

Proof. By Lemma 4.1, we may instead consider $\tilde{W}_t^{N,k}$, the proportion of queues with length at least k in a system that has the same arrivals and choices but has no customer departures.

For $0 \leq t \leq t_0$, let A_t be the event that the total number of customers arriving during $[0, t]$ is at most $2t\lambda N$. By inequality (4.2), $\mathbf{Pr}(\overline{A}_{t_0}) \leq e^{-\lambda t_0 N/3}$.

Fix $j \in \{1, \ldots, N\}$. For $0 \leq t \leq t_0$ let $X_t^{N,j}$ be the number of new arrivals at queue j during $[0, t]$. The probability that a particular customer joins queue j is at most d/N. It follows that on the event A_t, the number of new customers in queue j until time t is stochastically at most a binomial random variable $Bin(2\lambda t N, d/N)$. Then for each positive integer k

$$\mathbf{Pr}(X_t^{N,j} \geq k) \leq \mathbf{Pr}(\overline{A_t}) + \mathbf{Pr}(Bin(2\lambda t N, d/N) \geq k).$$

It follows that for N sufficiently large we have for each $k \leq f(N)$,

$$\mathbf{Pr}(X_{t_0}^{N,j} \geq k) \leq 2e^{-2d\lambda t_0} \frac{(2d\lambda t_0)^k}{k!}$$

so that

$$\mathbf{Pr}(X_{t_0}^{N,j} \geq k) \sim e^{-k \log k(1+o(1))}, \quad k \leq f(N), \quad \text{as } N \to \infty.$$

Thus for any constant $0 < \mu < 1$ there exists a constant $\alpha > 0$ such that for all $k \leq f(N)$,

$$\mathbf{Pr}(X_{t_0}^{N,j} \geq k) \leq \frac{1}{2}\alpha\mu^{2k}.$$

Now the $X_{t_0}^{N,j}$ for $j = 1, \ldots, N$ are independent, and so by inequality (4.2)

$$\mathbf{Pr}(\sum_{j=1}^{N} \mathbb{I}_{X_{t_0}^{N,j} \geq k} \geq N\alpha\mu^{2k}) \leq e^{-N\alpha\mu^{2k}/6}, \quad k \leq f(N).$$

To complete the proof we use the fact that if a queue has length at least k at time t_0 then it must have had at least $\lfloor k/2 \rfloor$ customers at time 0, or it has received at least $\lfloor k/2 \rfloor$ customers during $[0, t_0]$. □

We now consider the case $d \geq 2$.

Lemma 4.3. *Let $d \geq 2$. Then there exists a constant $0 < \mu_0 < 1$ and a constant $\alpha_0 > 0$ such that for $\alpha \geq \alpha_0$ and $\mu_0 \leq \mu < 1$ the following holds. Let $f(N)$ be a function such that $N\mu^{d^{\lceil f(N)/2 \rceil}} \to \infty$ as $N \to \infty$. If*

$$\mathbf{Pr}(\sup_{k \leq f(N)/2} \tilde{V}_0^{N,k}/\mu^{d^k} \geq \alpha) \leq p_N,$$

then

$$\mathbf{Pr}(\sup_{k \leq f(N)} \sup_{0 \leq t \leq t_0} \tilde{V}_t^{N,k}/\mu^{d^{\lfloor k/2 \rfloor}} \geq 2\alpha) \leq p_N + f(N)e^{-N\mu^{d^{\lceil f(N)/2 \rceil}}/\alpha}.$$

Proof. The results in [1] imply the existence of constants $\alpha_0 > 0$ and $0 < \mu_0 < 1$ such that if $k = k(N)$ satisfies $N\mu^{d^k} \to \infty$ as $N \to \infty$, then

$$\mathbf{Pr}(\sum_{j=1}^{N} \mathbb{I}_{X_{t_0}^{N,j} \geq k} \geq N\alpha_0\mu_0^{d^k}) \leq e^{-N\mu^{d^k}/\alpha_0}.$$

The proof may now be completed in the same way as the proof of Lemma 4.2. □

The bounds in Lemmas 4.2 and 4.3 can be improved by placing further restrictions on the initial conditions. However, for $d \geq 2$ one can do substantially better by working directly with the limiting deterministic process. This is demonstrated in the final lemma of this subsection, which bounds the decay of high-indexed components of the solution to the differential equation.

Lemma 4.4. *The following are true:*

1. *Let $d = 1$. Suppose $1 \geq v_0^{k-1} \geq v_0^k \geq 0$ for each $k \in \mathbb{N}$. Then $1 \geq v_t^{k-1} \geq v_t^k \geq 0$ for all $k \in \mathbb{N}$ and all $t > 0$. Furthermore, there exist constants $\beta > 0$ and $0 < \nu < 1$ such that, if $v_0^k \leq \beta\nu^{2k}$ for all $k \in \mathbb{N}$, then $v_t^k \leq 2\beta\nu^{k-1}$ for all $k \in \mathbb{N}$ and $t > 0$.*
2. *Let $d \geq 2$. Suppose $1 \geq v_0^{k-1} \geq v_0^k \geq 0$ for each $k \in \mathbb{N}$. Then $1 \geq v_t^{k-1} \geq v_t^k \geq 0$ for all $k \in \mathbb{N}$ and $t > 0$. If also for some constants $\tilde{\beta} > 0$ and $0 < \tilde{\nu} < 1$, we have $v_0^k \leq \tilde{\beta}\tilde{\nu}^{d^k}$ for all $k \in \mathbb{N}$, then there exist constants $\beta > 0$ and $0 < \nu < 1$ such that $v_t^k \leq \beta\nu^{d^k}$ for all $k \in \mathbb{N}$ and $t > 0$.*

Proof. The case $d = 1$ follows easily from the proof of Lemma 4.2, and the fact that v_t^k is equal to the expected number of queues of length at least k at time t.

Consider the case $d \geq 2$. The fact that $1 \geq v_t^{k-1} \geq v_t^k \geq 0$ for all $k \in \mathbb{N}$ and all $t > 0$ was proved in [26]. Let $\pi = (\pi^k)_{k \geq 0}$, where $\pi^k = \lambda^{\frac{d^k-1}{d-1}}$ for all $k \in \mathbb{N}$; thus π is the unique fixed point of the differential equation (4.3) subject to $\pi^0 = 1$ and $\sum_k \pi^k < \infty$. Let $R_t = \sup_{k \geq 0} (v_t^k/\pi^k)^{1/d^k}$. We first show that $R_t \leq R_0$ for all $t > 0$. To see this, notice that $v_t^k \leq \tilde{v}_t^k$ for all $t \geq 0$ and non-negative integers k, where $\tilde{v}_t^k$ solves the differential equation (4.3) subject to the initial condition $\tilde{v}_0^k = (R_0)^{d^k} \pi^k$ ($k \in \mathbb{N}$). This can be proved directly as in [26]; or it follows from the standard theory of *quasi-monotone* ordinary differential equations in infinite-dimensional spaces, see [4] for a thorough account of such equations.

But $\tilde{v}_0$ is a fixed point, so that $\tilde{v}_t^k = \tilde{v}_0^k$ for all $k \in \mathbb{N}$, $t > 0$, and so $R_t \leq \tilde{R}_t = \tilde{R}_0 = R_0$. Now, since $v_0^k \leq \min\{1, \tilde{\beta}\tilde{\nu}^{d^k}\}$ for all k, where $\tilde{\beta} > 0$ and $0 < \tilde{\nu} < 1$, we have that $R_0 < 1/\lambda^{1/(d-1)}$. Hence

$$v_t^k \leq \pi^k (R_0)^{d^k} = \lambda^{-1/(d-1)} (\lambda^{1/(d-1)} R_0)^{d^k} = \lambda^{-1/(d-1)} \nu^{d^k}$$

for some $0 < \nu < 1$. □

The upper bounds in Lemma 4.4 for $d \geq 2$ are significantly tighter than the bounds in Lemma 4.3; the former rather than the latter reflect the "true" behaviour of $(\tilde{V}_t^N)_{\geq 0}$. For an appropriate range of $\delta_{N,k}$ Lemma 4.4 alone will suffice for a law of large numbers - see Theorem 4.6 below. Many variants of the fluid limit established here are possible; some refinements are considered in [17].

4.2. Convergence to a fluid limit

In what follows, we use the phrase *asymptotically almost surely* (aas) to mean "with probability $\to 1$ as $N \to \infty$". Theorem 3.1 will be used to establish a quantitative law of large numbers.

We assume appropriate initial conditions $\tilde{V}_0^N$, so that Lemmas 4.2 and 4.4 may be applied. Let $t_0 > 0$. Observe that for all $k \in \mathbb{N}$, for each number $\eta > 0$

$$\sup_{||\theta|| \leq \eta} \left| \frac{\delta^2 m^N(v, N\theta)}{\delta \theta_k^2} \right| \leq \frac{e^\eta (\lambda (v^{k-1})^d + v^k)}{N},$$

and hence

$$\sup_{0 \leq t \leq t_0} \sup_{||\theta|| \leq N\eta} \left| \frac{\delta^2 m^N(\tilde{V}_t^N, \theta)}{\delta \theta_k^2} \right| \leq \frac{e^\eta (\lambda \sup_{0 \leq t \leq t_0} (\tilde{V}_t^{k-1,N})^d + \sup_{0 \leq t \leq t_0} \tilde{V}_t^{k,N})}{N}.$$

Let $\hat{S}^N = \{1, \dots, k^{**}\}$ for a suitable function $k^{**} = k^{**}(N)$, let $k^* = k^*(N) \leq k^{**}(N)$, and let $\eta > 0$. If $d = 1$, then by Lemma 4.2 for $k \leq k^*$ we can take $A_{N,k} = c\nu^k/N$, where $c = (\lambda + 1)e^\eta \beta$, and $\beta > 0$, $0 < \nu < 1$ are appropriate constants. For $k^* < k \leq k^{**}$ we put $A_{N,k} = A_{N,k^*}$. Let $d \geq 2$ and let $0 < \nu < 1$ be as in Lemma 4.4. Take a suitably large constant $\beta > 0$, and put $c = (\lambda + 1)e^\eta \beta$.

We define $A_{N,k} = c\nu^{d^k}/N$ for $k \leq k^*$, and $A_{N,k} = A_{N,k^*}$ for $k^* < k \leq k^{**}$. For each $N, k \in \mathbb{N}$ and $t \in [0, t_0]$ we let

$$\mathcal{A}_{N,k} = \{v \in E: \sup_{\|\theta\| \leq N\eta} \langle e_k, m''(v,\theta)e_k \rangle \leq A_{N,k}\},$$

$$\mathcal{A}_t^{N,k} = \{\tilde{V}_s^N \in \mathcal{A}_{N,k} \quad \forall s \in [0,t]\}, \quad \mathcal{A}_t^N = \wedge_{k \leq k^{**}} \mathcal{A}_t^{N,k}.$$

We now show one way to "cut off" the infinite-dimensional tail.

Lemma 4.5. *Let p_N be a sequence of positive numbers tending to 0 as $N \to \infty$. Then the following facts hold.*

1. *Let $d \geq 1$ and let the constant $0 < \mu < 1$ be as in Lemma 4.2. There exists a function $k^{**} = k^{**}(N) = O(\ln N)$ such that if*

$$\mathbf{Pr}(\sup_{k \leq k^{**}(N)/2} \tilde{V}_0^{N,k}/\mu^{2k} \geq \alpha) \leq p_N,$$

then

$$\sup_{k \geq k^{**}+1} \sup_{0 \leq t \leq t_0} \tilde{V}_t^{N,k} = 0 \quad aas.$$

2. *Let $d \geq 2$ and let the constant $0 < \mu < 1$ be as in Lemma 4.3. There exists a function $k^{**} = k^{**}(N) = \frac{2 \ln \ln N}{\ln d} + O(1)$ such that if*

$$\mathbf{Pr}(\sup_{k \leq k^{**}(N)/2} \tilde{V}_0^{N,k}/\mu^{d^k} \geq \alpha) \leq p_N,$$

then

$$\sup_{k \geq k^{**}+1} \sup_{0 \leq t \leq t_0} \tilde{V}_t^{N,k} = 0 \quad aas.$$

Proof. Let $d \geq 1$, let α, μ be as in Lemma 4.2, and let $f(N)$ be a function such that for N sufficiently large $2N\alpha\mu^{f(N)-1} < 1$. If the initial state satisfies the hypotheses of Lemma 4.2, then using Markov's inequality the probability that the maximum length of a queue ever exceeds $f(N) - 1$ during the interval $[0, t_0]$ is at most $p_N + \frac{1}{2}\alpha\mu^{f(N)-1}N$. This proves (1).

For $d \geq 2$ it follows from [1] that there exists a function $f(N) = \frac{2 \ln \ln N}{\ln d} + O(1)$ such that if the hypotheses of Lemma 4.3 hold, then the probability that the maximum length of a queue exceeds $f(N) - 1$ during the interval $[0, t_0]$ is $O(\ln^d N/N^{d-1})$ as $N \to \infty$. This proves (2). □

Let $d = 1$. Since we only consider a fixed-length interval $[0, t_0]$, we could (for a suitable choice of initial conditions) take $k^{**} = k^{**}(N) \sim \frac{\ln N}{\ln \ln N}$. However, in equilibrium the queue lengths are N independent geometric random variables each with parameter λ, and so aas the maximum length of a queue is around $\frac{\ln N}{\ln \lambda}$. Therefore it appears more natural to consider initial states where the maximum queue length is of order $\ln N$.

It is a consequence of Lemma 4.2 that if $k^{**}(N)$ grows subexponentially, and $k^{**}(N)/\ln N \to \infty$ as $N \to \infty$, then for all $d \geq 1$

$$\mathbf{Pr}(\sup_{k \geq k^{**}+1} \sup_{0 \leq t \leq t_0} \tilde{V}_t^{N,k} > 0) = e^{-\Omega(k^{**}(N))} \quad \text{as } N \to \infty.$$

This is the "cutoff" that will be used in Theorem 4.6 below for all positive integers d. Other cutoffs are possible and would result in slightly different variants of the fluid limit.

We now have all the ingredients needed for the law of large numbers. Let d be a positive integer, let v_0 satisfy the hypotheses of Lemma 4.4. Let $\ln N \le g(N) \le N^\epsilon$ (there is a fair amount of flexibility here), and let $0 < \nu < 1$ be as above. If $d = 1$, then we take $k^* = k^*(N)$ to be the largest integer for which $N\nu^{k^*} \ge g(N)$. If $d = 2$, then we take k^* be the largest integer for which $N\nu^{d^{k^*}} \ge g(N)$ Let $k^{**}(N)$ satisfy $k^{**}(N)/\log N \to \infty$ as $N \to \infty$. Let $\delta_{N,k}$ be a sequence of numbers such that $0 < \delta_{N,k} \le NA_{N,k}$ for each $k \le k^{**}(N)$, $N \in \mathbb{N}$; and for each constant $\alpha > 0$

$$\sum_{k \le k^{**}} e^{-\beta \delta_{N,k}^2 / A_{N,k}} \to 0 \quad \text{as } N \to \infty.$$

Assume further that $0 < \delta_{N,k} \le \delta_{n,k'}$ for every pair of integers $k > k^{**}$, $k' \le k^{**}$; and, with ν as in Lemma 4.4, in the limit as $N \to \infty$

$$\sup_{k \ge k^{**}} \nu^{-k} \delta_{N,k} \to \infty, \qquad \text{if } d = 1,$$

$$\sup_{k \ge k^{**}} \nu^{-d^k} \delta_{N,k} \to \infty, \qquad \text{if } d \ge 2.$$

Suppose that for each $\alpha > 0$,

$$\limsup_{N \to \infty} \sup_{k \le k^{**}} A_{N,k} \delta_{N,k}^{-2} \log \mathbf{Pr}(|\tilde{V}_0^{N,k} - v_0^k| \ge \alpha \delta_{N,k}) < 0,$$

and that

$$\limsup_{N \to \infty} [k^{**}(N)]^{-1} \log \mathbf{Pr}(\sup_{k \ge k^{**}} \frac{|\tilde{V}_0^{N,k} - v_0^k|}{\delta_{N,k}} \ge \alpha) < 0.$$

The law of large numbers that follows is just one in a large spectrum of such results.

Theorem 4.6. *Let $\alpha > 0$ and let $t_0 > 0$. Under the above assumptions, there exists a constant $\beta = \beta(t_0, \alpha)$ such that for all N sufficiently large*

$$\mathbf{Pr}(\sup_{k \in \mathbb{N}} \sup_{0 \le t \le t_0} \frac{|\tilde{V}_t^{N,k} - v_t^k|}{\delta_{N,k}} \ge \alpha) \le \sum_{k \le k^{**}} e^{-\beta \delta_{N,k}^2 / A_{N,k}} + e^{-\beta k^{**}(N)},$$

and hence tends to 0 as $N \to \infty$.

Proof. Let $(M_t^N)_{t \ge 0} = (M_t^{N,k} : k \in \mathbb{N})_{t \ge 0}$ be the martingale defined by

$$\tilde{V}_t^N = \tilde{V}_0^N + M_t^N + \int_0^t b^N(\tilde{V}_s^N) ds.$$

By Lemma 2.2 there exists $\epsilon_0 > 0$ such that for each $k \le k^{**}$ and $0 < \epsilon < \epsilon_0$,

$$\mathbf{Pr}\left(\mathcal{A}_{t_0}^{N,k} \wedge (\sup_{0 \le t \le t_0} |M_t^{N,k}| \ge \epsilon \delta_{N,k}) \right) \le 2e^{-\epsilon^2 \delta_{N,k}^2 / 2A_{N,k} t_0}.$$

Also by Lemma 4.4,

$$\limsup_{N\to\infty} \sup_{k\geq k^{**}} \sup_{t\leq t_0} \frac{1}{\delta_{N,k}} |v_t^k - v_0^k| = 0. \tag{4.4}$$

Finally, from the proof of Lemma 4.2,

$$\limsup_{N\to\infty} [k^{**}(N)]^{-1} \log \mathbf{Pr}(\sup_{k\geq k^{**}} \sup_{0\leq t\leq t_0} |\tilde{V}_t^{N,k} - \tilde{V}_0^{N,k}| > 0) < 0, \tag{4.5}$$

and the result follows from (3.11) in Theorem 3.1. □

5. Conclusions

We have proposed a quantitative law of large numbers and illustrated it with a queueing application. Similar techniques can be applied to the related model where each server has infinite capacity, and the arrival rate λ is any positive constant. An extended version of Theorem 3.1 will handle the "star" network, and many other network models. For details see the forthcoming papers [16, 17]. Furthermore, the exponential martingale method constitutes a simpler and more systematic alternative to the "differential equation" method in [28]; thus one challenge is to test it on combinatorial and random graph problems. Certainly, the power and versatility of this approach is yet to be fully explored.

Acknowledgement

I am grateful to James Norris for first introducing me to exponential martingales, and for many helpful discussions on the ideas presented here.

References

[1] Y. Azar, A. Broder, A. Karlin, and E. Upfal, *Balanced allocations.* SIAM J. Comput. **29** (1999), 180–200.

[2] S. Boucheron, F. Gamboa, and C. Léonard, *Bins and balls: large deviations of the empirical occupancy process.* Ann. Appl. Prob. **12** (2002), 607–636.

[3] R.W.R. Darling and J.R. Norris, *Structure of large random hypergraphs,* Preprint (2002).

[4] K. Deimling, *Ordinary Differential Equations in Banach Spaces.* Lecture Notes in Mathematics **596**, Springer-Verlag, 1977.

[5] D. L. Eager, E.D. Lazokwska, and J. Zahorjan, *Adaptive load sharing in homogeneous distributed systems.* IEEE Trans. Soft. Eng. **12** (1986), 662–675.

[6] S.N. Ethier and T.G. Kurtz, *Markov Processes: Characterization and Convergence.* Wiley, 1986.

[7] C. Graham, *Kinetic limits for large communication networks.* In Modelling in Applied Sciences: A Kinetic Theory Approach, N. Bellomo and M. Pulvirenti, eds., Birkhäuser, 1999.

[8] C. Graham and S. Méléard, *Propagation of chaos for a fully connected loss network with alternate routing.* Stoch. Proc. Appl. **44** (1993), 159–180.

[9] P.J. Hunt, *Limit Theorems for Stochastic Loss Networks.* PhD thesis, Cambridge University, 1990.

[10] P.J. Hunt, *Loss networks under diverse routing: the symmetric star network.* Adv. Appl. Prob. **25** (1995), 255–272.

[11] M.J. Luczak, *Probability, Algorithms and Telecommunication Systems.* DPhil thesis, Oxford University, 2000.

[12] M.J. Luczak, C. McDiarmid, and E. Upfal, *On-line routing of random calls in networks.* Prob. Ther. Rel. Fields **125** (2003).

[13] M.J. Luczak and C. McDiarmid, *On the power of two choices: balls and bins in continuous time.* Preprint (2003).

[14] M.J. Luczak and C. McDiarmid, *On the maximum queue length in the supermarket model.* In preparation (2003).

[15] M.J. Luczak and C. McDiarmid, *Dynamic balanced routing in networks.* In preparation (2003).

[16] M.J. Luczak and J.R. Norris, *A law of large numbers in almost finite dimensions.* In preparation (2003).

[17] M.J. Luczak and J.R. Norris, *Precise asymptotics of the supermarket model.* In preparation (2003).

[18] J.B. Martin and Y.M. Suhov, *Fast Jackson networks.* Ann. Appl. Prob. **9** (1998), 854–870.

[19] C. McDiarmid, *On the method of bounded differences.* In Surveys in Combinatorics, Siemons, J., ed., London Mathematical Society Lecture Notes Series 141, Cambridge University Press, 1989.

[20] C. McDiarmid, *Concentration.* In Probabilistic Methods for Algorithmic Discrete Mathematics, M. Habib, C. McDiarmid, J. Ramirez, B. Reed, eds., Springer-Verlag, 1998.

[21] M. Mitzenmacher, *The power of two choices in randomized load balancing.* PhD thesis, Berkeley, 1996.

[22] M. Mitzenmacher, *Studying balanced allocations with differential equations.* Comb. Prob. Comp. **8** (1999), 473–482.

[23] M. Mitzenmacher, A.W. Richa, and R. Sitaraman, *The power of two random choices: a survey of techniques and results.* In preparation (2002).

[24] A. Schwartz and A. Weiss, *Large deviations for performance analysis.* Chapman & Hall, 1995.

[25] S.R.E. Turner, *The effect of increasing routeing choice on resource pooling.* Prob. Eng. Inf. Sci. **12** (1998), 109–124.

[26] N.D. Vvedenskaya, R.L. Dobrushin, and F.I. Karpelevich, *Queueing system with selection of the shortest of two queues: an asymptotic approach.* Prob. Inf. Trans. **32** (1996), 15–27.

[27] W. Whitt, *Blocking when service is required from several facilities simultaneously* AT & T Tech. J. **64** (1985), 1807–1856.

[28] N.C. Wormald, *The differential equation method for random graph processes and greedy algorithms.* In Lectures on Approximation and Randomised Algorithms, M. Karonski and H.J. Prömel, eds., PWN, 1999.

[29] I.B. Ziedins and F.P. Kelly, *Limit theorems for loss networks with diverse routing.* Adv. Appl.Prob. **21** (1989), 804–830.

[30] S. Zhou, *A trace-driven simulation study of dynamic load balancing.* IEEE Trans. Soft. Eng. **14** (1988), 1327–1341.

Statistical Laboratory, Centre for Mathematical Sciences,
Wilberforce Road, Cambridge CB3 0WB, England
E-mail address: m.j.luczak@statslab.cam.ac.uk

Progress in Probability, Vol. 56, 113–128

Sufficient Conditions for Boundedness of Moving Average Processes

Michael B. Marcus and Jan Rosiński

Abstract. We show that the moving average process

$$\mathcal{X}_f(t) := \int_0^t f(t-s)\,dZ(s) \qquad t \in [0,T]$$

has a bounded version almost surely, when the kernel f has finite total 2-variation and Z is a symmetric Lévy process. We also obtain bounds for $E|\sup_{t\in[0,T]} \mathcal{X}_f(t)|$ and for uniform moduli of continuity of $\mathcal{X}_f(\cdot)$ and for the largest jump of $\mathcal{X}_f(\cdot)$ when it is not continuous. Similar results are obtained for forward average processes. The methods developed are also used to show that certain infinitely divisible random fields are bounded.

1. Introduction

We are interested in sample path properties of infinitely divisible moving average processes. Such processes occur naturally in applications. By a moving average process, we mean a process of the form

$$\mathcal{X}_f(t) := \int_0^t f(t-s)\,dZ(s) \qquad t \in [0,T] \tag{1.1}$$

where Z is a Lévy process and $f : [0,T] \mapsto R$ is a measurable function.

In [6] we study the continuity and boundedness of infinitely divisible processes in considerable detail. However, the results, which are often quite sharp in determining whether (1.1) has a continuous version, are not useful in determining whether it has a version which is bounded, except in some special cases. (See also the related work of Rootzén [10], Balkema and De Haan [1] on the continuity of stable moving averages). In this note we approach this question in a completely different, and much more elementary way, and obtain some simple conditions for (1.1) to be bounded almost surely.

2000 *Mathematics Subject Classification.* Primary 60G17, 60G51, 60G15; Secondary 60H05.
Key words and phrases. moving average processes, Lévy processes, infinitely divisible random fields.
This work was supported in part by grants from the National Science Foundation and PSC CUNY.

The results in this paper are given in terms of the total 2-variation. The total 2-variation of a function h on an interval I (bounded or unbounded) is defined by

$$[\![h]\!](I) = \sup\left\{\left(\sum_{i=1}^{n} |h(s_i) - h(s_{i-1})|^2\right)^{1/2} : s_0 < \cdots < s_n, s_i \in I, \quad n \geq 1\right\}. \tag{1.2}$$

We define $[\![f]\!](t) := [\![f]\!]([0,t])$. When f is increasing $[\![f]\!](t) = f(t) - f(0)$. When f has bounded variation, $[\![f]\!](t) \leq V(f)(t)$, where $V(f)(t)$ denotes the total variation of f on $[0,t]$. It is easy to construct examples where $[\![f]\!](t) < \infty$ but $V(f)(t) = \infty$.

We show that when certain obvious necessary conditions are satisfied and f in (1.1) has finite total 2-variation then $\mathcal{X}_f = \{\mathcal{X}_f(t), t \in [0,T]\}$ is bounded or continuous. One set of necessary conditions is that f must be continuous and $f(0) = 0$ for $\mathcal{X}_f$ to be continuous and f must be bounded for $\mathcal{X}_f$ to be bounded. Another is that $\mathcal{X}_f(t)$ must exist for all $t \in [0,T]$. To simplify the presentation we assume that the process Z is a symmetric Lévy process with finite first moment and cádlág trajectories, defined by

$$Ee^{i\lambda Z_s} = e^{-s\Psi(\lambda)} \tag{1.3}$$

where

$$\Psi(\lambda) = \int_0^\infty (1 - \cos \lambda x)\, d\theta(x) \tag{1.4}$$

with θ, a σ-finite measure on $(0,\infty)$, satisfying

$$\int_0^\infty (x^2 \wedge x)\ d\theta(x) < \infty. \tag{1.5}$$

Then $\mathcal{X}_f(t)$ in (1.1) is well-defined for all $t \in [0,T]$ if and only if

$$\int_0^T \int_0^\infty \left(x^2 f^2(s) \wedge 1\right)\ d\theta(x)\, ds < \infty, \tag{1.6}$$

see [7]. Actually, we also require that $E|\mathcal{X}_f(t)| < \infty$, which by [7] is equivalent to the condition

$$\int_0^T \int_0^\infty \left(x^2 f^2(s) \wedge x|f(s)|\right)\ d\theta(x)\, ds < \infty. \tag{1.7}$$

Using the results in [5] we can get concrete estimates for $E|\mathcal{X}_f(T)|$ in terms of θ and f. Let

$$\Phi(u) := \int_0^\infty \left((ux)^2 I(ux < 1) + (2ux - 1) I(ux \geq 1)\right)\, \theta(dx) \quad u \geq 0.$$

Φ is a convex function. Define the Orlicz norm of f by

$$\|f\|_\Phi = \inf\{c > 0 : \int_0^T \Phi(c^{-1}|f(s)|)\, ds \leq 1\}. \tag{1.8}$$

Then, by Theorem 2.1, [5]

$$0.125\|f\|_\Phi \leq E|\mathcal{X}_f(T)| \leq 2.125\|f\|_\Phi. \tag{1.9}$$

A simpler, although non-homogeneous, estimate is given by

$$0.25\min\{\xi,\sqrt{\xi}\}\le E|\mathcal{X}_f(T)|\le 2.125\max\{\xi,\sqrt{\xi}\} \tag{1.10}$$

where

$$\xi:=\int_0^T\int_0^\infty \left(x^2f^2(s)\wedge x|f(s)|\right)\theta(dx)ds. \tag{1.11}$$

See (16), [5].

We now give some sufficient conditions for the continuity and boundedness of the moving average process given in (1.1). Note that the term $E|\mathcal{X}_{[\![f]\!]}(T)|$ in (1.12) can be estimated using (1.9) or (1.10) with $[\![f]\!]$ instead of f.

Theorem 1.1. *Let $\mathcal{X}_f=\{\mathcal{X}_f(t),t\in[0,T]\}$ be as given in* (1.1) *and assume that $[\![f]\!](T)<\infty$. There exists a version $\mathcal{X}'_f$ of $\mathcal{X}_f$ with bounded sample paths. Furthermore,*

$$E\sup_{t\in T}\mathcal{X}'_f(t)\le KE|\mathcal{X}_{[\![f]\!]}(T)| \tag{1.12}$$

where $K=16\sqrt{2}\int_0^{\sqrt{2}/4}(\log 1/u)^{1/2}\,du$.

If f is continuous on $[0,T]$ and $f(0)=0$, then $\mathcal{X}'_f$ has continuous sample paths.

To obtain results for the modulus of continuity of $\mathcal{X}_f$ in (1.1) we define the τ-Lipschitz norm of a function. Let $\tau(s,t)$, $s,t\in[0,T]$ be a pseudo-metric on $[0,T]$

$$\|\tau(q)\|:=\sup_{s,t\in[0,T]}\frac{|q(s)-q(t)|}{\tau(s,t)}. \tag{1.13}$$

Theorem 1.2. *Let $\{\mathcal{X}_f(t),t\in[0,T]\}$ be as in Theorem* 1.1. *Assume that there exists a translation invariant pseudo-metric τ on $[0,T]$ such that $\|\tau([\![f]\!])\|<\infty$. Then there exists a version $\mathcal{X}'_f$ of $\mathcal{X}_f$ such that for all $0<\delta\le[\![f]\!](T)/\|\tau([\![f]\!])\|$,*

$$E(\sup_{\substack{s,t\in[0,T]\\ \tau(s,t)\le\delta}}\mathcal{X}'_f(t)-\mathcal{X}'_f(s))\le 26\sqrt{2}E|\mathcal{X}_{[\![f]\!]^{1/2}}(T)|\left(\widetilde{\delta}\log\frac{18\,[\![f]\!](T)}{\widetilde{\delta}}\right)^{1/2} \tag{1.14}$$

where $\widetilde{\delta}=\|\tau([\![f]\!])\|\,\delta$.

We now consider the largest jump $\Delta\mathcal{X}_f$ of $\mathcal{X}_f$ when f is not continuous.

Theorem 1.3. *Let $\{\mathcal{X}_f(t),t\in[0,T]\}$ be as in Theorem* 1.1. *Then there exists a version $\mathcal{X}'_f$ of $\mathcal{X}_f$ such that*

$$E(\Delta\mathcal{X}'_f)\le 26\sqrt{2}E|Z(T)|\left(\Delta[\![f]\!]^2\log\frac{15[\![f]\!](T)}{(\Delta[\![f]\!]^2)^{1/2}}\right)^{1/2} \tag{1.15}$$

where $\Delta[\![f]\!]^2=\lim_{\delta\to0}\sup_{|s-t|\le\delta}|[\![f]\!]^2(s)-[\![f]\!]^2(t)|$ is the largest jump of $[\![f]\!]^2$.

Theorems 1.1–1.3 have analogues for forward average processes. Let $p : R^+ \mapsto R$ be a measurable function. A forward average process is a process of the form

$$\mathcal{Y}_p(t) := \int_0^\infty p(t+s)\,dZ(s) \tag{1.16}$$

where Z is an independent increment process with cádlág trajectories. In general, for $\mathcal{Y}_p(t)$ to exist for fixed t, p must go to zero, in some sense, at infinity.

Define $[\![p[\![(t) := [\![p]\!]([t,\infty))$; (see (1.2)). Similarly to (1.6)–(1.7), $\mathcal{Y}_{[\![p[\![}(0)$ exists when

$$\int_0^\infty \int_0^\infty x^2 [\![p[\![^2(s) \wedge 1\,d\theta(x)\,ds < \infty \tag{1.17}$$

and $E|\mathcal{Y}_{[\![p[\![}(0)| < \infty$ when

$$\int_0^\infty \int_0^\infty x^2 [\![p[\![^2(s) \wedge x[\![p[\![(s)\,d\theta(x)\,ds < \infty. \tag{1.18}$$

To get estimates for $E|\mathcal{Y}_{[\![p[\![}(0)|$ take $f = [\![p[\![$ and $T = 0$ in (1.9) and (1.10) and set $T = \infty$ in (1.8) and (1.11).

Theorem 1.4. *Let $\mathcal{Y}_p = \{\mathcal{Y}_p(t), t \in R^+\}$ be as given in* (1.16) *and assume that $[\![p[\![(0) < \infty$. There exists a version $\mathcal{Y}'_p$ of $\mathcal{Y}_p$ with bounded sample paths. Furthermore, when $E|\mathcal{Y}_{[\![p[\![}(0)| < \infty$*

$$E \sup_{t\in R^+} \mathcal{Y}'_p(t) \le KE|\mathcal{Y}_{[\![p[\![}(0)| \tag{1.19}$$

where K is given in Theorem 1.1. *If p is continuous on R^+, $\mathcal{Y}'_p$ has continuous sample paths.*

Theorem 1.5. *Let $\{\mathcal{Y}_p(t), t \in R^+\}$ be as in Theorem* 1.4. *Assume that $E|\mathcal{Y}_{[\![p[\![^{1/2}}(0)| < \infty$. Then there exists a version $\mathcal{Y}'_p$ of $\mathcal{Y}_p$ such that for $0 < \delta \le [\![p[\![(0)/\|\tau([\![p]\!])\|$,*

$$E(\sup_{\substack{s,t\in R_+\\ \tau(s,t)\le\delta}} \mathcal{Y}'_p(t) - \mathcal{Y}'_p(s)) \le 26\sqrt{2}E|\mathcal{Y}_{[\![p[\![^{1/2}}(0)| \left(\widetilde{\delta}\log\frac{18[\![p[\![(0)}{\widetilde{\delta}}\right)^{1/2}. \tag{1.20}$$

where $\widetilde{\delta} = \|\tau([\![p]\!])\|\,\delta$.

2. Simple conditions on sample boundedness and continuity

Throughout this section (T,d) denotes a metric or pseudo-metric space. It is index set of the stochastic processes we consider. Elsewhere in this paper, T denotes a positive number. Recall that a second order centered process $G = \{G(t), t \in T\}$ is said to be sub-Gaussian if

$$E\exp(\lambda(G(s) - G(t)) \le \exp\left(\frac{\lambda^2}{2}\mathrm{Var}(G(s) - G(t))\right)$$

for all $\lambda \in R$ and $s, t \in T$. The class of sub-Gaussian processes includes all centered Gaussian processes and, most notably, Rademacher processes. The latter are defined as random series

$$G(t) = \sum_j \epsilon_j \phi_j(t)$$

where $\{\epsilon_j\}$ is a Rademacher sequence of random variables and $\{\phi_j\}$ is a sequence of deterministic functions such that $\sum_j \phi_j^2(t) < \infty$ for all $t \in T$.

We recall Dudley's sufficient condition for the continuity of a Gaussian process in terms of metric entropy. We state it as it appears in Theorem 5.1.2, [4]; see also 4.3.1. Théorème, [3].

Theorem 2.1. *Let $G = \{G(t), t \in T\}$ be a mean zero Gaussian or sub-Gaussian process and let*

$$(E[G(s) - G(t)]^2)^{1/2} \le d(s,t). \tag{2.1}$$

Assume that

$$\int_0^D (\log N(T,d,u))^{1/2}\, du < \infty \tag{2.2}$$

where D denotes the diameter of (T,d) and $N(T,d,u)$ is the smallest number of closed d-balls of radius u needed to cover T. Then there exists a version G' of G with bounded uniformly continuous sample paths on (T,d) such that

$$E \sup_{t\in T} G'(t) \le 4\sqrt{2} \int_0^{D/2} (\log N(T,d,u))^{1/2}\, du \tag{2.3}$$

and

$$E\left(\sup_{\substack{s,t\in T\\ d(s,t)\le\epsilon}} G'(t) - G'(s) \right) \le 18 \int_0^{\epsilon} (\log N(T,d,u))^{1/2}\, du. \tag{2.4}$$

We apply this theorem when the pseudo-metric d is dominated in a simple way.

Lemma 2.2. *Let $G = \{G(t), t \in T\}$ be a Gaussian or sub-Gaussian process. Suppose that*

$$E|G(s) - G(t)|^2 \le |Y(s) - Y(t)| =: \rho(s,t) \tag{2.5}$$

where Y is a bounded function on T. Then there exists a version G' of G with bounded uniformly continuous sample paths on (T,ρ) such that

$$E \sup_{t\in T} G'(t) \le K_0 \, \|Y\|_\infty^{1/2} \tag{2.6}$$

where

$$K_0 = 16 \int_0^{\sqrt{2}/4} \left(\log \frac{1}{u}\right)^{1/2} du. \tag{2.7}$$

Moreover, for every $0 < \delta \le \sqrt{2}\|Y\|_\infty^{1/2}$

$$E\left(\sup_{\substack{s,t\in T\\ |Y(t)-Y(s)|^{1/2}\le\delta}} G'(t)-G'(s)\right) \le 26\delta\left(\log\frac{6\|Y\|_\infty^{1/2}}{\delta}\right)^{1/2}. \tag{2.8}$$

Proof. Set

$$\eta(s) = \frac{G(s)}{\sqrt{2}\|Y\|_\infty^{1/2}}$$

and

$$h(s) = \frac{Y(s)-y_0}{2\|Y\|_\infty}$$

where $y_0 = \inf_{[0,T]} Y(t)$. By (2.5)

$$\left(E(\eta(s)-\eta(t))^2\right)^{1/2} \le |h(s)-h(t)|^{1/2} =: \rho_1(s,t). \tag{2.9}$$

Consider the sub-Gaussian process $\eta = \{\eta(t), t\in T\}$ on (T,ρ_1). Note that the diameter of (T,ρ_1) is less than or equal to one and [0,1] contains the range of h. Cover this interval with at most $\lceil 1/\epsilon^2\rceil$ closed intervals of length ϵ^2, where $\epsilon \le 1$ and $\lceil x\rceil$ denotes the smallest integer that is greater than or equal to x. Number these intervals from 1 to $\lceil 1/\epsilon^2\rceil$. Consider the k-th interval. If the range of h intersected with this interval is not empty, choose t_k so that $h(t_k)$ is in this interval. Let $S_k = \{s : |h(t_k)-h(s)| \le \epsilon^2\}$. Note that $T \subset \cup S_k$. In other words $N(T,\rho_1,\epsilon) \le \lceil 1/\epsilon^2\rceil \le 2/\epsilon^2$. Therefore, by (2.3) there exists a version η' of η such that

$$E\sup_{t\in T}\eta'(t) \le 8\int_0^{1/2}\left(\log\frac{\sqrt{2}}{u}\right)^{1/2} du = 8\sqrt{2}\int_0^{\sqrt{2}/4}\left(\log\frac{1}{u}\right)^{1/2} du \tag{2.10}$$

which gives us (2.6).

By (2.4) and Jensen's inequality we have for every $a\in(0,1]$

$$E\left(\sup_{\substack{s,t\in T\\ |h(s)-h(t)|^{1/2}\le a}} \eta'(t)-\eta'(s)\right) \le 18\sqrt{2}\int_0^a\left(\log\frac{\sqrt{2}}{u}\right)^{1/2} du$$

$$\le 18\sqrt{2}\,a\left(\int_0^a \log\frac{\sqrt{2}}{u}\frac{du}{a}\right)^{1/2} = 18\sqrt{2}\,a\left(\log\frac{\sqrt{2}e}{a}\right)^{1/2}.$$

We get (2.8) by taking $a = \delta/(\sqrt{2}\|Y\|_\infty^{1/2})$. □

Remark 2.3. Note that, trivially, Y is continuous on (T,ρ) but it may be discontinuous on (T,d). If Y is also continuous on (T,d), G' has bounded continuous sample paths on (T,d) and (2.8) gives a modulus of continuity for G'. If Y is not continuous on (T,d), G' has bounded sample paths on (T,d) but (2.8) is not too interesting. This is because, as δ goes to zero, the supremum on the left hand side of (2.8) excludes points around the discontinuities of Y. In the next lemma we say something about the jumps of G' in the case when Y is discontinuous on (T, d).

Let

$$\Delta Y := \lim_{\delta \to 0} \sup_{\substack{s,t \in T \\ d(s,t) \le \delta}} |Y(s) - Y(t)|$$

denote the largest oscillation of Y on T. If Y is a cádlág function on a interval, ΔY is simply the largest jump of Y.

Lemma 2.4. *Let $G = \{G(t), t \in T\}$ be a Gaussian or sub-Gaussian process satisfying* (2.5). *Then*

$$E(\Delta G') \le 26 \left(\Delta Y \log \frac{6\|Y\|_\infty^{1/2}}{(\Delta Y)^{1/2}} \right)^{1/2}. \tag{2.11}$$

Proof. By the definition of ΔY, for every $\eta > 0$ we can find a δ such that $\{s, t \in T : d(s,t) \le \delta\} \subset \{s, t \in T : |Y(s) - Y(t)| \le (1+\eta)^2 \Delta Y\}$. Hence, by (2.8), for a given $\eta > 0$

$$\begin{aligned} E\left(\sup_{\substack{s,t \in T \\ d(s,t) \le \delta}} G'(t) - G'(s) \right) &\le E\left(\sup_{\substack{s,t \in T \\ |Y(s)-Y(t)| \le (1+\eta)^2 \Delta Y}} G'(t) - G'(s) \right) \\ &\le 26(1+\eta) \left(\Delta Y \log \frac{6\|Y\|_\infty^{1/2}}{(\Delta Y)^{1/2}} \right)^{1/2}. \end{aligned}$$

The first term in above inequality dominates $E(\Delta G')$. Thus letting $\eta \to 0$ completes the proof. □

The condition in (2.5) is rather specialized. It is satisfied trivially when $G(t) = gf(t)$, where g is a standard normal random variable and f is a bounded function on T. It is also satisfied when

$$G(t) = \sum_{k=1}^{\infty} a_k \epsilon_k f_k(t) \qquad t \in [0, \infty] \tag{2.12}$$

where $\{\epsilon_k\}$ is a Rademacher sequence and the $\{f_k\}$ are increasing functions such that $\sum_k a_k^2 f_k^2(t) < \infty$ for every t. This is because for $t \ge s$

$$\begin{aligned} \left(E_\epsilon(G(t) - G(s))^2\right)^{1/2} &= \left(\sum a_k^2 \left(f_k(t) - f_k(s)\right)^2\right)^{1/2} \\ &\le \left(\sum a_k^2 \left(f_k^2(t) - f_k^2(s)\right)\right)^{1/2} \\ &=: (Y_1(t) - Y_1(s))^{1/2}. \end{aligned} \tag{2.13}$$

We can do better than this. Consider the function $[\![f]\!]^2(t)$ defined immediately following (1.2). Note that for $u \le v \le t$

$$[\![f]\!]^2([u,t]) \ge [\![f]\!]^2([u,v]) + [\![f]\!]^2([v,t]) \tag{2.14}$$

Thus, taking $u=0$, we see that

$$|f(t)-f(v)|^2 \le [\![f]\!]^2([v,t]) \le [\![f]\!]^2(t)-[\![f]\!]^2(v). \tag{2.15}$$

For G as in (2.12) we have

$$\begin{aligned}\left(E_\epsilon(G(s)-G(t))^2\right)^{1/2} &= \left(\sum a_k^2\left(f_k(t)-f_k(s)\right)^2\right)^{1/2}\\ &\le \left(\sum a_k^2\left([\![f_k]\!]^2(t)-[\![f_k]\!]^2(s)\right)\right)^{1/2} \qquad (2.16)\\ &=: \left(Y_2(t)-Y_2(s)\right)^{1/2}.\end{aligned}$$

Clearly, we must assume that $Y_2(t)=\sum a_k^2[\![f_k]\!]^2(t)<\infty$ for every t. We apply Lemmas 2.2 and 2.4 with Y of the form Y_2.

The following example shows that the right-hand side of (2.11), except for the constant, is best possible: Let $T=[0,N]$ and consider the Gaussian process $G=\{G(t),\, t\in T\}$ where

$$G(t)=\sum_{j=1}^{\lfloor t\rfloor}\xi_j \tag{2.17}$$

where $\{\xi_j\}$ are independent identically distributed normal random variables with mean zero and variance one, and $\lfloor\cdot\rfloor$ denotes the integer part. We have

$$E(G(t)-G(s))^2 \le |\lfloor t\rfloor-\lfloor s\rfloor|. \tag{2.18}$$

Considering (2.5) we see that $Y(t)=\lfloor t\rfloor$, $\|Y\|_\infty=N$ and $\Delta Y=1$. Therefore, the right–hand side of (2.11) is less than $C(\log N)^{1/2}$ for some constant C, which is the correct order of growth of the jumps of G. (See e.g. Lemma 4.5.5, [4].)

3. Proofs

Proof of Theorem 1.1 We extend f to $[-T,T]$ putting $f(s)=0$ for $s<0$. It is well known that $\mathcal{X}_f$ can be represented as an integral with respect to a Poisson point process on $[0,T]\times R$ with mean measure $\frac{1}{2}dt\bar{\theta}(dx)$, where $\bar{\theta}$ is a symmetric measure that coincides with θ of (1.4) on R^+. Let (U_j,X_j), $j=1,2,\dots$, be an enumeration of its points. In particular, one can take the $\{U_j\}$ to be independent identically distributed uniform random variables on $[0,T]$. The $\{X_j\}$ are defined by $X_j=\epsilon_jV_j$, where $\{\epsilon_j\}$ is a Rademacher sequence, and $V_j=\theta^{-1}(T^{-1}\Gamma_j)$. Here $\{\Gamma_j\}$ is a sequence of the partial sums of independent identically distributed standard exponential random variables and $\theta^{-1}(x)=\inf\{u>0:\theta([u,\infty))<x\}$. Furthermore, the sequences $\{U_j\}$, $\{\epsilon_j\}$, and $\{\Gamma_j\}$ are independent of each other. (See [9] for this and other ways to enumerate a Poisson point process and [8] for the corresponding representations of infinitely divisible processes). Consequently, we can represent $\mathcal{X}_f(t)$ as follows:

$$\mathcal{X}_f(t)=\sum\epsilon_jV_jf(t-U_j)\quad a.s. \tag{3.1}$$

In fact $\sum V_j^2 < \infty$ a.s. is the sum of the squares of jumps of Z on $[0,T]$. Let

$$Y_3(t) = \sum V_j^2 [\![f]\!]^2(t - U_j) \tag{3.2}$$

where we take $[\![f]\!](s) = 0$ for $s < 0$. Since $[\![f]\!]$ is bounded, the series (3.2) converges almost surely for each $t \in [0,T]$.

Consider $\mathcal{X}_f$ conditionally as a Rademacher process. Then, by (2.15), for $0 \le s < t \le T$,

$$\begin{aligned}
\left(E_\epsilon(\mathcal{X}_f(t) - \mathcal{X}_f(s))^2\right)^{1/2} &= \left(\sum V_j^2 \left(f(t-U_j) - f(s-U_j)\right)^2\right)^{1/2} \\
&\le \left(\sum V_j^2 \left([\![f]\!]^2(t-U_j) - [\![f]\!]^2(s-U_j)\right)\right)^{1/2} \qquad (3.3) \\
&=: (Y_3(t) - Y_3(s))^{1/2}.
\end{aligned}$$

It now follows from Lemma 2.2, (2.6) that almost surely with respect to the probability space that supports $\{V_j\}$ and $\{U_j\}$, the Rademacher process in (3.1) has a bounded version $\mathcal{X}'_f$ which satisfies

$$E_\epsilon \sup_{t \in T} \mathcal{X}'_f(t) \le K_0 |Y_3(T)|^{1/2}. \tag{3.4}$$

Since $Y_3(T)$ is finite almost surely, $\sup_{t\in T} \mathcal{X}'_f(t)$ is finite almost surely.

Taking the expectation of (3.4) with respect to the probability space that supports $\{V_j\}$ and $\{U_j\}$ and noting that

$$\begin{aligned}
E|Y_3(T)|^{1/2} &= E\left(\sum V_j^2 [\![f]\!]^2(T - U_j)\right)^{1/2} \qquad (3.5) \\
&\le \sqrt{2}\, E|\sum \epsilon_j V_j [\![f]\!](T - U_j)| \\
&= \sqrt{2}\, E|\int_0^T [\![f]\!](T-s)\, dZ(s)| \\
&= \sqrt{2}\, E|\mathcal{X}_{[\![f]\!]}(T)|
\end{aligned}$$

we get (1.12). Here we use Khintchine's inequality, see e.g. [2], and the same representation that gives (3.1). We have $E|\mathcal{X}_{[\![f]\!]}(T)| < \infty$ because $[\![f]\!](t) \le [\![f]\!](T) < \infty$ and $E|Z(T)| < \infty$ by the assumption (1.5).

Note that when f is continuous then so is $[\![f]\!]$ (since $f(t) = 0$ for $t < 0$, the assumption $f(0) = 0$ is needed here). Consequently, Y_3 given by (3.2) is continuous. By the argument in the first paragraph of this proof along with Lemma 2.2, $\mathcal{X}_f$ has a version with continuous sample paths. □

Proof of Theorem 1.2 Let $t \geq s$. By (3.3) and the fact that $a^2 - b^2 \leq 2(a \vee b)|a-b|$ we have

$$\begin{aligned}(Y_3(t) - Y_3(s))^{1/2} &= \Big(\sum V_j^2 \left([\![f]\!]^2(t-U_j) - [\![f]\!]^2(s-U_j)\right)\Big)^{1/2} \\ &\leq \Big(2\sum V_j^2 [\![f]\!](t-U_j)\Big\{[\![f]\!](t-U_j) - [\![f]\!](s-U_j)\Big\}\Big)^{1/2} \\ &\leq \Big(2\sum V_j^2 [\![f]\!](T-U_j)\|\tau([\![f]\!])\|\,\tau(s,t)\Big)^{1/2}. \qquad (3.6)\end{aligned}$$

Let

$$I_1 := (2\sum V_j^2 [\![f]\!](T-U_j))^{1/2} \qquad (3.7)$$

and

$$I_2 := \|Y_3\|_\infty^{1/2} = (\sum V_j^2 [\![f]\!]^2(T-U_j))^{1/2}. \qquad (3.8)$$

We now use (2.8) with $\eta = I_1\widetilde{\delta}^{1/2} \wedge \sqrt{2}I_2$ (in the place of δ) and (3.6) to get

$$\begin{aligned}E_\epsilon(\sup_{\substack{s,t\in[0,T]\\ \tau(s,t)\leq\delta}} \mathcal{X}'_f(t) - \mathcal{X}'_f(s)) &\leq E_\epsilon(\sup_{\substack{s,t\in[0,T]\\ |Y_3(t)-Y_3(s)|^{1/2}\leq\eta}} \mathcal{X}'_f(t) - \mathcal{X}'_f(s)) \\ &\leq 26 I_1\widetilde{\delta}^{1/2}\left(\log\frac{6I_2}{I_1\widetilde{\delta}^{1/2}\wedge\sqrt{2}I_2}\right)^{1/2}.\end{aligned}$$

Note that

$$\begin{aligned}I_2 &= [\![f]\!](T)\left(\sum V_j^2 \frac{[\![f]\!]^2(T-U_j)}{[\![f]\!]^2(T)}\right)^{1/2} \qquad (3.9) \\ &\leq [\![f]\!](T)\left(\sum V_j^2 \frac{[\![f]\!](T-U_j)}{[\![f]\!](T)}\right)^{1/2} \\ &= 2^{-1/2}[\![f]\!]^{1/2}(T) I_1.\end{aligned}$$

Using this and (3.7) we see that

$$E_\epsilon(\sup_{\substack{s,t\in[0,T]\\ \tau(s,t)\leq\delta}} \mathcal{X}'_f(t) - \mathcal{X}'_f(s)) \leq 26 I_1\left(\widetilde{\delta}\log\frac{6/\sqrt{2}}{\widetilde{\delta}^{1/2}[\![f]\!]^{-1/2}(T)\wedge 1}\right)^{1/2}. \qquad (3.10)$$

Note that by Khintchine's inequality $EI_1 \leq 2E_\epsilon|\mathcal{X}_{[\![f]\!]^{1/2}}(T)|$, see e.g. [2]. As at the end of the proof of Theorem 1.1 we argue that $E|\mathcal{X}_{[\![f]\!]^{1/2}}(T)| < \infty$. Thus, taking the expectation of (3.10) with respect to the probability space that supports $\{V_j\}$ and $\{U_j\}$ and noting the restriction on δ we get (1.14). □

Proof of Theorem 1.3 Using the representation of Y_3 given in (3.2) we see that

$$\|Y_3\|_\infty = \sum_{j=1}^{\infty} V_j^2 [\![f]\!]^2(T-U_j) \leq V_1^2[\![f]\!]^2(T) + \|\widetilde{Y}_3\|_\infty \qquad (3.11)$$

where $\widetilde{Y}_3(t) = \sum_{j=2}^{\infty} \widetilde{V}_j^2 [\![f]\!]^2(t - U_j)$ with $\widetilde{V}_j = \theta^{-1}(T^{-1}(\Gamma_j - \Gamma_1))$. Clearly $\|\widetilde{Y}_3\|_\infty$ has the same distribution as $\|Y_3\|_\infty$ but is independent of V_1. From (3.2) we also infer that $\Delta Y_3 \le V_1^2(\Delta[\![f]\!]^2) \le V_1^2[\![f]\!]^2(T)$. Thus

$$\begin{aligned}
\Delta Y_3 \log \frac{6\|Y_3\|_\infty^{1/2}}{(\Delta Y_3)^{1/2}} &\le \Delta Y_3 \log\left(6\frac{V_1[\![f]\!](T) + \|\widetilde{Y}_3\|_\infty^{1/2}}{(\Delta Y_3)^{1/2}}\right) \\
&\le V_1^2(\Delta[\![f]\!]^2)\log\left(6\frac{V_1[\![f]\!](T) + \|\widetilde{Y}_3\|_\infty^{1/2}}{V_1(\Delta[\![f]\!]^2)^{1/2}}\right)
\end{aligned}$$

because the function $x \mapsto x\log(6a^{1/2}x^{-1/2})$ is increasing for $x \in (0, a]$, $a > 0$.

Note that by Khintchine's inequality, $E\|\widetilde{Y}_3\|_\infty^{1/2} \le \sqrt{2}E|\mathcal{X}_{[\![f]\!]}(T)|$ and

$$\begin{aligned}
E|V_1| &\le E(\sum_j V_j^2)^{1/2} = E(E_\epsilon(\sum_j \epsilon_j V_j)^2)^{1/2} \\
&\le \sqrt{2}E|\sum_j \epsilon_j V_j| = \sqrt{2}E|Z(T)|.
\end{aligned} \tag{3.12}$$

Therefore, using Jensen's inequality twice, first on $\|\widetilde{Y}_3\|_\infty^{1/2}$ and then on V_1 we see that

$$\begin{aligned}
&E\left(\Delta Y_3 \log\frac{6\|Y_3\|_\infty^{1/2}}{(\Delta Y_3)^{1/2}}\right)^{1/2} \\
&\le (\Delta[\![f]\!]^2)^{1/2}E_{V_1}\left\{V_1\, E_{\widetilde{Y}_3}\left[\log\left(\frac{6[\![f]\!](T)}{(\Delta[\![f]\!]^2)^{1/2}} + \frac{6\|\widetilde{Y}_3\|_\infty^{1/2}}{V_1(\Delta[\![f]\!]^2)^{1/2}}\right)\right]^{1/2}\right\} \\
&\le (\Delta[\![f]\!]^2)^{1/2}E_{V_1}\left\{V_1\left[\log\left(\frac{6[\![f]\!](T)}{(\Delta[\![f]\!]^2)^{1/2}} + \frac{6\sqrt{2}E|\mathcal{X}_{[\![f]\!]}(T)|}{V_1(\Delta[\![f]\!]^2)^{1/2}}\right)\right]^{1/2}\right\} \\
&\le (\Delta[\![f]\!]^2)^{1/2}EV_1\left[\log\left(\frac{6[\![f]\!](T)}{(\Delta[\![f]\!]^2)^{1/2}} + \frac{6\sqrt{2}E|\mathcal{X}_{[\![f]\!]}(T)|}{EV_1(\Delta[\![f]\!]^2)^{1/2}}\right)\right]^{1/2} \\
&\le \sqrt{2}E|Z(T)|\,(\Delta[\![f]\!]^2)^{1/2}\left[\log\left(\frac{6[\![f]\!](T)}{(\Delta[\![f]\!]^2)^{1/2}} + \frac{6E|\mathcal{X}_{[\![f]\!]}(T)|}{E|Z(T)|(\Delta[\![f]\!]^2)^{1/2}}\right)\right]^{1/2}
\end{aligned} \tag{3.13}$$

In the application of Jensen's inequality on V_1 we used the fact that for each $a > e^{1/2}$ the function $x \mapsto x[\log(a + x^{-1})]^{1/2}$ is concave and increasing for $x > 0$.

By (3.1) and the fact that $[\![f]\!]$ is increasing we see that

$$\begin{aligned}
E|\mathcal{X}_{[\![f]\!]}(T)| &\le [\![f]\!](T)E(\sum_j V_j^2)^{1/2} \\
&\le \sqrt{2}[\![f]\!](T)E|Z(T)|
\end{aligned} \tag{3.14}$$

where, for the last inequality we use (3.12).

Substituting this into (3.13) and using Lemma 2.4 on the marginal Rademacher process gives (1.15). □

Proof of Theorem 1.4 As in the proof of Theorem 1.1 we represent $\mathcal{Y}_p$ as an integral with respect to a Poisson point process on $R^+ \times R$ with mean measure $\frac{1}{2} dt\theta(dx)$. Then one can write

$$\mathcal{Y}_p(t) = \sum \epsilon_j W_j p(t + X_j) \quad a.s. \tag{3.15}$$

where $\{X_k\}$ are independent identically distributed exponential random variables with mean one and $W_j = \theta^{-1}(e^{-X_j}\Gamma_j)$. (See [8]). $\{\epsilon_j\}$ is a Rademacher sequence, and the random sequences $\{X_j\}$, $\{\Gamma_j\}$, $\{\epsilon_j\}$ are independent of each other. By Fubini's theorem in (3.15)

$$\sum W_j^2 p^2(t + X_j) < \infty \qquad a.s. \tag{3.16}$$

Let

$$Y_4(t) = \sum W_j^2 [\![p[\![^2(t + X_j). \tag{3.17}$$

Similar to (2.15) we have that for $t \le v$

$$|p(t) - p(v)|^2 \le [\![p]\!]^2([t, v]) \le [\![p[\![^2(t) - [\![p[\![^2(v). \tag{3.18}$$

Therefore, similar to (3.3) we have, for $t \le s$, that

$$E_\epsilon(\mathcal{Y}_p(t) - \mathcal{Y}_p(s))^2 \le (Y_4(t) - Y_4(s))^{1/2}. \tag{3.19}$$

Similar to the proof of Theorem 1.1 we get

$$E_\epsilon \sup_{t \in R^+} \mathcal{Y}'_{[\![p[\![}(t) \le K_0 Y_4^{1/2}(0). \tag{3.20}$$

Taking the remaining expectations we get (1.19). □

Making obvious modifications in the proof of Theorem 1.2 we obtain Theorem 1.5.

Remark 3.1. Suppose that Z in (1.1) is Brownian motion. Then, for $u > v$,

$$\begin{aligned} E\left(\mathcal{X}_f(u) - \mathcal{X}_f(v)\right)^2 &= \int_v^u f^2(u-s)\,ds + \int_0^v (f(u-s) - f(v-s))^2\,ds \\ &\le \int_v^u [\![f]\!]^2(u-s)\,ds + \int_0^v ([\![f]\!]^2(u-s) - [\![f]\!]^2(v-s))\,ds \\ &= \int_0^u [\![f]\!]^2(u-s)\,ds - \int_0^v [\![f]\!]^2(v-s)\,ds. \end{aligned}$$

Therefore, the results in Section 2 apply with

$$Y(t) = \int_0^t [\![f]\!]^2(t-s)\,ds$$

in (2.5).

A similar observation applies when Z in (1.16) is Brownian motion.

4. Examples

Perhaps the most interesting application of these results is when the integrands in (1.1) and (1.16) are pure jump processes with the sum of the squares of their jumps finite. In this case we see that $\{\mathcal{X}_f(t), t \in [0,T]\}$ has a bounded version for all symmetric Lévy processes Z and when, in addition, (1.17) holds $\{\mathcal{Y}_f(t), t \in R^+\}$ has a bounded version for all symmetric Lévy processes Z.

A symmetric Lévy process on $[0,T]$ has the property that the sum of the squares of its jumps is finite. Therefore if Z_1 and Z_2 are independent Lévy process on $[0,T]$

$$\sup_{t\in[0,T]} \Big| \int_0^t Z_1(t-s)\, dZ_2(s) \Big| < \infty \qquad a.s. \tag{4.1}$$

Let $\{a_k\}$, $a_1 = 1$, be a sequence of real numbers that decreases to zero. Let $\{h(x), x \in [0,1]\}$, $h(0) = 0$, be an increasing function. Let $\{f(x), x \in [0,1]\}$ be the continuous function defined by: $f(a_j) = h(a_j)$, j odd; $f(a_j) = -h(a_j)$, j even and linear in between. Clearly

$$\sum_{j=1}^{\infty} h^2(a_j) \le [\![f]\!](1) \le 8 \sum_{j=1}^{\infty} h^2(a_j). \tag{4.2}$$

When $a_j = 1/j$ the sums in (4.2) are finite if and only if

$$\int_0^1 \frac{h^2(u)}{u^2}\, du < \infty. \tag{4.3}$$

When $a_j = e^{-j}$ the sums in (4.2) are finite if and only if

$$\int_0^1 \frac{h^2(u)}{u}\, du < \infty. \tag{4.4}$$

Note that for $|f|(1)$, the total variation of f form zero to one, to be finite we must replace h^2 by h in (4.3) and (4.4).

Consider the function f for which one or both of the integrals in (4.3) and (4.4) are finite. It follows from Theorem 1.1 that $\mathcal{X}_f = \{\mathcal{X}_f(t), t \in [0,T]\}$ in (1.1) has a continuous version for all independent increment processes Z with cádlág trajectories. One can obtain examples where the slope of f is extremely large near zero. This shows that uniform smoothness of f is not a necessary condition for $\mathcal{X}_f$ to be continuous.

5. Processes on R^d

At the conference Professor Werner Linde asked us if we had any results on the boundedness of Lévy sheets. We show here, using the technique of Gaussian randomization, that Lévy sheets are bounded almost surely.

Let M be a symmetric infinitely divisible independently scattered random measure on Borel subsets of $[0,1]^d$. The characteristic function of M can be written in the form

$$Ee^{i\lambda M(A)} = \exp\left(-\int_A \int_0^\infty (1-\cos(\lambda x))\,\theta(dx,s)m(ds)\right) \tag{5.1}$$

where m is a probability measure on $[0,1]^d$ and $\theta(\,\cdot\,,s)$ a measurable family of Lévy measures on R^+; see [7].

We consider the random field

$$I(\mathbf{t}) := I(t_1,\dots,t_d) = M([0,t_1]\times\cdots\times[0,t_d]) \qquad \mathbf{t}\in[0,1]^d. \tag{5.2}$$

Assume that $E|M([0,1]^d)|<\infty$. Let $I_j(t_j) := M([0,1]\times\cdots\times[0,t_j]\times\cdots\times[0,1])$ and note that I_j has independent increments. Therefore I_j has a version with cádlág trajectories and Lévy's inequality implies that $E\sup_{t\in[0,1]}|I_j(t)| \le 2E|I_j(1)| = 2E|M([0,1]^d)|$.

Lemma 5.1. *Let $I_j(t_j) := M([0,1]\times\cdots\times[0,t_j]\times\cdots\times[0,1])$. Then*

$$E\sup_{\mathbf{t}\in[0,1]^d} I(\mathbf{t}) \le 2\sqrt{2}\,d\,E|M([0,1]^d)|. \tag{5.3}$$

Proof. We give the proof in the case $d=2$. Let $I_{t_1,t_2} := I_{[0,t_1]\times[0,t_2]}$. It follows from [8] that $\{I(\mathbf{t}),\mathbf{t}\in[0,1]^2\}$ admits a series representation

$$I(\mathbf{t}) = \sum_{k=1}^\infty \epsilon_k V_k I_{t_1,t_2}(Z_{k,1},Z_{k,2})$$

where $\{(Z_{k,1},Z_{k,2})\}_{k\ge1}$ is an independent identically distributed sequence of random variables with the common distribution m, $\sum_k V_k^2<\infty$ almost surely and $\{\epsilon_k\}$ is a Rademacher sequence independent of $\{V_k,(Z_{k,1},Z_{k,2})\}$. Consider the Gaussian randomization $\widetilde{I}$ of I given by

$$\widetilde{I}(\mathbf{t}) = \sum_{k=1}^\infty \gamma_k V_k I_{t_1,t_2}(Z_{k,1},Z_{k,2})$$

where $\{\gamma_k\}$ is a standard normal sequence independent of $\{V_k,(Z_{k,1},Z_{k,2})\}$. We have

$$E_\gamma|\widetilde{I}(\mathbf{t})-\widetilde{I}(\mathbf{s})|^2 = \sum_{k=1}^\infty V_k^2|I_{t_1,t_2}(Z_{k,1},Z_{k,2}) - I_{s_1,s_2}(Z_{k,1},Z_{k,2})|^2. \tag{5.4}$$

Note that

$$\begin{aligned} |I_{t_1,t_2}-I_{s_1,s_2}|^2 &\le |I_{t_1,t_2}-I_{s_1,t_2}|^2 + |I_{s_1,t_2}-I_{s_1,s_2}|^2 \\ &\le |I_{t_1,1}-I_{s_1,1}|^2 + |I_{1,t_2}-I_{1,s_2}|^2. \end{aligned} \tag{5.5}$$

Let $\{\gamma_k^{(1)}\}$ and $\{\gamma_k^{(2)}\}$ be mutually independent copies of $\{\gamma_k\}$ which are also independent of $\{V_k, (Z_{k,1}, Z_{k,2})\}$. Consider the random field $\widetilde{J}(\mathbf{t}) := \widetilde{I}_1(t_1) + \widetilde{I}_2(t_2)$ where

$$\widetilde{I}_1(t) = \sum_{k=1}^{\infty} \gamma_k^{(1)} V_k I_{[0,t]}(Z_{k,1})$$

and

$$\widetilde{I}_2(t) = \sum_{k=1}^{\infty} \gamma_k^{(2)} V_k I_{[0,t]}(Z_{k,2}).$$

Equations (5.4) and (5.5) imply that

$$E_\gamma|\widetilde{I}(\mathbf{t}) - \widetilde{I}(\mathbf{s})|^2 \le E_\gamma|\widetilde{J}(\mathbf{t}) - \widetilde{J}(\mathbf{s})|^2.$$

Therefore, by the extended version of Slepian's Lemma; see e.g. Theorem 3.2.5, [3]

$$\begin{aligned} E_\gamma \sup_{\mathbf{t}\in[0,1]^2} \widetilde{I}(\mathbf{t}) &\le E_\gamma \sup_{\mathbf{t}\in[0,1]^2} |\widetilde{J}(\mathbf{t})| \qquad (5.6)\\ &\le E_\gamma \sup_{t\in[0,1]} |\widetilde{I}_1(t)| + E_\gamma \sup_{t\in[0,1]} |\widetilde{I}_2(t)|. \end{aligned}$$

Note that $\widetilde{I}_1$ and $\widetilde{I}_2$ are symmetric Lévy processes for which

$$\begin{aligned} E \sup_{t\in[0,1]} |\widetilde{I}_n(t)| &\le 2E|\widetilde{I}_n(1)| = 2EE_\gamma|\widetilde{I}_n(1)| \\ &= \sqrt{\frac{8}{\pi}} E(E_\gamma|\widetilde{I}_n(1)|^2)^{1/2} \le \frac{4}{\sqrt{\pi}} E|I_n(1)| \\ &= \frac{4}{\sqrt{\pi}} E|M([0,1]^d)|, \end{aligned}$$

$n = 1, 2$. (The last inequality follows from Khintchine's inequality.) Finally we note that

$$E_\epsilon \sup_{\mathbf{t}\in[0,1]^2} I(\mathbf{t}) \le \sqrt{\frac{\pi}{2}} E_\gamma \sup_{\mathbf{t}\in[0,1]^2} \widetilde{I}(\mathbf{t}). \qquad (5.7)$$

Substituting this into (5.6) and taking the remaining expectations completes the proof.

To obtain (5.7) we note that if $\zeta = \{\zeta_t, t \in T\}$ is a symmetric stochastic process

$$E\sup_{s,t}|\zeta_s - \zeta_t| = E\sup_{s,t}(\zeta_s - \zeta_t) = E(\sup_s \zeta_s + \sup_t(-\zeta_t)) = 2E\sup_s \zeta_s \qquad (5.8)$$

where the last step uses the fact that ζ is symmetric. Consequently, with ζ_t first equal to $\sum_{j=1}^{\infty} \gamma_j \phi_j(t)$ and then to $\sum_{j=1}^{\infty} \epsilon_j \phi_j(t)$, we see that

$$\begin{aligned}
2E\sup_t \sum_{j=1}^{\infty} \gamma_j \phi_j(t) &= E\sup_{s,t} |\sum_{j=1}^{\infty} \gamma_j(\phi_j(s) - \phi_j(t))| \\
&= E\sup_{s,t} |\sum_{j=1}^{\infty} \epsilon_j |\gamma_j|(\phi_j(s) - \phi_j(t))| \\
&\geq \sqrt{\frac{2}{\pi}} E\sup_{s,t} |\sum_{j=1}^{\infty} \epsilon_j(\phi_j(s) - \phi_j(t))| \\
&= 2\sqrt{\frac{2}{\pi}} E\sup_t \sum_{j=1}^{\infty} \epsilon_j \phi_j(t)
\end{aligned}$$

where $\{\phi_j(t)\}$ is any sequence of functions which are in ℓ^2 for each t. □

Acknowledgment. We are grateful to the referees for a very careful reading and for pointing out several errors and omissions.

References

[1] A.A. Balkema and L. De Haan. *Almost sure continuity of stable moving average processes with index less than one.* Annals Probab., **16**: 333–343, 1988.

[2] E. Giné and V. H. de la Peña, *Decoupling: from dependence to independence*, 1999, Springer-Verlag, New York.

[3] X. Fernique, *Fonctions Aléatoires Gaussiennes Vecteurs Aléatoires Gaussiennes*, 1997, CRM, Montreal.

[4] M. B. Marcus and J. Rosen, *Markov Processes and Gaussian processes*, book manuscript, in preparation.

[5] M.B. Marcus and J. Rosiński. *L^1–norms of stochastic integrals with respect to infinitely divisible random measures. Elect. Comm. in Probab.*, **6**:15–29, 2001.

[6] M. B. Marcus and J. Rosiński, *Continuity and boundedness of infinitely divisible processes: a Poisson point process approach*, submitted for publication

[7] B.S. Rajput and J. Rosiński. *Spectral representations of infinitely divisible processes.* Probab. Th. Rel. Fields, **82**: 451–487, 1989.

[8] J. Rosiński. *On path properties of certain infinitely divisible processes.* Stochastic Proc. Appl., **33**: 73–87, 1989.

[9] J. Rosiński. *Series representations of Lévy processes from the perspective of point processes*, in *Lévy Processes – Theory and Applications*, O.E. Barndorff-Nielsen et al (eds.), Birkhäuser, Boston, 401–415, 2001.

[10] H. Rootzén. *Extremes of moving averages of stable processes.* Annals Probab., **6**: 847–869, 1978.

Department of Mathematics, City College, New York, NY 10031, USA
E-mail address: `mbmarcus@earthlink.net`

Department of Mathematics, University of Tennessee, Knoxville, TN 39996, USA
E-mail address: `rosinski@math.utk.edu`

Progress in Probability, Vol. 56, 129–156

Notes on the Speed of Entropic Convergence in the Central Limit Theorem

Laurent Miclo

Abstract. In the context of the study of convergence speeds in the Central Limit Theorem, we investigate some consequences of a general Lipschitz contraction property of probability transition kernels with respect to relative entropy. This Markovian approach will enable us to discuss examples whose behavior is not covered by results recently obtained by several authors. More precisely, let $X_0, \ldots, X_n$ be IID real random variables, centered and normalized. It is known that if their law admits a positive spectral gap and a finite relative entropy with respect to ν, the standard Gaussian distribution, then the relative entropy of law of $(X_0 + \cdots + X_n)/\sqrt{n+1}$ with respect to ν goes to zero at least as $\mathcal{O}(1/(n+1))$, for large $n \in \mathbb{N}$. The two goals of this paper are: on one hand, for any fixed $p \in \mathbb{N}^*$, to find conditions insuring an entropic convergence faster than $\mathcal{O}(1/(n+1)^{p/2})$ and on the other hand to relax the spectral gap assumption, even at the cost of slower convergence bounds.

1. Introduction

The objective of this paper is to present a Markovian approach to the entropic convergence in Central Limit Theorem. Even if we are not yet completely satisfied with the results obtained, they enable to exhibit some interesting examples with respect to known behaviors.

We recall the existing results. Let $X_0, X_1, \ldots$ be a sequence of IID real random variables. We denote by μ their common law and we assume that it is centered and of variance 1. For $n \in \mathbb{N}$, consider m_n the law of $(X_0 + \cdots + X_n)/\sqrt{n+1}$. By Central Limit Theorem, for large $n \in \mathbb{N}$, the distribution m_n weakly converges to ν, the centered and normalized Gaussian law.
Under supplementary conditions, Barron & Johnson [14] and Artstein, Ball, Barthe & Naor [5, 3] studied quantitatively this convergence in a entropic sense. Recall

Received by the editors April 1, 2003.
2000 *Mathematics Subject Classification.* Primary 60F05; Secondary 60J05, 94A17, 49R50, 44A60.
Key words and phrases. Central Limit Theorem, Berry-Esséen type bounds, relative entropy, inhomogeneous Markovian chains, moments of Gibbs measures, modified logarithmic Sobolev inequalities, spectral gap, Fisher information, weighted Poincaré inequalities.

that the (relative) entropy of a probability measure m on $\mathbb{R}$ with respect to another one η is defined by

$$\mathrm{Ent}(m|\eta) \;:=\; \begin{cases} \int \ln\left(\frac{dm}{d\eta}\right)\, dm \;\leq\; +\infty, & \text{if } m \ll \eta, \\ +\infty, & \text{otherwise.} \end{cases}$$

Furthermore, we say that a law μ on $\mathbb{R}$ admits a spectral gap λ, if the quantity

$$\lambda \;:=\; \inf_{f\in\mathcal{C}_{\mathrm{b}}^1(\mathbb{R})\,:\, f\neq\mu[f],\, \mu-\text{a.s.}} \frac{\mu[(f')^2]}{\mu[(f-\mu[f])^2]}$$

is positive, where $\mathcal{C}_{\mathrm{b}}^1(\mathbb{R})$ is the set of $\mathcal{C}^1$ mappings from $\mathbb{R}$ to $\mathbb{R}$ which are bounded and whose derivative is equally bounded.

Theorem 1.1 ([14, 5, 3]). *With previous notations, assume that μ is of finite entropy with respect to ν and admits a spectral gap. Then there exists a constant $C_0 \geq 0$ (depending on $\mathrm{Ent}(\mu|\nu)$ and λ) such that for any $n \in \mathbb{N}$,*

$$\mathrm{Ent}(m_n|\nu) \;\leq\; \frac{C_0}{n+1}$$

and it was furthermore shown in [2] that the LHS is non-increasing with respect to the time parameter $n \in \mathbb{N}$.

One outcome of this theorem is that it enables us to strengthen Berry and Esséen bounds. Remember that one version (cf [12]) of this result asserts that if μ is supposed to admit a finite third absolute moment (namely $\int |x|^3\,\mu(dx) < +\infty$), then there exists a universal constant $0 < C_1 \leq 3$ such that for all $n \in \mathbb{N}$,

$$\sup_{x\in\mathbb{R}} |m_n[(-\infty,x]] - \nu[(-\infty,x]]| \;\leq\; \frac{C_1}{\sqrt{n+1}} \int |x|^3\,\mu(dx)$$

This order of convergence in n can be deduced from the above theorem, at least under the assumptions presented there. Indeed, this is a direct consequence of Csiszár–Kullback inequality (see e.g. [1]), saying that for any probabilities m, η, we are assured of

$$\|m-\eta\|_{\mathrm{tv}} \;\leq\; \sqrt{2}\sqrt{\mathrm{Ent}(m|\eta)}$$

where our convention for the total variation norm is that

$$\|m-\eta\|_{\mathrm{tv}} \;:=\; \sup_{f\in\mathcal{B}(\mathbb{R})\,:\,\|f\|_\infty\leq 1} |(m-\eta)[f]|$$

with $\mathcal{B}(\mathbb{R})$ the set of bounded measurable functions on $\mathbb{R}$ (but the previous bound is valid on any measurable space). In particular, we have

$$\begin{aligned} \sup_{x\in\mathbb{R}} |m_n[(-\infty,x]] - \nu[(-\infty,x]] \;&\leq\; \frac{\|m_n-\nu\|_{\mathrm{tv}}}{2} \\ &\leq\; \frac{1}{\sqrt{2}}\sqrt{\mathrm{Ent}(m_n|\nu)} \end{aligned}$$

Remark nevertheless that the spectral gap hypothesis implies that μ admits moments of all orders (even "small" exponential moments, see e.g. [15]), thus we are far away from the optimal validity condition of Berry and Esséen bounds, but it is also true that the total variation norm is much stronger than the supremum of the difference between distribution functions (consider for instance the situation where μ is a weighted sum of Dirac masses, then we have that $\mathrm{Ent}(m_n|\nu) = +\infty$ and $\|m_n - \nu\|_{\mathrm{tv}} = 2$ for all $n \in \mathbb{N}$, while Berry and Esséen bounds are valid).

Our purpose in this note is twofold: on one hand for any $p > 1$, we will find probabilities $\mu \neq \nu$ such that the corresponding quantity $\mathrm{Ent}(m_n|\nu)$ goes to zero faster than $1/(n+1)^p$ (Proposition 3.8) and on the other hand we will discuss the spectral gap hypothesis, showing ways to relax it in certain circumstances, in particular if one can content oneself with bad estimates of order $1/\sqrt{n+1}$ for the speed of convergence (Proposition 4.3). Finally let us mention that the ideas put forth here are relatively simple, so we hope numerous improvements are possible, see next section for a challenging direction.

2. A Markovian point of view

Our starting point is the observation that the sequence $((X_0+\cdots+X_n)/\sqrt{n+1})_{n\in\mathbb{N}}$ can be viewed as an inhomogeneous Markov chain. So one can try to apply general Markovian methods for the quantitative study of the relative entropy evolution to this particular context. Indeed, there are several possible choice for the reference measure; one could consider the instantaneous invariant probabilities (as it is customary in simulated annealing see for instance [17]) or the target probability ν (one can even think of intermediate solutions). We will follow here this latter approach, since we already have a lot of informations about the standard Gaussian law.

As in the introduction, let $X_0, \ldots, X_n, \ldots$ be a sequence of independent real random variables. For our purpose they no longer need to be identically distributed and for $n \in \mathbb{N}$, we will denote by μ_n the law of X_n, which we assume to be centered and of variance 1.
For fixed $n \in \mathbb{N}$, let K_{n+1} be the Markovian kernel describing the transition from $Y_n := (X_0 + \cdots + X_n)/\sqrt{n+1}$ to Y_{n+1}. More precisely, for any $x \in \mathbb{R}$, $K_{n+1}(x, \cdot)$ is the law of $\sqrt{\frac{n+1}{n+2}}x + \frac{X_{n+1}}{\sqrt{n+2}}$, so that this ensures:

$$\forall\, f \in \mathcal{B}(\mathbb{R}), \qquad \mathbb{E}[f(Y_{n+1})|Y_0, Y_1, \ldots, Y_n] \;=\; K_{n+1}[f](Y_n)$$

(as usual this equality is understood $\mathbb{P}$-a.s., with $\mathbb{P}$ the underlying probability).
From general results (cf [10]), we know that for every distribution m (and a priori for every probability ν, even if from now on, the latter will stand for the standard Gaussian law), we have

$$\mathrm{Ent}(mK_{n+1}|\nu K_{n+1}) \;\leq\; \mathrm{Ent}(m|\nu).$$

In our applications in this paper, we will only use this easy relation, but let us indicate how it can be quantified, since maybe the reader will be able to go further than us. Following [10], we first construct a "generalized" kernel $K^*_{n+1,\nu}$, namely a Markovian operator from $\mathbb{L}^1(\nu)$ to $\mathbb{L}^1(\nu K_{n+1})$, by requiring that

$$\forall\, f \in \mathbb{L}^1(\nu), \qquad K^*_{n+1,\nu}[f] \;:=\; \frac{d(f\nu)K_{n+1}}{d\nu K_{n+1}}$$

In order to compute it, let $g \in \mathcal{B}(\mathbb{R})$ be a test function. By definition and by an obvious change of variable, we have

$$\begin{aligned}
&((f\nu)K_{n+1})[g]\\
&= \int f(x)K_{n+1}[g](x)\,\nu(dx)\\
&= \int f(x)g\left(\sqrt{\frac{n+1}{n+2}}x + \frac{y}{\sqrt{n+2}}\right)\mu_{n+1}(dy)\nu(dx)\\
&= \int \mu_{n+1}(dy)\int dx\,\frac{\exp(-x^2/2)}{\sqrt{2\pi}}f(x)g\left(\sqrt{\frac{n+1}{n+2}}x + \frac{y}{\sqrt{n+2}}\right)\\
&= \int \mu_{n+1}(dy)\sqrt{\frac{n+2}{n+1}}\frac{1}{\sqrt{2\pi}}\int dx\exp\left(-\frac{n+2}{n+1}\frac{\left(x-y/\sqrt{n+2}\right)^2}{2}\right)\\
&\qquad f\left(\sqrt{\frac{n+2}{n+1}}\left(x-y/\sqrt{n+2}\right)\right)g(x)\\
&= \int dx\,F_n(x,f)g(x)
\end{aligned}$$

where for any $x \in \mathbb{R}$ and $f \in \mathbb{L}^1(\nu)$,

$$\begin{aligned}
F_n(x,f) \;:=\; &\sqrt{\frac{n+2}{n+1}}\frac{1}{\sqrt{2\pi}}\int \mu_{n+1}(dy)\exp\left(-\frac{n+2}{n+1}\frac{\left(x-y/\sqrt{n+2}\right)^2}{2}\right)\\
&f\left(\sqrt{\frac{n+2}{n+1}}\left(x-\frac{y}{\sqrt{n+2}}\right)\right)
\end{aligned}$$

(this expression is dx-a.s. finite), thus it appears that we can take

$$\begin{aligned}
K^*_{n+1,\nu}[f](x) &= \frac{F_n(x,f)}{F_n(x,\mathbb{1})}\\
&= \frac{1}{\varphi_{n+1}(x)}\int \mu_{n+1}(dy)\exp\left(\frac{\sqrt{n+2}}{n+1}xy - \frac{1}{2}\frac{y^2}{n+1}\right)\\
&\qquad f\left(\sqrt{\frac{n+2}{n+1}}\left(x-y/\sqrt{n+2}\right)\right)
\end{aligned}$$

with

$$\forall\, x \in \mathbb{R}, \qquad \varphi_{n+1}(x) \;=\; \mathbb{E}\left[\exp\left(\frac{\sqrt{n+2}}{n+1}xX_{n+1} - \frac{1}{2}\frac{X^2_{n+1}}{n+1}\right)\right]$$

In fact our main interest in kernel $K^*_{n+1,\nu}$ comes from the Dirichlet form $\mathcal{E}_{n+1,\nu}$ defined on $\mathbb{L}^2(\nu)$ by the following formula, for any $f, g \in \mathbb{L}^2(\nu)$,

$$\begin{aligned}
&\mathcal{E}_{n+1,\nu}(f,g)\\
&\quad := \nu[f(\mathrm{Id} - K_{n+1}K^*_{n+1,\nu})[g]]\\
&\quad = \int \nu(dx)\, f(x)(g(x) - K_{n+1}K^*_{n+1,\nu}[g](x))\\
&\quad = -\int \nu(dx)\, f(x)K_{n+1}K^*_{n+1,\nu}[g-g(x)](x)\\
&\quad = -\int \nu(dx)\mu_{n+1}(dy)\, f(x)K^*_{n+1,\nu}[g-g(x)]\left(\sqrt{\frac{n+1}{n+2}}x + \frac{y}{\sqrt{n+2}}\right)
\end{aligned}$$

This leads us to calculate for any $x, y \in \mathbb{R}$,

$$\begin{aligned}
&K^*_{n+1,\nu}[g-g(x)]\left(\sqrt{\frac{n+1}{n+2}}x + \frac{y}{\sqrt{n+2}}\right)\\
&\quad = \frac{1}{\varphi_{n+1}\left(\sqrt{\frac{n+1}{n+2}}x + \frac{y}{\sqrt{n+2}}\right)}\int \mu_{n+1}(dz)\exp\left(-\frac{z^2}{2(n+1)} + \frac{xz}{\sqrt{n+1}} + \frac{yz}{n+1}\right)\\
&\qquad\quad \times\left(g\left(x + y/\sqrt{n+1} - z/\sqrt{n+1}\right) - g(x)\right)
\end{aligned}$$

This suggests to consider the change of variable where x is replaced by $x - y/\sqrt{n+1}$, since then we end up with a rather symmetrical expression:

$$\begin{aligned}
&\mathcal{E}_{n+1,\nu}(f,g)\\
&\quad = \int \nu(dx)\mu_{n+1}(dy)\mu_{n+1}(dz)\frac{\exp\left(\frac{x(y+z)}{\sqrt{n+1}} - \frac{y^2}{2(n+1)} - \frac{z^2}{2(n+1)}\right)}{\varphi_{n+1}\left(\sqrt{\frac{n+1}{n+2}}x\right)} f(x - y/\sqrt{n+1})\\
&\qquad\quad \times(g(x - y/\sqrt{n+1}) - g(x - z/\sqrt{n+1}))\\
&\quad = \frac{1}{2}\int \nu(dx)\mu_{n+1}(dy)\mu_{n+1}(dz)\frac{\exp\left(\frac{x(y+z)}{\sqrt{n+1}} - \frac{y^2}{2(n+1)} - \frac{z^2}{2(n+1)}\right)}{\varphi_{n+1}\left(\sqrt{\frac{n+1}{n+2}}x\right)}\\
&\qquad \times(f(x - y/\sqrt{n+1}) - f(x - z/\sqrt{n+1}))(g(x - y/\sqrt{n+1}) - g(x - z/\sqrt{n+1})).
\end{aligned}$$

So we are led to introduce for $n \in \mathbb{N}$ and $x \in \mathbb{R}$, the distribution $M_{n+1,x}$ defined on $\mathbb{R}$ by

$$\begin{aligned}
&M_{n+1,x}[h]\\
&\quad := \frac{1}{\varphi_{n+1}\left(\sqrt{\frac{n+1}{n+2}}x\right)}\int \mu_{n+1}(dy)\exp\left(\frac{xy}{\sqrt{n+1}} - \frac{y^2}{2(n+1)}\right)h(x - y/\sqrt{n+1})
\end{aligned}$$

with $h \in \mathcal{B}(\mathbb{R})$ a test function, because we can rewrite

$$\mathcal{E}_{n+1,\nu}(f,g) = \frac{1}{2}\int \nu(dx)\varphi_{n+1}\left(\sqrt{\frac{n+1}{n+2}}x\right)\mathrm{Cov}(f,g;M_{n+1,x})$$

where $\mathrm{Cov}(f,g;M_{n+1,x})$ denote the covariance of f and g with respect to $M_{n+1,x}$. In particular, the modified logarithmic Sobolev constant α_n associated to ν and $\mathcal{E}_{n+1,\nu}$,

$$\alpha_n := \inf_{f\in\mathbb{L}^2(\nu)\setminus\mathrm{Vect}(\mathbf{1})} \frac{\mathcal{E}_{n+1,\nu}(f^2,\ln(f^2))}{\mathrm{Ent}(f^2;\nu)}$$

where $\mathrm{Ent}(f^2;\nu)$ is an abbreviation for $\mathrm{Ent}(f^2\nu/\nu[f^2]|\nu)$, can be expressed as

$$\begin{aligned}\alpha_n &= \frac{1}{2}\int \nu(dx)\varphi_{n+1}\left(\sqrt{\frac{n+1}{n+2}}x\right)\mathrm{Cov}(f^2,\ln(f^2);M_{n+1,x})\\ &\geq \inf_{f\in\mathbb{L}^2(\nu)\setminus\mathrm{Vect}(\mathbf{1})} \frac{\int \nu(dx)\varphi_{n+1}\left(\sqrt{\frac{n+1}{n+2}}x\right)\mathrm{Ent}(f^2;M_{n+1,x})}{\mathrm{Ent}(f^2;\nu)}\end{aligned}$$

(where we have used Jensen's inequality $\int \ln(f^2)\,dM_{n+1,x} \leq \ln\left(\int f^2\,dM_{n+1,x}\right)$ to traditionally deduce that $\mathrm{Cov}(f^2,\ln(f^2);M_{n+1,x}) \geq 2\mathrm{Ent}(f^2;M_{n+1,x})$).
The role of this ergodic coefficient α_n is particularly important, since we have shown in [10] that there exists a universal constant $0<\rho\leq 1$ such that

$$\mathrm{Ent}(mK_{n+1}|\nu K_{n+1}) \leq (1-\rho\alpha_n)\mathrm{Ent}(m|\nu) \tag{2.1}$$

Unfortunately, except in the case where $\mu_{n+1}=\nu$ (or more generally if $d\mu_{n+1}/d\nu$ is bounded above and below by positive constants), we don't know how to estimate α_n! Nevertheless, we remark that if $\mu_{n+1}=\nu$, then this ensures:

$$\mathrm{Ent}(mK_{n+1}|\nu K_{n+1}) = \mathrm{Ent}(mK_{n+1}|\nu) \leq \left(1-\frac{1}{n+2}\right)\mathrm{Ent}(m|\nu) \tag{2.2}$$

as it can be proved by directly using a classical Ornstein-Uhlenbeck process, which permits to go from m to mK_{n+1} in a continuous time interval of length $\ln((n+2)/(n+1))$ (cf also [10]).

Coming back to the general situation, we note that

$$\mathrm{Ent}(mK_{n+1}|\nu) = \mathrm{Ent}(mK_{n+1}|\nu K_{n+1}) + \int \ln\left(\frac{d\nu K_{n+1}}{d\nu}\right)dm_n$$

and we compute that ν-a.s. for $x\in\mathbb{R}$,

$$\frac{d\nu K_{n+1}}{d\nu}(x) = \sqrt{\frac{n+2}{n+1}}\exp\left(-\frac{x^2}{2(n+1)}\right)\varphi_{n+1}(x)$$

Taking into account that for each $n\in\mathbb{N}$, m_n is of variance 1, we get that

$$\mathrm{Ent}(m_{n+1}|\nu) \leq \mathrm{Ent}(m_n|\nu)+\frac{1}{2}\ln\left(\frac{n+2}{n+1}\right)-\frac{1}{2(n+1)}+\int \ln(\varphi_{n+1})\,dm_{n+1} \tag{2.3}$$

and our task in next section will be to evaluate this last term.

Remarks 2.1. a) The fact that the sequence $(\text{Ent}(m_n|\nu))_{n\geq 0}$ is non-increasing when all the distributions μ_n, $n \in \mathbb{N}$, are equal seems difficult to deduce from (2.3). Of course in general, when these laws are different, this monotonicity property is wrong, consider for instance the cases where $\mu_0 = \nu \neq \mu_1$.
b) The universal constant ρ appears in a very bad place in (2.1). Indeed, let be given a sequence $(E_n)_{n\in\mathbb{N}}$ of nonnegative reals verifying the inequalities

$$\forall\, n \in \mathbb{N}, \qquad E_{n+1} \leq (1 - a/(n+1))E_n + b/(n+1)^c \tag{2.4}$$

where $a, b, c > 0$ are fixed. Then by analogy with the corresponding differential inequality, it can be shown that there exists a constant $C > 0$ such that for all $n \in \mathbb{N}$, we are assured of

$$\begin{aligned} c - a < 1 \quad &\Rightarrow \quad E_n \leq \frac{C}{(n+1)^{c-1}} \\ c - a = 1 \quad &\Rightarrow \quad E_n \leq \frac{C\ln(n+2)}{(n+1)^a} \\ c - a > 1 \quad &\Rightarrow \quad E_n \leq \frac{C}{(n+1)^a} \end{aligned}$$

(and starting from the opposite inequalities in (2.4), one has similar reversed bounds, see for instance appendix A of [11]) so the coefficient a is quite crucial for the asymptotic behavior of the sequence $(E_n)_{n\in\mathbb{N}}$.
c) Nevertheless, we note that for fixed $x \in \mathbb{R}$ and $f \in \mathcal{C}_\mathrm{b}^1(\mathbb{R})$, we have, if f takes values in some compact subset of $(0, +\infty)$,

$$\begin{aligned} \lim_{n\to\infty} (n+1)\,\text{Cov}(f^2, \ln(f^2); M_{n+1,x}) \quad &= \quad 4(f'(x))^2 \\ \lim_{n\to\infty} \varphi_{n+1}\left(\sqrt{\frac{n+1}{n+2}}x\right) \quad &= \quad 1 \end{aligned}$$

so heuristically, we hope that for large $n \in \mathbb{N}$,

$$\begin{aligned} \alpha_n \quad &\sim \quad \frac{2}{n+1} \inf_{f\in\mathbb{L}^2(\nu)\setminus\text{Vect}(\mathbb{1})} \frac{\int (f'(x))^2\,\nu(dx)}{\text{Ent}(f^2;\nu)} \\ &= \quad \frac{1}{n+1} \end{aligned}$$

where for the last identity we have used the well-known fact that the logarithmic Sobolev constant associated to ν is $1/2$ (see for instance [1]).
More precisely, this expected behavior leads us to conjecture that under nice conditions on the family $(\mu_n)_{n\geq 0}$ (a uniform spectral gap assumption for instance?), we have directly for large $n \in \mathbb{N}$,

$$\inf_{f\in\mathbb{L}^2(\nu)\setminus\text{Vect}(\mathbb{1})} \frac{\text{Ent}(f^2;\nu) - \text{Ent}(K^*_{n+1,\nu}[f^2];\nu K_{n+1})}{\text{Ent}(f^2;\nu)} \quad \sim \quad \frac{1}{n+1}$$

(in [10] we have shown that the LHS always belongs to the interval $[\rho\alpha_n, \alpha_n]$ and this the reason of the apparition of the universal constant ρ in (2.1)), namely (2.2) would be asymptotically almost satisfied.

Of course, such a result would greatly help our approach of the subject, since via the difference inequalities of previous remark (b) and the considerations of next sections, we would end up with bounds close to that of Theorem 1.1, under appropriate assumptions (but surprisingly, it would not be possible by this method to get a convergence speed estimate better than $\mathcal{O}(1/(n+1))$). We hope to be able to develop such a study in the future.

Finally, let us remark that it is not really necessary to consider all the functions $f \in \mathbb{L}^2(\nu) \setminus \mathrm{Vect}(\mathbb{1})$ in the above infima, since we only need the corresponding inequalities with $f = f_n := \sqrt{dm_n/d\nu}$ and one can already have at his disposal some informations on this function (for instance a uniform spectral gap assumption on the family $(\mu_p)_{p\in\mathbb{N}}$ implies the same property for the family $(m_p)_{p\in\mathbb{N}} = (f_p^2\nu)_{p\in\mathbb{N}}$, see e.g. next section).

3. Examples of "fast" convergence

We will be interested here in properties of (subclasses of) $\mathcal{N}_p$, for $p \in \mathbb{N}$, the set of probability measures on $\mathbb{R}$ whose p first moments coincide with those of ν (by convention the moment of order 0 is the total mass, so $\mathcal{N}_0$ is just the set of all probabilities on $\mathbb{R}$).

These sets are quite natural in our setting, for instance our basic assumption in last section was that for every $n \in \mathbb{N}$, $\mu_n \in \mathcal{N}_2$. Then all the distributions m_n, $n \in \mathbb{N}$, also belong to $\mathcal{N}_2$. This kind of stability by appropriately weighted convolution is a general fact for the $\mathcal{N}_p$, $p \in \mathbb{N}$:

Lemma 3.1. *Let $p \in \mathbb{N}$, and $0 \le t \le 1$ be fixed. If we are given two probabilities $\mu, \mu' \in \mathcal{N}_p$, then we are assured that $m \in \mathcal{N}_p$, where m is the probability defined by*

$$\forall\, f \in \mathcal{B}(\mathbb{R}), \qquad m[f] \;=\; \int f(tx + \sqrt{1-t^2}y)\, \mu(dx)\mu'(dy).$$

Proof. Let $k \in \mathbb{N}$, $0 \le k \le p$, be given. We compute that

$$\begin{aligned}
\int x^k\, m(dx) &= \int (tx + \sqrt{1-t^2}y)^k\, \mu(dx)\mu'(dy) \\
&= \int \sum_{0\le l\le k} \binom{k}{l} t^l x^l (1-t^2)^{\frac{k-l}{2}} y^{k-l}\, \mu(dx)\mu'(dy) \\
&= \sum_{0\le l\le k} \binom{k}{l} t^l (1-t^2)^{\frac{k-l}{2}} \int x^l\, \mu(dx) \int y^{k-l}\, \mu'(dy) \\
&= \sum_{0\le l\le k} \binom{k}{l} t^l (1-t^2)^{\frac{k-l}{2}} \int x^l\, \nu(dx) \int y^{k-l}\, \nu(dy)
\end{aligned}$$

$$= \int (tx + \sqrt{1-t^2}y)^k \, \nu(dx)\nu(dy)$$
$$= \int x^k \, \nu(dx).$$
□

In particular, if for some $p \in \mathbb{N}$, we assume that for all $n \in \mathbb{N}$, $\mu_n \in \mathcal{N}_p$, then we also end up with $m_n \in \mathcal{N}_p$ for all $n \in \mathbb{N}$.
As mentioned at the end of last section, the spectral gap is also "preserved" by this kind of operation. In a certain manner, this observation (applied to discrete "carrés du champs") was at the heart of celebrated Gross' proof [13] of the logarithmic Sobolev inequality for the standard normal distribution.

Lemma 3.2. *Let $t \geq 0$ and two probabilities μ and μ' be given and define m as in the previous lemma. If we assume that μ and μ' admit respectively as spectral gaps $\lambda > 0$ and $\lambda' > 0$, then m also satisfies such an inequality and its spectral gap is larger than $\lambda \wedge \lambda'$.*

Proof. It is well-known (cf. for instance [1]) that $\mu \otimes \mu'$ admits a spectral gap larger than $\lambda \wedge \lambda'$, in the sense that

$$\forall\, g \in \mathcal{C}_b^1(\mathbb{R}^2), \qquad (\lambda \wedge \lambda')\mu \otimes \mu'[(g - \mu \otimes \mu'[g])^2] \;\leq\; \mu \otimes \mu'[(\partial_1 g)^2 + (\partial_2 g)^2]$$

(where ∂_1 and ∂_2 designate the partial derivatives with respect to the first and second variables). Let a function $f \in \mathcal{C}_b^1(\mathbb{R})$ be given and consider the mapping $g \in \mathcal{C}_b^1(\mathbb{R}^2)$ defined by

$$\forall\, (x,y) \in \mathbb{R}^2, \qquad g(x,y) \;:=\; f(tx + \sqrt{1-t^2}y).$$

Clearly, we have

$$\begin{aligned} \mu \otimes \mu'[g] &= m[f] \\ \mu \otimes \mu'[(g - \mu \otimes \mu'[g])^2] &= m[(f - \mu[f])^2] \end{aligned}$$

and since

$$\forall\, (x,y) \in \mathbb{R}^2, \qquad (\partial_1 g)^2(x,y) + (\partial_2 g)^2(x,y) \;=\; (f'(tx + \sqrt{1-t^2}y))^2$$

we are also assured of $\mu \otimes \mu'[(\partial_1 g)^2 + (\partial_2 g)^2] = m[(f')^2]$. Thus it appears that

$$(\lambda \wedge \lambda')\, m[(f - \mu[f])^2] \;\leq\; m[(f')^2]$$

and the above lemma follows at once. □

These two invariance properties lead us to introduce for any $p \in \mathbb{N}$ and any $\lambda > 0$, the class $\mathcal{N}_p(\lambda)$ of elements from $\mathcal{N}_p$ with a spectral gap larger than λ.
Our main task in this section will be to prove the following result which will be fundamental for our future estimations.

Proposition 3.3. *Let $p \in \mathbb{N}$ and $\lambda > 0$ be fixed. With the notations of the previous section, assume that all the distributions μ_n, $n \in \mathbb{N}$, belong to $\mathcal{N}_p(\lambda)$, then there exists a constant $C > 0$ (only depending on p and λ) such that*

$$\forall\, n \in \mathbb{N}, \left| \frac{1}{2} \ln\left(\frac{n+2}{n+1}\right) - \frac{1}{2(n+1)} + \int \ln(\varphi_{n+1})\, dm_{n+1} \right| \;\leq\; \frac{C}{(n+1)^{(p+1)/2}}.$$

The proof of this bound is based on classical Taylor expansions (thus in some sense, we are only recycling the idea underlying the simple proof of the Central Limit Theorem via characteristic functions), but we will take some care in justifying them in the next string of technical lemmas. We begin by introducing some notation.

Let a probability μ be fixed, we define for any $0 \leq t \leq 1$ and any $x, y \in \mathbb{R}$,

$$\begin{aligned} h(t) &:= t\sqrt{1+t^2} \\ U_t(x,y) &= h(t)xy - t^2\frac{y^2}{2} \\ F_x(t) &:= \ln\left(\int \exp(U_t(x,y))\, d\mu\right). \end{aligned}$$

The parameter t should be think of as $1/\sqrt{n+1}$, since one would have noticed that if $\mu = \mu_{n+1}$, then we get by definition,

$$\forall\, x \in \mathbb{R}, \qquad \varphi_{n+1}(x) \quad = \quad F_x\left(\frac{1}{\sqrt{n+1}}\right).$$

So we are interested in differentiating $F_x(t)$ with respect to small t to obtain, for large $n \in \mathbb{N}$, appropriate expansions of the expression considered in Proposition 3.3. Formally it is not very difficult, and the Gibbs probability

$$\mu_{t,x}(dy) \quad := \quad \frac{\exp(U_t(x,y))\,\mu(dy)}{\mu[\exp(U_t(x,\cdot))]} \tag{3.1}$$

appears to have a promising role (equally note that the probability $M_{n+1,x}$ introduced in the previous section can then be written $\mu_{\frac{1}{\sqrt{n+1}},\sqrt{\frac{n+1}{n+2}}x}$, if $\mu = \mu_{n+1}$). For instance and at least heuristically, we get

$$\partial_t F_x(t) \quad = \quad \int \partial_t U_t(x,y)\,\mu_{t,x}(dy).$$

Indeed this inequality is correct; the usual rule of differentiation under the integral is fulfilled, since we check that for any $0 \leq t \leq 1$ and $x, y \in \mathbb{R}$,

$$\begin{aligned} |\partial_t U_t(x,y)\exp(U_t(x,y))| &\leq \left|\frac{1+2t^2}{\sqrt{1+t^2}}xy - ty^2\right| \exp((1+t^2)x^2/2) \\ &\leq (3\,|x|\,|y| + y^2)\exp(x^2) \end{aligned}$$

which is integrable in y with respect to μ. Nevertheless, we need better estimations of $\partial_t F_x(t)$ than those deduced from this bound and the Jensen inequality $\mu[\exp(U_t(x,\cdot)] \geq \exp(\mu[U_t(x,\cdot)])$, because we shall rather differentiate in t integrals of $F_x(t)$ with respect to certain distributions of x (which will not necessarily integrate expression like $\exp(\epsilon x^2)$, for any $\epsilon > 0$; recall that the typical example of a probability on $\mathbb{R}$ having a spectral gap is the exponential law on $\mathbb{R}_+$). Before working in this direction, let us recall a general result, in fact valid on any measurable space.

Lemma 3.4. *Let η be a probability and V be a non-negative measurable function. Then for any $q \geq 0$, we have*

$$\int V^q \frac{\exp(-V)}{Z} d\eta \leq \int V^q \, d\eta$$

where $Z := \int \exp(-V)\, d\eta$ is the normalizing constant.

Proof. For $s > 0$, let $Z_s := \int \exp(-sV)\, d\eta$ and define η_s as the probability $\exp(-sV)\eta/Z_s$. Without any difficulty, we compute that for $s > 0$,

$$\begin{aligned}
\partial_s \int V^q \, d\eta_s &= -\int V^{q+1}\, d\eta_s + \int V^q \, d\eta_s \int V \, d\eta_s \\
&\leq -\int V^{q+1}\, d\eta_s + \left(\int V^{q+1}\, d\eta_s\right)^{q/(q+1)} \left(\int V^{q+1}\, d\eta_s\right)^{1/(q+1)} \\
&= 0.
\end{aligned}$$

So for any $0 < u \leq s$, we have

$$\int V^q \, d\eta_s \leq \int V^q \, d\eta_u$$

and the RHS is converging to $\eta[V^q] \leq +\infty$ when u goes to 0_+. □

This simple bound will be quite useful to deduce the next crucial one:

Lemma 3.5. *Assume that μ admits a spectral gap $\lambda > 0$, then for any $q \in \mathbb{N}$, there exists a finite constant $C(\lambda, q)$ such that*

$$\forall\, 0 \leq t \leq 1, \forall\, x \in \mathbb{R}, \qquad \int |y|^q \, \mu_{t,x}(dy) \leq C(\lambda, q)(1 + |x|^{3q}).$$

Proof. By classical approximation results, the bound

$$\int f^2 \, d\mu \leq \left(\int f \, d\mu\right)^2 + \lambda^{-1} \int (f')^2 \, d\mu$$

is extended to any function f of class $\mathcal{C}^1$ on $\mathbb{R}$, by allowing that the RHS can be infinite.
For fixed $t \geq 0$, $x \in \mathbb{R}$ and $p \in \mathbb{N}^*$, we apply this inequality with the mapping f defined by

$$f : \mathbb{R} \ni y \mapsto y^p \exp(U_t(x,y)/2)$$

so, after dividing by $\int \exp(U_t(x,y))\, d\mu$, we obtain that

$$\begin{aligned}
&\int y^{2p} \, \mu_{t,x}(dy) \\
&\leq \frac{\left(\int y^p \exp(U_t(x,y)/2)\, \mu(dy)\right)^2}{\int \exp(U_t(x,y))\, d\mu} + \int (py^{p-1} + y^p(h(t)x - t^2 y)/2)^2 \, \mu_{t,x}(dy) \\
&\leq \int y^{2p} \, \mu(dy) + 2p^2 \int y^{2p-2} \, \mu_{t,x}(dy) + \frac{1}{2} \int y^{2p}(h(t)x - t^2 y)^2 \, \mu_{t,x}(dy).
\end{aligned} \tag{3.2}$$

The two first terms of the RHS are quite easy to dispose of: due the spectral gap inequality verified by μ, we know there exists a finite constant $C_1(\lambda, 2p)$ bounding $\int y^{2p}\,\mu(dy)$ independently of such μ. In other respects, there exists a finite constant $C_2(\lambda, 2p)$ such that

$$\forall\, y \in \mathbb{R}, \qquad y^{2p-2} \;\leq\; C_2(\lambda, 2p) + \frac{1}{6p^2}y^{2p},$$

relation implying that

$$2p^2 \int y^{2p-2}\,\mu_{t,x}(dy) \;\leq\; 2p^2 C_2(\lambda, 2p) + \frac{1}{3}\int y^{2p}\,\mu_{t,x}(dy).$$

To treat the last term of (3.2), we rewrite it as

$$\begin{aligned}
&\frac{t^2}{2}\int y^{2p}(\sqrt{1+t^2}x - ty)^2\,\mu_{t,x}(dy)\\
&\leq \frac{t}{2}\int |y|^{2p-1}(\sqrt{1+t^2}\,|x| + |\sqrt{1+t^2}x - ty|)(\sqrt{1+t^2}x - ty)^2\,\mu_{t,x}(dy)\\
&\leq \frac{h(t)}{2}\,|x|\int |y|^{2p-1}(\sqrt{1+t^2}x - ty)^2\,\mu_{t,x}(dy)\\
&\quad + \frac{t}{2}\int |y|^{2p-1}\left|\sqrt{1+t^2}x - ty\right|^3\,\mu_{t,x}(dy).
\end{aligned} \tag{3.3}$$

Noting that we also have for any $t \geq 0$ and $x \in \mathbb{R}$,

$$\mu_{t,x}(dy) \;=\; \frac{\exp(-(\sqrt{1+t^2}x - ty)^2)\mu(dy)}{\int \exp(-(\sqrt{1+t^2}x - tz)^2)\,\mu(dz)}$$

we are led to apply twice Lemma 3.4 with reference probability and potential

$$\begin{aligned}
\eta(dy) &:= \frac{|y|^{2p-1}\,\mu(dy)}{\int |z|^{2p-1}\,\mu(dz)}\\
V(y) &:= (\sqrt{1+t^2}x - ty)^2
\end{aligned}$$

and respectively with $q = 1$ and $q = 3/2$. Thus we get for instance that for $0 \leq t \leq 1$,

$$\begin{aligned}
&\frac{h(t)}{2}\,|x|\int |y|^{2p-1}(\sqrt{1+t^2}x - ty)^2\,\mu_{t,x}(dy)\\
&\leq\; \frac{1}{\sqrt{2}}\,|x|\int |y|^{2p-1}(\sqrt{1+t^2}x - ty)^2\,\mu(dy)\int |y|^{2p-1}\,\mu_{t,x}(dy)\\
&\leq\; C_3(\lambda, p)(1 + |x|^3)\left(\int |y|^{2p}\,\mu_{t,x}(dy)\right)^{1-1/(2p)}
\end{aligned}$$

where $C_3(\lambda, p)$ is an appropriate constant (once again we have used that the quantity $\int |y|^{2p+3}\,\mu(dy)$ is uniformly bounded over probabilities μ with a spectral gap larger than λ). Now using a Young relation, we can find another finite constant $C_4(\lambda,p)$ such that the last RHS is bounded by

$$C_4(\lambda,p)(1+x^{6p})+\frac{1}{6}\int |y|^{2p}\,\mu_{t,x}(dy).$$

We can proceed in a similar way with the term $t\int |y|^{2p-1}\,|\sqrt{1+t^2}x-ty|^3\mu_{t,x}(dy)/2$ and combining all these estimates we end up with the bound stated in the above lemma if $q=2p$. The general case follows by suitable Hölder inequalities. □

In particular, these computations show that under the hypothesis of previous lemma, one can find a finite constant $C(\lambda)$ such that for any $0\le t\le 1$ and $x\in\mathbb{R}$,

$$\begin{aligned}|\partial_t F_x(t)| &= \left|\int \frac{1+2t^2}{\sqrt{1+t^2}}xy-ty^2\,\mu_{t,x}(dy)\right|\\ &\le 3\,|x|\int |y|\,\mu_{t,x}(dy)+\int y^2\,\mu_{t,x}(dy)\\ &\le C(\lambda)(1+x^6).\end{aligned}$$

The next result generalizes this kind of bound:

Lemma 3.6. *For any fixed $x\in\mathbb{R}$, the mapping $[0,1]\ni t\mapsto F_x(t)$ belongs to $\mathcal{C}^\infty([0,1])$ and if we assume that μ admits a spectral gap $\lambda>0$, then for any given $p\in\mathbb{N}^*$, there exists another finite constant $C(\lambda,p)$ such that*

$$\forall\, 0\le t\le 1,\ \forall\, x\in\mathbb{R},\qquad |\partial_t^p F_x(t)|\ \le\ C(\lambda,p)(1+x^{6p}).$$

Proof. If V is a polynomial function in one variable, it is not difficult to justify the following differentiation under the integral, as in the discussion before Lemma 3.4, for any $0\le t\le 1$ and $x\in\mathbb{R}$,

$$\begin{aligned}\partial_t\int V(y)\,\mu_{t,x}(dy) &= \int V(y)\partial_t U(x,y)\,\mu_{t,x}(dy)\\ &\quad-\int V(y)\,\mu_{t,x}(dy)\int \partial_t U_t(x,y)\,\mu_{t,x}(dy)\end{aligned}$$

so taking into account that h belongs to $\mathcal{C}^\infty(\mathbb{R}_+)$, it appears easily that $[0,1]\ni t\mapsto F_x(t)$ is equally of class $\mathcal{C}^\infty$. Indeed, if $p\in\mathbb{N}^*$ is given, $\partial_t^p F_x(t)$ appears as a weighted sum of products of expressions like

$$\int \partial_t^{\alpha_0}U_t(x,y)\partial_t^{\alpha_1}U_t(x,y)\cdots\partial_t^{\alpha_r}U_t(x,y)\,\mu_{t,x}(dy)$$

where $\alpha_0\ge\alpha_1\ge\cdots\ge\alpha_r>0$ are $r+1$ nonnegative integers. More precisely, let us denote $H_\alpha(t,x)$ this integral, where $\alpha=(\alpha_i)_{i\in\mathbb{N}}$ is a multi-index whose entries are non-increasing elements of $\mathbb{N}$ and $\alpha_l=0$ for $l>r$. Let $\mathcal{A}$ be the set of all such sequences (with varying $r\in\mathbb{N}\sqcup\{-1\}$, $r=-1$ corresponds to the element of $\mathcal{A}$ whose all entries are 0 and which we will also designate by 0, by traditional conventions $H_0=\mathbf{1}$) and for $\alpha=(\alpha_i)_{i\in\mathbb{N}}\in\mathcal{A}$, we note $|\alpha|=\sum_{i\in\mathbb{N}}\alpha_i\in\mathbb{N}$ (α is then sometimes called a partition of $|\alpha|$).

Let us go one step further and iterate this construction. First we put on $\mathcal{A}$ the lexicographical total order, namely for two given elements $\alpha = (\alpha_i)_{i\in\mathbb{N}}$ and $\alpha' = (\alpha'_i)_{i\in\mathbb{N}}$ of $\mathcal{A}$, we say that $\alpha \geq \alpha'$ if there exists $i_0 \in \mathbb{N} \sqcup \{\infty\}$ such that $\alpha_i = \alpha'_i$ for any $0 \leq i < i_0$ and $\alpha_{i_0} > \alpha'_{i_0}$ (of course this condition is void if $i_0 = \infty$, or equivalently if $\alpha = \alpha'$). Next we consider $\mathbb{A}$ the set of all sequences of non-increasing elements of $\mathcal{A}$ which are null after some rank. As before, the height of an element $A = (A_i)_{i\in\mathbb{N}} \in \mathbb{A}$ is the nonnegative integer $|A| := \sum_{i\in\mathbb{N}} |A_i|$ and we associate to A the mapping H_A defined on $[0,1] \times \mathbb{R}$ by

$$\forall\, 0 \leq t \leq 1,\, \forall\, x \in \mathbb{R}, \qquad H_A(t,x) \;:=\; \prod_{i\in\mathbb{N}} H_{A_i}(t,x).$$

Then it can be shown recursively that for any $A \in \mathbb{A}$, there exists an integer $N(A) \in \mathbb{Z}$, independent from the real numbers $0 \leq t \leq 1$, $x \in \mathbb{R}$, and from the underlying distribution μ, such that for any fixed $p \in \mathbb{N}^*$,

$$\forall\, 0 \leq t \leq 1,\, \forall\, x \in \mathbb{R}, \qquad \partial_t^p F_x(t) \;=\; \sum_{A\in\mathbb{A}\,:\,|A|=p} N(A) H_A(t,x)$$

(one would have noticed there is only a finite number of $A \in \mathbb{A}$ verifying $|A| = p$).

To compute the coefficients $N(A)$, for $A \in \mathbb{A}$, one can apply a sort of tree algorithm: if $A = (A_0, \ldots, A_r, 0, \ldots) \in \mathbb{A}$ is given, with $A_r \neq 0$, at height $|A|$, it gives birth to three types of sons, each of them of height $|A| + 1$:

- Let us denote for $0 \leq i \leq r$, $A_i = (\alpha_{i,0}, \alpha_{i,1}, \ldots, \alpha_{i,r_i}, 0, \ldots)$, with $\alpha_{i,r_i} > 0$. Then for any choice of $0 \leq i \leq r$ and $0 \leq j \leq r_i$, we obtain a son of A by replacing $\alpha_{i,j}$ by $\alpha_{i,j} + 1$ (all the other coordinates remaining the same) and by rearranging in a natural way the object thus obtained in order to ensure that it still belongs to $\mathbb{A}$ (i.e. that the monotonicity properties entering the definitions of $\mathcal{A}$ and $\mathbb{A}$ are fulfilled). Thus one has created $\sum_{0\leq i\leq r} r_i$ sons and some of them can be equal.
- Another type of son is obtained by choosing an index $0 \leq i \leq r$, replacing A_i by $(\alpha_{i,0}, \alpha_{i,1}, \ldots, \alpha_{i,r_i}, 1, 0, 0, \ldots)$ and proceeding to the necessary rearrangements. This operation creates r sons, as before not necessarily different.
- Finally, we add r new sons, all of them equal to $(A_0, A_1, \ldots, A_r, 1, 0, 0, \ldots)$, where 1 designate here the element of $\mathcal{A}$ given by $(1, 0, 0, \ldots)$.

If B is a son of A, we will write $A \rightharpoonup B$ if it was created by one of the two first procedures and $A \rightharpoondown B$ otherwise. Then we have that for any $B \in \mathbb{A}$,

$$N(B) \;=\; \sum_{A\in\mathbb{A}\,:\,A\rightharpoonup B} N(A) - \sum_{A\in\mathbb{A}\,:\,A\rightharpoondown B} N(A).$$

For instance, we compute that at height 3, for any $0 \leq t \leq 1$ and any $x \in \mathbb{R}$,

$$\begin{aligned} \partial_t^3 F_x(t) \;=\;& H_{A^{(0)}}(t,x) + 3H_{A^{(1)}}(t,x) + H_{A^{(2)}}(t,x) - 3H_{A^{(3)}}(t,x) \\ & -3H_{A^{(4)}}(t,x) + 2H_{A^{(5)}}(t,x) \end{aligned} \tag{3.4}$$

where

$$A^{(0)} := \begin{array}{|c|}\hline 3 \\ \hline\end{array} \qquad A^{(1)} := \begin{array}{|c|c|}\hline 2 & 1 \\ \hline\end{array} \qquad A^{(2)} := \begin{array}{|c|c|c|}\hline 1 & 1 & 1 \\ \hline\end{array}$$

$$A^{(3)} := \begin{array}{|c|}\hline 2 \\ \hline 1 \\ \hline\end{array} \qquad A^{(4)} := \begin{array}{|c|c|}\hline 1 & 1 \\ \hline 1 & \\ \cline{1-1}\end{array} \qquad A^{(5)} := \begin{array}{|c|}\hline 1 \\ \hline 1 \\ \hline 1 \\ \hline\end{array}$$

(with obvious notations, we have represented the elements of $\mathbb{A}$ as arrays, the lines corresponding to elements of $\mathcal{A}^*$, but the fact that above they appear under forms of Young tableaux is accidental).
Now let us come back for some given $\alpha = (\alpha_i)_{i\in\mathbb{N}} \in \mathcal{A}^* := \mathcal{A}\backslash\{0\}$ and for $0 \le t \le 1$, $x, y \in \mathbb{R}$, to the expression $\prod_{i\in\mathbb{N}} \partial_t^{\alpha_i} U_t(x,y)$, with the unusual convention that $\partial_t^0 U_t(x,y) \equiv 1$. Taking into account that

$$\begin{array}{rrcl} & \partial_t U_t(x,y) & = & h'(t)xy - ty^2 \\ & \partial_t^2 U_t(x,y) & = & h''(t)xy - y^2 \\ \forall\, k \in \mathbb{N},\, k \ge 3, & \partial_t^k U_t(x,y) & = & h^{(k)}(t)xy \end{array}$$

we see there exist coefficients $a(\alpha, k, t)$ depending on $\alpha \in \mathcal{A}^*$, $k \in \mathbb{N}$, $k \le r(\alpha) := \max\{i \in \mathbb{N} : \alpha_i \ne 0\}$ and $0 \le t \le 1$, such that this ensures the following polynomial expansion for any $x, y \in \mathbb{R}$,

$$\prod_{i\in\mathbb{N}} \partial_t^{\alpha_i} U_t(x,y) = \sum_{0\le k\le r(\alpha)} a(\alpha,k,t)x^k y^{2r(\alpha)-k}.$$

Furthermore it is quite clear that for any $p \in \mathbb{N}$, the quantity

$$b(p) := \sup_{\alpha\in\mathcal{A}\,:\,|\alpha|=p} \left(\sup_{0\le k\le r(\alpha),\, 0\le t\le 1} |a(\alpha,k,t)| \right)$$

is finite, so via Young inequalities, we can find a family of constants $(C(p))_{p\in\mathbb{N}^*}$ such that for any $\alpha = (\alpha_i)_{i\in\mathbb{N}} \in \mathcal{A}^*$,

$$\forall\, 0 \le t \le 1,\, \forall\, x, y \in \mathbb{R}, \qquad \left| \prod_{i\in\mathbb{N}} \partial_t^{\alpha_i} U_t(x,y) \right| \le C(|\alpha|)(1 + |x|^{2|\alpha|} + |y|^{2|\alpha|})$$

(one would have noticed the general fact $r(\alpha) \le |\alpha|$). The bound stated in Lemma 3.6 follows by putting together all these considerations and applying Lemma 3.5 and some more Young inequalities. □

We now have at our disposal all the ingredients needed for the proof of Proposition 3.3.
So let $p \in \mathbb{N}$ and $\lambda > 0$ be fixed as in the statement of this result. We make the hypothesis that we are given two probabilities $m, \mu \in \mathcal{N}_p(\lambda)$ and we define for $0 \le t \le 1$,

$$G(t) := \frac{1}{2}\ln(1+t^2) - \frac{1}{2}t^2 - \int F_x(t)\, m(dx) \tag{3.5}$$

which is just the expression we want to bound, if $m = m_{n+1}$, $\mu = \mu_{n+1}$ and $t = 1/\sqrt{n+1}$. We also notice that

$$G(t) \quad = \quad \int \ln\left(\frac{d\nu Q_{\mu,t}}{d\nu}\right) dm$$

where $Q_{\mu,t}$ is the Markovian kernel defined as K_{n+1}, but replacing μ_{n+1} by μ and $n+1$ by $1/t^2$ (namely for any $x \in \mathbb{R}$, $Q_{\mu,t}(x\cdot)$ is the law of $\frac{x}{\sqrt{1+t^2}} + \frac{tX}{\sqrt{1+t^2}}$, where X is distributed as μ). In particular G is identically equal to zero if $\mu = \nu$, since in this case the standard Gaussian distribution ν is invariant (even reversible) for $Q_{\nu,t}$, for any $t \geq 0$.
The positive spectral gap of m ensures that it admits moments of all order, thus estimates of Lemma 3.6 enable us to be convinced that $[0,1] \ni t \mapsto G(t)$ belongs to $\mathcal{C}^\infty([0,1])$ and that we can find a finite constant $C_5(\lambda,p)$, depending only on λ and p, bounding $G^{(p+1)}(t)$ uniformly in $t \in [0,1]$. Classical Taylor expansion up to the p-order with remainder term then show that for any $0 \leq t \leq 1$,

$$\left|G(t) - G(0) - G'(0)t - \cdots - G^{(p)}(0)\frac{t^p}{p!}\right| \quad \leq \quad C_5(\lambda,p)\frac{t^{p+1}}{(p+1)!}. \tag{3.6}$$

But let us return to the considerations developed in the proof of Lemma 3.6. Since we have for any $x, y \in \mathbb{R}$,

$$\begin{aligned}
&\partial_t U_t(x,y)|_{t=0} = h'(0)xy = xy \\
&\partial_t^2 U_t(x,y)\big|_{t=0} = h''(0)xy - y^2 = -y^2 \\
\forall\, k \in \mathbb{N},\, k \geq 3, \qquad &\partial_t^k U_t(x,y)\big|_{t=0} = h^{(k)}(0)xy
\end{aligned}$$

it appears that for any $\alpha = (\alpha_i)_{i\in\mathbb{N}} \in \mathcal{A}^*$, the expression $\prod_{i\in\mathbb{N}} \partial_t^{\alpha_i} U_t(x,y)\big|_{t=0}$ is a polynomial in x and y where the variable y (respectively x) appears at most at the power $|\alpha|$. Furthermore, for any $x \in \mathbb{R}$, $\mu_{0,x} = \mu$, so for any $k \in \mathbb{N}$, $0 < k \leq p$, the quantity $G^{(k)}(0)$ only depends on m and μ through their moments of order less or equal to p. By assumption, the latters coincide with those of the standard normal law and we know that if $\mu = \nu$ then all derivatives of G are null. Thus we conclude that in our case we equally have $G'(0) = \cdots = G^{(p)}(0) = 0 = G(0)$ (the latter equality is always true), so Proposition 3.3 follows from (3.6), once we allow the replacements $m = m_{n+1}$, $\mu = \mu_{n+1}$ and $t = 1/\sqrt{n+1}$ (taking into account Lemmas 3.1 and 3.2 to check that $m_{n+1} \in \mathcal{N}_p(\lambda)$).

Remarks 3.7. a) Because the laws we considered have moments of all order, we have not taken much care in exponents of the variable x. Nevertheless, let us indicate that it is possible to improve the $|x|^{3q}$ of Lemma 3.5 into x^{2q} (and by consequence the x^{6p} of Lemma 3.6 can be replaced by x^{4p}). To do so, instead of upper bounding $t\,|y|$ by $\sqrt{1+t^2}\,|x| + |\sqrt{1+t^2}x - ty|$ in (3.3), rather bound t^2y^2 by $(\sqrt{1+t^2}x - ty)^2 + 2\sqrt{1+t^2}\,|x|\,|\sqrt{1+t^2}x - ty| + (1+t^2)x^2$. One can then carry on with manipulations similar to those presented after (3.3) and conclude to the above mentioned improvement. But we are not sure that the power of x thus obtained is

the best possible one. Indeed, we conjecture that in Lemma 3.5, it is a term like $|x|^q$ which should appear. Note that it is the case if $\mu = \nu$, situation where exact computations can be conducted.
b) At the end of the proof of Proposition 3.3, the requirement that $m \in \mathcal{N}_p(\lambda)$ was a little too strong. In view of above remark, if $p \in \mathbb{N}^*$ is fixed, it is sufficient that m admits a moment of order $4(p+1)$ to obtain there exists a finite constant C depending on p, $\int x^{4(p+1)}\, m(dx)$ and $\lambda > 0$, such that $\mu \in \mathcal{N}_p(\lambda)$ implies that

$$\forall\, 0 \le t \le 1, \qquad |G(t)| \;\le\; Ct^{p+1}.$$

c) In next section we will also discuss about the hypothesis that μ admits a positive spectral gap. But let us already mention that as far as only bounds like (3.6) are concerned, there is an easy condition dispensing us from this assumption. It corresponds to the cases where μ has a compact support, say for instance that $\mu[(-\infty, -M) \sqcup (M, +\infty)] = 0$, where $M > 0$ is finite. In this situation we are assured in Lemma 3.5 of the obvious bound $\int |y|^q\, \mu_{t,x}(dy) \le M^q$, for any $q \in \mathbb{N}$, any $0 \le t \le 1$ and above all any $x \in \mathbb{R}$. Then reexamining the above computations, it appears that for any $p \in \mathbb{N}^*$, we can find a finite constant C depending only on p, M and $\int |x|^{p+1}\, m(dx)$ (assumed to be finite), insuring that for all such $\mu \in \mathcal{N}_p$, we have $|G(t)| \le Ct^{p+1}$. Recall that there exists such probabilities μ which are finite weighted sums of Dirac masses and thus are not admitting a positive spectral gap in the way we have defined it.
d) Using the computation made in (3.4), it appears at height 3 that for $\mu, m \in \mathcal{N}_2(\lambda)$, we have

$$G^{(3)}(0) = \int \partial_t^3 F_x(0) m(dx) = \int H_{A^{(2)}}(0,x)\, m(dx) = \int x^3\, m(dx) \int y^3 \mu(dy).$$

Thus if furthermore $\int y^3\, \mu(dy) \neq 0 \neq \int x^3\, m(dx)$, then the corresponding $G(t)$ is equivalent to $G^{(3)}(0)t^3/6$ for small $t > 0$ (the difference between these terms being at least of order $\mathcal{O}(t^4)$). Nevertheless a little miracle comes to our rescue when we apply this result: if for all $n \in \mathbb{N}$, $\mu_n \in \mathcal{N}_2$ and μ_n admits a moment of order 3, we get that for any $n \in \mathbb{N}$,

$$\begin{aligned}\int x^3\, m_n(dx) &= \mathbb{E}[Y_n^3] \\ &= \frac{1}{(n+1)^{3/2}} \sum_{0 \le i \le n} \mathbb{E}[X_i^3]\end{aligned}$$

so typically if $\int y^3\, \mu_n(dy)$ is bounded uniformly in $n \in \mathbb{N}$, then $\int x^3\, m_n(dx)$ is of order $1/\sqrt{n+1}$. In particular, assuming that for all $n \in \mathbb{N}$, $\mu_n \in \mathcal{N}_2(\lambda)$ for some fixed $\lambda > 0$, we end up with the existence of a constant $C(\lambda)$ depending only on λ, such that

$$\forall\, n \in \mathbb{N}, \qquad |\mathrm{Ent}(m_{n+1}|\nu) - \mathrm{Ent}(m_{n+1}|\nu K_{n+1})| \;\le\; \frac{C(\lambda)}{(n+1)^2}.$$

Thus taking into account the conjecture given in remark 2.1 (c), this estimate is quite promising, since we would obtain a general bound of order $\ln(e+n)/(n+1)$ (as already mentioned, the considerations of next section will indicate why the spectral gap assumption is not so crucial for the above arguments).

It is time now to present examples where the entropy goes to zero faster than what is predicted by Theorem 1.1. Of course, the (basic, i.e. not taking into account modified logarithmic Sobolev inequalities) Markovian considerations of section 2 and the estimates of Proposition 3.3 are not enough for this kind of result, since they will only offer bounds which are increasing with respect to time. So we need another trick; the convolution with the standard Gaussian law and rearrangements of random variables.
For fixed $p \in \mathbb{N} \setminus \{0,1\}$ and $\lambda > 0$, let us denote by $\widetilde{\mathcal{N}}_p(\lambda)$ the set of laws m which are constructed as in Lemma 3.1, with $\mu \in \mathcal{N}_p(\lambda)$, $\mu' = \nu$ and $t = 1/2$. It follows from Lemmas 3.1 and 3.2 that $\widetilde{\mathcal{N}}_p(\lambda) \subset \mathcal{N}_p(1 \wedge \lambda) = \mathcal{N}_p(\lambda)$, since the spectral gap of ν is just 1 and it is the largest possible spectral gap of elements of $\mathcal{N}_2$, as it can be checked by considering the identity as test function. Let us mention that such perturbed measures also lay at the heart of previous analysis of entropic convergence in the Central Limit Theorem by Linnik [16], Brown [8] and Barron [6]. Furthermore, part of the recent progresses of Johnson and Barron [14] and of Ball, Barthe and Naor [5] is to get rid of this necessity.

Proposition 3.8. *In the setting of Section 2, if all the distributions μ_n, $n \in \mathbb{N}$, belong to $\widetilde{\mathcal{N}}_p(\lambda)$, for some $p \in \mathbb{N} \setminus \{0,1\}$ and $\lambda > 0$, then there exist a finite constant $C(\lambda,p)$ depending only on those parameters, such that*

$$\forall\, n \in \mathbb{N}, \qquad \mathrm{Ent}(m_n|\nu) \;\leq\; \frac{C(\lambda,p)}{(n+1)^{\frac{p-1}{2}}}.$$

Proof. Let us return to probabilist notations. By definition of $\widetilde{\mathcal{N}}_p(\lambda)$, for each $n \in \mathbb{N}$, we can write $X_n = Z_n/\sqrt{2} + W_n/\sqrt{2}$, where Z_n and W_n are independent and whose respective law belongs to $\mathcal{N}_p(\lambda,p)$ for the former and is equal to ν for the latter. We can also assume that all the random variables Z_n, $n \in \mathbb{N}$, and $W_{n'}$, $n' \in \mathbb{N}$, are mutually independent (at least, these considerations are justified up to a possible modification of the underlying probability space).
Let a time $N \in \mathbb{N}^*$ be temporally fixed. We consider a new set of random variables $(\widetilde{X}_n)_{0\leq n\leq N}$ defined by

$$\widetilde{X}_n \;:=\; \begin{cases} W_{2n}/\sqrt{2} + W_{2n+1}/\sqrt{2}, & \text{if } 0 \leq n < \lfloor (N+1)/2 \rfloor, \\ W_{2n}/\sqrt{2} + Z_0/\sqrt{2}, & \text{if } \lfloor (N+1)/2 \rfloor \leq n < (N+1)/2, \\ Z_{2n-N-1}/\sqrt{2} + Z_{2n-N}/\sqrt{2}, & \text{if } (N+1)/2 \leq n \leq N. \end{cases}$$

For $0 \leq n \leq N$, let us also denote $\widetilde{m}_n$ the law of $(\widetilde{X}_0 + \cdots + \widetilde{X}_n)/\sqrt{n+1}$. In particular we have $\widetilde{m}_N = m_N$. But up to time $\lfloor (N+1)/2 \rfloor - 1$, we have $\widetilde{m}_n = \nu$ and after this time the difference of entropy $\mathrm{Ent}(\widetilde{m}_{n+1}|\nu) - \mathrm{Ent}(\widetilde{m}_n|\nu)$ is bounded above by $C_6(\lambda,p)/(n+1)^{\frac{p+1}{2}}$, for a certain finite constant $C_6(\lambda,p)$ depending only on λ and p, due to the fact that the law of $\widetilde{X}_{n+1}$ belongs to $\mathcal{N}_p(\lambda)$, according to the remark before the statement of Proposition 3.8. So we end up with the estimate

$$\begin{aligned}
\mathrm{Ent}(m_N|\nu) &= \mathrm{Ent}(\widetilde{m}_N|\nu) \\
&= \sum_{\lfloor (N+1)/2\rfloor -1\leq n<N} \mathrm{Ent}(\widetilde{m}_{n+1}|\nu)-\mathrm{Ent}(\widetilde{m}_n|\nu) \\
&\leq C_6(\lambda,p)\sum_{n\geq\lfloor (N+1)/2\rfloor -1}\frac{1}{(n+1)^{\frac{p+1}{2}}} \\
&\leq \frac{C(\lambda,p)}{(N+1)^{\frac{p-1}{2}}}
\end{aligned}$$

with for instance $C(\lambda,p) := \frac{p+1}{p-1}2^{p-1}C_6(\lambda,p)$.
For $N=0$, we directly get that $\mathrm{Ent}(m_0|\nu)$ is bounded above by a constant only depending on $\lambda>0$ and $p\in\mathbb{N}\setminus\{0,1\}$, by applying the inequality (3.6) with μ the law of Z_0, $m=\mu_0$ and $t=1/2$ (and resorting to the convention that $\widetilde{m}_{-1}:=\nu$ is the law of W_0). □

In Lemma 4.4 of the next section, we will see how to generalize this result to weaker convolutions.

4. Some bounds without spectral gap assumption

In the previous computations, the spectral gap hypothesis is not as crucial as it may seem at first view and we will discuss here ways to relax it.

At the end of last section, it was necessary to convolve with a Gaussian distribution to obtain our examples of fast convergence. We now recall how, to some extent, it is possible to "deconvolve", via an assumption of finite modified Fisher information. The latter is the quantity associated to any probability m on $\mathbb{R}$ by the formula

$$I(m) \;:=\; \begin{cases} \int \left|\nabla \ln\left(\frac{dm}{d\nu}\right)\right|^2 dm \;\leq\; +\infty, & \text{if } m\ll\nu \\ +\infty, & \text{otherwise.}\end{cases}$$

where ∇ is the weak derivative corresponding to Radon–Nikodym differentiation with respect to Lebesgue measure.
To see its relation with the weighted convolutions under study, let us consider $(P_t)_{t\geq 0}$ the Ornstein–Uhlenbeck semigroup (which has already made a discreet apparition in last section as $(Q_{\nu,\ln(1+t^2)})_{t\geq 0}$), which acts on nonnegative measurable functions f by

$$\forall\, t\geq 0, \forall\, x\in\mathbb{R},\qquad P_t[f](x) \;=\; \int f(\exp(-t/2)x+\sqrt{1-\exp(-t)}y)\,\nu(dy).$$

The next result is so standard in Markovian semigroup theory (see e.g. [4] or [1]), that we will not recall its proof.

Theorem 4.1. *Let $f \in \mathbb{L}^1(\nu)$ be a density of probability with respect to ν. For any $t \geq 0$, we denote $m(t) := P_t[f]\nu$. Then we have*

$$\forall\, t \geq 0, \qquad \partial_t \mathrm{Ent}(m(t)|\nu) \;=\; -\frac{I(m(t))}{2}$$

and the mapping

$$\mathbb{R}_+ \ni t \;\mapsto\; I(m(t))$$

is non-increasing (i.e. the mapping $\mathbb{R}_+ \ni t \mapsto \mathrm{Ent}(m(t)|\nu)$ is convex).

We are particularly interested in the following consequence, which enables "small" deconvolution:

$$\forall\, t \geq 0, \qquad \mathrm{Ent}(m|\nu) \;\leq\; \mathrm{Ent}(m(t)|\nu) + I(m)t/2. \tag{4.1}$$

In order to take advantage of this bound, let us reformalize the results obtained in section 3.

Definition 4.2. For any fixed constants $r, K \geq 0$, we define $\mathcal{M}_r^{(1)}(K)$ the set of probabilities m on $\mathbb{R}$ verifying $\int |x|^r \, m(dx) \leq K$. If furthermore $p \in \mathbb{N}$ is given, then let $\mathcal{M}_{p,r}^{(2)}(K)$ be the subset of $\mathcal{N}_p$ whose elements μ satisfy

$$\forall\, 0 \leq t \leq 1, \forall\, x \in \mathbb{R}, \qquad \int y^{2(p+1)} \, \mu_{t,x}(dy) \;\leq\; K(1 + |x|^r)$$

where the Gibbs distribution $\mu_{t,x}$ was defined in (3.1) with respect to μ.

The interest of these sets of measures is that if G is defined as in (3.5) with respect to $m \in \mathcal{M}^{(1)}_{r/2+(r/2)\vee(p+1)}(K_1)$ and $\mu \in \mathcal{M}^{(2)}_{p,r}(K_1)$, for some finite constants $p \in \mathbb{N}$ and $r, K_1, K_2 \geq 0$, then we have seen how to obtain a finite constant $C(p, r, K_1, K_2)$ depending only on its parameters, such that

$$\forall\, 0 \leq t \leq 1, \qquad |G(t)| \;\leq\; C(p, r, K_1, K_2)t^{p+1}. \tag{4.2}$$

Here is our main statement without any apparent convolution, result which is not very good, since the order $\mathcal{O}(1/n)$ is only asymptotically approached as p goes to infinity!

Proposition 4.3. *Consider once again the setting of Section 2. Let assume there exist constants $p \in \mathbb{N} \setminus \{0, 1\}$, $r, K_1, K_2, K_3 \geq 0$ such that*

$$\forall\, n \in \mathbb{N}, \qquad \begin{cases} m_n \in \mathcal{M}^{(1)}_{r/2+(r/2)\vee(p+1)}(K_1), \\ \mu_{n+1} \in \mathcal{M}^{(2)}_{p,r}(K_2), \\ I(m_n) \leq K_3. \end{cases}$$

Then there exists another finite constant $C(p, r, K_1, K_2, K_3) \geq 0$ depending only on the previous ones, such that

$$\forall\, n \in \mathbb{N}, \qquad \mathrm{Ent}(m_n|\nu) \;\leq\; \frac{C(p, r, K_1, K_2, K_3)}{(n+1)^{(p-1)/(p+1)}}.$$

The proof is based on the following extension of Proposition 3.8.

Lemma 4.4. *Assume that the sequence $(\mu_n)_{n\in\mathbb{N}}$ is as in previous proposition, except for the requirement of bounded modified Fisher information, and as usual let $(X_n)_{n\in\mathbb{N}}$ be independent variables distributed according to these laws. Let $0<s<1$ be fixed and $(W_n)_{n\in\mathbb{N}}$ be IID standard Gaussian variables, also independent from $(X_n)_{n\in\mathbb{N}}$. We consider for $n\in\mathbb{N}$, $\widetilde{X}_n=\sqrt{1-s^2}X_n+sW_n$ and we denote $\widetilde{m}_n$ the law of*

$$\widetilde{Y}_n \quad := \quad \frac{\widetilde{X}_0+\cdots+\widetilde{X}_n}{\sqrt{n+1}}.$$

Then there exists a finite constant $C(p,r,K_1,K_2)\geq 0$, not depending on $0<s<1$, such that

$$\forall\, n\in\mathbb{N},\qquad s^2(n+1)\geq 2 \quad\Longrightarrow\quad \mathrm{Ent}(\widetilde{m}_n|\nu) \;\leq\; \frac{C(p,r,K_1,K_2)}{[s^2(n+1)]^{(p-1)/2}}.$$

Proof. As in the proof of Lemma 3.8, we begin by fixing a time horizon $N\in\mathbb{N}$, and we consider the new sequence of random variables $(\widehat{Y}_n)_{-1\leq n\leq N}$ defined by the iteration

$$\widehat{Y}_{-1} := \frac{W_0+\cdots+W_N}{\sqrt{N+1}}$$

and for any $-1\leq n<N$

$$\widehat{Y}_{n+1} := \sqrt{\frac{s^2(N+1)+(1-s^2)(n+1)}{s^2(N+1)+(1-s^2)(n+2)}}\widehat{Y}_n+\frac{\sqrt{1-s^2}}{\sqrt{s^2(N+1)+(1-s^2)(n+2)}}X_{n+1}.$$

Indeed, we have for any $0\leq n\leq N$,

$$\widehat{Y}_n = \frac{sW_0+\cdots+sW_N+\sqrt{1-s^2}X_0+\cdots+\sqrt{1-s^2}X_n}{\sqrt{s^2(N+1)+(1-s^2)(n+1)}}$$

and we denote by $\widehat{m}_n$ its law. In particular it appears that $\widehat{m}_N=\widetilde{m}_N$. Since we have for any $0\leq n\leq N$,

$$\forall\; f\in\mathcal{B}(\mathbb{R}),\qquad \widehat{m}_n[f] \;=\; \int f(s_{N,n}x+\sqrt{1-s_{N,n}^2}y)\,\nu(dx)m_n(dy)$$

where $s_{N,n}:=\sqrt{N+1}s/\sqrt{(N+1)s^2+(1-s^2)(n+1)}$, we show without difficulty that the $\widehat{m}_n$, $0\leq n\leq N$, belong to $\mathcal{M}_{r'}^{(1)}(K_1(r'))$, with $r':=r/2+(r/2)\vee(p+1)$ and $K_1(r):=2^{r'/2}\max\{\int|x|^{r'}\,\nu(dx),K_1\}$. Thus we would like to apply bound (4.2) with $0\leq t\leq 1$ defined by

$$\frac{t}{\sqrt{1+t^2}} \;=\; \widetilde{t} := \frac{\sqrt{1-s^2}}{\sqrt{s^2(N+1)+(1-s^2)(n+2)}}$$

or equivalently $t=\widetilde{t}/\sqrt{1-\widetilde{t}^2}$. That is where the condition $s^2(n+1)\geq 2$ is useful, since it ensures that $\widetilde{t}\leq 1/\sqrt{2}$ and $t\leq 1$, and then that $t\leq\sqrt{2}\,\widetilde{t}$, thus via (4.2) we obtain

$$\mathrm{Ent}(\widehat{m}_{n+1}|\nu)\leq\mathrm{Ent}(\widehat{m}_n|\nu)+C_1(p,r,K_1,K_2)\left(\frac{\sqrt{1-s^2}}{\sqrt{s^2(N+1)+(1-s^2)(n+2)}}\right)^{p+1}$$

for an appropriate finite constant $C_1(p,r,K_1,K_2)\geq 0$ not depending on $0<s<1$. So summing these inequalities we end up with

$$\begin{aligned}\mathrm{Ent}(\widetilde{m}_N|\nu) &\leq C_1(p,r,K_1,K_2)\sum_{-1\leq n\leq N-1}\frac{(1-s^2)^{(p+1)/2}}{(s^2(N+1)+(1-s^2)(n+2))^{\frac{p+1}{2}}}\\ &= C_1(p,r,K_1,K_2)(1-s^2)^{(p+1)/2}\left(\frac{1}{(s^2(N+1)+(1-s^2))^{\frac{p+1}{2}}}\right.\\ &\quad\left.+\sum_{1\leq n\leq N}\frac{1}{(s^2(N+1)+(1-s^2)(n+1))^{\frac{p+1}{2}}}\right).\end{aligned}$$

We bound above the last sum by

$$\begin{aligned}&\int_1^N\frac{1}{(s^2(N+1)+(1-s^2)u)^{\frac{p+1}{2}}}\,du\\ &\leq \int_1^{+\infty}\frac{1}{(s^2(N+1)+(1-s^2)u)^{\frac{p+1}{2}}}\,du\\ &= \frac{2}{(1-s^2)(1-p)}\left[(s^2(N+1)+(1-s^2)u)^{\frac{1-p}{2}}\right]_1^{+\infty}\\ &= \frac{2}{(1-s^2)(p-1)}\frac{1}{(s^2(N+1)+(1-s^2))^{\frac{p-1}{2}}},\end{aligned}$$

so we get, since $p\geq 2$,

$$\begin{aligned}\mathrm{Ent}(\widetilde{m}_N|\nu) &\leq C_1(p,r,K_1,K_2)\left(\frac{1}{(s^2(N+1)+(1-s^2))^{\frac{p+1}{2}}}\right.\\ &\quad\left.+\frac{2}{p-1}\frac{1}{(s^2(N+1)+(1-s^2))^{\frac{p-1}{2}}}\right)\\ &\leq 3C_1(p,r,K_1,K_2)\frac{1}{(s^2(N+1)+(1-s^2))^{\frac{p-1}{2}}}\\ &\leq \frac{3C_1(p,r,K_1,K_2)}{(s^2(N+1))^{\frac{p-1}{2}}}\end{aligned}$$

which is the required result. □

We can now come to the

Proof of Proposition 4.3. Using the reversibility of the Ornstein–Uhlenbeck semigroup with respect to ν, we obtain, with the notation of the above lemma and of Theorem 4.1,

$$\forall\, n\in\mathbb{N},\qquad \widetilde{m}_n = m_nP_{\ln(1/(1-s^2))} = m_n(\ln(1/(1-s^2)))$$

thus taking into account bound (4.1), our assumptions imply that if $0<s\leq 1/\sqrt{2}$ and if $(n+1)s^2\geq 2$, then

$$\begin{aligned}
\operatorname{Ent}(m_n|\nu) &\leq \operatorname{Ent}(\widetilde{m}_n|\nu) + \frac{1}{2} I(m_n) \ln\left(\frac{1}{1-s^2}\right) \\
&\leq \frac{C(p,r,K_1,K_2)}{[s^2(n+1)]^{(p-1)/2}} + \frac{K_3}{2}\frac{s^2}{1-s^2} \\
&\leq \frac{C(p,r,K_1,K_2)}{[s^2(n+1)]^{(p-1)/2}} + K_3 s^2.
\end{aligned}$$

It remains now to optimize in the parameter $0 < s \leq 1/\sqrt{2}$ to be convinced of the validity of the proposition. Indeed, the minimizing s is of order $(n+1)^{\frac{1-p}{2(p+1)}}$, so the condition $2/(n+1) \leq s^2 \leq 1/2$ is satisfied for large enough $n \in \mathbb{N}$. For the other $n \in \mathbb{N}$, note that the logarithmic Sobolev inequality verified by ν directly gives us

$$\forall\, n \in \mathbb{N}, \qquad \operatorname{Ent}(m_n|\nu) \;\leq\; \frac{1}{2} I(m_n) \;\leq\; \frac{K_3}{2}. \qquad \square$$

Of course, the problem with the hypotheses of Proposition 4.3 is that they are not immediate to verify, since they use the distributions m_n, for $n \in \mathbb{N}$. So we will give below simple sufficient criteria enabling to check them only in terms of the given family $(\mu_n)_{n\in\mathbb{N}}$.

To treat the bounded modified Fisher information condition, let us come back to the "true" Fisher information of a probability m on $\mathbb{R}$ which is defined as the quantity

$$J(m) \;:=\; \begin{cases} \int \left|\nabla \ln\left(\frac{dm}{d\lambda}\right)\right|^2 dm \;\leq\; +\infty, & \text{if } m \ll \lambda, \\ +\infty, & \text{otherwise.} \end{cases}$$

where λ is the traditional Lebesgue measure on $\mathbb{R}$. So if m is a probability absolutely continuous with respect to λ, we have

$$\begin{aligned}
I(m) &= \int \left|\nabla \ln\left(\frac{dm}{d\lambda}\right) + \nabla \ln\left(\frac{d\lambda}{d\nu}\right)\right|^2 dm \\
&= \int \left|\nabla \ln\left(\frac{dm}{d\lambda}\right)(x) + x\right|^2 m(dx) \\
&= J(m) + 2\int x \nabla \ln\left(\frac{dm}{d\lambda}\right)(x)\, m(dx) + \int x^2\, m(dx) \\
&= J(m) + 2\int x \nabla \frac{dm}{d\lambda}(x)\, \lambda(dx) + \int x^2\, m(dx) \\
&= J(m) - 2\int (\nabla x) \frac{dm}{d\lambda}(x)\, \lambda(dx) + \int x^2\, m(dx) \\
&= J(m) - 2\int m(dx) + \int x^2\, m(dx) \\
&= J(m) - 2 + \int x^2\, m(dx)
\end{aligned}$$

(these computations are justified if $\sqrt{dm/d\lambda}$ is $\mathcal{C}^1$ with compact support, but the relation $I(m) = J(m) - 2 + \int x^2\, m(dx)$ can next be extended to the general case by traditional approximation procedures).

The advantage of J is that it is also an "invariant" quantity in the sense of Lemma 3.1, namely we have the following well-known bound of Blachman and Stam (see for instance [9] or [1]):

Theorem 4.5. *Let m be a distribution constructed on $\mathbb{R}$ as in the statement of Lemma 3.1, starting from $0 \leq t \leq 1$ and from two probabilities μ, μ'. Then we have*

$$J(m) \;\leq\; t^2 J(\mu) + (1-t^2) J(\mu').$$

Remark 4.6. Still with the notations of the above result, we have if μ or μ' are centered and have the same variance,

$$\begin{aligned}
I(m) &= J(m) - 2 + \int x^2\, m(dx) \\
&\leq t^2 J(\mu) + (1-t^2) J(\mu') + t^2 \!\int x^2\, \mu(dx) + (1-t^2) \!\int x^2\, \mu'(dx) - 2(t^2 + 1 - t^2) \\
&= t^2 I(\mu) + (1-t^2) I(\mu') \leq I(\mu) \vee I(\mu').
\end{aligned}$$

But this relation is no longer necessarily true if μ and μ' are not centered, consider for instance the case where $\mu = \mu'$ is the normalized Gaussian distribution of mean 1.

Thus to ensure that for every $n \in \mathbb{N}$, we have $I(m_n) \leq K_3$, for some constant $K_3 \geq 0$, it is sufficient to ask that for all $n \in \mathbb{N}$, $\mu_n \in \mathcal{N}_2$ and $J(\mu_n) \leq K_3 + 1$. To deal with the belonging of the m_n, $n \in \mathbb{N}$, to some $\mathcal{M}_r^{(1)}(K_1)$, for fixed constants $r, K_1 \geq 0$, we will use another stability property. More precisely, if $r \in \mathbb{N}$ and $M \geq 1$ are given, let $\widetilde{\mathcal{S}}_r(M)$ be the set of probabilities μ on $\mathbb{R}$ which are symmetrical with respect to the origin and such that

$$\forall\, 0 \leq k \leq r, \qquad \int x^{2k}\, \mu(dx) \;\leq\; M^k \int x^{2k}\, \nu(dx).$$

Lemma 4.7. *For any r and M as above, the set $\widetilde{\mathcal{S}}_r(M)$ is stable by the operations described in Lemma 3.1, in the sense that if m is constructed in this way from $0 \leq t \leq 1$ and $\mu, \mu' \in \widetilde{\mathcal{S}}_r(M)$, then we also have $m \in \widetilde{\mathcal{S}}_r(M)$.*

Proof. The preservation of the symmetry with respect to zero is clear, so we just compute that for $0 \leq k \leq r$,

$$\begin{aligned}
\int x^{2k}\, m(dx) &= \int \sum_{0 \leq l \leq 2k} \binom{2k}{l} t^l x^l (1-t^2)^{\frac{2k-l}{2}} y^{2k-l}\, \mu(dx)\mu'(dy) \\
&= \sum_{0 \leq l \leq k} \int \binom{2k}{2l} t^{2l} x^{2l} (1-t^2)^{k-l} y^{2(k-l)}\, \mu(dx)\mu'(dy) \\
&\leq M^k \sum_{0 \leq l \leq k} \int \binom{2k}{2l} t^{2l} x^{2l} (1-t^2)^{k-l} y^{2(k-l)}\, \nu(dx)\nu(dy) \\
&= M^k \int x^{2k}\, \nu(dx).
\end{aligned}$$

□

Remark 4.8. This result can be slightly improved by requiring not the symmetry of the elements μ of $\widetilde{\mathcal{S}}_r(M)$, but only that for every $1 \le k \le r$,

$$\int x^{2k-1}\, m(dx) \quad = \quad 0$$

(note that this property is also stable by the operations described in Lemma 3.1.

For two given as above constants $r \in \mathbb{N}$ and $K \ge 1$, let $\mathcal{S}_r(K)$ be the set of symmetrical distributions μ verifying $\int x^{2r}\, \mu(x) \le K$. Denoting

$$M_r(K) \quad := \quad \sup_{0 \le k \le r} \left(\frac{K^{k/r}}{\int x^{2k}\, \nu(dx)} \right)^{1/k} \quad < +\infty$$

we easily check that $\mathcal{S}_r(K) \subset \widetilde{\mathcal{S}}_r(M_r(K))$. Thus if we assume there exist $r \in \mathbb{N}$ and a finite constant $K_4 \ge 0$ such that for all $n \in \mathbb{N}$, $\mu_n \in \mathcal{S}_r(K_4)$, then we are assured that for every $n \in \mathbb{N}$, $m_n \in \widetilde{\mathcal{S}}_r(M_r(K_4)) \subset \mathcal{M}^{(1)}_{2r}(K_2)$, with $K_2 := M_r^r(K_4) \int x^{2r}\, \nu(dx)$.
So for instance we get the following result, where hypotheses are only made on the family $(\mu_n)_{n\in\mathbb{N}}$, but which remains not very good in view of Theorem 1.1 (nevertheless we will check below that the hypothesis of spectral gap is not necessary here, nor in Proposition 4.3).

Corollary 4.9. *Still in the context of Section 2, assume there exist $r \in \mathbb{N}$ and a finite constant $K \ge 0$ such that for all $n \in \mathbb{N}$,*

$$\begin{aligned} \int x^2\, \mu_n(dx) &= 1, \\ \mu_n &\in \mathcal{S}_{r+4\vee r}(K), \\ \mu_n &\in \mathcal{M}^{(3)}_{3,2r}(K), \\ J(\mu_n) &\le K, \end{aligned}$$

where for any $p \in \mathbb{N}$ and $r, K \ge 0$, $\mathcal{M}^{(3)}_{p,r}(K)$ is the set of probabilities μ on $\mathbb{R}$ verifying the second point of (4.2), then one can find an appropriate constant $C(r,K)$, depending only on its parameters r and K, such that

$$\forall\, n \in \mathbb{N}, \qquad \mathrm{Ent}(m_n|\nu) \quad \le \quad \frac{C(r,K)}{\sqrt{n}}.$$

Proof. One would have furthermore noticed that the two first above conditions imply that $\mu_n \in \mathcal{N}_3$, so we can apply Proposition 4.3 with $p = 3$. □

To finish, let us present two kinds of examples without spectral gap satisfying the previous conditions. For simplicity we will only consider constant sequences $(\mu_n)_{n\in\mathbb{N}}$ and denote $\mu := \mu_0$ that we assume to be symmetrical. This probability will have moments of all orders and for any $p \in \mathbb{N}$, there will exist $r, K \ge 0$ such that $\mu \in \mathcal{M}^{(3)}_{p,r}(K)$. We will also check that there is no obstruction to the conditions that $\int x^2\, \mu(dx) = 1$ and that μ admits a finite Fisher information.

• The first kind of examples is based on remark 3.7 (c). To ensure that μ has no positive spectral gap, we ask for the existence of two constants $0 < M_1 < 1 < M_2$ such that $\mu[(-M_2, -M_1) \sqcup (M_1, M_2)] = 1$ (indeed, considering a smooth function f on $\mathbb{R}$ verifying $f \equiv 0$ on $(-\infty, 0]$ and $f \equiv 1$ on $[M_1, +\infty)$, we get that the spectral gap has to be zero). To obtain a finite Fisher information for μ, we impose that $\mu \ll \lambda$ and that $\sqrt{d\mu/d\lambda}$ is smooth on $\mathbb{R}$ (since one can also write $J(\mu) = 4\int |\nabla\sqrt{d\mu/d\lambda}|^2\, d\lambda$), for instance it is sufficient that the density $d\mu/d\lambda$ is positive inside (M_1, M_2) and that in a right neighborhood of M_1 (respectively a left neighborhood of M_2), $d\mu/d\lambda(x)$ is proportional to $\exp(-1/(x - M_1))$ (resp. to $\exp(-1/(M_2 - x))$). Then it is quite clear that one find such probabilities μ having furthermore 1 for variance.

• At the opposite of the previous examples, our second type of probabilities μ will have tails heavier that those of exponential distributions and this feature will equally forbid a positive spectral gap. We will nevertheless resort to some weighted Poincaré's inequalities (besides it would be interesting to elaborate more general conditions for the belonging to sets like $\mathcal{M}^{(3)}_{p,r}(K)$, for fixed constants $p, r, K \geq 0$, for instance we are wondering which kind of functional inequalities can serve as criteria).
So let assume there exist two constants $0 < \epsilon < 2$ and $C > 0$ such that the symmetrical probability μ verifies that for any absolutely continuous mapping f on $\mathbb{R}$,

$$\int (f - \mu[f])^2\, d\mu \;\; \leq \;\; C \int (f'(y))^2 (1 + |y|^{2-\epsilon})\, \mu(dy) \tag{4.3}$$

If one considers for test function f the power mapping $\mathbb{R} \ni y \mapsto y^q$, with $q \in \mathbb{N}$, then it appears easily that $\int y^{2q}\, \mu(dy)$ has to be bounded by a quantity depending only on q, ϵ and C, thus μ admits moments of all orders. Then returning to the proof of Lemma 3.5 and in particular to the bound (3.3), where we can directly use a slight variant of the trick mentioned in remark 3.7 (a), $(ty)^2 \leq 2(\sqrt{1+t^2}x - ty)^2 + 2(1+t^2)x^2$, it appears that for any $p \in \mathbb{N}$, μ belongs to $\mathcal{M}^{(3)}_{p,r(p)}(K(\epsilon, C, p))$, with $r(p) := 8(p+1)/\epsilon$ and for some appropriate finite constant $K(\epsilon, C, p) > 0$, as usual depending only on ϵ, C and p.
In other respects, the symmetry of μ and Hardy's inequalities (cf [18, 7] or [1]) enable to obtain a simple criterion for the validity of (4.3). More precisely, for fixed $\epsilon > 0$, the best possible constant C in (4.3) satisfies $B/2 \leq C \leq 4B$, with

$$B \;\; := \;\; \sup_{t>0} \int_0^t \frac{1}{(1 + |y|^{2-\epsilon})\, \frac{d\mu}{d\lambda}(y)}\, \lambda(dy)\; \mu[[t, +\infty)]$$

(where $d\mu/d\lambda$ is a priori the Radon–Nikodym–Lebesgue derivative of the part of μ which is absolutely continuous with respect to λ). But if we assume that $\mu \ll \lambda$ and that $d\mu/d\lambda(y)$ is proportional to $\exp(-y^\alpha)$ for some given $\alpha > 0$ and for $y > 0$

large enough, then we get that for t large enough, $\mu[[t,+\infty)]$ is proportional to

$$\int_t^{+\infty} \exp(-y^\alpha)\,dy = \frac{1}{\alpha}\int_{t^\alpha}^{+\infty} y^{\frac{1-\alpha}{\alpha}} \exp(-y)\,dy$$

and it is well-known that such an integral is equivalent for large $t>0$ to the expression $\alpha^{-1}t^{1-\alpha}\exp(-t^\alpha)$. Similar computations show that up to a multiplicative constant, the term $\int_0^t \frac{\exp(-y^\alpha)}{1+|y|^{2-\epsilon}}\,dy$ is equivalent to $t^{\epsilon-\alpha-1}\exp(t^\alpha)$, for large $t>0$. Thus if $\alpha \geq \epsilon/2$, the inequality (4.3) is satisfied. Of course, if furthermore we impose that $d\mu/d\lambda$ is positive and smooth everywhere, then its above behavior at infinity implies that $J(\mu) < +\infty$. There is also no difficulty in locally deforming this density to ensure that it admits 1 as variance. But if we take $\alpha = \epsilon/2 < 1$, then the tail $\mu[[t,+\infty)]$ does not decrease exponentially fast in large $t>0$, so the spectral gap has to be null (cf for instance [15]), result which in the above situation can also be deduced by another application of Hardy's inequalities.

Remark 4.10. Let $p \in \mathbb{N}\setminus\{0,1\}$ be given, it is possible to find examples of the above second type which furthermore belong to $\mathcal{N}_p$, because one has a lot of freedom for the form of the density $d\mu/d\lambda$ on compact subsets. Then performing a convolution with a Gaussian distribution as in Proposition 3.8, we end up with a convergence speed of order $\mathcal{O}(1/n^{(p-1)/2})$. But note that the probabilities thus obtained keep the heavy tail property and thus do not admit a spectral gap. So one can find probabilities μ without spectral gap whose entropic convergence speed is at least of order $\mathcal{O}(1/n^{(p-1)/2})$. This feature leads us to think that the spectral gap assumption in Theorem 1.1 is maybe not very natural.

References

[1] C. Ané, S. Blachère, D. Chafaï, P. Fougères, I. Gentil, F. Malrieu, C. Roberto, and G. Scheffer. *Sur les inégalités de Sobolev logarithmiques.* Société Mathématique de France, Paris, 2000. With a preface by D. Bakry and M. Ledoux.

[2] S. Artstein, K. Ball, F. Barthe, and A. Naor. Entropy growth for sums of independent random variables. Preprint, 2002.

[3] S. Artstein, K. Ball, F. Barthe, and A. Naor. More on entropy production. Preprint in preparation, personal communication, 2002.

[4] D. Bakry. L'hypercontractivité et son utilisation en théorie des semigroupes. In P. Bernard, editor, *Lectures on Probability Theory. Ecole d'Eté de Probabilités de Saint-Flour XXII-1992*, Lecture Notes in Mathematics **1581**. Springer-Verlag, 1994.

[5] K. Ball, F. Barthe, and A. Naor. Entropy jumps in the presence of a spectral gap. Preprint, 2002.

[6] A. R. Barron. Entropy and the central limit theorem. *Ann. Probab.*, **14**(1):336–342, 1986.

[7] S. G. Bobkov and F. Götze. Exponential integrability and transportation cost related to logarithmic Sobolev inequalities. *J. Funct. Anal.*, **163**(1):1–28, 1999.

[8] L. D. Brown. A proof of the Central Limit Theorem motivated by the Cramér-Rao inequality. In *Statistics and probability: essays in honor of C. R. Rao*, pages 141–148. North-Holland, Amsterdam, 1982.

[9] E. A. Carlen. Superadditivity of Fisher's information and logarithmic Sobolev inequalities. *J. Funct. Anal.*, **101**(1):194–211, 1991.

[10] P. Del Moral, M. Ledoux, and L. Miclo. About supercontraction properties of Markov kernels. Preprint, Publications du Laboratoire de Statistique et Probabilités, no 2001-01, CNRS and Université Paul Sabatier (Toulouse III), France, 2001.

[11] P. Del Moral and L. Miclo. On convergence of chains with occupational self-interactions. Preprint, to appear in *Royal Society Special Edition on Stochastic Analysis*, 2003.

[12] W. Feller. *An introduction to probability theory and its applications, volume II.* Wiley Series in Probability and Mathematical Statistics. John Wiley and Sons, New York, 1966.

[13] L. Gross. Logarithmic Sobolev inequalities. *American Journal of Mathematics*, **97**(4):1061–1083, 1976.

[14] O. Johnson and A. Barron. Fisher information inequalities and the Central Limit Theorem. Preprint in submission, Statslab Research Report 2001-17, PDF file downloadable at arXiv:math.PR/0111020 v1 2 Nov 2001, 2003.

[15] M. Ledoux. Concentration of measure and logarithmic Sobolev inequalities. In *Séminaire de Probabilités, XXXIII*, pages 120–216. Springer, Berlin, 1999.

[16] Ju. V. Linnik. An information-theoretic proof of the central limit theorem with Lindeberg conditions. *Theor. Probability Appl.*, **4**:288–299, 1959.

[17] L. Miclo. Remarques sur l'hypercontractivité et l'évolution de l'entropie pour des chaînes de Markov finies. In J. Azéma, M. Emery, and M. Yor, editors, *Séminaire de Probabilités XXXI*, Lecture Notes in Mathematics 1655, pages 136–167. Springer-Verlag, Berlin, 1997.

[18] B. Muckenhoupt. Hardy's inequality with weights. *Studia Mathematica*, XLIV:31–38, 1972.

Acknowledgments

This paper was motivated by Franck Barthe's talk given at the conference on Stochastic Inequalities and their Applications (Barcelona, June 17–21, 2002), where he presented some of the results [2, 5, 3], recently obtained with his co-authors. I am indebted to him for all the supplementary explanations he kindly gave me. I am also very grateful to Michel Ledoux and Paul-Marie Samson for all the discussions we had around this subject and to the referee for complementary informations and references.

Laboratoire de Statistique et Probabilités, UMR 5583,
CNRS et Université Paul Sabatier,
118, route de Narbonne,
31062 Toulouse cedex 4, France,
E-mail address: `miclo@math.ups-tlse.fr`

Progress in Probability, Vol. 56, 157–168

On a Nonsymmetric Version of the Khinchine–Kahane Inequality

Krzysztof Oleszkiewicz

Abstract. We prove a new version of the Khinchine–Kahane inequality in which Bernoulli random variables no longer need to be symmetric. The constant in the inequality is optimal up to some universal factor. The proof uses hypercontractive methods and the optimal hypercontractivity constant for a mean-zero Bernoulli random variable is found. A simple observation generalizing Pisier's Rademacher projection norm estimate is added.

1. Introduction and notation

Throughout the paper given a random variable Z we will denote by $\|Z\|_p$ its p-th norm $(E|Z|^p)^{1/p}$; if Z is bounded then $\|Z\|_\infty = \operatorname{ess\,sup} |Z|$. If Z is vector-valued then $\|Z\|_p$ denotes the p–th norm of $\|Z\|$.

The classical Khinchine inequality [9] states that for $p > 0$, real numbers $a_1, a_2, \ldots, a_n$ and independent symmetric Bernoulli random variables $r_1, r_2, \ldots, r_n$ there is

$$A_p\|S\|_2 \leq \|S\|_p \leq B_p\|S\|_2,$$

where $S = \sum_{k=1}^n a_k r_k$ and the constants $A_p, B_p > 0$ depend only on p. Kahane [8] proved that the inequality remains valid (with possible change of constants) if the real coefficients are replaced by vectors of a normed linear space and the absolute value is replaced by the norm. The optimal constants in the Khinchine inequality were established by Whittle [23] ($p \geq 3$), Stečkin [20] (p even), Szarek [21] ($p = 1$) and Haagerup [6] (general p) whereas the optimal constants in the Khinchine–Kahane inequality in most cases are not known (see [16] and [12] for exceptions; also, it is obvious that the optimal A_p is equal to 1 if $p > 2$ and the optimal B_p is equal to 1 if $p < 2$). It is known however that for $p > 2$ the Khinchine–Kahane inequality holds with $B_p = \sqrt{p-1}$ and that $B_p p^{-1/2}$ must be uniformly bounded away from zero (see [14] for the proof and related results).

2000 *Mathematics Subject Classification.* Primary 46B09.
Key words and phrases. Bernoulli distribution, hypercontractivity, moment inequality, random vector, Rademacher projection.
The author was supported by the Polish KBN grant 2 P03A 027 22.

In this note we establish a new version of the Khinchine–Kahane inequality in which Bernoulli random variables need to be mean-zero but not necessarily symmetric. The constants will now depend on both Bernoulli's distribution and p.

In what follows let $\alpha \in (0, 1/2)$ and $\beta = 1 - \alpha$. Some formulas make sense also if $\alpha = \beta = 1/2$, usually a limit $\alpha \to 1/2$ can be taken. Let $X, X_1, X_2, \dots$ be i.i.d. real random variables with $P(X = -\alpha) = \beta$ and $P(X = \beta) = \alpha$, so that $EX = 0$. We denote by $Y, Y_1, Y_2, \dots$ i.i.d. random variables with $Y \sim X_1 - X_2$, i.e. $P(Y = 1) = P(Y = -1) = \alpha\beta$, $P(Y = 0) = \alpha^2 + \beta^2$. Let $v_1, v_2, \dots$ be vectors of a separable Banach space F. In fact F can be just a normed linear space if we consider finite sums only. By S we will denote a random sum $\sum_{k=1}^n X_k v_k$. Let $p > 2, q = p' = \frac{p}{p-1} \in (1, 2)$ and

$$K_p(\alpha) = \begin{cases} (1/\alpha)^{1/2-1/p} & \text{if } p \le \ln(1/\alpha), \\ \sqrt{\frac{1/\alpha}{\ln(1/\alpha)}}\sqrt{p} & \text{if } p \ge \ln(1/\alpha). \end{cases}$$

By $\gtrsim$ and $\simeq$ we mean inequality and equality up to some universal factor, respectively.

2. Main result

Under the above notation the following theorem holds true:

Theorem 2.1. *There exists a universal positive constant C such that for any positive integer n and any choice of vectors $v_1, v_2, \dots, v_n$ we have*

$$\|S\|_p \le CK_p(\alpha)\|S\|_2.$$

The constant $B_p(\alpha) = CK_p(\alpha)$ is optimal up to a universal factor.

An easy and standard symmetrization argument implies that this theorem can be equivalently stated for $S = \sum_{k=1}^n Y_k v_k$. It suffices to note that for $p \ge 1$ and any random variable Z with $EZ = 0$ there is $\|Z\|_p \le \|Z - Z'\|_p \le 2\|Z\|_p$, where Z' is an independent copy of Z. This follows by the Jensen inequality and the triangle inequality.

Proof. Let us start the proof by showing that the constant $CK_p(\alpha)$ is essentially optimal even if $F = \mathbb{R}$. Indeed, if $p \le \ln(1/\alpha)$ then one can just take $n = 1$ and any $v_1 \ne 0$. If $p \ge \ln(1/\alpha)$ then consider $n = [p/\ln(1/\alpha)]$ and $v_1 = v_2 = \dots = v_n = 1$. In this case one can easily compute

$$\|S\|_2 = \sqrt{n}\|X\|_2 \simeq \sqrt{n\alpha}$$

and estimate

$$\|S\|_p \ge \left((n\beta)^p P(S = n\beta)\right)^{1/p} = n\beta\alpha^{n/p} \ge \frac{1}{2}n\alpha^{2/\ln(1/\alpha)} \simeq n,$$

so that

$$\|S\|_p/\|S\|_2 \gtrsim \sqrt{n/\alpha} \simeq \sqrt{\frac{1/\alpha}{\ln(1/\alpha)}}\sqrt{p}.$$

To prove the upper bound we will adopt the approach of [18]. Let $H = L_2(\{-\alpha,\beta\}, \beta\delta_{-\alpha}+\alpha\delta_\beta)$. Consider a linear operator $L = Id - E$, where E denotes the expectation taken with respect to the measure $\mu_\alpha = \beta\delta_{-\alpha} + \alpha\delta_\beta$ (the probability distribution of X). One can easily check that L is a self-adjoint operator on H. Remark 2 of [17] states (after some reformulation) that for any function $f : \{-\alpha,\beta\} \to \mathbb{R}$ and any $q \in (1,2)$ the inequality

$$Ef^2 - (E|f|^q)^{2/q} \leq \frac{C_q(\alpha)}{\alpha\beta} EfLf \tag{2.1}$$

holds with

$$C_q(\alpha) = \frac{\alpha^{1-2/q} - \beta^{1-2/q}}{\alpha^{-2/q} - \beta^{-2/q}}$$

where the constant is optimal.

The above inequality is a strengthening of the logarithmic Sobolev inequality on the two-point probability space which can be recovered (with the optimal constant due to Rothaus [unpublished], Higuchi and Yoshida [7] [without proof], and Diaconis and Saloff-Coste [5]) as the limit case $q \to 2$. Note that $EfLf = \alpha\beta(f(\beta) - f(-\alpha))^2$, so that the energy functional on the right hand side of the inequality coincides with the so-called discrete gradient squared.

The optimality of $C_q(\alpha)$ implies that

$$C_q(\alpha) = \alpha\beta \sup_f \frac{Ef^2 - (E|f|^q)^{2/q}}{EfLf},$$

where supremum is taken over all non-constant functions. The function $q \mapsto (E|f|^q)^{2/q}$ is non-decreasing on $(1,2)$, hence we deduce that $C_q(\alpha) \leq C_1(\alpha) = \alpha\beta$. Therefore (2.1) can be equivalently stated as

$$\|(Id - \frac{C_q(\alpha)}{\alpha\beta} L)^{1/2} f\|_2 \leq \|f\|_q$$

for all $f \in H$. Let $t_q(\alpha)$ be a positive number such that $\left(1 - \frac{C_q(\alpha)}{\alpha\beta}\right)^{1/2} = e^{-t_q(\alpha)}$, then based on the fact that $\sigma(L) = \{0,1\}$ we can reformulate the above inequality to the following form:

$$\|e^{-t_q(\alpha)L}\|_{L_q \to L_2} \leq 1.$$

Let $\mathbf{H} = L_2(\{-\alpha,\beta\}^n, \mu_\alpha^{\otimes n})$ and let $\mathbf{L} : \mathbf{H} \to \mathbf{H}$ be given by

$$\mathbf{L} = L \otimes Id \otimes \ldots \otimes Id + Id \otimes L \otimes Id \otimes Id \ldots \otimes Id + \ldots + Id \otimes Id \otimes \ldots \otimes Id \otimes L.$$

Standard reasoning implies that for any $f \in \mathbf{H}$ we have

$$\|e^{-t_q(\alpha)\mathbf{L}}\|_{L_q \to L_2} \leq 1,$$

where L_q and L_2 are taken with respect to $\mu_\alpha^{\otimes n}$ (see [12] and [18]). By duality we have

$$\|e^{-t_q(\alpha)\mathbf{L}}\|_{L_2 \to L_p} \leq 1 \tag{2.2}$$

for all $f \in \mathbf{H}$ and $p > 2$ ($q = \frac{p}{p-1}$ is the Hölder conjugate of p). We will apply the above inequality to some particular functions (a similar trick has been used in the proof of Lemma 2.2, [19]). Let $g(x) = \|\sum_{k=1}^n x_k v_k\|$ and $v^*(x) = \sum_{k=1}^n x_k v^*(v_k)$, where $v^* \in F^*$. Clearly, $g = \sup_{v^*:\, \|v^*\|=1} v^*$. Note that $(e^{-t\mathbf{L}})_{t>0}$ is a non-negativity preserving semigroup (this can be proved in many ways, see [18] or note that it is given by convolution with some non-negative function if we identify $\{-\alpha, \beta\}^n$ with the group $\mathbb{Z}_2 \times \mathbb{Z}_2 \times \ldots \times \mathbb{Z}_2$), so that $e^{-t_q(\alpha)\mathbf{L}} g \geq e^{-t_q(\alpha)\mathbf{L}} v^* = e^{-t_q(\alpha)} v^*$ for all $v^* \in F^*$ with $\|v^*\| = 1$. Taking supremum over all v^* with $\|v^*\| = 1$ we get that

$$e^{-t_q(\alpha)\mathbf{L}} g \geq e^{-t_q(\alpha)} g \geq 0,$$

which together with (2.2) yields

$$\|g\|_p \leq e^{t_q(\alpha)} \|g\|_2$$

or equivalently

$$\|S\|_p \leq \left(1 - \frac{C_q(\alpha)}{\alpha\beta}\right)^{-1/2} \|S\|_2.$$

To finish the proof it suffices to show that $K_p(\alpha) \gtrsim \left(1 - \frac{C_q(\alpha)}{\alpha\beta}\right)^{-1/2}$. To this end, notice that

$$\left(1 - \frac{C_q(\alpha)}{\alpha\beta}\right)^{-1/2} = \sqrt{\frac{\beta\alpha^{\frac{2}{p}-1} - \alpha\beta^{\frac{2}{p}-1}}{\beta^{2/p} - \alpha^{2/p}}}$$

and that $\varphi(\alpha) = (1-\alpha)^{2/p} - \alpha^{2/p}$ is a non-increasing function with $\varphi(e^{-p}) \geq 1 - 2e^{-2}$ for $p > 2$. Therefore for $\alpha \leq e^{-p}$ we have

$$\left(1 - \frac{C_q(\alpha)}{\alpha\beta}\right)^{-1/2} \leq \sqrt{\frac{\alpha^{\frac{2}{p}-1}}{\varphi(e^{-p})}} \simeq (1/\alpha)^{1/2-1/p}.$$

The second case is more complicated. First note that for $\gamma \geq 1$ there is

$$\psi(\gamma) = \frac{p(\gamma^{2/p} - 1)}{1 - \gamma^{-2/q}} \gtrsim \max(1, \ln \gamma)$$

uniformly in p. Then observe that for $\alpha \geq e^{-p}$ we have

$$K_p(\alpha) \simeq \sqrt{\frac{1/\alpha}{\max(1, \ln(\beta/\alpha))}} \sqrt{p} \gtrsim \sqrt{\frac{p \cdot \beta/\alpha}{\psi(\beta/\alpha)}} = \left(1 - \frac{C_q(\alpha)}{\alpha\beta}\right)^{-1/2}. \qquad \square$$

It is easy to generalize Theorem 2.1 to the case where X_k's are independent Bernoulli random variables with different parameters $\alpha_k \in (0, 1/2]$. Then the theorem holds with the constant $K_p(\alpha)$ being replaced by $K_p(\min_k \alpha_k) \simeq \max_k K_p(\alpha_k)$ and essentially the same proof works.

One can also reformulate the assertion of Theorem 2.1 to the form

$$\|S\|_p \lesssim K_{p,n}(\alpha) \|S\|_2$$

where

$$K_{p,n}(\alpha) = \begin{cases} (1/\alpha)^{1/2-1/p} & \text{if } p \leq \ln(1/\alpha), \\ \sqrt{\frac{1/\alpha}{\ln(1/\alpha)}} \sqrt{p} & \text{if } \ln(1/\alpha) \leq p \leq n \ln(1/\alpha), \\ \sqrt{n/\alpha} & \text{if } p \geq n \ln(1/\alpha), \end{cases}$$

with the constant optimal (up to a universal factor) for given $p > 2$ and positive integer n. To see it note that for $p \geq n\ln(1/\alpha)$ there is $\|\cdot\|_p \simeq \|\cdot\|_\infty$ for the norms taken with respect to the measure $\mu_\alpha^{\otimes n}$ since the supremum of every function $f : \{-\alpha, \beta\}^n \to \mathbb{R}$ is attained on a set of measure not less than α^n. Therefore for $p \geq n\ln(1/\alpha)$ there is also $\|\cdot\|_p \simeq \|\cdot\|_{n\ln(1/\alpha)}$ and the result follows. The cases indicating the optimality (up to a universal factor) of the constants are the same as before, with some additional zero vectors if necessary.

Constants $A_q(\alpha) = \sqrt{1 - \frac{C_q(\alpha)}{\alpha\beta}}$ for $q \in (1, 2)$ can be obtained in a similar way (just skip the duality argument in the proof), but they are far from optimal for q close to 1. Some standard method to avoid this problem is presented in the next section of the paper. For $q = 1$ a slightly stronger result is true (see [18] for proof):

Theorem 2.2. *Let $Z_1, Z_2, \ldots, Z_n$ be independent symmetric bounded real random variables and let $v_1, v_2, \ldots v_n$ be vectors of some normed linear space. Then for $S = \sum_{k=1}^n Z_k v_k$ we have*

$$\|S\|_2 \leq \max_{1\leq k\leq n} (\sqrt{2}, \frac{\|Z_k\|_\infty}{\|Z_k\|_2})\|S\|_1.$$

This fact used for the random variables $Z_k = Y_k$ for $k = 1, 2, \ldots, n$ easily gives the non-symmetric Khinchine–Kahane inequality with $A_q(\alpha) \geq \kappa_\varepsilon \alpha^{1/q-1/2}$ for $q \in (\varepsilon, 1]$ for any $\varepsilon \in (0, 1)$; however the positive constant κ_ε may tend to zero for $\varepsilon \to 0$. The standard argument involving use of the Hölder inequality can be found in [16]. In fact it gives the optimal (not only optimal up to some universal factor) constants in the Kahane-type inequality

$$\|\sum_{k=1}^n Y_k v_k\|_p \leq (2\alpha\beta)^{1/p-1/q}\|\sum_{k=1}^n Y_k v_k\|_q$$

for all $p \in (0, 2]$ and $q \in (0, 1]$ such that $p \geq q$. One can use a modified symmetrization trick (decsribed below Theorem 2.1) to obtain a slightly weaker version of the above inequality for $S = \sum_{k=1}^n X_k v_k$.

Theorem 2.1 implies a simple tail bound:

Corollary 2.3. *Under notation of Theorem 2.1 for any $t \geq 2Ce\sqrt{1/\alpha}$ we have*

$$P(\|S\| \geq t\|S\|_2) \leq \exp\big(-\frac{\alpha\ln(1/\alpha)t^2}{e^2C^2}\big).$$

Proof. It suffices to use the Chebyshev inequality for $p = \frac{\alpha\ln(1/\alpha)t^2}{e^2C^2}$. □

Clearly one has also $P(\|S\| \geq t\|S\|_2) \leq t^{-2}$ for any $t \geq 1$. Both tail bounds are quite close to optimal. Indeed, let n be a positive integer and put $S = X_1 + X_2 + \ldots + X_n$, so that $\|S\|_2 = \sqrt{n\alpha\beta}$. Then for $t = \sqrt{n\beta/\alpha}$ we have

$$P(\|S\| \geq t\|S\|_2) = P(X_1 + X_2 + \ldots + X_n \geq n\beta) =$$
$$\alpha^n \geq \exp(-2\alpha\ln(1/\alpha)t^2),$$

whereas for $n \leq \frac{1}{2\alpha}$ and $t = \frac{1}{2\sqrt{n\alpha\beta}}$ we have

$$P(\|S\| \geq t\|S\|_2) \geq P(X_1 + X_2 + \ldots + X_n \geq 1/2) \geq$$

$$n\alpha\beta^{n-1} \geq \frac{\beta^n}{4t^2} \geq \frac{1}{4t^2}(1-\alpha)^{\frac{1}{2\alpha}} \geq \frac{1}{8t^2},$$

since $\beta + (n-1)(-\alpha) \geq 1/2$.

Note that Theorem 2.1 reveals some new phenomenon which was not present in the case of the classical Khinchine–Kahane inequality — the equality cases depend strongly both on p and α. Indeed, if one defines a symmetric real random variable T_α by setting $P(|T_\alpha| \geq t)$ to be equal to $\sup_S P(\|S\| \geq t)$ (more precisely, a left-continuous modification of this function of the parameter t), where supremum is taken over all sums $S = \sum_k X_k v_k$ with $\|S\|_2 \leq 1$ and arbitrary n, for a fixed $\alpha \in (0, 1/2]$, then $\lim_{\alpha \to 0} ET_\alpha^2 = \infty$, so that no tight stochastic majorization in terms of the second moment is possible for the class of Bernoulli sums with small parameter α (in the class of Rademacher sums the role of the majorant was played by Gaussian distribution). On the other hand, using the above tail bounds one can easily prove that $\|T_\alpha\|_2 \simeq \sqrt{\ln(1/\alpha)}$, which means that the majorization problems are serious only for *very* small values of α.

The case of real coefficients is much better understood. If $a_k \in \mathbb{R}$ for $k = 1, 2, \ldots, n$ then $\|\sum_{k=1}^n a_k Y_k\|_2 = \sqrt{2\alpha\beta}\,(\sum_{k=1}^n a_k^2)^{1/2}$ and there are also some very precise formulas for $\|\sum_{k=1}^n a_k Y_k\|_p$ involving certain Orlicz norms, see [15]. They are, however, quite complicated.

3. Hypercontractivity constants and chaos moment estimates

Let us turn to a slightly different proof of Theorem 2.1. The functional inequality (2.1) implies that for all vectors v and w of a normed linear space the inequality

$$\|v + \sigma_{2,q}(\alpha) X w\|_2 \leq \|v + Xw\|_q \tag{3.1}$$

holds true, where $q \in (1,2)$ and

$$\sigma_{2,q}(\alpha) = \sqrt{1 - \frac{C_q(\alpha)}{\alpha\beta}} = \alpha^{-1/2}\beta^{-1/2}\sqrt{\frac{\beta^{2-2/q} - \alpha^{2-2/q}}{\alpha^{-2/q} - \beta^{-2/q}}},$$

and the constant cannot be improved (replaced by a greater number). This means that X is $(2,q)$-hypercontractive and its optimal $(2,q)$-hypercontractivity constant is equal to $\sigma_{2,q}(\alpha)$ - see [11] for the definition of hypercontractivity of a random variable and [13, 14] for some related results. The optimal hypercontractivity constants for the symmetric Bernoulli variables (case $\alpha = \beta = 1/2$) were established by Bonami [3] and Beckner [1]. Our proof in some aspects follows [1] (this argument was also presented in [14]). Indeed, put $f(-\alpha) = \|v - \alpha w\|$ and $f(\beta) = \|v + \beta w\|$. Note that $\sigma = \sigma_{2,q}(\alpha) \in (0,1)$,

$$v + \sigma\beta w = (\alpha + \sigma\beta)(v + \beta w) + (1-\sigma)\beta(v - \alpha w)$$

and

$$v - \sigma\alpha w = (1-\sigma)\alpha(v+\beta w) + (\beta+\sigma\alpha)(v-\alpha w).$$

Therefore by the triangle inequality we get $\|v+\sigma\beta w\| \le (\alpha+\sigma\beta)f(\beta) + (1-\sigma)\beta f(-\alpha)$ and $\|v-\sigma\alpha w\| \le (1-\sigma)\alpha f(\beta) + (\beta+\sigma\alpha)f(-\alpha)$. Hence

$$\begin{aligned}
&E\|v+\sigma X w\|^2 \\
&\quad\le \alpha\big((\alpha+\sigma\beta)f(\beta)+(1-\sigma)\beta f(-\alpha)\big)^2+\beta\big((1-\sigma)\alpha f(\beta)+(\beta+\sigma\alpha)f(-\alpha)\big)^2 \\
&\quad= \alpha f(\beta)^2+\beta f(-\alpha)^2-\alpha\beta(1-\sigma^2)\big(f(\beta)-f(-\alpha)\big)^2 = Ef^2-\frac{C_q(\alpha)}{\alpha\beta}EfLf.
\end{aligned}$$

By (2.1) the last expression can be estimated from above by

$$(E|f|^q)^{2/q} = \big(\alpha f(\beta)^q+\beta f(-\alpha)^q\big)^{2/q} = (E\|v+Xw\|^q)^{2/q},$$

which proves our claim. There is equality in (3.1) for $w = \frac{\beta^{2/q}-\alpha^{2/q}}{\beta\alpha^{2/q}+\alpha\beta^{2/q}}v$.

To get a similar estimate for the p-th norm ($p>2$) consider the functions $f, \varphi_{v^*} : \{-\alpha,\beta\} \to \mathbb{R}$ given by $f(x) = \|v+xw\|$ and $\varphi_{v^*}(x) = v^*(v+xw) = v^*(v)+v^*(w)x$ for $v^* \in F^*$. Since $e^{-t_q(\alpha)L}$ preserves non-negativity there is

$$(e^{-t_q(\alpha)L}f)(x) \ge (e^{-t_q(\alpha)L}\varphi_{v^*})(x) = v^* + e^{-t_q(\alpha)}v^*(w)x = v^*(v+e^{-t_q(\alpha)}xw).$$

Taking supremum over $v^* \in F^*$ with $\|v^*\|=1$ we get

$$(e^{-t_q(\alpha)L}f)(x) \ge \|v+e^{-t_q(\alpha)}xw\|$$

which together with (2.2) yields

$$\|v+e^{-t_q(\alpha)}Xw\|_p \le \|v+Xw\|_2.$$

Therefore for all vectors v and w of a normed linear space the inequality

$$\|v+\sigma_{p,2}(\alpha)Xw\|_p \le \|v+Xw\|_2 \tag{3.2}$$

holds true, with the optimal $(p,2)$-hypercontractivity constant

$$\sigma_{p,2}(\alpha) = e^{-t_q(\alpha)} = \sigma_{2,q}(\alpha) = \sqrt{\frac{\beta^{2/p}-\alpha^{2/p}}{\beta\alpha^{\frac{2}{p}-1}-\alpha\beta^{\frac{2}{p}-1}}}.$$

There is equality in (3.2) for

$$w = \frac{(\beta^{2/p}-\alpha^{2/p})^{1/2}(\beta\alpha^{\frac{2}{p}-1}-\alpha\beta^{\frac{2}{p}-1})^{1/2}}{\beta\alpha^{2/p}+\alpha\beta^{2/p}}v.$$

Some tedious but elementary calculations were omitted.

Both (3.2) and (3.1) in a standard way imply Kahane-type inequalities with constants $B_p(\alpha) = \sigma_{p,2}(\alpha)^{-1}$ for $p>2$ and $A_q(\alpha) = \sigma_{2,q}(\alpha)$ for $q \in (1,2)$ respectively (see [13] or [14]). Note that the constants $A_q(\alpha)$ obtained in this way can be far from optimal if q is close to 1. Let us discuss briefly a standard way to deal with this problem. Let $q \in (0,2)$. For any $\varepsilon \in (0,1)$ with $\varepsilon < 2-q$ by the Hölder inequlity we get

$$E\|S\|^{2-\varepsilon} = E\|S\|^{\frac{2(2-\varepsilon-q)}{2-q}}\|S\|^{\frac{q\varepsilon}{2-q}} \le (E\|S\|^2)^{\frac{2-\varepsilon-q}{2-q}}(E\|S\|^q)^{\frac{\varepsilon}{2-q}},$$

so that

$$\|S\|_q/\|S\|_2 \geq (\|S\|_{2-\varepsilon}/\|S\|_2)^{\frac{(2-q)(2-\varepsilon)}{q\varepsilon}} \geq \sigma_{2,2-\varepsilon}(\alpha)^{\frac{(2-q)(2-\varepsilon)}{q\varepsilon}} \to \left((\alpha/\beta)^{\frac{1}{\beta-\alpha}}\right)^{1/q-1/2}$$

as $\varepsilon \to 0$. Now it suffices to note that $(\alpha/\beta)^{\frac{1}{\beta-\alpha}} \simeq \alpha$ to get the following

Theorem 3.1. *There exists a universal positive constant D such that for any $q \in (0,2)$, any positive integer n and any choice of vectors $v_1, v_2, \ldots, v_n$ we have*

$$(D\alpha)^{1/q-1/2}\|S\|_2 \leq \|S\|_q,$$

where $S = \sum_{k=1}^n X_k v_k$.

Obviously, the same theorem holds for $S = \sum_{k=1}^n Y_k v_k$ since $Y \sim X_1 - X_2$.

Let us turn to multilinear random forms. For $A \subset \{1,2,\ldots,n\}$ we can define a function $w_A : \{-\alpha,\beta\}^n \to \mathbb{R}$ by

$$w_A(x) = \prod_{k\in A} x_k.$$

It will be called an α-biased Walsh function and $|A|$ will be called its degree. The family $(w_A)_{A\subset\{1,2,\ldots,n\}}$ is a complete orthogonal system in $L^2(\{-\alpha,\beta\}^n, \mu_\alpha^n)$ and a natural generalization of the classical Walsh system. It has been used by Talagrand [22] and Kindler and Safra [10]. Any linear combination of the α-biased Walsh functions (with vector coefficients from some normed space) will be called an α-biased chaos. It is easy to check that

$$e^{-t\mathbf{L}}w_A = e^{-t|A|}w_A.$$

Using this fact (or an approach based on the hypercontractivity of Bernoulli random variables) one easily gets that for any $p > 2$ and any vectors $(v_A)_{A\subset\{1,2,\ldots,n\}}$ of a normed linear space there is

$$\|\sum_{A\subset\{1,2,\ldots,n\}} \sigma_{p,2}(\alpha)^{|A|}v_A w_A\|_p \leq \|\sum_{A\subset\{1,2,\ldots,n\}} v_A w_A\|_2,$$

where L_2 and L_p norms are taken with respect to the measure $\mu_\alpha^{\otimes n}$. In particular for any $p > 2$ and any real numbers $(a_A)_{A\subset\{1,2,\ldots,n\}}$ we have

$$\|S_d\|_p \leq \sigma_{p,2}(\alpha)^{-d}\|S_d\|_2, \tag{3.3}$$

where $S_d = \sum_{A\subset\{1,2,\ldots,n\}} a_A w_A$ is an α-biased chaos of degree not exceeding d. Here by the degree of chaos we understand the supremum of degrees of all Walsh functions that appear in it with non-zero coefficients. This kind of estimate is standard by now and was established in the case of the homogeneous chaos with vector coefficients by Borell [4] (a slightly different approach can be found for example in [12]; Bonami [2] proved a similar result in the symmetric setting with real coefficients). The inequality (3.3) becomes clear when we notice that $\|S_d\|_2$ is non-decreasing in $|a_A|$ for every $A \subset \{1,2,\ldots,n\}$. As a direct conclusion we get the following

Corollary 3.2. *Let $Z_1, Z_2, \ldots, Z_n$ be independent indicator random variables with $P(Z_k = 1) = \alpha$ and $P(Z_k = 0) = \beta$ for $k = 1, 2, \ldots, n$, where $\alpha \in (0, 1/2)$ and $\beta = 1 - \alpha$. Then for any polynomial $Q \in \mathbb{R}[x_1, x_2, \ldots, x_n]$ and any $p > 2$ we have*

$$\|Q(Z_1, Z_2, \ldots, Z_n)\|_p \le \sigma_{p,2}(\alpha)^{-\deg Q} \|Q(Z_1, Z_2, \ldots, Z_n)\|_2.$$

Recall that $\sigma_{p,2}(\alpha)^{-1} \simeq K_p(\alpha)$. The corollary can have some combinatorial applications since expressions of this type appear naturally for example in small subgraph counts in random graphs. On the other hand the combinatorial estimates it yields usually seem to be far from the best possible.

Proof. Note that $(Z_k)^m \sim Z_k$ for any positive integer m. Therefore we can replace all powers higher than 1 appearing in Q by first powers and then after a natural linear change of variables ($X_k = Z_k - \alpha$) we get an α-biased chaos of order that does not exceed $\deg Q$, so that we can use (3.3) to finish the proof. □

4. Bernoulli projection

Let us finish by extending Pisier's bound on the norm of the Rademacher projection to the nonsymmetric case. The proof given below follows closely [19], with a slight change of setting and notation.

Let F be a normed linear space. For any function $f : \{-\alpha, \beta\}^n \to F$ there exists the unique representation

$$f = \sum_{A \subset \{1,2,\ldots,n\}} w_A v_A,$$

where w_A are α-biased Walsh functions and the vector coefficients v_A can be obtained as

$$v_A = Ef(X_1, X_2, \ldots, X_n) w_A(X_1, X_2, \ldots, X_n).$$

Let us define linear operators Q_k for $k = 0, 1, \ldots, n$ by

$$Q_k f = \sum_{A \subset \{1,2,\ldots,n\} : |A| = k} w_A v_A.$$

We will call $Q_1 : L_2(F) \to L_2(F)$ the Bernoulli projection and we will denote by $d(F, H)$ the Banach–Mazur distance between F and Hilbert space H of the same dimension. The following theorem is a nonsymmetric counterpart to Theorem 2.5 of [19].

Theorem 4.1. *Under the above notation*

$$\|Q_1\|_{L_2(F) \to L_2(F)} \lesssim \begin{cases} d(F, H) & \text{if } d(F, H) \le 1/\alpha, \\ \frac{1/\alpha}{\ln(1/\alpha)} \ln d(F, H) & \text{if } d(F, H) \ge 1/\alpha. \end{cases}$$

Proof. Since the operator Q_1 is a contraction in L_2 when acting on real-valued functions (i.e. $\|Q_1\|_{L_2(\mathbb{R})\to L_2(\mathbb{R})} \le 1$), by Lemma 2.2 of [19] one has $\|Q_1\|_{L_2(F)\to L_2(F)} \le d(F,H)$. Hence we can reduce our considerations to the case $d(F,H) \ge 1/\alpha$. Let us define $T(\varepsilon) : L_2(\{-\alpha,\beta\}) \to L_2(\{-\alpha,\beta\})$ by

$$T(\varepsilon) = Q_0 + \varepsilon Q_1$$

and $\mathbf{T}(\varepsilon) : L_2(\{-\alpha,\beta\}^n) \to L_2(\{-\alpha,\beta\}^n)$ by

$$\mathbf{T}(\varepsilon) = \sum_{k=0}^{n} \varepsilon^k Q_k.$$

Note that $\mathbf{T}(e^{-t}) = e^{-t\mathbf{L}}$. One can easily check that $T(\varepsilon)$ preserves non-negativity for $\varepsilon \in [-\alpha/\beta, 1]$ and therefore by an easy tensorization argument $\mathbf{T}(\varepsilon)$ also preserves non-negativity for $\varepsilon \in [-\alpha/\beta, 1]$ (when acting on real-valued functions). Also, it is easy to prove that $\mathbf{T}(\varepsilon)$ is a contraction in L_2 when acting on real-valued functions, for any $\varepsilon \in [-1,1]$. Hence, again by Lemma 2.2 of [19], $\|\mathbf{T}(\varepsilon)\|_{L_2(F)\to L_2(F)} \le 1$ for any $\varepsilon \in [-\alpha/\beta, 1]$ (in particular for all $\varepsilon \in [-\alpha,\alpha]$). Consider a polynomial $P(\varepsilon) = \sum_{k=0}^{m} x_k \varepsilon^k$ (m will be chosen later) and a trigonometric polynomial $Q(t) = P(\alpha \sin t)$. By the classical Bernstein inequality there is $\|Q'\|_\infty \le m\|Q\|_\infty$, so that

$$|x_1| = |P'(0)| = |Q'(0)|/\alpha \le m\|Q\|_\infty/\alpha \le \frac{m}{\alpha} \sup_{\varepsilon\in[-\alpha,\alpha]} |P(\varepsilon)|.$$

Next, consider $\mathbf{P}(\varepsilon) = \sum_{k=0}^{m} \varepsilon^k Q_k$ (now the polynomial has operator coefficients). Using the above estimate for $P = \varphi(\mathbf{P})$, where $\varphi : Lin\big(L_2(F), L_2(F)\big) \to \mathbb{R}$ is a bounded linear functional, and taking the supremum over all φ with $\|\varphi\| = 1$ one gets

$$\|Q_1\|_{L_2(F)\to L_2(F)} \le \frac{m}{\alpha} \sup_{\varepsilon\in[-\alpha,\alpha]} \|\mathbf{P}(\varepsilon)\|_{L_2(F)\to L_2(F)}.$$

Therefore

$$\begin{aligned} &\|Q_1\|_{L_2(F)\to L_2(F)} \\ &\le \frac{m}{\alpha}\Big(\sup_{\varepsilon\in[-\alpha,\alpha]} \|\mathbf{T}(\varepsilon)\|_{L_2(F)\to L_2(F)} + \sup_{\varepsilon\in[-\alpha,\alpha]} \|\sum_{k>m} \varepsilon^k Q_k\|_{L_2(F)\to L_2(F)}\Big) \\ &\le \frac{m}{\alpha}\big(1 + \alpha^{m+1} d(F,H)\big) \end{aligned}$$

because for $|\varepsilon| \le 1$ the operator $\sum_{k>m} \varepsilon^{k-m-1} Q_k$ is a contraction in L_2 when acting on real-valued functions (use Lemma 2.2 of [19] one more time). Choosing $m = \lceil \frac{\ln d(F,H)}{\ln(1/\alpha)} \rceil$ we finish the proof. □

References

[1] W. Beckner, *Inequalities in Fourier analysis*, Ann. Math. **102** (1975), 159–182.

[2] A. Bonami, *Construction d'opérateurs de convolution sur le groupe D^∞*, C. R. Acad. Sci. Paris Sér. A-B **266** (1968), A655–A657

[3] A. Bonami, *Étude des coefficients de Fourier des fonctions de $L^p(G)$* Ann. Inst. Fourier (Grenoble), **20** 1970 fasc. 2 (1971), 335–402

[4] C. Borell, *On the integrability of Banach space valued polynomials*, in *Séminaire de Probabilités, XIII (Univ. Strasbourg, Strasbourg, 1977/78)*, 1–3, Lecture Notes in Math. **721**, Springer, Berlin, 1979

[5] P. Diaconis, L. Saloff–Coste, *Logarithmic Sobolev inequalities for finite Markov chains*, Ann. Appl. Prob. **6** (1996), 695–750

[6] U. Haagerup, *The best constants in the Khinchine inequality*, Studia Math. **70** (1981), 231–283

[7] Y. Higuchi, N. Yoshida, *Analytic conditions and phase transition for Ising models*, Lecture Notes in Japanese (1995)

[8] J.-P. Kahane, *Sur les sommes vectorielles* $\sum \pm u_n$, C. R. Acad. Sci. Paris **259** (1964), 2577–2580

[9] A. Khintchine, *Über dyadische Brüche*, Math. Z. **18** (1923), 109–116

[10] G. Kindler and S. Safra, *Noise-Resistant Boolean-Functions are Juntas* (personal communication)

[11] W. Krakowiak and J. Szulga, *Hypercontraction principle and random multilinear forms*, Probab. Theory Related Fields **77** (1988), 325–342

[12] S. Kwapień, R. Latała and K. Oleszkiewicz, *Comparison of moments of sums of independent random variables and differential inequalities*, J. Funct. Anal. **136** (1996), 258–268

[13] S. Kwapień and J. Szulga, *Hypercontraction methods in moment inequalities for series of independent random variables in normed spaces*, Ann. Probab. **19** (1991), 369–379

[14] S. Kwapień and W. A. Woyczyński, *Random series and stochastic integrals: single and multiple*, Probability and its Applications, Birkhäuser, 1992

[15] R. Latała, *Estimation of moments of sums of independent real random variables*, Ann. Probab. **25** (1997), 1502–1513

[16] R. Latała and K. Oleszkiewicz, *On the best constant in the Khinchine–Kahane inequality* Studia Math. **109** (1994), 101–104

[17] R. Latała and K. Oleszkiewicz, *Between Sobolev and Poincaré*, in *Geometric aspects of functional analysis,* 147–168, Lecture Notes in Math. **1745**, Springer, 2000

[18] K. Oleszkiewicz, *Comparison of moments via Poincaré-type inequality*, in *Advances in stochastic inequalities (Atlanta, GA, 1997),* 135–148, Contemp. Math. **234**, Amer. Math. Soc., 1999

[19] G. Pisier, *The volume of convex bodies and Banach space geometry,* Cambridge Tracts in Math. **94**, Cambridge University Press, 1989

[20] S. B. Stečkin, *On the best lacunary systems of functions,* (in Russian) Izv. Akad. Nauk SSSR Ser. Mat. **25** (1961), 357–366

[21] S. J. Szarek, *On the best constants in the Khinchine inequality,* Studia Math. **58** (1976), 197–208

[22] M. Talagrand *On Russo's approximate zero-one law,* Ann. Prob. **22** (1994), 1576–1587

[23] P. Whittle, *Bounds for the moments of linear and quadratic forms in independent variables,* Teor. Verojatnost. i Primenen. **5** (1960), 331–335

Acknowledgement

The result concerning the Bernoulli projection was inspired by a question posed by Prof. Vitali Milman and it was obtained while I was visiting the Univeristy of Missouri–Columbia. I would like to thank my MU host, Prof. Alexander Koldobsky for his hospitality and creating excellent research conditions.

Institute of Mathematics, Warsaw University, Banacha 2, 02-097 Warsaw, Poland
E-mail address: koles@mimuw.edu.pl

Progress in Probability, Vol. 56, 169–185

Dimensionality Reduction in Extremal Problems for Moments of Linear Combinations of Vectors with Random Coefficients

Iosif Pinelis

Abstract. It is shown that

$$\sup\left\{\mathsf{E}\varphi\left(\left\|\sum_{i=1}^{n}\xi_i x_i\right\|^2\right) : x_i \in H,\ \sum_{i=1}^{n}\|x_i\|^2 = B^2\right\}$$

does not depend on $\dim H \geq 1$, where $(H, \|\cdot\|)$ is a Hilbert space, φ is any convex function, and $\xi_1, \dots, \xi_n$ are any (real-valued) random variables. An immediate corollary is the following vector extension of the Whittle–Haagerup inequality: let $\varepsilon_1, \dots, \varepsilon_n$ be independent Rademacher random variables, and let $x_1, \dots, x_n$ be vectors in H; then

$$\mathsf{E}\left\|\sum_{i=1}^{n}\varepsilon_i x_i\right\|^p \leq \mathsf{E}\,|\nu|^p\left(\sum_{i=1}^{n}\|x_i\|^2\right)^{p/2} \qquad \forall p \geq 2,$$

where $\nu \sim N(0,1)$. Dimensionality reduction in the case when all the lengths $\|x_i\|$ are fixed is also considered. Open problems are stated.

1. Introduction

Let $\varepsilon_1, \varepsilon_2, \dots$ be independent Rademacher random variables (r.v.'s), so that

$$\mathsf{P}(\varepsilon_i = 1) = \mathsf{P}(\varepsilon_i = -1) = 1/2$$

for all i. The first exact versions of the Khinchine inequality [8] are apparently due to Whittle [22], who proved that, for any $p \geq 3$, any natural n, any real numbers $a_1, \dots, a_n$, and

$$B := \left(a_1^2 + \cdots + a_n^2\right)^{1/2},$$

one has

$$\mathsf{E}\left|\sum_{i=1}^{n}\varepsilon_i a_i\right|^p \leq \mathsf{E}\left|\sum_{i=1}^{n}\varepsilon_i \frac{B}{\sqrt{n}}\right|^p, \tag{1.1}$$

2000 *Mathematics Subject Classification.* Primary: 60E15; Secondary: 60B99, 60G50.
Key words and phrases. Khinchine inequality, exact probability inequalities, extremal problems, extreme points, exact comparison inequalities, Rademacher sums, linear combinations of vectors with random coefficients, moments.

which implies

$$\mathsf{E}\left|\sum_{i=1}^{n} \varepsilon_i a_i\right|^p \leq B^p\, \mathsf{E}\,|\nu|^p, \tag{1.2}$$

where ν is a standard normal r.v. By the exactness, say of inequality (1.1), we mean that the right-hand side of inequality (1.1) is the exact upper bound on its left-hand side, for any given values of the parameters on which the bound depends: $p \geq 3$, a natural n, and $B \geq 0$; similarly, inequality (1.2) is exact (given any $p \geq 3$ and $B \geq 0$). Haagerup [7] proved (1.2) for all $p \geq 2$. The case $2 < p < 3$ is the most difficult one; the only known proof in this case is due to Haagerup [7]. In contrast, the exact upper bound, B^p, for the case $0 < p \leq 2$ is trivial.

Note that, while (1.2) holds for all $p \geq 2$, the same cannot be said of (1.1). Indeed, e.g., if $C(p, n, B)$ stands for the right-hand side of (1.1), then $C(5/2, 2, 1) > C(5/2, 3, 1)$.

Analogues of the Khinchine inequality for the maximum of the first n cumulative sums of weighted Rademacher r.v.'s were considered in Paley and Zygmund [10], Graversen and Peskir [6], and Pinelis [17].

The only condition on the moment function $g_p(u) = |u|^p$ used in the proof of inequality (1.1) in [22] was that the second derivative g_p'' is convex, for $p \geq 3$. Therefore, one has

$$\mathsf{E} g\left(\sum_{i=1}^{n} \varepsilon_i a_i\right) \leq \mathsf{E} g\left(\sum_{i=1}^{n} \varepsilon_i \frac{B}{\sqrt{n}}\right) \tag{1.3}$$

and

$$\mathsf{E} g\left(\sum_{i=1}^{n} \varepsilon_i a_i\right) \leq \mathsf{E} g(B\nu) \tag{1.4}$$

for all (generalized moment functions) $g \in C^2_{\mathsf{conv}}$, where C^2_{conv} stands for the set of all functions $g\colon \mathbb{R} \to \mathbb{R}$ whose second derivative g'' is convex. The richness of the class C^2_{conv} allows one to extract from (1.4) the following tail comparison inequality, whose multivariate analogue was obtained in Pinelis [13, 14]:

$$\mathsf{P}\left(\sum_{i=1}^{n} \varepsilon_i a_i \geq u\right) \leq c\,\mathsf{P}(B\nu \geq u) \quad \forall u \in \mathbb{R}, \tag{1.5}$$

where $c := 2e^3/9 = 4.46\ldots$. The proof of this inequality and its mentioned multivariate analogue was simplified in Pinelis [15]. A very simple proof of (1.5) was given in Bobkov, Götze, and Houdré [1], albeit with a larger constant c. For other developments concerning inequality (1.5), see Dufour and Hallin [3] and Pinelis [19].

A "discrete" counterpart of inequality (1.5) was obtained in Pinelis [16]: if $a_1^2 + \cdots + a_n^2 = n$, then for all $u \in 2\mathbb{Z} - n$

$$\mathsf{P}\left(\sum_{i=1}^{n} \varepsilon_i a_i \geq u\right) \leq c\,\mathsf{P}\left(\sum_{i=1}^{n} \varepsilon_i \geq u\right)$$

with the same $c = 2e^3/9$. Using the central limit theorem, it is easy to see that the latter inequality implies (1.5).

A number of other exact probability inequalities for generalized moment functions $g \in C^2_{\mathsf{conv}}$ were given in Cox and Kemperman [2], Eaton [4, 5], Utev [20], and Pinelis [14, 18].

In particular, Utev [20] proved that

$$\mathsf{E}g\left(\left\|\sum_{i=1}^{n}\varepsilon_i x_i\right\|\right) \le \mathsf{E}g\left(\sum_{i=1}^{n}\varepsilon_i\|x_i\|\right) \quad \text{for all } \textit{even } g \in C^2_{\mathsf{conv}}. \tag{1.6}$$

Here and in what follows, $x_1, \ldots, x_n$ are arbitrary vectors in any given Hilbert space H, equipped with a scalar product $(\cdot,\cdot)$ and the corresponding Hilbert norm $\|\cdot\|$.

In Pinelis [14], the following multidimensional generalization of (1.2) was obtained:

$$\mathsf{E}g\left(\left\|\sum_{i=1}^{n}\eta_i x_i\right\|\right) \le \mathsf{E}g\left(\left\|\sum_{i=1}^{n}\xi_i x_i\right\|\right) \quad \text{for all } \textit{even } g \in C^2_{\mathsf{conv}}, \tag{1.7}$$

where $\eta_1, \ldots, \eta_n$ are independent r.v.'s with $\mathsf{E}\eta_i = 0$ and $\mathsf{P}(|\eta_i| \le 1) = 1$ and $\xi_1, \ldots, \xi_n$ are independent symmetric r.v.'s with $\mathsf{E}\xi_i^2 = 1$, for all i. In particular, letting $H = \mathbb{R}$, $\eta_i = \varepsilon_i$, and $\xi_i \sim N(0,1)$, one has (1.4) (cf. Eaton [4, 5]). Thus, (1.7) may be considered as a generalized Khinchine inequality.

In Pinelis [18], two classes of spherically symmetric functions $g\colon H \to \mathbb{R}$, generalizing the class C^2_{conv}, are characterized, and applications to exact probability inequalities are given.

2. Statement of results

Consider the hypothetical inequality

$$\mathsf{E}\varphi\left(\left\|\sum_{i=1}^{n}\varepsilon_i x_i\right\|^2\right) \le \mathsf{E}\varphi\left(\left(\sum_{i=1}^{n}\varepsilon_i\|x_i\|\right)^2\right), \tag{2.1}$$

where $\varphi\colon [0,\infty) \to [0,\infty)$ is any convex function. The latter inequality would imply (2.3) indeed for all $p \ge 2$ (take here $\varphi(u) = u^{p/2}$).

However, inequality (2.1) is not true in general. For example, consider the situation when $n = 3$, $\|x_1\| = \|x_2\| = \|x_3\| = 1$, $(x_1, x_2) = (x_1, x_3) = (x_2, x_3) = -1/2$, and $\varphi(u) = \max(0, u-1)$.

If now $n > 3$, then one can complement these x_1, x_2, x_3 by sufficiently small $x_4, \ldots, x_n$ so that inequality (2.1) still does not hold.

Alternatively, one may generalize the above example from $n = 3$ to any odd $n > 1$. Indeed, let $x_i := \sqrt{\frac{n}{n-1}}(e_i - \overline{e})$ for $i \in \{1, \ldots, n\}$, where $(e_1, \ldots, e_n)$ is any orthonormal system of n vectors in H and $\overline{e} := (e_1 + \cdots + e_n)/n$. Let again $\varphi(u) =$

$\max(0, u-1)$. Then the right-hand side of (2.1) is $n-1+\mathsf{P}\left(\sum_{i=0}^{n} \varepsilon_i = 0\right) = n-1$, while the left-hand side is $n-1+\mathsf{P}\left(\left|\sum_{i=0}^{n} \varepsilon_i\right| = n\right) > n-1$.

In view of such examples, here we shall obtain exact inequalities, which are weaker in some aspects than (2.1) and yet imply a vector version of the Haagerup inequality - see Theorem 2.2 below. At the same time, those exact inequalities will hold for random variables more general than independent Rademacher.

Theorem 2.1. *For any Hilbert space H with* $\dim H \geq 1$, *any convex function φ, any natural n, any (real-valued) r.v.'s $\xi_1, \ldots, \xi_n$, and any $B > 0$, one has*

$$\sup\left\{\mathsf{E}\varphi\left(\left\|\sum_{i=1}^{n} \xi_i x_i\right\|^2\right) : x_i \in H,\ \sum_{i=1}^{n} \|x_i\|^2 = B^2\right\}$$
$$= \sup\left\{\mathsf{E}\varphi\left(\left(\sum_{i=1}^{n} \xi_i a_i\right)^2\right) : a_i \in \mathbb{R},\ \sum_{i=1}^{n} a_i^2 = B^2\right\}. \qquad (2.2)$$

The proof of this theorem, as well as other proofs, is deferred to the next section.

Theorem 2.1 is not very hard to prove. Moreover, since it is true for any r.v.'s $\xi_1, \ldots, \xi_n$ whatsoever, one might think that the statement of Theorem 2.1 would easily follow from the special case of the statement when the ξ_i's are non-random. However, that is not so, because sup is a convex (and not concave) functional.

Theorem 2.1 states that the left-hand side of (2.2) does not depend on the dimension of the Hilbert space H. Hence, it can be considered as a dimensionality reduction device, as well as (1.6).

Theorem 2.2. *For* all $p \geq 2$,

$$\mathsf{E}\left\|\sum_{i=1}^{n} \varepsilon_i x_i\right\|^p \leq \mathsf{E}\,|\nu|^p \left(\sum_{i=1}^{n} \|x_i\|^2\right)^{p/2}. \qquad (2.3)$$

This follows immediately from Theorem 2.1 and the Whittle–Haagerup inequality (1.2).

Note that inequalities (1.6) and (1.2) together imply (2.3) - but only for $p \geq 3$.

Remark. A referee suggested the following alternative proof of Theorem 2.2. Let here $\xi_1, \xi_2, \ldots$ be i.i.d. $N(0,1)$ r.v.'s defined on a probability space with a probability measure μ. Then, for any $p > 0$ and any separable Hilbert space H with an orthonormal basis $e_1, e_2, \ldots$, one has the classical isometric embedding

$$H \ni x \mapsto \left(\mathsf{E}|\nu|^p\right)^{-1/p} \sum_j (x, e_j)\xi_j$$

of H into $L_p(\mu)$. Therefore, it is enough to establish (2.3) with $\|\cdot\|_p$ in place of the Hilbert norm $\|\cdot\|$ and with arbitrary functions $y_i \in L_p(\mu)$ in place of x_i. But, using the Haagerup and Minkowski inequalities, one has for all $p \geq 2$

$$\begin{aligned}
\mathsf{E}\left\|\sum_i \varepsilon_i y_i\right\|_p^p &= \int \mathsf{E}\left|\sum_i \varepsilon_i y_i(t)\right|^p \mu(dt) \leq \mathsf{E}\,|\nu|^p \int \left(\sum_i y_i(t)^2\right)^{p/2} \mu(dt) \\
&= \mathsf{E}\,|\nu|^p \left\|\sum_i y_i^2\right\|_{p/2}^{p/2} \leq \mathsf{E}\,|\nu|^p \left(\sum_i \|y_i^2\|_{p/2}\right)^{p/2} \\
&= \mathsf{E}\,|\nu|^p \left(\sum_i \|y_i\|_p^2\right)^{p/2}.
\end{aligned}$$

As the referee indicated, a reasoning similar to this was used as early as in the 70s by Szarek [21, Theorem 1b] to extend his lower bound

$$\mathsf{E}\left|\sum_{i=1}^n \varepsilon_i a_i\right| \geq 2^{-1/2}\left(a_1^2 + \cdots + a_n^2\right)^{1/2}$$

to L_1 and hence to any Banach space which is isometrically isomorphic to a subspace of L_1. In turn, Szarek ascribed the corresponding ideas to Orlicz [9] and also seemingly to B. Tomaszewski.

However, in contrast with Theorem 2.1, the just described method of dimensionality reduction seems to work only for power moment functions φ.

Let us now turn back to inequality (2.1), which, as we saw, does not hold in general. One may ask: To what extent dimensionality reduction is possible for $\mathsf{E}\varphi\left(\|\sum_{i=1}^n \varepsilon_i x_i\|^2\right)$ when one fixes all the lengths $\|x_1\|, \ldots, \|x_n\|$ rather than only the sum of their squares? We shall show that it is then possible to reduce the dimension from n to $\lfloor(\sqrt{1+8n}-1)/2\rfloor \sim \sqrt{2n}$.

Fix an arbitrary n-tuple of strictly positive real numbers

$$\mathbf{a} := (a_1, \ldots, a_n)$$

and denote by $K_n(\mathbf{a})$ the set of all n-tuples $\mathbf{x} := (x_1, \ldots, x_n) \in H^n$ with the given lengths $\|x_1\| = a_1, \ldots, \|x_n\| = a_n$:

$$K_n(\mathbf{a}) := \{\mathbf{x} \in H^n \colon \|x_1\| = a_1, \ldots, \|x_n\| = a_n\}.$$

Further, let $K_{n,r}(\mathbf{a})$ be the set of all n-tuples $(x_1, \ldots, x_n) \in K_n(\mathbf{a})$ of rank r:

$$K_{n,r}(\mathbf{a}) := \{\mathbf{x} \in K_n(\mathbf{a}) \colon \operatorname{rank}\mathbf{x} = r\};$$

as usual, by the rank of an n-tuple $\mathbf{x} = (x_1, \ldots, x_n) \in H^n$ we mean the dimension of its linear hull.

Theorem 2.3. *For any Hilbert space H with* $\dim H \geq 1$, *any convex function φ, any natural n, any (real-valued) r.v.'s $\xi_1, \ldots, \xi_n$, and any $\mathbf{a} := (a_1, \ldots, a_n) \in (0, \infty)^n$,* *one has*

$$\sup\left\{\mathsf{E}\varphi\left(\left\|\sum_{i=1}^n \xi_i x_i\right\|^2\right) : \mathbf{x}\in K_n(\mathbf{a})\right\}$$
$$= \sup\left\{\mathsf{E}\varphi\left(\left\|\sum_{i=1}^n \xi_i x_i\right\|^2\right) : \mathbf{x}\in K_n(\mathbf{a}), \operatorname{rank}\mathbf{x}\le r_*(n)\right\}, \quad (2.4)$$

where

$$r_*(n) := \left\lfloor \frac{\sqrt{1+8n}-1}{2} \right\rfloor. \quad (2.5)$$

Statements below detail that of Theorem 2.3. To present them, we need some preliminaries.

For any r.v.'s $\eta_1,\dots,\eta_m$, let us say that they are linearly independent if $\mathsf{P}(c_1\eta_1+\dots+c_m\eta_m\ne 0)\ne 0$ for any real constants $c_1,\dots,c_m$ not all of which are zero.

Fix any natural n, any $\mathbf{a}=(a_1,\dots,a_n)\in(0,\infty)^n$, and any r.v.'s $\xi_1,\dots,\xi_n$.

For any given convex function φ, let us say that an n-tuple $\mathbf{x}=(x_1,\dots,x_n)\in K_n(\mathbf{a})$ is a *φ-maximal system* if

$$\mathsf{E}\varphi\left(\left\|\sum_{i=1}^n \xi_i x_i\right\|^2\right) \ge \mathsf{E}\varphi\left(\left\|\sum_{i=1}^n \xi_i y_i\right\|^2\right) \quad \forall (y_1,\dots,y_n)\in K_n(\mathbf{a}).$$

For any $\mathbf{x}=(x_1,\dots,x_n)\in K_{n,r}(\mathbf{a})$ (so that the rank of $\mathbf{x}$ is r) and for any given orthonormal basis (o.b.) $\mathbf{e}:=(e_1,\dots,e_r)$ of $\mathbf{x}$, let $x_{i,k}$ denote the coordinates of x_i in the o.b. $\mathbf{e}$, so that

$$x_i = x_{i,1}e_1+\dots+x_{i,r}e_r \quad \forall i\in\{1,\dots,n\}.$$

For any $\mathbf{x}\in K_{n,r}(\mathbf{a})$, consider the following condition:

(A) for *every* o.b. of $\mathbf{x}$, the $n\times r$ matrix $\left(x_{i,k}^2\right)$ (consisting of the squares of the coordinates) is of (full) rank r.

Definition 2.4. Let $\operatorname{ext} K_n(\mathbf{a})$ stand for the set of all $\mathbf{x}\in K_n(\mathbf{a})$ that satisfy condition (A).

Here ext stands for "extreme". This notation is justified by

Proposition 2.5.

(i) For any convex function φ such that $\mathsf{E}\varphi\left(\left(\sum_{i=1}^n|\xi_i|a_i\right)^2\right)<\infty$, there exists a φ-maximal system in $\operatorname{ext} K_n(\mathbf{a})$.

(ii) If, moreover, the r.v.'s $\xi_i\xi_j$ $(1\le i<j\le n)$ are linearly independent and the function φ is strictly convex, then all *φ-maximal systems are in* $\operatorname{ext} K_n(\mathbf{a})$.

Thus, in a certain sense, Proposition 2.5 reduces the consideration of the set of the φ-maximal systems for convex functions φ to the consideration of the set $\operatorname{ext} K_n(\mathbf{a})$. As for the condition of linear independence of the r.v.'s $\xi_i\xi_j$ ($1 \le i < j \le n$), it is satisfied if e.g. the ξ_i's are (stochastically) independent and have zero means and nonzero variances (for then the $\xi_i\xi_j$'s are mutually orthogonal); in particular, Rademacher ε_i's will obviously do for the ξ_i's.

Theorem 2.6.

(i) *All* $\mathbf{x} \in \operatorname{ext} K_n(\mathbf{a})$ *are of rank* $r \le r_*(n)$; *that is,*

$$\operatorname{ext} K_n(\mathbf{a}) \subseteq \bigcup_{r=1}^{r_*(n)} K_{n,r}(\mathbf{a}),$$

where $r_*(n)$ *is given by* (2.5).

(ii) *For every natural* $r \le r_*(n)$, *almost all* $\mathbf{x} \in K_{n,r}(\mathbf{a})$ *are in* $\operatorname{ext} K_n(\mathbf{a})$. *Here, "almost all" can be understood in almost any reasonable sense, as will be clear from the proof.*

Thus, speaking somewhat loosely, Proposition 2.5 and Theorem 2.6 together say that all φ-maximal systems $\mathbf{x} = (x_1, \dots, x_n)$ are of rank $\le r_*(n)$, and almost all n-tuples $\mathbf{x} = (x_1, \dots, x_n)$ of any rank $\le r_*(n)$ are φ-maximal systems for some convex functions φ.

On the other hand, the following proposition shows that for $r \in \{3, \dots, r_*(n)\}$ the set $K_{n,r}(\mathbf{a}) \setminus \operatorname{ext} K_n(\mathbf{a})$ necessarily contains systems $\mathbf{x}$ "in a general position".

Proposition 2.7. *Let* r *be a natural number and let* $n \ge r(r+1)/2$, *so that* $r \le r_*(n)$.

(i) *If* $r = 1$, *then* $K_{n,r}(\mathbf{a}) \setminus \operatorname{ext} K_n(\mathbf{a}) = \emptyset$.

(ii) *If* $r = 2$, *then every system* $(x_1, \dots, x_n) \in K_{n,r}(\mathbf{a}) \setminus \operatorname{ext} K_n(\mathbf{a})$ *contains a pair of collinear vectors.*

(iii) *If* $r \ge 3$, *then there exists a system* $(x_1, \dots, x_n) \in K_{n,r}(\mathbf{a}) \setminus \operatorname{ext} K_n(\mathbf{a})$ *"in a general position", i.e. such that every subsystem of* $(x_1, \dots, x_n)$ *consisting of* r *vectors is of (full) rank* r.

Some of the above results were announced without proof in [12].

We conclude this section with

Open problems.

(i) It follows from (1.6) that

$$\mathsf{E}\left\|\sum_{i=1}^n \varepsilon_i x_i\right\|^p \le \mathsf{E}\left|\sum_{i=1}^n \varepsilon_i \|x_i\|\right|^p \tag{2.6}$$

for all $p \ge 3$, which reduces the dimension of the linear hull of the x_i's from n (in general) to 1. However, for $p \in (2,3)$ we have so far been able to reduce the dimension only to $r_*(n) \sim \sqrt{2n}$, by Theorem 2.3. Does inequality (2.6) hold for $p \in (2,3)$? (Comparison with the Whittle–Haagerup inequality suggests that the case $p \in (2,3)$ may be much more difficult concerning (2.6) as well.) More generally, let us recall that (2.1) does not hold for all convex functions φ. However, one can ask whether (2.1) holds for all functions φ of the form $\varphi(t) = f(\sqrt{t})$, where

$$f \in C^1_{\text{conv}} := \{f\colon\ f'(0+) = 0 \text{ and } f' \text{ is convex on } [0, \infty)\}.$$

(All such functions φ are convex, because $f'(u)/u$ is non-decreasing in $u > 0$ for all $f \in C^1_{\text{conv}}$.) Note that, if $f(u) = u^p$ for some $p \geq 2$ and all $u \geq 0$, then $f \in C^1_{\text{conv}}$ and the corresponding function φ is given by $\varphi(t) = t^{p/2}$. Therefore, to prove (2.6) for all $p \geq 2$, it suffices to prove (2.1) just for the functions φ of the form $\varphi(t) = f_a(\sqrt{t})$, where $f_a(u) := \max(0, u-a)^2$, for all $a > 0$; indeed, every $f \in C^1_{\text{conv}}$ admits the Taylor expansion

$$f(u) = f(0) + \frac{u^2}{2}\, f''(0) + \frac{1}{2}\int_{0+}^{\infty} f_a(u)\, df''(a) \quad \forall u \geq 0,$$

where f'' is the (non-decreasing) right derivative of the convex function f'.

(ii) Can the bound $r \leq r_*(n)$ in Theorem 2.3 be improved? A claim that this bound is the best possible in a certain weaker sense was made in [12]. However, I have since found an error in the proof of that claim.

3. Proofs

Proof of Theorem 2.1. Take any system of vectors $(x_1, \dots, x_n) \in H^n$ such that

$$\sum_{i=1}^{n} \|x_i\|^2 = B^2.$$

The $n \times n$ Gram matrix

$$\alpha := ((x_i, x_j))$$

is symmetric and nonnegative-definite. Therefore, it has a spectral decomposition of the form

$$\alpha = \tau\lambda\tau^T,$$

where the $n \times n$ matrices $\tau = (\tau_{i,j})$ and $\lambda = (\lambda_{i,j})$ are orthogonal and diagonal, respectively; τ^T stands for the transpose of τ; moreover, $\lambda_{i,i} \geq 0$ for all i and

$$\sum_{i=1}^{n} \lambda_{i,i} = \sum_{i=1}^{n} \alpha_{i,i} = \sum_{i=1}^{n} \|x_i\|^2 = B^2.$$

Hence,

$$\left\| \sum_{i=1}^{n} \xi_i x_i \right\|^2 = \sum_{i,j=1}^{n} \alpha_{i,j}\xi_i\xi_j = \sum_{i,j,k=1}^{n} \lambda_{k,k}\tau_{i,k}\tau_{j,k}\xi_i\xi_j = \sum_{k=1}^{n} \frac{\lambda_{k,k}}{B^2} \left(\sum_{i=1}^{n} \xi_i B\tau_{i,k} \right)^2.$$

Using now the convexity of φ, one has

$$\mathsf{E}\varphi\left(\left\|\sum_{i=1}^n \xi_i x_i\right\|^2\right) \le \sum_{k=1}^n \frac{\lambda_{k,k}}{B^2}\mathsf{E}\varphi\left(\left(\sum_{i=1}^n \xi_i B\tau_{i,k}\right)^2\right)$$
$$\le \sup\left\{\mathsf{E}\varphi\left(\left(\sum_{i=1}^n \xi_i a_i\right)^2\right) : \sum_{i=1}^n a_i^2 = B^2\right\},$$

because $\sum_{i=1}^n (B\tau_{i,k})^2 = B^2\ \forall k$, in view of the orthogonality of τ. Now the statement of the theorem follows. □

Proof of Theorem 2.3. This follows immediately from part (i) of Proposition 2.5 and part (i) of Theorem 2.6, which will be proved below. □

Along with condition (A) stated on page 174, let us introduce the following conditions:

(B) for *every* o.b. of $\mathbf{x}$, the system of homogeneous linear equations

$$\sum_{1\le k\le \ell\le r} x_{i,k}x_{i,\ell}\rho_{k,\ell} = 0 \quad \forall i \in \{1,\dots,n\} \tag{3.1}$$

for the unknowns $\rho_{k,\ell}$ $(1 \le k \le \ell \le r)$ has no nontrivial solutions;

(C) the same, for *some* o.b. of $\mathbf{x}$.

Lemma 3.1. *The conditions (A), (B), and (C) are mutual equivalent.*

Proof. The statement of the lemma will follow from the implications

$$(\neg\mathrm{A}) \implies (\neg\mathrm{B}) \implies (\neg\mathrm{C}) \implies (\neg\mathrm{A}),$$

where $\neg$ stands for the negation.

$(\neg\mathrm{A}) \implies (\neg\mathrm{B})$. If (A) is false, then for some o.b. of $\mathbf{x}$ the system of equations

$$\sum_{k=1}^r x_{i,k}^2\nu_k = 0 \quad \forall i \in \{1,\dots,n\}$$

has a nontrivial solution for the unknowns ν_k; letting now $\rho_{k,\ell} := \nu_k\delta_{k,\ell}$, one sees that (B) is also false. Here and in the sequel, $\delta_{k,\ell} := I\{k = \ell\}$ denotes the Kronecker symbol, as usual.

$(\neg\mathrm{C}) \implies (\neg\mathrm{A})$. Suppose that (C) is false; that is, for every (and hence for some) o.b. $\mathbf{e} = (e_1,\dots,e_r)$ of $\mathbf{x} = (x_1,\dots,x_n)$, system (3.1) has a nontrivial solution $(\rho_{k,\ell})$. For such $\rho_{k,\ell}$'s, let $\mu = (\mu_{k,\ell})$ be the $r \times r$ matrix with

$$\mu_{k,\ell} = \mu_{\ell,k} = \rho_{k,\ell}/(2 - \delta_{k,\ell})$$

whenever $1 \le k \le \ell \le r$. Note that the matrix μ is nonzero and symmetric. Next, let here (not to be confused with the standard normal r.v. ν in the Introduction)

$$\nu := x\mu x^T,$$

where

$$x = (x_{i,j}),$$

the $n \times r$ matrix of the coordinates of the vectors x_i in the o.b. $\mathbf{e}$. Then (3.1) means that

$$\nu_{i,i} = 0 \quad \forall i \in \{1, \dots, n\}. \tag{3.2}$$

Further, let

$$\mu = \tau \lambda \tau^T$$

be a spectral decomposition of μ, where $r \times r$ matrices τ and λ are, respectively, orthogonal and (nonzero) diagonal. Let also

$$\tilde{x} := x\tau;$$

note that the elements $\tilde{x}_{i,k}$ of the matrix $\tilde{x}$ are the coordinates of x_i in the o.b.

$$\tilde{e}_j := \tau_{1,j} e_1 + \dots + \tau_{r,j} e_r, \quad j \in \{1, \dots, r\}. \tag{3.3}$$

Moreover, one has

$$\nu = \tilde{x} \lambda \tilde{x}^T.$$

It follows by (3.2) that $\sum_{k=1}^r \tilde{x}_{i,k}^2 \lambda_{k,k} = 0$, while λ is diagonal and nonzero. Thus, (A) is false.

($\neg$B) $\implies$ ($\neg$C). Suppose that (B) is false. Then, for some o.b. $\mathbf{e} = (e_1, \dots, e_r)$ of $\mathbf{x}$, there exists a nonzero symmetric $r \times r$ matrix μ such that the matrix $\nu = x\mu x^T$ satisfies condition (3.2). Let now $\tilde{\mathbf{e}} = (\tilde{e}_1, \dots, \tilde{e}_r)$ be any other o.b. of $\mathbf{x}$. Then there exists an orthogonal $r \times r$ matrix $\tau = (\tau_{i,j})$ such that (3.3) holds. If $\tilde{x} = (\tilde{x}_{i,k})$ is the matrix of the coordinates of the vectors x_i in the o.b. $\tilde{\mathbf{e}}$, then $x = \tilde{x}\tau^T$, whence $\nu = x\mu x^T = \tilde{x}\tilde{\mu}\tilde{x}^T$, where $\tilde{\mu} := \tau^T \mu \tau$ is a nonzero symmetric matrix. This implies that (C) is false. □

Remark. The just proved Lemma 3.1 implies, in particular, that conditions (B) and (C) are equivalent to each other. In contrast, it is clear from part (i) of Theorem 2.6 that one cannot replace the term "every" in condition (A) by "some".

Proof of Proposition 2.5. Introduce the set $\mathcal{K}_n(\mathbf{a})$ of all symmetric nonnegative-definite $n \times n$ real matrices $\alpha = (\alpha_{i,j})$ with $\alpha_{i,i} = a_i$ for all $i \in \{1, \dots, n\}$.

This set is convex and compact relative to the natural linear and topological structure on $\mathcal{K}_n(\mathbf{a})$.

Let $\operatorname{ext} \mathcal{K}_n(\mathbf{a})$ denote the set of all extreme points of $\mathcal{K}_n(\mathbf{a})$.

Let us identify the systems of vectors $\mathbf{x} = (x_1, \dots, x_n)$ that can be obtained from one another via an orthogonal transformation. Then the formula

$$\alpha_{i,j} = (x_i, x_j) \quad \text{for all } i \text{ and } j \text{ in } \{1, \dots, n\} \tag{3.4}$$

defines a 1-to-1 correspondence between $K_n(\mathbf{a})$ and $\mathcal{K}_n(\mathbf{a})$.

Lemma 3.2. *With the stated identification, formula* (3.4) *also carries a* 1-*to*-1 *correspondence between* $\operatorname{ext} K_n(\mathbf{a})$ *and* $\operatorname{ext} \mathcal{K}_n(\mathbf{a})$.

Proof. One has

$$\alpha = xx^T, \tag{3.5}$$

where $x = (x_{i,k})$ is the $n \times r$ matrix of the coordinates of vectors $x_1, \dots, x_n$ in any o.b. of the system $\mathbf{x} = (x_1, \dots, x_n)$. Since matrix x is of full rank, one can extend x to a non-singular $n \times n$ matrix. Denoting this extended matrix also by x, one can rewrite (3.5) as

$$\alpha = x\lambda x^T, \tag{3.6}$$

where $\lambda = (\lambda_{i,j})$ is an $n \times n$ matrix with

$$\lambda_{i,j} := \delta_{i,j} I\{i \le r\}.$$

If $\alpha \in \mathcal{K}_n(\mathbf{a}) \setminus \operatorname{ext} \mathcal{K}_n(\mathbf{a})$, then there exists a nonzero symmetric $n \times n$ matrix $\nu = (\nu_{i,j})$ such that $\alpha - \nu \in \mathcal{K}_n(\mathbf{a})$ and $\alpha + \nu \in \mathcal{K}_n(\mathbf{a})$, so that $\alpha - \nu$ and $\alpha + \nu$ are symmetric nonnegative-definite matrices and $\nu_{i,i} = 0$ for all $i \in \{1, \dots, n\}$. Introduce now

$$\mu := x^{-1}\nu \left(x^{-1}\right)^T \ne 0,$$

so that

$$\nu = x\mu x^T.$$

Then the matrices $\lambda \pm \mu = x^{-1}(\alpha \pm \nu)\left(x^{-1}\right)^T$ are symmetric and nonnegative-definite. It follows that $\mu_{k,k} = 0$ for all $k > r$, because otherwise one of the matrices $\lambda - \mu$ or $\lambda + \mu$ would have a negative diagonal entry. But then one also has $\mu_{k,\ell} = 0$ whenever $\max(k, \ell) > r$ (otherwise, for $k \ne \ell$, the matrices $\lambda \pm \mu$ would have a negative principal minor of order 2). Therefore

$$0 = \nu_{i,i} = \sum_{k,\ell=1}^{r} x_{i,k} x_{i,\ell} \mu_{k,\ell} \quad \forall i \in \{1, \dots, n\},$$

which can be rewritten as

$$\sum_{1 \le k \le \ell \le r} x_{i,k} x_{i,\ell} \left(2 - \delta_{k,\ell}\right) \mu_{k,\ell} = 0 \quad \forall i \in \{1, \dots, n\}, \tag{3.7}$$

so that condition (C) is violated. Now Lemma 3.1 and definition 2.4 imply that $\mathbf{x} \in K_n(\mathbf{a}) \setminus \operatorname{ext} K_n(\mathbf{a})$.

Vice versa, let $\mathbf{x} = (x_1, \dots, x_n) \in K_n(\mathbf{a}) \setminus \operatorname{ext} K_n(\mathbf{a})$, so that the system of equations (3.7) has a nonzero solution. For the corresponding symmetric $n \times n$ matrix $\mu \ne 0$ with $\mu_{k,\ell} = 0$ whenever $\max(k, \ell) > r$, both matrices $\lambda - t\mu$ and $\lambda + t\mu$ are nonnegative-definite for all small enough real t. Indeed, there is a positive real number C such that for all real numbers $z_1, \dots, z_n$ one has

$$\left| \sum_{i,j=1}^{n} \mu_{i,j} z_i z_j \right| = \left| \sum_{i,j=1}^{r} \mu_{i,j} z_i z_j \right| \le C \left(z_1^2 + \cdots + z_r^2\right),$$

whence

$$\sum_{i,j=1}^{n} \left(\lambda_{i,j} + t\mu_{i,j}\right) z_i z_j \ge (1 - |t|C) \left(z_1^2 + \cdots + z_r^2\right) \ge 0$$

if $|t| \le 1/C$. It follows that the matrices $\alpha \pm t\nu = x(\lambda \pm t\mu)x^T \in \mathcal{K}_n(\mathbf{a})$, where $\nu := x\mu x^T \neq 0$. Thus, $\alpha \in \mathcal{K}_n(\mathbf{a}) \setminus \operatorname{ext} \mathcal{K}_n(\mathbf{a})$.

Lemma 3.2 is proved. □

Let us now complete the proof of Proposition 2.5.

Let φ be a convex function such that $\mathsf{E}\varphi\left(\left(\sum_{i=1}^n |\xi_i| a_i\right)^2\right) < \infty$. Introduce

$$\Phi(\alpha) := \mathsf{E}\varphi\left(\sum_{i,j=1}^n \alpha_{i,j}\xi_i\xi_j\right) = \mathsf{E}\varphi\left(\left\|\sum_{i=1}^n \xi_i x_i\right\|^2\right).$$

Since Φ is a convex function as well, it attains a maximum value on $\mathcal{K}_n(\mathbf{a})$ at some $\alpha \in \operatorname{ext} \mathcal{K}_n(\mathbf{a})$, by a classic result due to Minkowski; see e.g. [11]. Hence and from Lemma 3.2 one obtains part (i) of Proposition 2.5.

Moreover, suppose that the function φ is strictly convex, a symmetric $n \times n$ matrix $\beta = (\beta_{i,j})$ with $\beta_{i,i} = 0\ \forall i$ is nonzero, and the r.v.'s $\xi_i\xi_j$ $(1 \le i < j \le n)$ are linearly independent; then the inequality

$$\varphi\left(\sum_{i,j=1}^n (\alpha_{i,j} - \beta_{i,j})\xi_i\xi_j\right) + \varphi\left(\sum_{i,j=1}^n (\alpha_{i,j} + \beta_{i,j})\xi_i\xi_j\right) \ge 2\varphi\left(\sum_{i,j=1}^n \alpha_{i,j}\xi_i\xi_j\right)$$

is strict with a nonzero probability, because

$$\mathsf{P}\left(\sum_{i,j=1}^n \beta_{i,j}\xi_i\xi_j \neq 0\right) > 0.$$

Taking now the expectations, we see that function Φ is strictly convex. In view of the same result by Minkowski, one obtains part (ii) of Proposition 2.5. □

Proof of Theorem 2.6. (i) Note that inequalities $r \le r_*(n)$ and $r(r+1)/2 \le n$ are equivalent. Therefore, for any $r > r_*(n)$, the number $r(r+1)/2$ of the unknowns in the homogeneous linear system (3.1) is greater than the number n of the equations. Hence, (3.1) has a nontrivial solution. This reasoning and Lemma 3.1 imply part (i) of Theorem 2.6.

(ii) This part of Theorem 2.6 is based on

Lemma 3.3. *For every $r \in \{1, \dots, r_*(n)\}$, there exists some $\mathbf{y} = (y_1, \dots, y_n) \in K_{n,r}(\mathbf{a}) \cap \operatorname{ext} K_n(\mathbf{a})$.*

Proof. Let us assume first that $n = r(r+1)/2$. Let $\xi_{i,k}$ with $i \in \{1, \dots, n\}$ and $k \in \{1, \dots, r\}$ be independent identically distributed r.v.'s with $\mathsf{E}\xi_{i,k} = 0$, $\mathsf{Var}\,\xi_{i,k} = 1$, and a finite $\gamma := \mathsf{E}\xi_{i,k}^4$. Let us enumerate arbitrarily all the n pairs (k, ℓ) such that $1 \le k \le \ell \le r$ and denote by $(k(j), \ell(j))$ the jth pair, $j = 1, \dots, n$. Let us assume, without loss of generality, that

$$k(j) = \ell(j) \iff j \in \{1, \dots, r\}. \tag{3.8}$$

Consider the random $n \times n$ matrix $\zeta = (\zeta_{i,j})$ with

$$\zeta_{i,j} := \xi_{i,k(j)}\xi_{i,\ell(j)}.$$

Then

$$\mathsf{E}(\det\zeta)^2 = \sum_{\pi,\sigma}(-1)^{|\pi|+|\sigma|}E_{\pi,\sigma}, \quad \text{where} \quad E_{\pi,\sigma} := \prod_{i=1}^{n} E_{i;\pi,\sigma} \quad \text{and}$$

$$E_{i;\pi,\sigma} := \mathsf{E}\xi_{i,k(\pi(i))}\xi_{i,\ell(\pi(i))}\xi_{i,k(\sigma(i))}\xi_{i,\ell(\sigma(i))};$$

here π and σ stand for arbitrary permutations of the set $\{1,\dots,n\}$, and $|\pi| := I\{\pi \text{ is odd}\}$. Because the $\xi_{i,k}$'s are independent and zero-mean, one has

$$E_{i;\pi,\sigma} \neq 0 \iff i \in C_{\pi,\sigma} := C^{(1)}_{\pi,\sigma} \cup C^{(2)}_{\pi,\sigma} \cup C^{(3)}_{\pi,\sigma},$$

where

$$C^{(1)}_{\pi,\sigma} := \{i\colon k(\pi(i)) = \ell(\pi(i)) \ \&\ k(\sigma(i)) = \ell(\sigma(i))\};$$
$$C^{(2)}_{\pi,\sigma} := \{i\colon k(\pi(i)) = k(\sigma(i)) \ \&\ \ell(\pi(i)) = \ell(\sigma(i))\};$$
$$C^{(3)}_{\pi,\sigma} := \{i\colon k(\pi(i)) = \ell(\sigma(i)) \ \&\ \ell(\pi(i)) = k(\sigma(i))\}.$$

In view of (3.8),

$$C^{(1)}_{\pi,\sigma} = \{i\colon \pi(i) \le r \ \&\ \sigma(i) \le r\}.$$

By the definition of the enumeration of the pairs (k,ℓ),

$$C^{(2)}_{\pi,\sigma} = \{i\colon \pi(i) = \sigma(i)\}.$$

Also, if $i \in C^{(3)}_{\pi,\sigma}$, then the condition $k(j) \le \ell(j)$ for all j implies that $\ell(\pi(i)) \ge k(\pi(i)) = \ell(\sigma(i)) \ge k(\sigma(i)) = \ell(\pi(i))$. It follows that

$$C^{(3)}_{\pi,\sigma} = C^{(1)}_{\pi,\sigma} \cap C^{(2)}_{\pi,\sigma} =: C^{(0)}_{\pi,\sigma},$$

and so,

$$C_{\pi,\sigma} = C^{(1)}_{\pi,\sigma} \cup C^{(2)}_{\pi,\sigma}.$$

Moreover,

$$E_{i;\pi,\sigma} = \begin{cases} \gamma & \text{if } i \in C^{(0)}_{\pi,\sigma}; \\ 1 & \text{if } i \in C_{\pi,\sigma} \setminus C^{(0)}_{\pi,\sigma}; \\ 0 & \text{if } i \notin C_{\pi,\sigma}. \end{cases}$$

Hence,

$$E_{\pi,\sigma} = \gamma^p I\{C_{\pi,\sigma} = \{1,\dots,n\}\}, \quad \text{where} \quad p := p_{\pi,\sigma} = |C^{(0)}_{\pi,\sigma}|;$$

here $|\cdot|$ stands for the cardinality.

Note that $p_{\pi,\sigma} \le |C^{(1)}_{\pi,\sigma}| \le r$. Moreover, if $\pi = \sigma$, then $C_{\pi,\sigma} = \{1,\dots,n\}$ and $p_{\pi,\sigma} = r$. Vice versa is also true. Indeed, assume that $C_{\pi,\sigma} = \{1,\dots,n\}$, $p_{\pi,\sigma} = r$, and $\pi \neq \sigma$. Then there exists some $i_0 \in \{1,\dots,n\} = C_{\pi,\sigma}$ such that $\pi(i_0) \neq \sigma(i_0)$.

In fact, $i_0 \in C^{(1)}_{\pi,\sigma}$, because $\pi(i) = \sigma(i)$ for all $i \in C^{(2)}_{\pi,\sigma}$. Hence, $\pi(i_0) \le r$, while $\pi(i_0) \neq \sigma(i_0)$. But $|\{i\colon \pi(i) \le r\}| = r$. It follows that

$$r = p_{\pi,\sigma} = |C^{(0)}_{\pi,\sigma}| = |\{i\colon \sigma(i) = \pi(i) \le r\}| \le |\{i\colon \pi(i) \le r\} \setminus \{i_0\}| = r - 1 < r,$$

which is a contradiction.

We see that

$$\mathsf{E}(\det\zeta)^2 = A_{r,r}\gamma^r + A_{r,r-1}\gamma^{r-1} + \cdots + A_{r,0},$$

where the coefficients $A_{r,k}$ depend only on their indices, r and k, and

$$A_{r,r} = n! = (r(r+1)/2)! > 0.$$

Hence, $\mathsf{E}(\det\zeta)^2 \to \infty$ as $\gamma \to \infty$. Therefore, there exists an $n \times r = (r(r+1)/2) \times r$ matrix $(x^{(0)}_{i,k})$ such that the $n \times n$ matrix $(x^{(0)}_{i,k(j)}x^{(0)}_{i,\ell(j)})$ is nonsingular.

If now $n > r(r+1)/2$, then let us extend the just constructed $(r(r+1)/2) \times r$ matrix $(x^{(0)}_{i,k})$ to an $n \times r$ matrix $x = (x_{i,k})$ arbitrarily by adding $n - r(r+1)/2$ nonzero rows. Note that

$$\operatorname{rank}\left(x_{i,k(j)}x_{i,\ell(j)}\right) \ge \operatorname{rank}\left(x^{(0)}_{i,k(j)}x^{(0)}_{i,\ell(j)}\right) = r(r+1)/2.$$

Letting the $x_{i,k}$'s be the coordinates of some vectors $x_1, \ldots, x_n$ in some o.b. $(e_1, \ldots, e_r)$, we see that condition (C) is satisfied. By the construction, all the x_i's are nonzero vectors. Let

$$y_i := \frac{a_i}{\|x_i\|}\,x_i \quad \forall i \in \{1, \ldots, n\}.$$

Then, in view of Lemma 3.1, the system of vectors $\mathbf{y} = (y_1, \ldots, y_n)$ belongs to $\operatorname{ext} K_n(\mathbf{a})$.

To complete the proof of Lemma 3.3, it remains to notice that $\operatorname{rank}\mathbf{y}(= \operatorname{rank}\mathbf{x}) = r$, so that $\mathbf{y} \in K_{n,r}(\mathbf{a})$. Indeed, suppose that there exist real numbers $c_1, \ldots, c_r$, not all of which are zero, such that

$$\sum_{k=1}^{r} c_k x_{i,k} = 0 \quad \forall i \in \{1, \ldots, n\}.$$

Then, letting $\rho_{k,\ell} := (1 - \delta_{k,\ell}/2)(c_k + c_\ell)$, one has

$$\sum_{1\le k\le \ell\le r} x_{i,k}x_{i,\ell}\rho_{k,\ell} = \sum_{k,\ell=1}^{r} c_k x_{i,k}x_{i,\ell} = 0.$$

for all $i \in \{1, \ldots, n\}$. But this contradicts condition (B), because at least one of the numbers $\rho_{k,k} = c_k$ is nonzero.

Lemma 3.3 is thus proved. □

Now we are ready to prove part (ii) of Theorem 2.6. To be specific, let us show that, for every natural n and every natural $r \le r_*(n)$, the set $K_{n,r}(\mathbf{a}) \setminus \operatorname{ext} K_n(\mathbf{a})$ is nowhere dense in $K_{n,r}(\mathbf{a})$. Here, the topology on $K_{n,r}(\mathbf{a})$ is induced by the embedding

$$K_{n,r}(\mathbf{a}) \ni (x_1, \ldots, x_n) \mapsto (x_{i,k})$$

of the set $K_{n,r}(\mathbf{a})$ into the set of all $n \times r$ matrices, where, as before, the $x_{i,k}$'s are the coordinates of the x_i's relative to an arbitrary fixed o.b. $\mathbf{e} = (e_1, \dots, e_r)$.

Let $\mathbf{y} = (y_1, \dots, y_n)$ be the system of vectors constructed in the proof of Lemma 3.3. Take any $\mathbf{z} = (z_1, \dots, z_n) \in K_{n,r}(\mathbf{a})$. Let then

$$u_i(t) := z_i + ty_i \quad \forall i \in \{1, \dots, n\} \ \forall t \in \mathbb{R}.$$

Next, let the $u_{i,k}(t)$'s stand for the coordinates of the $u_i(t)$'s relative to the o.b. $\mathbf{e}$. Let us use here as well the enumeration of the pairs (k, ℓ) such that $1 \le k \le \ell \le r$, as described in the proof of Lemma 3.3. Consider the $(r(r+1)/2) \times (r(r+1)/2)$ matrices $\big(u_{i,k(j)}(t)u_{i,\ell(j)}(t)\big)$ and $\big(y_{i,k(j)}y_{i,\ell(j)}\big)$ with $i, j \in \{1, \dots, r(r+1)/2\}$. Then

$$D(t) := \det\big(u_{i,k(j)}(t)u_{i,\ell(j)}(t)\big) = t^{r(r+1)} \det\big(y_{i,k(j)}y_{i,\ell(j)}\big) + f(t),$$

where $f(t)$ is a polynomial in t of degree $< r(r+1)$.

Since $\det\big(y_{i,k(j)}y_{i,\ell(j)}\big) \neq 0$, the polynomial $D(t)$ has only finitely many zeroes. Hence, for all but finitely many real t, condition (C) with the $u_{i,k}(t)$'s in place of the $x_{i,k}$'s is satisfied. Moreover, $u_i(t) = 0$ only for finitely many values of t. Let now

$$x_i(t) := \frac{a_i}{\|u_i(t)\|}\, u_i(t) \quad \forall i \in \{1, \dots, n\};$$

then $\mathbf{x}(t) := (x_1(t), \dots, x_n(t)) \in \operatorname{ext} K_n(\mathbf{a})$ except for finitely many values of t; moreover, as seen from the proof of Lemma 3.3, one has $\operatorname{rank} \mathbf{x}(t) = r$.

It follows that the set $\operatorname{ext} K_n(\mathbf{a})$ is everywhere dense in $K_{n,r}(\mathbf{a})$, for every $r \le r_*(n)$. By Lemma 3.1, one has $(x_1, \dots, x_n) \in K_{n,r}(\mathbf{a}) \setminus \operatorname{ext} K_n(\mathbf{a})$ iff $(x_1, \dots, x_n) \in K_{n,r}(\mathbf{a})$ and all the $(r(r+1)/2) \times (r(r+1)/2)$ minors of the $n \times (r(r+1)/2)$ matrix $\big(x_{i,k(j)}x_{i,\ell(j)}\big)$ are zero. Therefore, the set $K_{n,r}(\mathbf{a}) \setminus \operatorname{ext} K_n(\mathbf{a})$ is closed in $K_{n,r}(\mathbf{a})$. Hence, it is nowhere dense in $K_{n,r}(\mathbf{a})$. □

Proof of Proposition 2.7.

(i) Part (i) of Proposition 2.7 is trivial.

(ii) First, note that the conditions $r = 2$ and $n \ge r(r+1)/2$ imply $n \ge 3$. Take now any system $\mathbf{x} = (x_1, \dots, x_n) \in K_{n,r}(\mathbf{a}) \setminus \operatorname{ext} K_n(\mathbf{a})$. Then condition (A) on page 174 is violated. That is, for some o.b. of $\mathbf{x}$, the rank of the $n \times 2$ matrix $(x_{i,k}^2)$ does not exceed 1, so that all the rows of this matrix are proportional to one another. Moreover, since the number of rows is $n \ge 3$, there are two different rows, say the ith and jth ones, of the $n \times 2$ matrix $(x_{i,k})$ such that $x_{i,1}x_{i,2}x_{j,1}x_{j,2} \ge 0$. This and the proportionality of the rows of matrix $(x_{i,k}^2)$ imply the collinearity of the vectors x_i and x_j.

(iii) Take any o.b. $\mathbf{e} = (e_1, \dots, e_r)$. Let

$$x_i := e_1 + t_i e_2 + \cdots + t_i^{r-2} e_{r-1} + (1 + t_i^2 + \cdots + t_i^{2(r-2)})^{1/2} e_r$$

for all $i \in \{1, \dots, n\}$. Here the t_i's are some real numbers, to be constructed by induction.

Note that here the matrix $(x_{i,k}^2)$ is of rank $< r$, because its rth column is the sum of the other ones, so that condition (A) is violated. It remains to select the t_i's so that every subsystem of $(x_1, \dots, x_n)$ consisting of r vectors is of (full) rank r. We shall do that by induction. Namely, we shall show that for every $m \in \{1, \dots, n\}$ there are real numbers $t_1, \dots, t_m$ such that every subsystem of the system $(x_1, \dots, x_m)$ consisting of r vectors is of (full) rank r. For $m < r$ this statement is trivial for any $t_1, \dots, t_m$, because then the system $(x_1, \dots, x_m)$ contains no subsystems consisting of r vectors. Let now $m \geq r - 1$ and suppose that the induction hypothesis is true, for some *distinct* real numbers $t_1, \dots, t_m$. Let $A(i_1, \dots, i_s)$ denote the minor of the matrix $x = (x_{i,k})$ corresponding to the first s columns of x and its rows labelled by $i_1, \dots, i_s$. By induction, one has $A(i_1, \dots, i_r) \neq 0$ whenever $1 \leq i_1 < \dots < i_r \leq m$. It suffices to show that, for some choice of a number t_{m+1} distinct from any of the numbers $t_1, \dots, t_m$, one has $A(i_1, \dots, i_{r-1}, m+1) \neq 0$ whenever $1 \leq i_1 < \dots < i_{r-1} \leq m$. But

$$A(i_1, \dots, i_{r-1}, m+1) = A(i_1, \dots, i_{r-1}) f(t_{m+1})^{1/2} + g(t_{m+1}),$$

where $f(t) := 1 + t^2 + \dots + t^{2(r-2)}$ and $g(t)$ is some polynomial (with coefficients possibly depending on $t_1, \dots, t_m$). Note that $A(i_1, \dots, i_{r-1})$ is a Vandermonde determinant and hence nonzero. Therefore, if $A(i_1, \dots, i_{r-1}, m+1)$ were zero for infinitely many values of t_{m+1}, then one would have $f(t) = h(t)^2$ for some polynomial $h(t)$ (namely, $h(t) = -g(t)/A(i_1, \dots, i_{r-1})$) and all real t. This would imply that all the roots of $f(t)$ are of even multiplicity, which is absurd, because the roots of $f(t)$, given by the formula $t = \exp\left(\sqrt{-1}\pi k/(r-1)\right)$ for $k \in \{1, \dots, 2r-3\} \setminus \{r-1\}$, are each of multiplicity 1 (for $r \geq 3$). Thus, one can find some value of t_{m+1}, which is distinct from any of the numbers $t_1, \dots, t_m$ and makes $A(i_1, \dots, i_{r-1}, m+1)$ nonzero. □

References

[1] S. G. Bobkov, F. Götze, and C. Houdré, *On Gaussian and Bernoulli covariance representations*, Bernoulli **7** (2001), 439–451.

[2] D. C. Cox and J. H. B. Kemperman, *Sharp bounds on the absolute moments of a sum of two i.i.d. random variables*, Ann. Probab. **11** (1983), 765–771.

[3] J.-M. Dufour and M. Hallin, *Improved Eaton bounds for linear combinations of bounded random variables, with statistical applications*, JASA, **88** (1993), 1026–1033.

[4] M. L. Eaton, *A note on symmetric Bernoulli random variables*, Ann. Math. Statist. **41** (1970), 1223–1226.

[5] M. L. Eaton, *A probability inequality for linear combinations of bounded random variables*, Ann. Statist. **2** (1974), 609–614.

[6] S. E. Graversen and G. Peskir, *Extremal problems in the maximal inequalities of Khinchine*, Math. Proc. Cambridge Philos. Soc., **123**, No. 1 (1995), 169–177.

[7] U. Haagerup, *The best constants in the Khinchine inequality*, Studia Math., **70** (1982), 231–283.

[8] A. Khinchin, *Über dyadische Brüche*, Math. Z., **18** (1923), 109–116.

[9] W. Orlicz, *Über unbedingte Konvergentz in Funktionenräumen I*, Studia Math., **4** (1933), 33–37.

[10] R. E. A. C. Paley and A. Zygmund, *On some series of functions,* (1), Proc. Cambridge Philos. Soc., **26** (1930), 337–357.

[11] R. R. Phelps, *Lectures on Choquet's Theorem,* (1966), Van Nostrand, Princeton.

[12] I. F. Pinelis, *Some extremal probability problems in Hilbert space,* Theory Probab. Appl., **29** (1984), 419–420.

[13] I. Pinelis, *Extremal probabilistic problems and Hotelling's T^2 test under symmetry condition,* Preprint (1991).

[14] I. Pinelis, *Extremal probabilistic problems and Hotelling's T^2 test under a symmetry condition,* Ann. Statist., **22** (1994), 357–368.

[15] I. Pinelis, *Optimal tail comparison based on comparison of moments,* In: High dimensional probability (Oberwolfach, 1996), Progr. Probab., **43** (1998), Birkhäuser, Basel, 297–314.

[16] I. Pinelis, *Fractional sums and integrals of r-concave tails and applications to comparison probability inequalities,* In: Advances in stochastic inequalities (Atlanta, GA, 1997), Contemp. Math., **234** (1999), Amer. Math. Soc., Providence, 149–168.

[17] I. Pinelis, *On exact maximal Khinchine inequalities,* In: High dimensional probability, II (Seattle, WA, 1999), Progr. Probab., **47** (2000), Birkhäuser, Boston, 49–63.

[18] I. Pinelis, *Spherically symmetric functions with a convex second derivative and applications to extremal probabilistic problems,* Math. Ineq. & Appl. **5** (2002), 7–26.

[19] I. Pinelis, *L'Hospital type rules for monotonicity: applications to probability inequalities for sums of bounded random variables,* JIPAM. J. Inequal. Pure Appl. Math. **3** (2002), Article 20, 8 pp. (electronic). `http://jipam.vu.edu.au/v3n1/013_01.html`.

[20] S. A. Utev, *Extremal problems in moment inequalities. (Russian)* In: Limit theorems of probability theory, 56–75, 175, Trudy Inst. Mat., **5** "Nauka" Sibirsk. Otdel., Novosibirsk, 1985.

[21] S. J. Szarek, *On the best constants in the Khinchin inequality,* Studia Math. **58** (1976), 197–208.

[22] P. Whittle, *Bounds for the moments of linear and quadratic forms in independent variables,* Teor. Verojatnost. i Primenen. **5** (1960), 331–335.

Department of Mathematical Sciences, Michigan Technological University, Houghton, Michigan 49931
E-mail address: `ipinelis@mtu.edu`

Part III

Empirical Processes

Progress in Probability, Vol. 56, 189–212

Moderate Deviations of Empirical Processes

Miguel A. Arcones

Abstract. We give necessary and sufficient conditions for the moderate deviations of empirical processes and of sums of i.i.d. random vectors with values in a separable Banach space. Our approach is based on a characterization of the large deviation principle using the large deviations of the finite dimensional distributions plus an asymptotic exponential equicontinuity condition.

1. Introduction

We study the moderate deviations for different types of sequences of empirical processes $\{U_n(t) : t \in T\}$, where T is an index set. We also consider the moderate deviations of sums of i.i.d. random vectors with values in a separable Banach space. Our results are stated as functional large deviations with a Gaussian rate function.

General references on (functional) large deviations are Bahadur [4]; Varadhan [23]; Deuschel and Stroock [9] and Shwartz and Weiss [21]. We consider stochastic processes as elements of $l_\infty(T)$, where T is an index set. $l_\infty(T)$ is the Banach space consisting of the bounded functions defined in T with the norm $\|x\|_\infty = \sup_{t\in T} |x(t)|$. We will use the following definition:

Definition 1.1. Given a sequence of stochastic processes $\{U_n(t) : t \in T\}$, a sequence of positive numbers $\{\epsilon_n\}_{n=1}^\infty$ such that $\epsilon_n \to 0$, and a function $I : l_\infty(T) \to [0, \infty]$, we say that $\{U_n(t) : t \in T\}$ satisfies the LDP (large deviation principle) with speed ϵ_n^{-1} and with a good rate function I if:

(i) For each $0 \le c < \infty$, $\{z \in l_\infty(T) : I(z) \le c\}$ is a compact set of $l_\infty(T)$.

(ii) For each set $A \in l_\infty(T)$,

$$\begin{aligned} -I(A^o) &\le \liminf_{n\to\infty} \epsilon_n \log(\Pr{}_*\{\{U_n(t) : t \in T\} \in A\}) \\ &\le \limsup_{n\to\infty} \epsilon_n \log(\Pr{}^*\{\{U_n(t) : t \in T\} \in A\}) \le -I(\bar{A}), \end{aligned}$$

where for $B \subset l_\infty(T)$, $I(B) = \inf\{I(z) : z \in B\}$.

By Theorem 3.2 in Arcones [1], this definition is equivalent to the large deviations of the finite dimension distributions plus an asymptotic equicontinuity

2000 *Mathematics Subject Classification.* Primary 62E20; Secondary 62F12.
Key words and phrases. Moderate deviations, empirical processes, Banach space valued r.v.'s.

condition. This will allow us to obtain necessary and sufficient conditions for the moderate deviations of the considered stochastic processes.

We consider stochastic proceses $\{U_n(t) : t \in T\}$ satisfying the large deviation principle with a Gaussian rate function. This rate function is related with a covariance function on T. By a covariance function R on T, we mean a function $R : T \times T \to \mathbb{R}$ such that for each $s, t \in T$ $R(s,t) = R(t,s)$, and for each $\lambda_1, \dots, \lambda_m \in \mathbb{R}$ and each $t_1, \dots, t_m \in T$, $\sum_{j,k=1}^{m} \lambda_j \lambda_k R(t_j, t_k) \geq 0$. By Theorem II.3.1 in Doob [10], a function $R : T \times T \to \mathbb{R}$ is a covariance function if and only if there exists a (Gaussian) process $\{Z(t) : t \in T\}$ with mean zero and covariance given by $E[Z(s)Z(t)] = R(s,t)$, for each $s, t \in T$. In the considered situations, the rate function of the LDP of $\{(U_n(t_1), \dots, U_n(t_m))\}$ is

$$I_{t_1,\dots,t_m}(u_1, \dots, u_m) = \sup_{\lambda_1,\dots,\lambda_m} \left(\sum_{j=1}^{m} \lambda_j u_j - 2^{-1} \sum_{j,k=1}^{\infty} \lambda_j \lambda_k R(t_j, t_k) \right), \tag{1.1}$$

where $u_1, \dots, u_m \in \mathbb{R}$. This is the rate function of the LDP of the finite dimensional distributions of a Gaussian process. If $\{Z(t) : t \in T\}$ is a Gaussian process with mean zero and covariance function R, then for each $t_1, \dots, t_m$, $(n^{-1/2}Z(t_1), \dots, n^{-1/2}Z(t_m))$ satisfies the LDP with speed n and the rate function in (1.1).

For sums of i.i.d.r.v.'s, the moderate deviations can be defined as follows. Let $\{X_j\}_{j=1}^{\infty}$ be a sequence of nondegenerate i.i.d.r.v.'s such that for some $\lambda > 0$, $E[e^{\lambda |X_1|}] < \infty$. Let $\{a_n\}_{n=1}^{\infty}$ be a sequence of real numbers such that $a_n \to \infty$ and $a_n^{-1} n^{1/2} \to \infty$, it follows from the results in Petrov [19] that for each $t \geq 0$,

$$\lim_{n \to \infty} a_n^{-2} \log(\Pr\{|\sum_{j=1}^{n} (X_j - E[X_j])| \geq t n^{1/2} a_n\}) = -t^2/(2\mathrm{Var}(X_1)).$$

Cramér [8] obtained the previous result assuming the extra condition that $a_n^{-1} n^{1/2} \log n \to \infty$. We obtain necessary and sufficient conditions for the moderate deviations of sums of i.i.d.r.v.'s which apply to r.v.'s which may not have finite second moment. In particular, we obtain that $\{n^{-1/2} a_n^{-1} \sum_{j=1}^{n} X_j\}$ satisfies the LDP with speed a_n^2 and a Gaussian rate if and only if $E[X] = 0$, $E[X^2] < \infty$ and

$$\lim_{n \to \infty} a_n^{-2} \log(n \Pr\{|X| \geq n^{1/2} a_n\}) = -\infty.$$

We also give necessary and sufficient conditions for the moderate deviations with a Gaussian rate of empirical processes and for sums of i.i.d.r.v.'s with values in a separable Banach space. The partial sums processes obtained from the processes above satisfy the LDP under the same conditions as the regular sums do. Moderate deviations for empirical processes have been studied by Borovkov and Mogul'skiĭ[5, 6], Ledoux [15] and Wu [24, 25], among other authors. In other situations, moderate deviations may have not a Gaussian rate.

We will use the usual multivariate notation. For example, given $u = (u_1, \dots, u_d)' \in \mathbb{R}^d$ and $v = (v_1, \dots, v_d)' \in \mathbb{R}^d$, $u'v = \sum_{j=1}^d u_j v_j$ and $|u| = (\sum_{j=1}^n u_j^2)^{1/2}$. Whenever, we consider a sequence of i.i.d.r.v.'s $\{X_j\}$, X will denote a copy of X_1. c will denote an arbitrary constant which may vary from occurrence to occurrence.

2. Moderate deviations of empirical processes

The basis of our work is the following theorem:

Theorem 2.1. *(Theorem 3.2 in Arcones [1]) Let $\{U_n(t) : t \in T\}$ be a sequence of stochastic processes, let $\{\epsilon_n\}$ be a sequence of positive numbers that converges to zero. Let $I : l_\infty(T) \to [0, \infty]$ and let $I_{t_1,\dots,t_m} : \mathbb{R}^m \to [0, \infty]$ be a function, where $t_1, \dots, t_m \in T$. Let d be a pseudometric in T. Consider the conditions:]*

(a.1) *(T, d) is totally bounded.*
(a.2) *For each $t_1, \dots, t_m \in T$, $(U_n(t_1), \dots, U_n(t_m))$ satisfies the LDP with speed ϵ_n and good rate function $I_{t_1,\dots,t_m}$.*
(a.3) *For each $\tau > 0$,*

$$\lim_{\eta \to 0} \limsup_{n \to \infty} \epsilon_n \log \left(\Pr{}^* \left\{ \sup_{d(s,t) \le \eta} |U_n(t) - U_n(s)| \ge \tau \right\} \right) = -\infty.$$

(b) *$\{U_n(t) : t \in T\}$ satisfies the LDP in $l_\infty(T)$ with speed ϵ_n^{-1} and with good rate function I.*

If the set of conditions (a) is satisfied for some pseudometric d, then (b) holds with

$$I(z) = \sup\{I_{t_1,\dots,t_m}(z(t_1), \dots, z(t_m)) : t_1, \dots, t_m \in T, m \ge 1\}.$$

If (b) is satisfied, then the set of conditions (a) holds with

$$I_{t_1,\dots,t_m}(u_1, \dots, u_m) = \inf\{I(z) : z \in l_\infty(T), (z(t_1), \dots, z(t_m)) = (u_1, \dots, u_m)\}$$

and the pseudometric $\rho(s,t) = \sum_{k=1}^\infty k^{-2} \min(\rho_k(s,t), 1)$, where $\rho_k(s,t) = \sup\{|u_2 - u_1| : I_{s,t}(u_1, u_2) \le k\}$.

First, we see how to express the rate function I, when the rate function for the finite dimensional distributions is given by (1.1).

Theorem 2.2. *Let T be a parameter set and let R be covariance function on T. Let $\{f(\cdot, t) : t \in T\}$ be a class of measurable functions on the same measure space $(\Omega, \mathcal{F}, \mu)$ such that for each $t \in T$, $\int f(x,t)\, d\mu(x) = 0$ and $\int (f(x,t))^2\, d\mu(x) < \infty$, and for each $s, t \in T$, $\int f(x,s) f(x,t)\, d\mu(x) = R(s,t)$. Then,*

(i) *For each* $t_1, \dots, t_m \in T$, *and each* $u_1, \dots, u_m \in \mathbb{R}$,

$$\sup\left\{\sum_{j=1}^{m} \lambda_j u_j - 2^{-1} \sum_{j,k=1}^{\infty} \lambda_j \lambda_k R(t_j, t_k) : \lambda_1, \dots, \lambda_m\right\}$$

$$= \inf\left\{2^{-1} \int \gamma^2(x)\, d\mu(x) : \right.$$

$$\left. \gamma \in L_2,\ \int \gamma(x) f(x, t_j)\, d\mu(x) = z(t_j) \text{ for each } 1 \le j \le m\right\}.$$

Besides, if the infimum above is finite, there exists a function γ *attaining the infimum.*

(ii) *If* $\{f(\cdot, t) : t \in T\}$ *is a separable subset of* L_2, *then for each* $z \in l_\infty(T)$,

$$\sup\{I_{t_1,\dots,t_m}(z(t_1), \dots, z(t_m)) : t_1, \dots, t_m \in T, m \ge 1\}$$

$$= \inf\left\{2^{-1} \int \gamma^2(x)\, d\mu(x) : \gamma \in L_2,\right.$$

$$\left. \int \gamma(x) f(x, t)\, d\mu(x) = z(t) \text{ for each } t \in T\right\},$$

where

$$I_{t_1,\dots,t_m}(u_1, \dots, u_m)$$

$$= \inf\left\{2^{-1} \int \gamma^2(x)\, d\mu(x) : \gamma \in L_2,\right.$$

$$\left. \int \gamma(x) f(x, t_j)\, d\mu(x) = z(t_j) \text{ for each } 1 \le j \le m\right\}.$$

As before, if the infimum above is finite, there exists a function γ *attaining the infimum.*

Proof. Part (i) follows from Lemma 4.1 in Arcones [2] with $\Phi(x) = \Psi(x) = 2^{-1}x^2$, $x \in \mathbb{R}$. Since a bounded set of L_2 is weakly compact, the infimum in part (i) is attained.

Let

$$I^{(1)}(z) := \sup\{I_{t_1,\dots,t_m}(z(t_1), \dots, z(t_m)) : t_1, \dots, t_m \in T, m \ge 1\}$$

and let

$$I^{(2)}(z) := \inf\{2^{-1} \int \gamma^2(x)\, d\mu(x) : \int \gamma(x) f(x, t)\, d\mu(x) = z(t) \text{ for each } t \in T\}.$$

It is obvious that for each $z \in l_\infty(T)$, $I^{(2)}(z) \ge I^{(1)}(z)$. Next, we show that $I^{(1)}(z) \ge I^{(2)}(z)$. We may assume that $I^{(2)}(z) < \infty$. Take a sequence $\{t_n\}$ in T such that

$$I^{(1)}(z) = \lim_{n\to\infty} I_{t_1,\dots,t_n}(z(t_1), \dots, z(t_n)).$$

Given $r_1, \dots, r_m, s_1, \dots, s_k \in T$, we have that

$$I_{r_1,\dots,r_m}(z(r_1), \dots, z(r_m)) \le I_{r_1,\dots,r_m,s_1,\dots,s_k}(z(r_1), \dots, z(r_m), z(s_1), \dots, z(s_k)).$$

Hence, we may assume that $\{f(\cdot, t_n) : n \ge 1\}$ is dense in L_2. Let $\gamma_n \in L_2$ be such that

$$2^{-1} \int \gamma_n^2(x)\, d\mu(x) = I_{t_1,\dots,t_n}(z(t_1), \dots, z(t_n)).$$

Then, there exists a subsequence γ_{n_k} and $\gamma \in L_2$ such that γ_{n_k} converges weakly to γ. This implies that for each $m \ge 1$,

$$z(t_m) = \lim_{k\to\infty} \int \gamma_{n_k}(x) f(x, t_m)\, d\mu(x) = \int \gamma(x) f(x, t_m)\, d\mu(x).$$

Since $I^{(2)}(z) < \infty$, $z : (T, \|\cdot\|_2) \to \mathbb{R}$ is continuous. From this and the fact that $\{f(\cdot, t_m) : m \ge 1\}$ is dense in L_2, we get that $z(t) = \int \gamma(x) f(x,t)\, d\mu(x)$, for each $t \in T$. □

Given a covariance function R, there exists a Gaussian process $\{Z(t) : t \in T\}$ with mean zero and covariance R. Let $\mathcal{L}$ be the closed vector space of L_2 generated by $\{Z(t) : t \in T\}$. If $\{U_n(t) : t \in T\}$ satisfies the LDP with speed ϵ_n^{-1} and for each $t_1, \dots, t_m \in T$, the rate function of the LDP of $\{(U_n(t_1), \dots, U_n(t_m))\}$ is

$$I_{t_1,\dots,t_m}(u_1, \dots, u_m) = \sup_{\lambda_1,\dots,\lambda_m} \left(\sum_{j=1}^m \lambda_j u_j - 2^{-1} \sum_{j,k=1}^m \lambda_j \lambda_k R(t_j, t_k) \right), \tag{2.1}$$

then, by Theorems 2.1 and 2.2, the rate function of the LDP of $\{U_n(t) : t \in T\}$ is

$$I(z) = \inf\{2^{-1} E[\gamma^2] : \gamma \in \mathcal{L}, E[\gamma Z(t)] = z(t) \text{ for each } t \in T\}. \tag{2.2}$$

It follows that if $\sup_{t\in T} R(t,t) > 0$, then for each $\lambda \ge 0$,

$$\inf\{I(z) : \sup_{t\in T} |z(t)| \ge \lambda\} = \frac{\lambda^2}{2 \sup_{t\in T} R(t,t)}.$$

If $\sup_{t\in T} R(t,t) = 0$, then

$$I(z) = \begin{cases} 0 & \text{if } \sup_{t\in T} |z(t)| = 0 \\ \infty & \text{if } \sup_{t\in T} |z(t)| > 0. \end{cases}$$

So, if $\sup_{t\in T} R(t,t) > 0$, then for each $\lambda \ge 0$,

$$\lim_{n\to\infty} \epsilon_n \log \left(\Pr\{\sup_{t\in T} |U_n(t)| \ge \lambda\} \right) = -\frac{\lambda^2}{2 \sup_{t\in T} R(t,t)}.$$

If $\sup_{t\in T} R(t,t) = 0$, then for each $\lambda > 0$,

$$\lim_{n\to\infty} \epsilon_n \log \left(\Pr\{\sup_{t\in T} |U_n(t)| \ge \lambda\} \right) = -\infty.$$

We also have that

$$\rho_k(s,t) := \sup\{|u_2 - u_1| : I_{s,t}(u_1,u_2) \le qk\}$$
$$= \sup\{|E[\gamma(Z(s) - Z(t))]| : \gamma \in \mathcal{L}, 2^{-1}E[\gamma^2] \le k\} = (2k)^{1/2}\|Z(s) - Z(t)\|_2.$$

So, the LDP implies that $\{Z(t) : t \in T\}$ is a totally bounded set of L_2.

The rates above appear in the large deviations of Gaussian processes. If $\{Z(t) : t \in T\}$ is a Gaussian process with mean zero and covariance R, then the finite dimensional distributions of $\{n^{-1/2}Z(t) : t \in T\}$ satisfy the LDP with speed n and the rate function in (2.1). If $\sup_{t\in T}|Z(t)| < \infty$ a.s., then $\{n^{-1/2}Z(t) : t \in T\}$ satisfy the LDP with speed n and with the rate in (2.2).

To get the LDP for the finite dimensional distributions, we will apply the following lemma.

Lemma 2.3. *Let $\{X_{n,j} : 1 \le j \le n\}$ be a triangular array of independent r.v.'s with values in $\mathbb{R}^d$ and mean zero. Let $\{a_n\}_{n=1}^\infty$ be a sequence of real numbers converging to infinity. Suppose that:*

(i) The following limit exists and it is finite: $\lim_{n\to\infty}\sum_{j=1}^n E[X_{n,j}X'_{n,j}] =: \Sigma$.

(ii) There exists a constant τ such that for each $1 \le j \le n$, $|X_{n,j}| \le \tau a_n^{-1}$ a.s.

(iii) For each $\delta > 0$, $a_n^{-2}\sum_{j=1}^n \Pr\{|X_{n,j}| \ge \delta a_n^{-1}\} \to 0$.

Then, $a_n^{-1}\sum_{j=1}^n X_{n,j}$ satisfies the LDP with speed a_n^2 and rate function $I(u) = \sup_\lambda(\lambda' u - 2^{-1}\lambda\Sigma\lambda')$.

Proof. By Theorem II.2 in Ellis [12], it suffices to prove that for each $\lambda \in \mathbb{R}^d$,

$$\lim_{n\to\infty} a_n^{-2}\log(E[\exp(a_n\sum_{j=1}^n \lambda' X_{n,j})]) = 2^{-1}\lambda'\Sigma\lambda.$$

First, we prove that by a Taylor expansion, we have that

$$a_n^{-2}\sum_{j=1}^n E[\exp(a_n\lambda' X_{n,j}) - 1] \to 2^{-1}\lambda'\Sigma\lambda. \tag{2.3}$$

Since $a_n|X_{n,j}|$, $1 \le j \le n$, are uniformly bounded, we need to prove that

$$a_n^{-2}\sum_{j=1}^n E[|a_n\lambda' X_{n,j}|^3] \to 0.$$

We have that for any $\delta > 0$,

$$a_n^{-2}\sum_{j=1}^n E[|a_n\lambda' X_{n,j}|^3]$$
$$= a_n^{-2}\sum_{j=1}^n E[|a_n\lambda' X_{n,j}|^3 I_{a_n|X_{n,j}|\le\delta}] + a_n^{-2}\sum_{j=1}^n E[|a_n\lambda' X_{n,j}|^3 I_{a_n|X_{n,j}|>\delta}]$$
$$\le \delta|\lambda|\sum_{j=1}^n E[\lambda' X_{n.j}X'_{n,j}\lambda] + |\lambda|^3\tau^3 a_n^{-2}\sum_{j=1}^n \Pr\{|X_{n,j}| \ge \delta a_n^{-1}\}.$$

Hence, $\limsup_{n\to\infty} a_n^{-2}\sum_{j=1}^n E[|a_n\lambda' X_{n,j}|^3] \le \delta|\lambda|\lambda'\Sigma\lambda$. Since δ is arbitrary, (2.3) follows.

Again, using that $a_n|X_{n,j}|$, $1 \le j \le n$, are uniformly bounded, we have that

$$a_n^{-2}\sum_{j=1}^n |\log(E[\exp(a_n\lambda' X_{n,j})]) - E[\exp(a_n\lambda' X_{n,j}) - 1]|$$
$$\le ca_n^{-2}\sum_{j=1}^n |E[\exp(a_n\lambda' X_{n,j}) - 1]|^2$$
$$= ca_n^{-2}\sum_{j=1}^n |E[\exp(a_n\lambda' X_{n,j}) - 1 - a_n\lambda' X_{n,j}]|^2$$
$$\le ca_n^{-2}\sum_{j=1}^n (E[|a_n\lambda' X_{n,j}|^2])^2$$
$$\le ca_n^{-2}\sum_{j=1}^n E[|a_n\lambda' X_{n,j}|^4] \to 0.$$

□

Next, we consider the moderate deviations of sums of real valued i.i.d.r.v.'s. We present a general theorem which applies to r.v.'s which may not have finite second moment.

Theorem 2.4. *Let $\{X_j\}_{j=1}^\infty$ be a sequence of i.i.d.r.v.'s. Let $\{a_n\}_{n=1}^\infty$ and let $\{c_n\}_{n=1}^\infty$ be two sequences of real numbers such that $a_n \nearrow \infty$ and $a_n^{-1}c_n^{-1}n \nearrow \infty$ and $\{n^{-1}c_n^2\}$ is nondecreasing. Then, the following sets of conditions ((a), (b), (c)) are equivalent:*

(a.1) $\lim_{n\to\infty} a_n^{-2}\log(n\Pr\{|X| \ge c_n a_n\}) = -\infty$.
(a.2) $a_n^{-1}c_n^{-1}nE[XI(|X| \le a_n c_n)] \to 0$.
(a.3) $c_n^{-2}n\mathrm{Var}(XI(|X| \le a_n^{-1}c_n))$ *converges to a finite limit* σ^2.
(b) $\{c_n^{-1}a_n^{-1}\sum_{j=1}^n X_j\}$ *satisfies the LDP with speed* a_n^2 *and a rate function* I *such that* $\lim_{|\lambda|\to\infty}\lambda^{-1}I(\lambda) = \infty$ *and for each* $\delta > 0$, $\inf\{I(z) : |z| \ge \delta\} > 0$.

(c) $\{c_n^{-1}a_n^{-1}\sum_{j=1}^{[nu]} X_j : 0 \le u \le 1\}$ *satisfies the LDP in* $l_\infty([0,1])$ *with speed* a_n^2 *and a rate function* I *such that* $\lim_{\lambda\to\infty}\lambda^{-1}\inf\{I(z) : \sup_{0\le u\le 1}|z(u)| \ge \lambda\} = \infty$ *and for each* $\delta > 0$, $\inf\{I(z) : \sup_{0\le u\le 1}|z(u)| \ge \delta\} > 0$.

Moreover, the rate function in (b) is given by

$$I(z) = \frac{z^2}{2\sigma^2}, \text{ if } \sigma^2 > 0. \tag{2.4}$$

and

$$I(z) = \begin{cases} 0 & \text{if } z = 0 \\ \infty & \text{if } z \ne 0 \end{cases} \tag{2.5}$$

if $\sigma^2 = 0$.
The rate function in (c) is given by

$$I(z) = \begin{cases} 2^{-1}\sigma^{-2}\int_0^1 (z'(u))^2\,du & \text{if } z(0) = 0 \text{ and } z \text{ is absolutely continuous} \\ \infty & \text{else} \end{cases} \tag{2.6}$$

if $\sigma^2 > 0$, *and*

$$I(t) = \begin{cases} 0 & \text{if } \sup_{0\le u\le 1}|z(u)| = 0 \\ \infty & \text{if } \sup_{0\le u\le 1}|z(u)| > 0 \end{cases} \tag{2.7}$$

if $\sigma^2 = 0$.

Proof. First, we prove that (a) implies (b). Observe that if $n^{-1/2}c_n \to M < \infty$, then $E[X^2] < \infty$ and $E[X] = 0$. If $n^{-1/2}c_n \to \infty$, then $E[X^2] = \infty$. For each $\delta > 0$,

$$a_n^{-2}\log(\Pr\{|c_n^{-1}a_n^{-1}\sum_{j=1}^n X_j I_{|X_j|\ge a_n c_n}| \ge \delta\}) \le a_n^{-2}\log(n\Pr\{|X| \ge a_n c_n\}) \to -\infty.$$

Next, we prove that for each $\delta > 0$,

$$a_n^{-2}\log\Pr\{|c_n^{-1}a_n^{-1}\sum_{j=1}^n X_j I_{a_n^{-1}c_n \le |X_j| < a_n c_n}| \ge \delta\} \to -\infty. \tag{2.8}$$

We have that for each $\lambda > 0$,

$$\begin{aligned} &a_n^{-2}\log\Pr\{|c_n^{-1}a_n^{-1}\sum_{j=1}^n X_j I_{a_n^{-1}c_n \le |X_j| < a_n c_n}| \ge \delta\} \\ &\quad\le -\lambda\delta + na_n^{-2}\log E[\exp(\lambda c_n^{-1}a_n|X|I_{a_n^{-1}c_n\le|X_j|<a_nc_n})]. \end{aligned}$$

Hence, if we show that for each $0 < \lambda < \infty$,

$$na_n^{-2}E[\exp(\lambda c_n^{-1}a_n|X|)I_{a_n^{-1}c_n\le|X_j|<a_nc_n} - 1] \to 0, \tag{2.9}$$

then (2.8) follows by letting $\lambda \to \infty$. Given $1 < M < \infty$,

$$na_n^{-2} E[\exp(\lambda c_n^{-1} a_n |X|) I_{a_n^{-1} c_n \le |X| < a_n c_n} - 1]$$
$$= na_n^{-2} \int_0^\infty \Pr\{\exp(\lambda c_n^{-1} a_n |X|) I_{a_n^{-1} c_n \le |X| < a_n c_n} - 1 \ge t\}\, dt$$
$$\le (e^M - 1) na_n^{-2} \Pr\{|X| \ge a_n^{-1} c_n\}$$
$$+ na_n^{-2} \int_{e^M - 1}^\infty \Pr\{\exp(\lambda c_n^{-1} a_n |X|) \ge t+1, a_n c_n \ge |X|\}\, dt =: I + II.$$

By the change of variables $t + 1 = e^u$, we have that

$$II \le a_n^{-2} n \int_M^{\lambda a_n^2} \Pr\{|X| \ge u\lambda^{-1} c_n a_n^{-1}\} e^u\, du.$$

Take $m = m(n,u)$ such that $a_m c_m \le u\lambda^{-1} c_n a_n^{-1} < a_{m+1} c_{m+1}$. Observe $m(n,1) \to \infty$ as $n \to \infty$. So, given $c > 1 + 2\lambda$, for m large

$$m \Pr\{|X| \ge a_m c_m\} \le -e^{-ca_m^2}.$$

Hence, for $u \ge 1$, and n large enough,

$$na_n^{-2} \Pr\{|X| \ge u\lambda^{-1} c_n a_n^{-1}\} e^u \qquad \le nu^{-2} \lambda^2 c_n^{-2} a_{m+1}^2 c_{m+1}^2 \Pr\{|X| \ge c_m a_m\} e^u$$
$$\le n\lambda^2 c_n^{-2} a_{m+1}^2 c_{m+1}^2 m^{-1} e^{-ca_m^2 + u}.$$

We also have that for $m \ge 3$, $a_{m+1}^2 c_{m+1}^2 \le a_m^2 c_m^2 (m+1)^2 m^{-2} \le 2a_m^2 c_m^2$. Since $u < \lambda a_n^2$, $a_m c_m \le u\lambda^{-1} c_n a_n^{-1} < a_n c_n$, for n large, $m \le n$. So, $c_n^{-2} c_m^2 n m^{-1} \le 1$,

$$nc_n^{-2} a_{m+1}^2 c_{m+1}^2 m^{-1} \le 2nc_n^{-2} a_m^2 c_m^2 m^{-1} \le 2a_m^2$$

and

$$u \le \lambda c_n^{-1} a_n a_{m+1} c_{m+1} \le 2\lambda c_n^{-1} a_n a_m c_m \le 2\lambda a_m^2.$$

From these estimations, for n large enough,

$$na_n^{-2} \Pr\{|X| \ge \lambda^{-1} u a_n^{-1} c_n\} e^u \le 2\lambda^2 a_m^2 e^{-ca_m^2 + u}$$
$$\le 2\lambda^2 e^{-(c-1)a_m^2 + u} \le 2\lambda^2 e^{-((c-1)2^{-1}\lambda^{-1} - 1)u}.$$

Hence,

$$\limsup_{n\to\infty} na_n^{-2} \int_M^{\lambda a_n^2} \Pr\{|X| \ge \lambda^{-1} c_n a_n^{-1} u\} e^u\, du$$
$$\le 2\lambda^2 ((c-1)2^{-1}\lambda^{-1} - 1)^{-1} e^{-((c-1)2^{-1}\lambda^{-1} - 1)M}.$$

Since M can be made arbitrarily large, $II \to 0$. By the previous estimations with $u = \lambda$, we get that $I \le (e^M - 1) 2\lambda^2 e^{-(c-1)a_m^2 + \lambda} \to 0$. Hence, (2.9) follows.

It follows from (2.9) that

$$na_n^{-1} c_n^{-1} E[|X| I_{a_n^{-1} c_n \le |X_j| < a_n c_n}] \to 0$$

By the previous estimations, it suffices to show that $\{c_n^{-1}a_n^{-1}\sum_{j=1}^n(X_jI_{|X_j|<a_n^{-1}c_n} - E[X_jI_{|X_j|<a_n^{-1}c_n}])\}$ satisfies the LDP with speed a_n^2 and rate function $I(t) = \sup_{u\in\mathbb{R}}(ut - 2^{-1}u^2\sigma^2)$. This follows from Lemma 2.3. Therefore, we got that (a) implies (b).

Next, we prove that (b) implies (a). Let $J(t) = \inf\{I(u) : |u| \geq t\}$. Then, for each $t > 0$,

$$\limsup_{n\to\infty} a_n^{-2}\log(\Pr\{a_n^{-1}c_n^{-1}|\sum_{j=1}^n X_j| \geq t\}) \leq -J(t).$$

By Lemma 2.1 in Arcones [3], for each $t > 0$,

$$\limsup_{n\to\infty} a_n^{-2}\log(n\Pr\{|X| \geq 3a_nc_nt\}) \leq -J(t).$$

Given $t > 1$, $m = m(n,t)$ such that $a_mc_m \leq 3^{-1}t^{-1}c_na_n < a_{m+1}c_{m+1}$. Observe that for each $t > 0$, $m(n,t) \to \infty$ as $n \to \infty$. Since $t > 1$, $m \leq n$. For n large,

$$a_nc_n \leq 3ta_{m+1}c_{m+1} \leq 6ta_mc_m \leq 6ta_nc_m \leq 6tn^{-1/2}m^{1/2}a_nc_n.$$

So, $m^{-1}n \leq 36t^2$. We also have that

$$a_n^2 \leq 36t^2c_n^{-2}a_m^2c_m^2 \leq 36t^2n^{-1}a_m^2m \leq 36t^2a_n^{-1}c_n^{-1}a_m^3c_m \leq 12ta_m^2.$$

From these estimations, for n large enough,

$$\begin{aligned} n\Pr\{|X| \geq a_nc_n\} &\leq n\Pr\{|X| \geq 3ta_mc_m\} \\ &\leq m^{-1}ne^{-2^{-1}J(t)a_m^2} \leq 36t^2e^{-2^{-5}t^{-1}J(t)a_n^2}. \end{aligned}$$

Since, $\lim_{t\to\infty} t^{-1}J(t) = \infty$, (a.1) follows.

Since for each $\delta > 0$, $\inf\{I(t) : |t| \geq \delta\} > 0$, $c_n^{-1}a_n^{-1}\sum_{j=1}^n X_j \overset{\Pr}{\to} 0$. By the necessary and sufficient conditions for the weak law of the large numbers (see for example Corollary 10.1.3 in Chow and Teicher [7]), (a.2) follows.

Since (a.1) and (a.2) hold, by the proof (a) implies (b), we have that the sum of the r.v.'s for $|X_j| \geq a_n^{-1}c_n$ is asymptotically exponentially negligible. So, we have that for each $t > 0$,

$$\limsup_{n\to\infty} a_n^{-2}\log(\Pr\{|c_n^{-1}a_n^{-1}\sum_{j=1}^n(X_jI_{|X_j|<a_n^{-1}c_n} - E[X_jI_{|X_j|<a_n^{-1}c_n}])| \geq t\}) \leq -z(t).$$

Let $\{X_n'\}$ be an independent copy of $\{X_n\}$. Then, for each $t > 0$,

$$\limsup_{n\to\infty} a_n^{-2}\log(\Pr\{|c_n^{-1}a_n^{-1}\sum_{j=1}^n(X_jI_{|X_j|<a_n^{-1}c_n} - X_j'I_{|X_j'|<a_n^{-1}c_n})| \geq 2t\}) \leq -z(t).$$

By Proposition 6.14 in Ledoux and Talagrand [16],

$$E[\exp(\lambda c_n^{-1} a_n |\sum_{j=1}^{n} (X_j I_{|X_j|<a_n^{-1}c_n} - X_j' I_{|X_j'|<a_n^{-1}c_n})|)] \tag{2.10}$$
$$\leq e^{2\lambda a_n^2 t} + 2e^{2\lambda a_n^2 (t+a_n^{-2}) - 2z(t)a_n^2}$$
$$\times E[\exp(\lambda c_n^{-1} a_n |\sum_{j=1}^{n} (X_j I_{|X_j|<a_n^{-1}c_n} - X_j' I_{|X_j'|<a_n^{-1}c_n})|)].$$

Hence, for $0 < \lambda < 2^{-1}t^{-1}z(t)$, and n large enough.

$$E[\exp(\lambda c_n^{-1} a_n |\sum_{j=1}^{n} (X_j I_{|X_j|<a_n^{-1}c_n} - X_j' I_{|X_j'|<a_n^{-1}c_n})|)] \leq 2e^{2\lambda a_n^2 t}.$$

Now, Lemma 10.2.1 in Chow and Teicher [7] implies that for $0 < \lambda < 2^{-1}$,

$$\exp(\lambda^2 a_n^2 s_n^2 g(-2\lambda)(1 - 2\lambda^2))$$
$$\leq E[\exp(\lambda c_n^{-1} a_n \sum_{j=1}^{n} (X_j I_{|X_j|<a_n^{-1}c_n} - X_j' I_{|X_j'|<a_n^{-1}c_n}))],$$

where $g(x) = x^{-2}(e^x - 1 - x)$ and $s_n^2 = c_n^{-2} n \mathrm{Var}(X I_{|X|<a_n^{-1}c_n})$. Hence, $\limsup_{n\to\infty} s_n^2 < \infty$. If $\lim_{n\to\infty} s_n^2$ does not exist, we have sequences such that (a.1)–(a.3) are satisfied with different σ^2. But, this implies that there are different rates functions for the LDP of the whole sequence, in contradiction. Therefore, (a.3) holds. Hence, (b) implies (a).

The contraction principle implies that (c) implies (b).

Finally, we prove that (b) implies (c). We use Theorem 2.1. Let

$$U_n(u) = c_n^{-1} a_n^{-1} \sum_{j=1}^{[nu]} (X_j I_{|X_j|<a_n^{-1}c_n} - E[X_j I_{|X_j|<a_n^{-1}c_n}]), 0 \leq u \leq 1.$$

We need to prove that $\{U_n(u) : 0 \leq u \leq 1\}$ satisfies the LDP in $l_\infty([0,1])$ with speed a_n^2. The convergence of the finite dimensional distributions follows from Lemma 2.3. The corresponding covariance function is

$$R(u_1, u_2) = \lim_{n\to\infty} \mathrm{Cov}(U_n(u_1), U_n(u_2)) = \min(u_1, u_2)\sigma^2. \tag{2.11}$$

To prove (a.3) in Theorem 2.1, it suffices to prove that for each $\tau, M > 0$, there exists a function $\pi : [0,1] \to [0,1]$ with finite range such that

$$\limsup_{n\to\infty} a_n^{-2} \log \left(\Pr\{ \sup_{0\leq u\leq 1} |U_n(u) - U_n(\pi(u))| \geq \tau\} \right) \leq -M. \tag{2.12}$$

Let

$$U_n'(u) = c_n^{-1} a_n^{-1} \sum_{j=1}^{[nu]} (X_j' I_{|X_j'|<a_n^{-1}c_n} - E[X_j' I_{|X_j'|<a_n^{-1}c_n}]), 0 \leq u \leq 1.$$

Given a positive integer m, let $\pi : [0,1] \to [0,1]$ given by $\pi(u) = \frac{[mu]}{m}$. By symmetrization (see Lemma 1.2.1 in Giné and Zinn [13]) and the Lévy inequality, we have that

$$\begin{aligned}
&\Pr\{\sup_{0\le u\le 1} |U_n(u) - U_n(\pi(u))| \ge \tau\}\\
&\quad = \Pr\{\max_{1\le j\le m} \max_{(j-1)/m<u<j/m} |U_n(u) - U_n((j-1)/m)| \ge \tau\}\\
&\quad \le m \Pr\{\max_{0<u<1/m} |U_n(u)| \ge \tau\}\\
&\quad \le 2m \Pr\{\max_{0<u<1/m} |U_n(u) - U_n'(u)| \ge 2^{-1}\tau\}\\
&\quad \le 4m \Pr\{|U_n(1/m) - U_n'(1/m)| \ge 2^{-1}\tau\}\\
&\quad \le 8m \Pr\{|U_n(1/m)| \ge 2^{-2}\tau\}
\end{aligned}$$

By the Prokhorov inequality (Prokhorov [20]),

$$\limsup_{n\to\infty} a_n^{-2} \log(\Pr\{|U_n(1/m)| \ge 2^{-2}\tau\} \le -2^{-3}\tau \mathrm{arcsinh}(2^{-2}\sigma^{-2}m\tau).$$

Hence, (2.13) follows.

The rate function in (b) is $I(z) = \sup_{t\in R}(zt - t^2\sigma^2/2)$, which is given by (2.3) and (2.4).

The rate function in (c) is determined by the covariance function in (2.13). The class of functions $\{\sigma I(x \le t) : 0 \le t \le 1\}$ in the probability space $([0,1], \mathcal{B}, dx)$ has this covariance function. So, the rate function is

$$I(z) = \inf\{2^{-1}\int_0^1 \gamma^2(x)\,dx : z(t) = \sigma \int_0^1 \gamma(x) I(x \le t)\,dx \text{ for each } 0 \le t \le 1\}.$$

This rate function is given by (2.5) and (2.6). □

To obtain (a) implies in (b) in the previous theorem it suffices that $a_n \to \infty$, $a_n^{-1}c_n^{-1}n \to \infty$ and $\inf_{n\ge 1} n^{-1}c_n^2 > 0$.

Next, we will present two examples, where Theorem 2.4 applies to r.v.'s with infinite second moment.

Let $\{X_j\}_{j=1}^\infty$ be a sequence of symmetric i.i.d.r.v.'s with $\Pr\{|X| \ge t\} = t^{-2}$, for each $t \ge 1$. Let $\{a_n\}_{n=1}^\infty$ be a sequence of real numbers such that $a_n \to \infty$ and $a_n^{-2}\log\log n \to \infty$. By Theorem 2.4, $\{n^{-1/2}(\log n)^{-1/2}a_n^{-1}\sum_{j=1}^n X_j\}$ satisfies the LDP with speed a_n^2 and rate function $I(t) = 2^{-1}t^2$. Observe that $\{n^{-1/2}(\log n)^{-1/2}\sum_{j=1}^n X_j\}$ converges in distribution to a normal r.v. with mean zero and variance 1.

Given $2 > p > 1$, let $\{X_j\}_{j=1}^\infty$ be a sequence of symmetric i.i.d.r.v.'s with $\Pr\{|X| \ge t\} = t^{-p}$, for each $t \ge 1$. Let $\{a_n\}_{n=1}^\infty$ and $\{c_n\}_{n=1}^\infty$ be two sequences of real numbers such that $a_n \to \infty$ and $a_n^{-1}c_n^{-1}n \to \infty$ and $a_n^{-2}\log(n^{-1}c_n^p) \to \infty$. Then, Theorem 2.4 gives that $\{c_n^{-1}a_n^{-1}\sum_{j=1}^n X_j\}$ satisfies the LDP with speed a_n^2 and the rate function I in (2.4). It is well known that $n^{-1/p}\sum_{j=1}^n X_j$ converges in distribution to a symmetric stable r.v. of order p.

To get (b) implies (a), we need that the rate function $I(t)$ satisfies $\lim_{|t|\to\infty} t^{-1}I(t) = \infty$. There are examples of sequence of i.i.d.r.v.'s $\{X_n\}$ such that $\{a_n^{-1}n^{-1/2}\sum_{j=1}^n X_j\}$ satisfies the LDP with speed a_n^2 and a rate function $I(t) = a|t|^p$, for some $a > 0$ and some $0 < p < 1$ (see for example, Nagaev [17, 18]).

When $c_n = n^{1/2}$, the conditions in the previous theorem simplify.

Theorem 2.5. *Let $\{X_j\}_{j=1}^\infty$ be a sequence of i.i.d.r.v.'s. Let $\{a_n\}_{n=1}^\infty$ be a sequence of real numbers such that $a_n \nearrow \infty$ and $n^{-1/2}a_n \searrow 0$. Then, the following conditions are equivalent:*

(a) *$E[X] = 0$, $E[X^2] < \infty$ and*

$$\lim_{n\to\infty} a_n^{-2}\log(n\Pr\{|X| \geq n^{1/2}a_n\}) = -\infty. \tag{2.13}$$

(b) *$\{a_n^{-1}n^{-1/2}\sum_{j=1}^n X_j\}$ satisfies the LDP with speed a_n^2 and a rate I such that $\lim_{|\lambda|\to\infty}\lambda^{-1}I(\lambda) = \infty$.*

(c) *$\{a_n^{-1}n^{-1/2}\sum_{j=1}^{[nu]} X_j : 0 \leq u \leq 1\}$ satisfies the LDP in $l_\infty([0,1])$ with speed a_n^2 and a rate function I such that*

$$\lim_{\lambda\to\infty}\lambda^{-1}\inf\{I(z) : |z(1)| \geq \lambda\} = \infty.$$

Moreover, the rate functions in (b) and (c) are given by (2.3)–(2.5).

Proof. Assume (a). Condition (a) implies condition (a) in Theorem 2.4. So, (b) and (c) follow. (c) implies (b) trivially.

Assume (b). The proof is similar to that of Theorem 2.4. The difference is that we do not assume that the rate function satisfies that for each $\delta > 0$, $\inf\{I(z) : |z| \geq \delta\} > 0$. The argument in the proof of Theorem 2.4 implies that

$$\lim_{n\to\infty} a_n^{-2}\log(n\Pr\{|X| \geq n^{1/2}a_n\}) = -\infty.$$

So, $\lim_{t\to\infty} t^2\Pr(|X| \geq t) = 0$. In particular, for each $0 < p < 2$, $E[|X|^p] < \infty$ and $\lim_{t\to\infty} t^{2-p}E[|X|^pI(|X| \geq t)] = 0$.

Since for M large enough, $\inf\{I(z) : |z| \geq M\} > 0$, $a_n^{-1}n^{-1/2}\sum_{j=1}^n X_j$ is bounded in probability. This implies that $n^{-1}\sum_{j=1}^n X_j \overset{\Pr}{\to} 0$. Since $E[|X|] < \infty$, we have that $E[X] = 0$. Hence,

$$\lim_{n\to\infty} n^{1/2}a_n^{-1}E[XI(|X| \leq a_nn^{1/2}] = \lim_{n\to\infty} n^{1/2}a_n^{-1}E[XI(|X| > a_nn^{1/2})] = 0.$$

we got that conditions (a.1) and (a.2) in Theorem 2.4 hold. Proceeding as in the proof of Theorem 2.4, we get that $E[X^2] < \infty$. □

To obtain (a) implies in (b) in the previous theorem it suffices that $a_n \to \infty$ and $n^{-1/2}a_n \to 0$.

By the results in Cramér [8] and Petrov [19], if for some $\lambda > 0$, $E[e^{\lambda|X|}] < \infty$, then (b) in Corollary 2.4 holds. But, Corollary 2.4 applies to r.v.'s whose moment generating function is not defined in a neighborhood of zero. It follows from Corollary 2.4 that if X is symmetric and for some $0 < p < 1$,

$$\lim_{t\to\infty} t^{-p}\log(\Pr\{|X| \geq t\}) = -1,$$

and $\{a_n\}$ a sequence of positive numbers such that $n^{-p/2}a_n^{2-p} \to \infty$ and $n^{-1/2}a_n \to 0$, then $\{a_n^{-1}n^{-1/2}\sum_{j=1}^n X_j\}$ satisfies the LDP with speed a_n^2 and with rate function $I(t) = \frac{t^2}{2E[X^2]}$. It is easy to see that if for some $\lambda > 0$, $E[e^{\lambda|X|}] < \infty$, then (2.14) holds.

We will obtain the LDP for empirical processes, using a general lemma dealing with triangular array of empirical processes. Let $(\Omega_n, \mathcal{A}_n, Q_n)$, $n \geq 1$, be a sequence of probability spaces and let $(S_{n,j}, \mathcal{S}_{n,j})$, $1 \leq j \leq k_n$, $n \geq 1$, be measurable spaces. For each $n \geq 1$, let $\{X_{n,j} : 1 \leq j \leq k_n\}$ be independent r.v.'s defined on $(\Omega_n, \mathcal{A}_n, Q_n)$ and with values in $(S_{n,j}, \mathcal{S}_{n,j})$, $1 \leq j \leq k_n$. Let $f_{n,j}(\cdot, t) : (S_{n,j}, \mathcal{S}_{n,j}) \to \mathbb{R}$ be a measurable function for each $1 \leq j \leq k_n$, each $n \geq 1$ and each $t \in T$. Let $U_n(t) := \sum_{j=1}^{k_n} f_{n,j}(X_{n,j}, t)$. To avoid measurability problems, we will assume that $\Omega_n := \prod_{j=1}^{k_n} S_{n,j}$, $\mathcal{A}_n := \prod_{j=1}^{k_n} \mathcal{S}_{n,j}$, $Q_n := \prod_{j=1}^{k_n} Q_{n,j}$ and that for each $1 \leq j \leq k_n$, $\{f_{n,j}(x,t) : t \in T\}$ is an image admissible Suslin class of functions (see page 80 in Dudley [11]). By an abuse of notation, we denote by Pr to Q_n.

Lemma 2.6. *Let d be a pseudometric in T. Assume that:*

(i) (T, d) is totally bounded.

(ii) For each $0 < M < \infty$,

$$\lim_{n\to\infty} \epsilon_n \log(\sum_{j=1}^{k_n} \Pr\{F_{n,j}(X_{n,j}) > M\}) = -\infty,$$

where $F_{n,j}(x) = \sup_{t\in T} |f_{n,j}(x,t)|$.

(iii) For each $0 < a, M, \lambda < \infty$,

$$\lim_{n\to\infty} \epsilon_n \log\left(E[\exp(\epsilon_n^{-1}\lambda\sum_{j=1}^{k_n} F_{n,j}(X_{n,j})I_{M\geq F_{n,j}(X_{n,j})>a\epsilon_n})]\right) = 0.$$

(iv) For some $a > 0$,

$$E[\sup_{t\in T}|\sum_{j=1}^{k_n}(f_{n,j}(X_{n,j},t)I_{F_{n,j}(X_{n,j})\leq a\epsilon_n} - E[f_{n,j}(X_{n,j},t)I_{F_{n,j}(X_{n,j})\leq a\epsilon_n}])|] \to 0.$$

(v) For some $a > 0$,

$$\lim_{\eta\to 0}\limsup_{n\to\infty}\sup_{d(s,t)\leq\eta} \epsilon_n^{-1}\sum_{j=1}^{k_n} \mathrm{Var}((f_{n,j}(X_{n,j},s) - f_{n,j}(X_{n,j},t))I_{F_{n,j}(X_{n,j})\leq a\epsilon_n}) = 0.$$

(vi) For a fixed $0 < a < \infty$, for each $s, t \in T$,

$$\lim_{n\to\infty} \epsilon_n^{-1} \sum_{j=1}^{k_n} \mathrm{Cov}(f_{n,j}(X_{n,j}, s) I_{F_{n,j}(X_{n,j}) \le a\epsilon_n}, f_{n,j}(X_{n,j}, t) I_{F_{n,j}(X_{n,j}) \le a\epsilon_n})$$

exists.
Then, for each $a > 0$,

$$\{\sum_{j=1}^{k_n} (f_{n,j}(X_{n,j}, t) - E[f_{n,j}(X_{n,j}, t) I_{F_{n,j}(X_{n,j}) \le a\epsilon_n}]) : t \in T\}$$

satisfies the LDP with speed ϵ_n and the rate function in (2.2) with respect to the covariance function

$$R(s,t) := \lim_{n\to\infty} \epsilon_n^{-1} \sum_{j=1}^{k_n} \mathrm{Cov}(f_{n,j}(X_{n,j}, s) I_{F_{n,j}(X_{n,j}) \le a\epsilon_n}, \\ f_{n,j}(X_{n,j}, t) I_{F_{n,j}(X_{n,j}) \le a\epsilon_n}),\ s, t \in T.$$

Proof. First, we prove that condition (iii) implies that if (iv)–(vi) hold for some $a > 0$, then they hold for each $a > 0$. Note that condition (iii) implies that for each $0 < a, M, \lambda < \infty$,

$$\lim_{n\to\infty} \epsilon_n E[\exp(\epsilon_n^{-1} \lambda \sum_{j=1}^{k_n} F_{n,j}(X_{n,j}) I_{M \ge F_{n,j}(X_{n,j}) > a\epsilon_n}) - 1] = 0.$$

This implies that for each $0 < a, M < \infty$,

$$\lim_{n\to\infty} E[\sum_{j=1}^{k_n} F_{n,j}(X_{n,j}) I_{M \ge F_{n,j}(X_{n,j}) > a\epsilon_n})] = 0 \tag{2.14}$$

and

$$\lim_{n\to\infty} E[\sum_{j=1}^{k_n} F_{n,j}^2(X_{n,j}) I_{M \ge F_{n,j}(X_{n,j}) > a\epsilon_n})] = 0.$$

Hence, we may assume that (iv)–(vi) hold for each $a > 0$.

Fixed $a > 0$, we use Theorem 2.1 to prove that $\{U_n(t) : t \in T\}$ satisfies the LDP, where

$$U_n(t) := \sum_{j=1}^{k_n} (f_{n,j}(X_{n,j}, t) - E[f_{n,j}(X_{n,j}, t) I_{F_{n,j}(X_{n,j}) \le a\epsilon_n}]).$$

Condition (i) implies (a.1) in Theorem 2.1.

Let

$$V_n(t) := \sum_{j=1}^{k_n} f_{n,j}(X_{n,j}, t) I_{F_{n,j}(X_{n,j}) > a\epsilon_n}.$$

First, we prove that $\{V_n(t) : t \in T\}$ is uniformly exponentially asymptotically negligible, i.e. for each $\tau > 0$,

$$\lim_{n\to\infty} \epsilon_n \log \left(\Pr\{\sup_{t\in T} |V_n(t)| \geq \tau\} \right) = -\infty. \tag{2.15}$$

By (ii) and (iii), we may take $\lambda, M > 0$ such that $\lambda\tau \geq 4c$,

$$\limsup_{n\to\infty} \epsilon_n \log(\sum_{j=1}^{k_n} \Pr\{F_{n,j}(X_{n,j}) > M\}) < -c$$

and

$$\limsup_{n\to\infty} \epsilon_n \log E[\exp(\epsilon_n^{-1}\lambda \sum_{j=1}^{k_n} F_{n,j}(X_{n,j}) I_{M \geq F_{n,j}(X_{n,j}) > a\epsilon_n})] < 4^{-1}\lambda\tau.$$

Then, for n large enough,

$$\begin{aligned}
&\epsilon_n \log(\Pr\{\sup_{t\in T} |\sum_{j=1}^{k_n} f_{n,j}(X_{n,j}, t) I_{F_{n,j}(X_{n,j}) > M}| \geq 2^{-1}\tau\}) \\
&\leq \epsilon_n \log(\sum_{j=1}^{k_n} \Pr\{F_{n,j}(X_{n,j}) > M\}) \leq -c,
\end{aligned}$$

and

$$\begin{aligned}
&\Pr\{\sup_{t\in T} |\sum_{j=1}^{k_n} f_{n,j}(X_{n,j}, t) I_{M > F_{n,j}(X_{n,j}) \geq a\epsilon_n}| \geq 2^{-1}\tau\} \\
&\leq \Pr\{\sum_{j=1}^{k_n} F_{n,j}(X_{n,j}) I_{M > F_{n,j}(X_{n,j}) \geq a\epsilon_n} \geq 2^{-1}\tau\} \\
&\leq e^{-2^{-1}\epsilon_n^{-1}\lambda\tau} E[\exp(\epsilon_n^{-1}\lambda \sum_{j=1}^{k_n} F_{n,j}(X_{n,j}) I_{M > F_{n,j}(X_{n,j}) \geq a\epsilon_n})] \leq e^{-c\epsilon_n^{-1}}.
\end{aligned}$$

Hence,

$$\limsup_{n\to\infty} \epsilon_n \log \left(\Pr\{\sup_{t\in T} |V_n(t)| \geq \tau\} \right) \leq -c$$

and (2.17) follows.

By (2.17), to get the LDP for the finite dimensional distributions, we need to prove that for each $t_1, \dots, t_m \in T$, $(W_n(t_1), \dots, W_n(t_m))$ satisfies the LDP, where

$$W_n(t) = \sum_{j=1}^{k_n} (f_{n,j}(X_{n,j}, t) I_{F_{n,j}(X_{n,j}) \leq a\epsilon_n} - E[f_{n,j}(X_{n,j}, t) I_{F_{n,j}(X_{n,j}) \leq a\epsilon_n}]).$$

We claim that Lemma 2.3 implies the LDP for the finite dimensional distributions of the process $\{W_n(t) : t \in T\}$. Condition (i) in Lemma 2.3 follows from (vi).

Condition (ii) in Lemma 2.3 is obviously satisfied. As to condition (iii) in Lemma 2.3, given $t \in T$ and $0 < \delta < a$,

$$\epsilon_n \sum_{j=1}^{k_n} \Pr\{|f_{n,j}(X_{n,j},t)I_{F_{n,j}(X_{n,j})\le\epsilon_n a} - E[f_{n,j}(X_{n,j},t)I_{F_{n,j}(X_{n,j})\le\epsilon_n a}]| \ge \epsilon_n\delta\}$$

$$\le \epsilon_n \sum_{j=1}^{k_n} \Pr\{|f_{n,j}(X_{n,j},t)I_{2^{-2}\epsilon_n\delta<F_{n,j}(X_{n,j})\le\epsilon_n a}$$

$$- E[f_{n,j}(X_{n,j},t)I_{2^{-2}\epsilon_n\delta<F_{n,j}(X_{n,j})\le\epsilon_n a}]| \ge 2^{-1}\epsilon_n\delta\}$$

$$\le 4\delta \sum_{j=1}^{k_n} E[F_{n,j}(X_{n,j},t)I_{2^{-2}\epsilon_n\delta<F_{n,j}(X_{n,j})\le\epsilon_n a}] \to 0,$$

by (2.16).

By (2.17), to prove (a.3) in Theorem 2.1, it suffices to consider $\{Z_n(t) : t \in T\}$, where

$$Z_n(t) = \sum_{j=1}^{k_n}(f_{n,j}(X_{n,j},t)I_{F_{n,j}(X_{n,j})<a\epsilon_n} - E[f_{n,j}(X_{n,j},t)I_{F_{n,j}(X_{n,j})<a\epsilon_n}]).$$

It is easy to see that by Theorem 1.4 in Talagrand [22], condition (v) imply that for each $\tau > 0$,

$$\lim_{\eta\to 0}\limsup_{n\to\infty} \epsilon_n \log\left(\Pr\{|\sup_{d(s,t)\le\eta}|Z_n(s)-Z_n(t)| \right.$$

$$\left. -E[\sup_{d(s,t)\le\eta}|Z_n(s)-Z_n(t)|]| \ge \tau\}\right) = -\infty.$$

Since condition (iv) implies that $E[\sup_{t\in T}|Z_n(t)|]| \to 0$, (a.3) in Theorem 2.1 follows. □

Next, we present necessary and sufficient conditions for moderate deviations of empirical processes. We will use the following set-up. Let $(S, \mathcal{S}, \nu)$ be a probability space. Let $\Omega = S^{\mathbb{N}}$, $\mathcal{A} = \mathcal{S}^{\mathbb{N}}$, and $Q = \nu^{\mathbb{N}}$. Let X_n be the n-th projection from Ω into S. Then, $\{X_n\}_{n=1}^{\infty}$ is a sequence of i.i.d.r.v.'s with values in S. Let $\{f(\cdot,t) : t \in T\}$ be an image admissible Suslin class of measurable functions from S into $\mathbb{R}$. F will denote an envelope of $\{f(\cdot,t) : t \in T\}$, i.e. a measurable function on S, such that $F(X) \ge \sup_{t\in T}|f(X,t)|$ a.s.

Theorem 2.7. *With the notation above, let $\{a_n\}_{n=1}^{\infty}$ and $\{c_n\}_{n=1}^{\infty}$ be two sequences of real numbers such that $a_n \nearrow \infty$ and $a_n^{-1}c_n^{-1}n \nearrow \infty$ and $\{n^{-1}c_n^2\}$ is nondecreasing. Then, the following sets of conditions ((a), (b) and (c)) are equivalent:*

(a.1) $\lim_{n\to\infty} a_n^{-2}\log(n\Pr\{F(X) \ge a_nc_n\}) = -\infty$.

(a.2) $a_n^{-1}c_n^{-1}\sup_{t\in T}|\sum_{j=1}^n f(X_j,t)| \xrightarrow{\Pr} 0$.

(a.3) *For each $s,t \in T$, the following limit exists:*

$$\lim_{n\to\infty} nc_n^{-2}\mathrm{Cov}(f(X,s)I_{F(X)<a_n^{-1}c_n}, f(X,t)I_{F(X)<a_n^{-1}c_n})$$

(a.4) *(T,d) is totally bounded, where*

$$d^2(s,t) = \lim_{n\to\infty} nc_n^{-2}\mathrm{Var}((f(X,s)-f(X,t))I_{F(X)<a_n^{-1}c_n}).$$

(a.5) $\lim_{\eta\to 0}\lim_{n\to\infty}\sup_{d(s,t)\le\eta} nc_n^{-2}\mathrm{Var}((f(X,s)-f(X,t))I_{F(X)<a_n^{-1}c_n}) = 0.$

(b) *$\{a_n^{-1}c_n^{-1}\sum_{j=1}^n f(X_j,t) : t \in T\}$ satisfies the LDP in $l_\infty(T)$ with speed a_n^2 and a rate function I such that $\lim_{\lambda\to\infty}\lambda^{-1}\inf\{I(z) : \sup_{t\in T}|z(t)| \ge \lambda\} = \infty$ and for each $\delta > 0$, $\inf\{I(z) : \sup_{t\in T}|z(t)| \ge \delta\} > 0$.*

(c) *$\{c_n^{-1}a_n^{-1}\sum_{j=1}^{[nu]} f(X_j,t) : 0 \le u \le 1, t \in T\}$ satisfies the LDP in $l_\infty([0,1]\times T)$ with speed a_n^2 and a rate function I such that*

$$\lim_{\lambda\to\infty}\lambda^{-1}\inf\{I(z) : \sup_{t\in T}|z(1,t)| \ge \lambda\} = \infty$$

and for each $\delta > 0$, $\inf\{I(z) : \sup_{t\in T}|z(1,t)| > \delta\} > 0$.

Moreover, in (b) the rate function is given by (2.2) with respect to the covariance function

$$R(s,t) = \lim_{n\to\infty} nc_n^{-2}E[f(X,s)f(X,t)I_{F(X)\le a_nc_n}].$$

In (c), the rate function is

$$I(\alpha) = \inf\Big\{2^{-1}\int_0^1\int_\Omega \xi^2(v,\omega)\,d\nu(\omega)\,dv : \text{ where } \xi : [0,1]\times\Omega \to \mathbb{R} \text{ is a measurable function such that } \alpha(u,t) = \int_0^u\int_\Omega \xi(v,\omega)Z(t,\omega)\,d\nu(\omega)\,dv, \text{ for each } u\in[0,1], t\in T\Big\},$$

where $(\Omega,\mathcal{F},\nu)$ is probability space, $\{Z(t,\omega) : t \in T\}$ is a Gaussian process defined on Ω with zero means and covariance given by

$$\int_\Omega Z(s,\omega)Z(t,\omega)\,d\nu(\omega) = \lim_{n\to\infty} nc_n^{-2}E[f(X,s)f(X,t)I_{F(X)\le a_nc_n}],$$

for each $s,t \in T$.

Proof. To prove that (a) implies (b) we apply Lemma 2.6. Conditions (i), (ii), (v) and (vi) in Lemma 2.6 are obviously satisfied. By the arguments used in (2.9), we have that for each $0 < a, M, \lambda < \infty$,

$$na_n^{-2}E[(\exp(\lambda c_n^{-1}a_nF(X)) - 1)\, I_{aa_n^{-1}c_n<F(X)\le a_nc_nM}] \to 0. \tag{2.16}$$

This implies condition (iii) in Lemma 2.6. From (a.1) and (2.18), we have that for each $a > 0$,

$$\sup_{t\in T} a_n^{-1}c_n^{-1}\sum_{j=1}^n F(X_j)I_{F(X_j)>aa_n^{-1}c_n} \xrightarrow{\Pr} 0.$$

This limit and (a.2) imply that for each $a > 0$,

$$\sup_{t\in T} a_n^{-1}c_n^{-1}|\sum_{j=1}^n f(X_j)I_{F(X_j)\le a_n^{-1}c_n}| \overset{\Pr}{\to} 0.$$

So, by the Hoffmann–Jørgensen inequality (see for example Proposition 6.8 in Ledoux and Talagrand [16]), condition (iv) in Lemma 2.6 follows.

Next, we prove that (b) implies (a). From Lemma 2.1 in Arcones [3], for each $\lambda > 0$,

$$\limsup_{n\to\infty} a_n^{-2}\log(n\Pr\{F(X)\ge 2\lambda a_n c_n\}) \le -\inf\{I(z) : |z|_\infty \ge \lambda\}.$$

This estimation and the argument in (2.11) imply (a.1). Since for each $\delta > 0$, $\inf\{I(z) : \sup_{t\in T}|z(t)| \ge \delta\} > 0$, (a.2) holds. Since (a.1) holds, the stochastic process

$$\{a_n^{-1}c_n^{-1}\sum_{j=1}^n (f(X_j,t)I_{F(X_j)<a_n^{-1}c_n} - E[f(X_j,t)I_{F(X_j)<a_n^{-1}c_n}]) : t\in T\}$$

satisfies the LDP. This fact and Theorem 2.4 imply (a.3). By the remark after Theorem 2.2, (a.4) holds. (a.5) follows from the argument in (2.12).

Finally, we prove that (a) implies (c). We use Theorem 2.1. By (2.18), it suffices to prove that $\{U_n(u,t) : 0\le u\le 1, t\in T\}$ satisfies the LDP, where

$$U_n(u,t) := c_n^{-1}a_n^{-1}\sum_{j=1}^{[nu]}(f(X_j,t)I_{F(X_j)<a_n^{-1}c_n} - E[f(X_j,t)I_{F(X_j)<a_n^{-1}c_n}]).$$

The LDP for the finite dimensional distributions follows from Lemma 2.3 applied to the independent random vectors

$$\begin{aligned} X_{n,j} = {} & (Y_j^{(1,1)}, Y_j^{(1,2)}, \dots, Y_j^{(1,m)}, \dots, Y_j^{(m,1)}, Y_j^{(m,2)}, \dots, Y_j^{(m,m)}) \\ & - E[(Y_j^{(1,1)}, Y_j^{(1,2)}, \dots, Y_j^{(1,m)}, \dots, Y_j^{(m,1)}, Y_j^{(m,2)}, \dots, Y_j^{(m,m)})], \quad 1\le j\le n \end{aligned}$$

where

$$Y_j^{(k,l)} = c_n^{-1}f(X_j,t_l)I(F(X_j,t_l) < a_n^{-1}c_n, j\le [nu_k]),$$

$t_1,\dots,t_m \in T$ and $u_1,\dots,u_m \in [0,1]$. Observe that the (k,l) coordinate of $\sum_{j=1}^n X_{n,j}$ is

$$c_n^{-1}\sum_{j=1}^{[nu_k]}\left(f(X_j,t_l)I_{F(X_j)<a_n^{-1}c_n} - E[f(X_j,t_l)I_{F(X_j)<a_n^{-1}c_n}]\right).$$

We have that the covariance of

$$(c_n^{-1}\sum_{j=1}^{[nu_1]} f(X_j,t_1)I_{F(X_j)<a_n^{-1}c_n}, c_n^{-1}\sum_{j=1}^{[nu_2]} f(X_j,t_2)I_{F(X_j)<a_n^{-1}c_n})$$

converges to

$$\min(u_1,u_2)\lim_{n\to\infty} nc_n^{-2}\mathrm{Cov}(f(X,t_1)I_{F(X)<a_n^{-1}c_n}, f(X,t_2)I_{F(X)<a_n^{-1}c_n}).$$

Next, we prove the asymptotic exponential equicontinuity. Given $\tau, M>0$, there exists a function $\pi_2 : T\to T$ with finite range such that

$$\limsup_{n\to\infty} a_n^{-2}\log(\Pr\{\sup_{t\in T}|V_n(1,t)-V_n(1,\pi_2(t))|\ge 2^{-3}\tau\})\le -M. \quad (2.17)$$

Let $\{X_j'\}$ be a independent copy of $\{X_j\}$. Let

$$V_n'(u,t) := c_n^{-1}a_n^{-1}\sum_{j=1}^{[nu]}(f(X_j',t)I_{F(X_j')<a_n^{-1}c_n} - E[f(X_j',t)I_{F(X_j')<a_n^{-1}c_n}]).$$

By symmetrization (see Lemma 1.2.1 in Giné and Zinn [13]) and the Lévy inequality, we have that

$$\begin{aligned}
&\Pr\{\sup_{t\in T,\ 0\le u\le 1}|V_n(u,t)-V_n(u,\pi_2(t))|\ge 2^{-1}\tau\} \quad (2.18)\\
&\le 2\Pr\{\sup_{t\in T,\ 0\le u\le 1}|V_n(u,t)-V_n(u,\pi_2(t))-(V_n'(u,t)-V_n'(u,\pi_2(t)))|\ge 2^{-2}\tau\}\\
&\le 4\Pr\{\sup_{t\in T}|V_n(1,t)-V_n(1,\pi_2(t))-(V_n'(1,t)-V_n'(1,\pi_2(t)))|\ge 2^{-2}\tau\}\\
&\le 8\Pr\{\sup_{t\in T}|V_n(1,t)-V_n(1,\pi_2(t))|\ge 2^{-3}\tau\}.
\end{aligned}$$

Let m be a positive integer such that $M\le 2^{-4}\tau\mathrm{arcsinh}(2^{-3}(\sup_{t\in\pi_2(T)}R(t,t))^{-2}m\tau)$. Let $\pi_2(u)=\frac{[mu]}{m}$, $0\le u\le 1$. Using the Prokhorov inequality as in the proof of the Theorem 2.4, we have that

$$\begin{aligned}
&\limsup_{n\to\infty} a_n^{-2}\log(\Pr\{\sup_{t\in\pi_2(T),\ 0\le u\le 1}|V_n(u,t)-V_n(\pi_1(u),t)|\ge 2^{-1}\tau\}) \quad (2.19)\\
&\le -2^{-4}\tau\mathrm{arcsinh}\left(2^{-3}(\sup_{t\in\pi_2(T)}R(t,t))^{-2}m\tau\right)\le -M.
\end{aligned}$$

From (2.19)–(2.21), we get that

$$\limsup_{n\to\infty} a_n^{-2}\log(\Pr\{\sup_{t\in T,\ 0\le u\le 1}|V_n(u,t)-V_n(\pi_1(u),\pi_2(t))|\ge\tau\})\le -M. \qquad \square$$

When $c_n = n^{1/2}$, the previous theorem gives the following:

Theorem 2.8. *Let $\{a_n\}$ be a sequence of real numbers such that $a_n\nearrow\infty$ and $n^{-1/2}a_n\searrow 0$. Then, the following sets of conditions are equivalent:*

(a.1) $\lim_{n\to\infty} a_n^{-2}\log(n\Pr\{\sup_{t\in T}|f(X,t)-E[f(X,t)]|\ge n^{1/2}a_n\})=-\infty.$

(a.2) $\sup_{t\in T}|a_n^{-1}n^{-1/2}\sum_{j=1}^n(f(X_j,t)-E[f(X_j,t)])|\xrightarrow{\Pr} 0.$

(a.3) *For each* $t\in T$, $E[f^2(X,t)]<\infty$.

(a.4) (T,d) *is totally bounded, where*

$$d^2(s,t)=\mathrm{Var}(f(X,s)-f(X,t)).$$

(b) $\{a_n^{-1}n^{-1/2}\sum_{j=1}^n(f(X_j,t)-E[f(X_j,t)]) : t\in T\}$ *satisfies the LDP in* $l_\infty(T)$ *with speed* a_n^2 *and a rate function* I *such that*

$$\lim_{\lambda\to\infty}\lambda^{-1}\inf\{I(z):\sup_{t\in T}|z(t)|\geq\lambda\}=\infty$$

and for each $\delta>0$, $\inf\{I(z):\sup_{t\in T}|z(t)|\geq\delta\}>0$.

(c) $\{a_n^{-1}n^{-1/2}\sum_{j=1}^{[nu]}(f(X_j,t)-E[f(X_j,t)]) : 0\leq u\leq 1, t\in T\}$ *satisfies the LDP in* $l_\infty([0,1]\times T)$ *with speed* a_n^2 *and a rate function* I *such that* $\lim_{\lambda\to\infty}\lambda^{-1}\inf\{I(z) : \sup_{t\in T}|z(1,t)|\geq\lambda\}=\infty$ *and for each* $\delta>0$, $\inf\{I(z):\sup_{t\in T}|z(1,t)|>\delta\}>0$.

Moreover, the rate function in (b) is

$$I(z)=\inf\{2^{-1}E[\alpha^2(X)] : z(t)=E[\alpha(X)(f(X,t)-E[f(X,t)])] \text{ for each } t\in T\}.$$

Moreover, the rate function in (c) is

$$\begin{aligned}I(z)= \quad &\inf\Big\{2^{-1}\int_0^1E[\gamma^2(u,X)]\,du : \text{ where } \gamma:[0,1]\times S\to\mathbb{R}\\ &\text{is a measurable function such that for each } u\in[0,1], t\in T,\\ &z(u,t)=\int_0^uE[\gamma(v,X)(f(X,t)-E[f(X,t)])]\,dv\Big\}.\end{aligned}$$

Ledoux [15] and Wu [24] proved the part (a) implies (b) in the previous theorem assuming a little different conditions. For example, they assume that for some $c>1$ and some $|\delta|<1/2$, $a_{nk}\leq ck^\delta a_n$, for each $n,k\geq 1$. These conditions do not apply to sequences that very close to n. For example, it does not apply to sequences of the form $a_n=n(\log n)^{-\alpha}$, where $\alpha>0$, for example. Condition (a.4) above is stated differently in Ledoux [15] and Wu [24].

For classical empirical processes Theorem 2.8 gives the following:

Corollary 2.9. *Let* $\{X_j\}_{j=1}^\infty$ *be a sequence of i.i.d.r.v.'s. Let* $\{a_n\}$ *be a sequence of real numbers such that* $a_n\to\infty$ *and* $n^{-1/2}a_n\to 0$. *Then,*

(i) $\{a_n^{-1}n^{-1/2}\sum_{j=1}^n(I(X_j\leq t)-P\{X_j\leq t\}) : t\in\mathbb{R}\}$ *satisfies the LDP in* $l_\infty(\mathbb{R})$ *with speed* a_n^2 *and rate function*

$$I(z)=\inf\{2^{-1}E[\alpha^2(X)] : z(t)=E[\alpha(X)(I(X\leq t)-P(X\leq t))] \text{ for all } t\in T\}.$$

(ii) $\{a_n^{-1}n^{-1/2}\sum_{j=1}^{[nu]}(I(X_j\leq t)-P\{X_j\leq t\}) : t\in\mathbb{R}, 0\leq u\leq 1\}$ *satisfies the LDP in* $l_\infty([0,1]\times\mathbb{R})$ *with speed* a_n^2 *and rate function*

$$\begin{aligned}I(z)= \quad &\inf\Big\{2^{-1}\int_0^1E[\gamma^2(u,X)]\,du where\gamma : [0,1]\times\mathbb{R} \text{ is a measurable function}\\ &\text{such that, for each } (u,t)\in[0,1]\times\mathbb{R},\\ &z(u,t)=\int_0^uE[\gamma(v,X)(I(X\leq t)-P(X\leq t))]\,dv\Big\}.\end{aligned}$$

In the previous corollary, if X has a positive density f_X, then the rate function in part (i) can be written as

$$I(z)=\begin{cases}\int_{-\infty}^\infty\frac{(z'(t))^2}{2f_X(t)}\,dt & \text{if } z \text{ is absolutely continuous and } \lim_{|t|\to\infty}z(t)=0.\\ \infty & \text{else .}\end{cases}$$

By taking T equal to the unit ball of a Banach space, the previous theorems give necessary and sufficient conditions for the moderate deviations of Banach space valued random vectors. In particular, we have:

Theorem 2.10. *Let $\{X_j\}_{j=1}^{\infty}$ be a sequence of i.i.d.r.v.'s with values in a separable Banach space B. Let $\{a_n\}_{n=1}^{\infty}$ and $\{c_n\}_{n=1}^{\infty}$ be two sequences of real numbers such that $a_n \nearrow \infty$ and $a_n^{-1}c_n^{-1}n \nearrow \infty$ and $\{n^{-1}c_n^2\}$ is nondecreasing. Then, the following sets of conditions ((a), (b) and (c)) are equivalent:*

(a.1) $\lim_{n\to\infty} a_n^{-2} \log(n \Pr\{|X| \geq c_n a_n\}) = -\infty$.

(a.2) $a_n^{-1}c_n^{-1}\sum_{j=1}^{n} X_j \stackrel{\Pr}{\to} 0$.

(a.3) *For each $f_1, f_2 \in B^*$, the following limit exists:*

$$\lim_{n\to\infty} nc_n^{-2}E[f_1(X)f_2(X)I_{|X|\leq a_n^{-1}c_n}],$$

where B^* is the dual of B.

(a.4) *(B_1^*, d) is totally bounded, where*

$$d^2(f_1, f_2) = \lim_{n\to\infty} nc_n^{-2}\mathrm{Var}((f_1(X) - f_2(X))I_{|X|\leq a_n^{-1}c_n})$$

and B_1^* is the unit ball of the dual of B.

(a.5) $\lim_{\eta\to 0}\lim_{n\to\infty} \sup_{f_1,f_2\in B_1^*, d(f_1,f_2)\leq\eta} nc_n^{-2}\mathrm{Var}((f_1(X) - f_2(X))I_{|X|\leq a_n^{-1}c_n}) = 0$.

(b) *$\{c_n^{-1}a_n^{-1}\sum_{j=1}^{n} X_j\}$ satisfies the LDP in $l_\infty(B)$ with speed a_n^2 and a good rate function such that $\lim_{\lambda\to\infty}\lambda^{-1}\inf\{I(z) : |z| \geq \lambda\} = \infty$ and for each $\delta > 0$, $\inf\{I(z) : |z| \geq \delta\} > 0$.*

(c) *$\{c_n^{-1}a_n^{-1}\sum_{j=1}^{[nu]} X_j : 0 \leq u \leq 1\}$ satisfies the LDP in $l_\infty([0,1], B)$ with speed a_n^2 and a good rate function such that $\lim_{\lambda\to\infty}\lambda^{-1}\inf\{I(z) : |z(1)| \geq \lambda\} = \infty$ and for each $\delta > 0$, $\inf\{I(z) : |z(1)| \geq \delta\} > 0$.*

When $c_n = n^{1/2}$, the conditions (a) in the previous theorem simplify considerably:

Theorem 2.11. *Let $\{X_j\}_{j=1}^{\infty}$ be a sequence of i.i.d.r.v.'s with values in a separable Banach space B. Let $\{a_n\}_{n=1}^{\infty}$ be a sequence of real numbers such that $a_n \nearrow \infty$ and $n^{-1/2}a_n \searrow 0$. Then, the following sets of conditions ((a), (b) and (c)) are equivalent:*

(a.1) $\lim_{n\to\infty} a_n^{-2}\log(n\Pr\{|X| \geq a_n n^{1/2}\}) = -\infty$.

(a.2) $a_n^{-1}n^{-1/2}\sum_{j=1}^{n}(X_j - E[X_j]) \stackrel{\Pr}{\to} 0$.

(a.3) *$\{f(X) - E[f(X)] : f \in B_1^*\}$ is a totally bounded set of L_2.*

(b) *$\{a_n^{-1}n^{-1/2}\sum_{j=1}^{n}(X_j - E[X_j])\}$ satisfies the LDP in $l_\infty(B)$ with speed a_n^2 and a good rate function such that $\lim_{t\to\infty}\lambda^{-1}\inf\{I(z) : |z| \geq \lambda\} = \infty$ and for each $\delta > 0$, $\inf\{I(z) : |z| \geq \delta\} > 0$.*

(c) *$\{a_n^{-1}n^{-1/2}\sum_{j=1}^{[nu]}(X_j - E[X_j]) : 0 \leq u \leq 1\}$ satisfies the LDP in $l_\infty([0,1], B)$ with speed a_n^2 and a good rate function such that $\lim_{\lambda\to\infty}\lambda^{-1}\inf\{I(z) : |z(1)| \geq \lambda\} = \infty$ and for each $\delta > 0$, $\inf\{I(z) : |z(1)| \geq \delta\} > 0$.*

Hu and Lee [14] obtained the LDP for the stochastic processes in (c) in the previous theorem assuming that for some $\lambda > 0$, $E[e^{\lambda|X|}] < \infty$.

References

[1] M. A. Arcones, *The large deviation principle for stochastic processes I.* To appear in Theor. Probab. Applic. (2002).

[2] M. A. Arcones, *The large deviation principle for stochastic processes II.* To appear in Theor. Probab. Applic. (2002).

[3] M. A. Arcones, *Large deviations of empirical processes.* To appear in *High Dimensional Probability III*, Progress in Probability, Birkhäuser (2003).

[4] R. R. Bahadur, *Some Limit Theorems in Statistics.* SIAM, Philadelphia, PA. 1971.

[5] A. A. Borovkov, and A. A. Mogul'skiĭ, *Probabilities of large deviations in topological spaces. I.* Siber. Math. J. **19** (1978), 988–1004.

[6] A. A. Borovkov, and A. A. Mogul'skiĭ, *Probabilities of large deviations in topological spaces. II.* Siber. Math. J. **21** (1980), 12–26.

[7] Y. S. Chow, and H. Teicher, *Probability Theory. Independence, Interchangeability, Martingales.* Springer, 1978.

[8] H. Cramér, *Sur un nouveau théorèm limite de la théorie des probabilities.* Actualités Sci. Ind. **736** (1938), 5–23.

[9] J. D. Deuschel, and D. W. Stroock, *Large Deviations.* Academic Press, 1989.

[10] J. L. Doob, *Stochastic Processes.* John Wiley, New York. 1953.

[11] R. M. Dudley, *Uniform Central Limit Theorems.* Cambridge University Press, 1999.

[12] R. S. Ellis, *Large deviations for a general class of random vectors.* Ann. Probab. **12** (1984), 1–12.

[13] E. Giné, and J. Zinn, Lectures on the central limit theorem for empirical processes. Lect. Notes in Math. **1221** (1986), 50–112. Springer.

[14] Y. Hu and T. Y. Lee, *Moderate deviation principles for trajectories of sums of independent Banach space valued random variables.* Trans. Amer. Math. Soc. (2001). To appear.

[15] M. Ledoux, Sur les déviation modérées de sommes de variable aléatories vectorialles independantes de même loi. *Ann. Inst. Henri Poincaré.* **28** (1992), 267–280.

[16] M. Ledoux, and M. Talagrand, *Probability in Banach Spaces.* Springer, 1991.

[17] A. V. Nagaev, *Integral limit theorems with regard to large deviations when Cramér's condition is not satisfied.* Theor. Probability Appl. **14** (1969), 51–64.

[18] A. V. Nagaev, *Integral limit theorems with regard to large deviations when Cramér's condition is not satisfied. II.* Theor. Probability Appl. **14** (1969), 193–208.

[19] V. V. Petrov, *A generalization of Cramér's limit theorem.* Select. Transl. Math. Statist. Probab. **6** (1966), 1–8, American Mathematical Society, Providence, R.I.

[20] Yu. V. Prokhorov, *An extremal problem in probability theory.* Theor. Probability Appl. **4** (1959), 211–214.

[21] A. Shwartz and A. Weiss, *Large Deviations for Performance Analysis. Queues, Communications, and Computing.* Chapman & Hall, London, 1995.

[22] M. Talagrand, *New concentration inequalities in product spaces.* Inventiones Math. **126** (1996), 505–563.

[23] S. R. S. Varadhan, *Large Deviations and Applications.* SIAM, Philadelphia, (1984).

[24] L. Wu, *Large deviations, moderate deviations and LIL for empirical processes.* Ann. Probab. **22** (1994), 17–27.

[25] L. Wu, *Moderate deviations of dependent random variables related to CLT.* Ann. Probab. **23** (1995), 420–445.

Department of Mathematical Sciences, Binghamton University,
Binghamton, NY 13902

E-mail address: arcones@math.binghamton.edu

Progress in Probability, Vol. 56, 213–247

Concentration Inequalities for Sub-Additive Functions Using the Entropy Method

Olivier Bousquet

Abstract. We obtain exponential concentration inequalities for sub-additive functions of independent random variables under weak conditions on the increments of those functions, like the existence of exponential moments for these increments. As a consequence of these general inequalities, we obtain refinements of Talagrand's inequality for empirical processes and new bounds for randomized processes associated to unbounded random variables. These results are obtained by further developing the entropy method introduced by Ledoux.

1. Introduction

Sums of independent real-valued random variables have been considerably studied. In 1947, Cramér already stated that this problem is at the heart of probability theory (see [9], page 170)

> [The study of the distribution of a sum of independent random variables] may be regarded as the very starting point of a large number of those investigations by which modern Theory of Probability was created.

This was quoted in Bennett's work [2] in 1962 where the sharpest bounds (up to the improvements brought by Hoeffding in [10]) were obtained on the deviations of sums of independent real-valued random variables.

Naturally, a lot of the subsequent work in probability theory has been focusing on extensions of these results. One possible direction is to consider the dependent case (sampling without replacement, r-dependence, martingales ...).

The other direction is to consider norms of sums of independent vector-valued random variables. The study of such objects can be considered as 'the very starting point of a large number of investigations' in Probability in Banach Spaces and in Empirical Processes Theory. One of the outcomes of such investigations is the development of general purpose concentration inequalities.

Key words and phrases. Concentration Inequalities, Empirical Processes, Talagrand's Inequality, Entropy Method.

Our goal here is to consider more general functions of independent random variables than sums or norms of sums. We will consider sub-additive functions which we define as follows

Definition 1.1. A function f of n variables is *sub-additive* if there exists a sequence of functions f_k of $n-1$ variables such that for all $x_1, \dots, x_n$,

$$\sum_{k=1}^{n} f(x_1, \dots, x_n) - f_k(x_1, \dots, x_{k-1}, x_{k+1}, \dots, x_n) \le f(x_1, \dots, x_n)\,.$$

A function of n independent random variables $X_1, \dots, X_n$ is *sub-additive* if $Z = f(X_1, \dots, X_n)$ is $\sigma(X_1, \dots, X_n)$-measurable, $Z_k = f_k(X_1, \dots, X_{k-1}, X_{k+1}, \dots, X_n)$ is $\sigma(X_1, \dots, X_{k-1}, X_{k+1}, \dots, X_n)$-measurable for all $k = 1, \dots, n$ and the above inequality is satisfied almost surely, i.e.

$$\sum_{k=1}^{n} (Z - Z_k) \le Z \text{ a.s.}\,.$$

In view of the above definition, the random variable of interest Z can be considered to behave as a sum of real valued random variables $Z - Z_k$. These differences $Z - Z_k$ will be called the *increments* of Z. Our goal will be to obtain results on Z from knowledge on the distribution of the increments.

Notice that sums ($Z = \sum X_k$) obviously satisfy the sub-additivity requirement and the increments are simply $Z - Z_k = X_k$. So that they are independent. Also, it can be easily verified that norms of sums of vectors are sub-additive functions. More generally, suprema of empirical processes are sub-additive (see Lemma C.1). This fact will allow us to obtain various refinements of Talagrand's inequality for empirical processes.

We want to obtain concentration inequalities, but as an intermediate step, we will bound the moment generating function of the random variables involved. Of course even if we have a sharp bound on that moment generating function, we may loose some sharpness when we apply Markov's inequality to get the deviation bound. However, we will not address this issue here and rather focus on obtaining sharp bounds on the moment generating function.

As a measure of how good the bounds we obtain are, we will compare them to what can be obtained for sums of real-valued independent random variables.

In the case of a sum of independent random variables, a bound on the Laplace transform is trivially obtained from a bound on the Laplace transforms of the individual random variables. Indeed, one has

$$\log \mathbb{E}\left[e^{\lambda Z}\right] \le \sum_{k=1}^{n} \log \mathbb{E}\left[e^{\lambda X_k}\right]\,.$$

Writing this in terms of the increments we obtain, when Z is a sum of independent random variables and Z_k is the sum without the k-th term, for all $\lambda \in \mathbb{R}$,

$$\log \mathbb{E}\left[e^{\lambda Z}\right] \leq \sum_{k=1}^{n} \log \mathbb{E}\left[e^{\lambda(Z-Z_k)}\right]. \tag{1.1}$$

In a more complicated case, one can use martingale techniques. A general result (used for example in [16]) can be written as follows,

$$\log \mathbb{E}\left[e^{\lambda(Z-\mathbb{E}[Z])}\right] \leq \sum_{k=1}^{n} \log \left\|\mathbb{E}_k\left[e^{\lambda(Z-Z_k)-\lambda\mathbb{E}_k[Z-Z_k]}\right]\right\|_{\infty},$$

where $\mathbb{E}_k[\cdot]$ means the conditional expectation, conditioned on the random variables $X_1, \ldots, X_{k-1}, X_{k+1}, \ldots, X_n$.

However, we will not follow this path but rather use the so-called *entropy method.* We use the entropy of the function $\lambda \mapsto e^{\lambda Z}$ and its tensorization property for functions of independent random variables. Denote $\mathbb{H}\left[e^{\lambda Z}\right] := \mathbb{E}\left[\lambda Z e^{\lambda Z}\right] - \mathbb{E}\left[e^{\lambda Z}\right] \log \mathbb{E}\left[e^{\lambda Z}\right]$. Then this function of λ is defined on the interior of the domain of definition of the Laplace transform of Z, $\lambda \mapsto \mathbb{E}\left[e^{\lambda Z}\right]$. The key property of this function is the following

$$\mathbb{H}\left[e^{\lambda Z}\right] \leq \mathbb{E}\left[\sum_{k=1}^{n} \mathbb{H}_k\left[e^{\lambda Z}\right]\right] = \mathbb{E}\left[\sum_{k=1}^{n} e^{\lambda Z_k} \mathbb{H}_k\left[e^{\lambda(Z-Z_k)}\right]\right]. \tag{1.2}$$

where $\mathbb{H}_k[\cdot]$ is the conditional entropy, conditioned on $X_1, \ldots, X_{k-1}, X_{k+1}, \ldots, X_n$ (in other words, it is the entropy with respect to X_k only). Notice that the above inequality is tight for sums of independent real-valued random variables since it implies (1.1).

We thus have to bound the entropy of the moment generating function of the increments. Thus, given conditions on the increments, we will obtain bounds on the moment generating function of Z. The sub-additivity condition will allow to bound the individual entropies in terms of the entropy of Z itself, which will in many cases improve the bounds.

2. ϕ-Entropy

In this section we will introduce a method based on a generalized entropy function which allows to connect properties of a function of independent random variables to properties of its increments. We shall first introduce some notation and then describe the main properties of the ϕ-entropy. It will then be shown that the moment generating function and the variance can be seen as entropy functions.

2.1. Notation

We consider a sequence of independent random variables $X_1, \ldots, X_n$ with values in some polish space $\mathcal{X}$. We introduce the following σ-algebras generated by these random variables

$$\begin{aligned} \mathcal{A} &:= \sigma(X_1, \dots, X_n) \\ \mathcal{A}_k &:= \sigma(X_1, \dots, X_{k-1}, X_{k+1}, \dots, X_n) \text{ for all } k = 1, \dots, n\,. \end{aligned}$$

Given a function $f : \mathcal{X}^n \to \mathbb{R}$ and n functions $f_k : \mathcal{X}^{n-1} \to \mathbb{R}$, we will define, assuming f is $\mathcal{A}$-measurable and f_k is $\mathcal{A}_k$-measurable, the random variables

$$\begin{aligned} Z &:= f(X_1, \dots, X_n)\,, \\ Z_k &:= f_k(X_1, \dots, X_{k-1}, X_{k+1}, \dots, X_n) \text{ for all } k = 1, \dots, n\,. \end{aligned} \tag{2.1}$$

We denote by $\mathbb{E}_k[\cdot]$ the expectation taken conditionally to $\mathcal{A}_k$ and by $\mathsf{Var}_k[\cdot]$ the variance taken conditionally to $\mathcal{A}_k$ (i.e. $\mathsf{Var}_k[Z] = \mathbb{E}_k\left[(Z - \mathbb{E}_k[Z])^2\right]$).

We are interested in the conditions that the random variable Z should satisfy in order to be concentrated around its expectation. We will express those conditions in terms of the *increments* of Z, i.e. the differences $Z - Z_k$.

In the remainder, when we consider a random variable denoted by Z, we will implicitly mean that Z is a defined as above and that a collection of n auxiliary random variables $Z_1, \dots, Z_n$ satisfying the above measurability conditions are given. The notation Y_k will always denote an auxiliary $\mathcal{A}$-measurable random variable.

In certain situations, it will be convenient to introduce independent copies of the X_k. We will denote them by X'_k and use the notation

$$Z'_k := f(X_1, \dots, X_{k-1}, X'_k, X_{k+1}, \dots, X_n)\,. \tag{2.2}$$

The following real-valued functions will appear often in the computation, we will thus give them a name

$$\begin{aligned} \psi(x) &= e^{-x} - 1 + x \\ \nu(x) &= 1 - (1+x)e^{-x}\,. \end{aligned} \tag{2.3}$$

Also, in the exponential rates of deviation, the following two functions will often appear

$$\begin{aligned} h_1(x) &= (1+x)\log(1+x) - x\,, \\ h_2(x) &= 1 + x - \sqrt{1+2x}\,. \end{aligned} \tag{2.4}$$

2.2. The Entropy Method

The entropy method was introduced by Ledoux [12] in order to obtain concentration inequalities for empirical processes. It was based on modified logarithmic Sobolev inequalities. Later the method was refined by Massart [14] and Rio [16] among others. In particular, many applications of this method have been exposed in [6, 7]. More recently, the method was generalized (see e.g. [13, 11, 8, 5]) to yield other types of results.

For a function ϕ defined on $\mathbb{R}^+$ which is convex, we define the entropy function $\mathbb{H}_\phi[\cdot]$ as

$$\mathbb{H}_\phi[T] = \mathbb{E}[\phi(T)] - \phi(\mathbb{E}T)\,,$$

for any non-negative random variable T such that T and $\phi(T)$ are integrable. When $\phi(x) = x\log x$, we simply denote $\mathbb{H}[T] = \mathbb{H}_\phi[T]$ this is the original definition of entropy and it explains the name (in that case, it relates to the information theoretic entropy of the distribution of T).

The key property of the entropy function is that it happens to be convex when ϕ is convex and $1/\phi''$ is concave. This in turn implies a tensorization inequality which relates the entropy of the function to the entropies with respect to each independent component [11].

When $\phi(x)$ is chosen to be equal to $x \log x$, then one can consider the random variable $T = e^{\lambda Z}$. This gives

$$\mathbb{H}\left[e^{\lambda Z}\right] = \mathbb{E}\left[\lambda Z e^{\lambda Z}\right] - \mathbb{E}\left[e^{\lambda Z}\right] \log \mathbb{E}\left[e^{\lambda Z}\right],$$

which is a function of the moment generating function of Z and its derivative since, denoting $F(\lambda) = \mathbb{E}\left[e^{\lambda Z}\right]$ we have $\mathbb{H}\left[e^{\lambda Z}\right] = \lambda F'(\lambda) - F(\lambda) \log F(\lambda)$. The main property of this function is summarized in the following lemma.

Lemma 2.1. *Let Z be a random variable defined as in (2.1). Then for all $\lambda \in \mathbb{R}$,*

$$\begin{aligned} \mathbb{H}\left[e^{\lambda Z}\right] &\leq \mathbb{E}\left[\sum_{k=1}^{n} \mathbb{H}_k\left[e^{\lambda Z}\right]\right] \\ &= \mathbb{E}\left[\sum_{k=1}^{n} e^{\lambda Z_k} \mathbb{H}_k\left[e^{\lambda(Z-Z_k)}\right]\right] \\ &= \mathbb{E}\left[\sum_{k=1}^{n} \inf_{u \in \mathbb{R}} \mathbb{E}_k\left[\psi(\lambda(Z - Z_k - u)) e^{\lambda Z}\right]\right]. \end{aligned}$$

A proof of this result can be found in [14]. Using this Lemma, if one has conditions on the increments, such as boundedness or bounds of certain moments, then one can obtain a differential inequality involving the moment generating function of Z. Solving this differential inequality will give the desired bound.

Before getting into the details of how this can be done, we will illustrate some more the generality of the entropy method. First we will show that the moment generating function itself can be seen as an entropy, as well as the variance. Then we will give bounds on the moment generating functions of sums of independent random variables and on the variance of sub-additive functions.

2.3. Moment Generating Function as an Entropy

Notice that $\log \mathbb{E}\left[e^{\lambda(Z-\mathbb{E}Z)}\right] = \mathbb{H}_\phi\left[e^{\lambda X}\right]$ with $\phi(x) = -\log x$. However, since $1/\phi''$ is not concave, tensorization does not happen unless one considers specific functions like sums. The following lemma allows to obtain bounds for sums from bounds on the individual Laplace transforms.

Lemma 2.2. *Let Z be a sum of n independent random variables $X_1, \ldots, X_n$ and $Z_k = \sum_{j \neq k} X_j$. We have for all $\lambda \in \mathbb{R}$,*

$$\begin{aligned}\mathbb{H}_\phi\left[e^{\lambda Z}\right] &= \sum_{k=1}^n \mathbb{H}_{\phi,k}\left[e^{\lambda(Z-Z_k)}\right] \\ &= \sum_{k=1}^n \inf_{u \in \mathbb{R}} \mathbb{E}_k\left[\psi(-\lambda(Z - Z_k - u))\right].\end{aligned}$$

This lemma is stated in a way that makes the comparison easy with Lemma 2.1. The main difference is that it only applies to specific functions Z. However, the involved quantities are quite similar which means that we will use similar techniques to obtain bounds in both cases.

2.4. Variance as an Entropy

The variance function is indeed a ϕ-entropy. It is obtained with the function $\phi(x) = x^2$. It satisfies a tensorization inequality.

Theorem 2.3 (Efron–Stein–Steele Inequality). *Let Z be a random variable defined as in* (2.1).

$$\begin{aligned}\mathsf{Var}[Z] &\leq \mathbb{E}\left[\sum_{k=1}^n \mathsf{Var}_k[Z]\right] \\ &= \mathbb{E}\left[\sum_{k=1}^n \mathsf{Var}_k[Z - Z_k]\right] \\ &= \mathbb{E}\left[\sum_{k=1}^n \inf_{u \in \mathbb{R}} \mathbb{E}_k\left[(Z - Z_k - u)^2\right]\right]. \qquad (2.5)\end{aligned}$$

Because of the way we have written the above result, it again seems similar to Lemma 2.1. As a consequence, similar techniques might be used to bound the variance.

3. Bounds for Sums

In this section we recall some bounds on the moment generating functions that can be obtained from simple conditions, such as boundedness or conditions on the moments. The main reason for presenting these results is to provide a comparison with the general inequalities that will be presented later on. Indeed, all the inequalities for sub-additive functions apply as a special case to sums of independent random variables, so in some sense, we cannot do better that the following results. The question is whether we can match them.

3.1. Bounds for Individual Real-Valued Random Variables

One way to ensure that the moment generating function of a random variable X is nicely bounded is when the random variable is bounded (almost surely) on both sides by some constants.

Lemma 3.1. *Assume $a \leq X \leq b$ almost surely. Then for any $\lambda \in \mathbb{R}$,*

$$\begin{aligned} \log \mathbb{E}\left[e^{\lambda X}\right] &\leq \log\left(\frac{(\mathbb{E}X-a)e^{\lambda b}+(b-\mathbb{E}X)e^{\lambda a}}{b-a}\right) && (3.1) \\ &\leq \frac{(b-a)^2\lambda^2}{8}+\lambda\mathbb{E}X\,, && (3.2) \end{aligned}$$

and for any $\lambda \geq 0$,

$$\log \mathbb{E}\left[e^{\lambda X}\right] \leq \frac{(\mathbb{E}X-a)(e^{\lambda(b-\mathbb{E}X)}-1-\lambda(b-\mathbb{E}X))}{b-\mathbb{E}X}+\lambda\mathbb{E}X\,, \quad (3.3)$$

and for any $\lambda \leq 0$,

$$\log \mathbb{E}\left[e^{\lambda X}\right] \leq \frac{(b-\mathbb{E}X)(e^{-\lambda(\mathbb{E}X-a)}-1+\lambda(\mathbb{E}X-a))}{\mathbb{E}X-a}+\lambda\mathbb{E}X\,. \quad (3.4)$$

If one has a bound on the variance, one can actually relax the boundedness on the left to obtain a tight control on the moment generating function from this bound as shows next lemma.

Lemma 3.2 (Bennett [2]). *Suppose $X \leq b$ a.s., $\mathbb{E}X = 0$ and $\mathsf{Var}X \leq \sigma^2$. Then for any $\lambda \geq 0$,*

$$\begin{aligned} \log \mathbb{E}\left[e^{\lambda X}\right] &\leq \log\left(\frac{b^2e^{-\lambda\sigma^2/b}+\sigma^2e^{\lambda b}}{b^2+\sigma^2}\right) && (3.5) \\ &\leq \frac{\sigma^2}{b^2}\left(e^{\lambda b}-1-\lambda b\right) && (3.6) \\ &\leq \frac{\sigma^2\lambda^2}{2(1-\lambda b/3)}\,, && (3.7) \end{aligned}$$

where the last inequality holds only if $\lambda \in [0, 3/b)$.

We notice that we presented four types of bounds. The first one, in (3.1) and (3.5) corresponds to the moment generating function of a Bernoulli random variable (a random variable that takes only two different values). The second one, in (3.3), (3.4) and (3.6), corresponds to the moment generating function of a Poisson random variable. This form gives so-called Bennett type deviation inequalities, i.e. the probability of deviations will have an exponential rate of the form $(1+t)\log(1+t)-t$. The third type, in (3.7) gives Bernstein type deviation inequalities (with exponential rate of the form $t^2/(1+t)$). Finally, in (3.2) the moment generating function of a Gaussian random variable appears.

The first type of inequality is tight since the bound is achieved when X has the distribution

$$\mathbb{P}[X=a]=\frac{b-\mathbb{E}X}{b-a} \text{ and } \mathbb{P}[X=b]=\frac{\mathbb{E}X-a}{b-a},$$

for (3.1) and

$$\mathbb{P}\left[X=-\frac{\sigma^2}{b}\right]=\frac{b^2}{b^2+\sigma^2} \text{ and } \mathbb{P}[X=b]=\frac{\sigma^2}{b^2+\sigma^2},$$

for (3.5).

In the remainder we will try to obtain similar bounds for sub-additive functions whose increments are bounded on both sides or bounded on the right. We will actually consider several possible conditions.

1. There exists a r.v. Y and $b \in \mathbb{R}$ such that $Y \leq X \leq b$ a.s.
   (a) $Y \leq a$ a.s., $\mathbb{E}Y^2 \leq \sigma^2$
   (b) $Y \geq a$ a.s., $\mathbb{E}Y^2 \leq \sigma^2$
   (c) $\mathbb{E}Y \geq 0$, $\|Y\|_\infty \leq b$, $\mathbb{E}\left[|Y|^2\right] \leq \sigma^2$
   (d) $\mathbb{E}Y \geq 0$, for all $p \geq 2$, $\mathbb{E}\left[|Y|^p\right] \leq p!b^{p-2}\sigma^2/2$
2. $\mathbb{E}X \geq 0$, $\|X\|_\infty \leq b$, $\mathbb{E}\left[|X|^2\right] \leq \sigma^2$
3. $\mathbb{E}X \geq 0$, $\mathbb{E}\left[|X|^p\right] \leq p!b^{p-2}\sigma^2/2$

Notice that condition 3 is the least we can require since it is necessary for the existence of the moment generating function on an open interval containing zero.

We next generalize the Poisson type bounds of the above two lemmas. The generalization we propose consists in replacing the second moment of X by the second moment of some other random variable Y which is smaller X. The price for that is that the variance factor we obtain involves the second moment of Y and the first moment of X. This will be very useful in applications (e.g. to suprema of empirical processes) as we shall see later.

Lemma 3.3. *Let X be a random variable and define $L(\lambda)=\log\mathbb{E}\left[e^{\lambda(X-\mathbb{E}[X])}\right]$ (for those λ for which it is defined). If X satisfies condition* 1.(a), *then for all $\lambda \geq 0$,*

$$L(\lambda) \leq \frac{\mathbb{E}Y^2-(\mathbb{E}X)^2+(a+b)\mathbb{E}[X-Y]}{(b-\mathbb{E}X)^2}\left(e^{\lambda(b-\mathbb{E}X)}-1-\lambda(b-\mathbb{E}X)\right).$$

If X satisfies condition 1.(b) *then for all $\lambda \leq 0$,*

$$L(\lambda) \leq \frac{\mathbb{E}Y^2-(\mathbb{E}X)^2+2b\mathbb{E}[X-Y]}{(\mathbb{E}X-a)^2}\left(e^{-\lambda(\mathbb{E}X-a)}-1+\lambda(\mathbb{E}X-a)\right).$$

If X satisfies condition 1.(c), *then for all $\lambda \leq 0$,*

$$L(\lambda) \leq \sigma^2(e^{-\lambda b}-1+\lambda b)-\lambda(1-e^{\lambda})\mathbb{E}X.$$

If X satisfies condition 1.(d), *then for all $\lambda \in]-1/b,0]$,*

$$L(\lambda) \leq \sigma^2\frac{\lambda^2}{2(1+b\lambda)}-\lambda(1-e^{\lambda})\mathbb{E}X.$$

If instead of the above boundedness assumption we simply have a control on all the moments of the random variable we obtain the following lemma.

Lemma 3.4. *If X satisfies condition 2, then for all $\lambda \in \mathbb{R}$,*

$$\log \mathbb{E}\left[e^{\lambda(X-\mathbb{E}X)}\right] \leq \frac{\sigma^2}{b^2}\left(e^{|\lambda| b} - 1 - |\lambda| b\right).$$

If X satisfies condition 3, then for all $\lambda \in (-1/b, 1/b)$,

$$\log \mathbb{E}\left[e^{\lambda(X-\mathbb{E}X)}\right] \leq \frac{\sigma^2\lambda^2}{2(1-b|\lambda|)}.$$

3.2. Consequences for Sums

From the results presented above, it is easy to derive bounds for sums of independent random variables.

Lemma 3.5 (Hoeffding [10], Bennett [2], Bernstein [3]). *Suppose $X_k \leq b$ a.s., $\mathbb{E}X_k = 0$ and $\mathsf{Var}X_k \leq \sigma_k^2$. Then for any $\lambda \geq 0$, denoting $\sigma^2 = n^{-1}\sum_{k=1}^n \sigma_k^2$,*

$$\log \mathbb{E}\left[e^{\lambda Z}\right] \leq n \log\left(\frac{b^2 e^{-\lambda\sigma^2/b} + \sigma^2 e^{\lambda b}}{b^2+\sigma^2}\right) \tag{3.8}$$

$$\leq \frac{n\sigma^2}{b^2}\left(e^{\lambda b} - 1 - \lambda b\right) \tag{3.9}$$

$$\leq \frac{n\sigma^2\lambda^2}{2(1-\lambda b/3)}, \tag{3.10}$$

where the last inequality holds only if $\lambda \in [0, 3/b)$.

Inequality (3.8) is due to Hoeffding. It is tight since if X_k has the distribution

$$\mathbb{P}\left[X_k = -\frac{\sigma^2}{b}\right] = \frac{b^2}{b^2+\sigma^2}, \mathbb{P}\left[X_k = b\right] = \frac{\sigma^2}{b^2+\sigma^2},$$

then the bound is achieved. Inequality (3.9) is due to Bennett and Inequality (3.10) is due to Bernstein.

4. Variance of Sub-Additive Functions

Before looking at the moment generating function of sub-additive functions, we start by looking at their variance. This will allow us to have an idea of which terms should appear in the bounds we want to obtain.

We will thus try to obtain bounds on the variance of sub-additive functions from Theorem 2.3. In some sense, since it relies on tensorization, we cannot hope to do better than that when we will use the tensorization of the entropy.

Lemma 4.1. *Let Z be a sub-additive function defined as in (2.1). Assume $Y_k \leq Z - Z_k \leq 1$ a.s. and $Y_k \leq b$ a.s., then if $\mathbb{E}[Z] \leq n(1+b)/2$,*

$$\mathsf{Var}[Z] \leq \sum_{k=1}^{n} \mathbb{E}\left[Y_k^2 - (1+b)Y_k\right] + \mathbb{E}[Z](1 + b - \mathbb{E}[Z]/n),$$

and if $\mathbb{E}[Z] \geq n(1+b)/2$,

$$\mathsf{Var}[Z] \leq \sum_{k=1}^{n} \left(\mathbb{E}\left[Y_k^2 - (1+b)Y_k\right] + (1+b)^2/4\right).$$

If additionally, $\mathbb{E}_k[Y_k] \leq 0$ a.s. then when $\mathbb{E}[Z] \geq n(1+b)/2$,

$$\mathsf{Var}[Z] \leq \sum_{k=1}^{n} \left(\mathbb{E}[\mathsf{Var}_k[Y_k]] - (1+b)\mathbb{E}[Y_k]/2 + (1+b)^2/4\right).$$

Proof. We start from Theorem 2.3 which gives

$$\begin{aligned}
\mathsf{Var}[Z] &\leq \sum_{k=1}^{n} \left(\mathbb{E}\left[(Z-Z_k)^2\right] - \mathbb{E}\left[(\mathbb{E}_k[Z-Z_k])^2\right]\right) \\
&= \sum_{k=1}^{n} \mathbb{E}\left[(Z-Z_k)^2 - (1+b)(Z-Z_k)\right] \\
&\quad + \sum_{k=1}^{n} \mathbb{E}\left[(1+b)\mathbb{E}_k[Z-Z_k] - (\mathbb{E}_k[Z-Z_k])^2\right] \\
&\leq \sum_{k=1}^{n} \left(\mathbb{E}\left[Y_k^2 - (1+b)Y_k\right] + \mathbb{E}\left[(1+b)\mathbb{E}_k[Z-Z_k] - (\mathbb{E}_k[Z-Z_k])^2\right]\right),
\end{aligned}$$

where we used, in the third line, the fact that $x \mapsto x^2 - (1+b)x$ is non-increasing for $x \leq b$ and is smaller than its value at b between b and 1.

If $\mathbb{E}[Z] \leq n(1+b)/2$, applying Jensen's inequality to the concave function $x \mapsto (1+b)x - x^2$ we obtain

$$\mathsf{Var}[Z] \leq \sum_{k=1}^{n} \mathbb{E}\left[Y_k^2 - (1+b)Y_k\right] + \left(1 + b - \frac{1}{n}\mathbb{E}\left[\sum_{k=1}^{n}(Z-Z_k)\right]\right)\mathbb{E}\left[\sum_{k=1}^{n}(Z-Z_k)\right].$$

Now, since $0 \leq \mathbb{E}[\sum_{k=1}^{n}(Z-Z_k)] \leq \mathbb{E}[Z] \leq n(1+b)/2$, and the function $x \mapsto (1+b)x - x^2/n$ is non-decreasing for $x \leq n(1+b)/2$, we obtain

$$\mathsf{Var}[Z] \leq \sum_{k=1}^{n} \mathbb{E}\left[Y_k^2 - (1+b)Y_k\right] + \mathbb{E}[Z]((1+b) - \mathbb{E}[Z]/n).$$

If $\mathbb{E}[Z] \geq n(1+b)/2$, using $(1+b)x - x^2 \leq (1+b)^2/4$ for all x, we obtain

$$\mathsf{Var}[Z] \leq \sum_{k=1}^{n} \left(\mathbb{E}\left[Y_k^2 - (1+b)Y_k\right] + (1+b)^2/4\right).$$

Using $0 \le \mathbb{E}_k[Y_k] \le b$ and the fact that $x \mapsto x^2 - (1+b)/2x$ is non-positive in $[0,b] \subset [0, 1+b/2]$, we have

$$\mathbb{E}_k[Y_k]^2 - \frac{1+b}{2}\mathbb{E}_k[Y_k] \le 0\,,$$

which gives the last inequality. □

Note that the first bound is tight when Z is the sum of independent Bernoulli random variables. For example if $Z = \sum_{k=1}^n X_k$ with X_k a Bernoulli random variable with parameter p. Then $\mathsf{Var}[Z] = np(1-p)$ which is obtained in the bound by taking $Z_k = Z - X_k$ and $Y_k = 0$.

We give here a result on the variance of randomized processes (which we will study in more details in a later section).

Lemma 4.2. *Let $\mathcal{F}$ be a set of n-dimensional vectors and define*

$$Z = \sup_{f \in \mathcal{F}} \sum_{k=1}^n X_k f_k\,.$$

Assume also that $\mathbb{E}[X_k] = 0$ and $\mathbb{E}\left[X_k^2\right] \le \sigma^2$. Then

$$\begin{aligned}\mathsf{Var}[Z] &\le \sigma^2 \sup_{f\in\mathcal{F}} \|f\|_2^2 + \mathbb{E}\left[\sup_{f\in\mathcal{F}} \sum_{k=1}^n X_k^2 f_k^2\right] \\ &\le \left(\sigma^2 + \mathbb{E}\left[\max_{k=1,\ldots,n} X_k^2\right]\right) \sup_{f\in\mathcal{F}} \|f\|_2^2\,.\end{aligned}$$

Proof. We introduce the variables Z'_k as defined in (2.2) and use the following version of Efron–Stein–Steele inequality

$$\mathsf{Var}[Z] \le \mathbb{E}\left[\sum_{k=1}^n \mathbb{E}\left[(Z - Z'_k)_+^2\right]\right]$$

Denoting by g the vector where the supremum is reached in Z, we obtain

$$\begin{aligned}\mathsf{Var}[Z] &\le \mathbb{E}\left[\sum_{k=1}^n (X_k - X'_k)^2 g_k^2\right] \\ &\le \mathbb{E}\left[\sum_{k=1}^n \sigma^2 g_k^2\right] + \mathbb{E}\left[\sum_{k=1}^n X_k^2 g_k^2\right]\,.\end{aligned}$$

□

Note that Bobkov [4] obtained the bound

$$\mathsf{Var}[Z] \le (1+\sigma^2) \sup_{f\in\mathcal{F}} \|f\|_2^2\,,$$

in the case $X_k \in [-1,1]$. This can be recovered as a consequence of the above lemma.

5. Previous Results

We present in this section some of the known results for sub-additive functions of independent random variables.

For the case of increments bounded on sides, Boucheron, Lugosi and Massart [6] obtained Bennett type bounds.

Theorem 5.1 (Boucheron, Lugosi and Massart [6]). *Let Z be sub-additive and defined as in* (2.1). *Moreover assume that for all $k = 1, \ldots, n$,*

$$0 \leq Z - Z_k \leq 1\,,$$

Then for all $\lambda \in \mathbb{R}$,

$$\log \mathbb{E}\left[e^{\lambda Z}\right] \leq \mathbb{E}\left[Z\right](e^{\lambda} - 1)\,.$$

Notice that this result is only useful when $\mathbb{E}\left[Z\right]/n \leq 1/4$, otherwise one can use McDiarmid's inequality [15] to get

$$\log \mathbb{E}\left[e^{\lambda Z}\right] \leq \lambda \mathbb{E}\left[Z\right] + n\lambda^2/8\,.$$

More recently, Rio obtained a refined version of the above theorem for the case $\mathbb{E}\left[Z\right]/n \leq 1/2$.

Theorem 5.2 (Rio [16]). *Under the conditions of Theorem* 5.1, *if additionally $\mathbb{E}\left[Z\right] < n/2$ then for $\lambda \geq 0$,*

$$\log \mathbb{E}\left[e^{\lambda Z}\right] \leq \frac{\mathbb{E}\left[Z\right]}{1 - \mathbb{E}\left[Z\right]/n}\left(e^{\lambda(1-\mathbb{E}[Z]/n)} - 1\right),$$

and for $\lambda \leq 0$,

$$\log \mathbb{E}\left[e^{\lambda Z}\right] \leq n \log\left(1 + \frac{\mathbb{E}\left[Z\right]}{n}(e^{\lambda} - 1)\right).$$

The above refinement gives the correct variance factor $\mathbb{E}\left[Z\right](1 - \mathbb{E}\left[Z\right]/n)$.

Another older result, due to van der Vaart and Wellner [19] gives a concentration inequality for functions which have a different sub-additivity property. The technique used to prove it relies on concentration inequalities for product measures, unlike the above results which rely on the entropy method.

Theorem 5.3 (van der Vaart and Wellner [19], page 455). *Let $S_n : \mathcal{X}^n \to [0, n]$ be a permutation symmetric function satisfying $\forall x \in \mathcal{X}^n$, $y \in \mathcal{X}^m$,*

$$S_n(x) \leq S_{n+m}(x, y)\,,$$

and

$$S_{n+m}(x, y) \leq S_n(x) + S_m(y)\,.$$

Consider a sequence of i.i.d. random variables $X_1, \ldots, X_n$ and denote $Z = S_n(X)$. Then for every $t > 0$

$$\mathbb{P}\left[Z \geq t\right] \leq e^{-\frac{1}{2} t \log \frac{t}{12(\mathbb{E}[Z] \vee 1)}}\,.$$

In order to see how the assumptions of the above theorem relate to our sub-additivity condition, denote

$$Z_k = S_{n-1}(X_1, \ldots, X_{k-1}, X_{k+1}, \ldots, X_n).$$

Then the conditions of Theorem 5.3 imply

$$0 \leq Z - Z_k \leq 1,$$

and

$$S_{(n-1)*n}(X, \ldots, X) \leq \sum_{k=1}^{n} Z_k.$$

So that

$$\sum_{k=1}^{n} (Z - Z_k) \leq nZ - S_{(n-1)*n}(X, \ldots, X).$$

Hence, if one has $S_{kn}(X, \ldots, X) = kS_n(X)$ we recover our sub-additivity condition and Theorem 5.2 applies. Thus, to some extent, the conditions of Theorem 5.3 refer to a more general notion of sub-additivity. However, it seems that in most applications our definition is enough.

Finally, it is possible to extend above results (with worse constants) to a more general situation without sub-additivity.

Theorem 5.4 (Boucheron, Lugosi and Massart [7]). *Let Z, Z_k and Z'_k be defined as in* (2.1) *and* (2.2). *If V_+ satisfies*

$$\sum_{k=1}^{n} \mathbb{E}\left[(Z - Z'_k)_+^2 | \mathcal{A}\right] \leq V_+,$$

and V_- satisfies

$$\sum_{k=1}^{n} \mathbb{E}\left[(Z - Z'_k)_-^2 | \mathcal{A}\right] \leq V_-,$$

then for all $\theta > 0$ and all $\lambda \in (0, 1/\theta)$,

$$\log \mathbb{E}\left[e^{\lambda(Z - \mathbb{E}[Z])}\right] \leq \frac{\lambda\theta}{1 - \lambda\theta} \log \mathbb{E}\left[e^{\lambda V_+/\theta}\right],$$

and for all $\theta \in (0, 1)$ and all $\lambda \in (-1, 0)$,

$$\log \mathbb{E}\left[e^{\lambda(Z - \mathbb{E}[Z])}\right] \leq \frac{-\lambda\theta}{1 + \lambda\theta} \log \mathbb{E}\left[e^{-\lambda V_-/\theta}\right].$$

6. Main Results

After these lengthy preliminaries we are now ready to state the main results of this work. We will give bounds on the moment generating function of sub-additive functions from conditions on their increments.

The first theorem applies to functions with bounded increments (on one side or on both sides).

Theorem 6.1. *Assume Z is sub-additive, $Y_k \leq Z - Z_k \leq 1$ and $\mathbb{E}_k [Y_k] \geq 0$. Also, let σ^2 be a real number such that*

$$\sigma^2 \geq \frac{1}{n} \sum_{k=1}^{n} \mathbb{E}_k \left[Y_k^2 \right] .$$

If for all $k = 1, \ldots, n$ there exists some $b > 0$ such that

$$Y_k \leq b \ \ a.s. ,$$

then let $v = (1+b)\mathbb{E}[Z] + n\sigma^2$. We obtain for all $\lambda \geq 0$,

$$\log \mathbb{E} \left[e^{\lambda(Z - \mathbb{E}[Z])} \right] \leq \psi(-\lambda) v ,$$

where ψ is defined in (2.1).

This theorem can be considered as a generalization of (the first part of) Theorem 5.2 since it gives a Bennett type concentration inequality for Z under less restrictive conditions. In particular the non-negativity does not have to hold almost surely but only in expectation. This can be extremely useful in applications. As an example, we will show that our result applies to suprema of empirical processes while Theorem 5.2 did not. If we apply Theorem 6.1 to a sum Z of independent random variables X_k such that $X_k \in [-1, 1]$, $\mathbb{E}[X_k] = 0$ and $\mathbb{E}\left[X_k^2\right] = \sigma_k^2$, denoting $\sigma^2 = n^{-1} \sum \sigma_k^2$, we have for $\lambda \geq 0$

$$\log \mathbb{E} \left[e^{\lambda Z} \right] \leq n\sigma^2 \psi(-\lambda) ,$$

which is exactly Bennett's inequality (see inequality (3.9) in Lemma 3.5). Another remark concerning Theorem 6.1 is that the variance factor we obtain $v = (1 + b)\mathbb{E}[Z] + n\sigma^2$ is slightly sub-optimal since Lemma 4.1 gives a better variance under similar conditions.

The second main result we present applies to functions whose increments have exponential moments.

Theorem 6.2. *Assume Z is sub-additive, and $\mathbb{E}_k [Z] \geq Z_k$.*

1. *If for all $k = 1, \ldots, n$, $\|Z - \mathbb{E}_k [Z]\|_\infty \leq 1$ and there exists a real σ_k such that,*
$$\mathbb{E}_k \left[(Z - \mathbb{E}_k [Z_k])^2 \right] \leq \sigma_k^2 ,$$
then let $v = \mathbb{E}[Z] + \sum \sigma_k^2$. We obtain for all $\lambda \in [-\log 2, 1/2)$,
$$\log \mathbb{E} \left[e^{\lambda(Z - \mathbb{E}[Z])} \right] \leq v\psi(-|\lambda|) ,$$
2. *If for all $k = 1, \ldots, n$ there exists a real σ such that for all $p \geq 2$,*
$$\mathbb{E}_k \left[|Z - \mathbb{E}_k [Z_k] |^p \right] \leq \frac{p! \sigma_k^2}{2} ,$$
then let $v = 2\mathbb{E}[Z] + \sum \sigma_k^2$. We obtain for all $\lambda \in [1/\sqrt{2} - 1, 1/2)$,
$$\log \mathbb{E} \left[e^{\lambda(Z - \mathbb{E}[Z])} \right] \leq \frac{v\lambda^2}{2(1 - |\lambda|)} .$$

Also the same results hold true for $\lambda \geq 0$ (within the above specified intervals) if in the conditions, $Z - \mathbb{E}_k[Z]$ is replaced by $Z - Z_k$.

To better understand the meaning of this result, let us apply it to a sum Z of random variables X_k such that $\mathbb{E}[X_k] = 0$ and $\mathbb{E}[|X_k|^p] \leq \sigma^2$. Then $Z - \mathbb{E}_k[Z] = X_k$ so that we obtain for all $\lambda \in [0, 1)$

$$\log \mathbb{E}\left[e^{\lambda Z}\right] \leq n\sigma^2 \psi(-\lambda),$$

which is again Bennett's inequality (see Lemma 3.4). Also, under the assumption $\mathbb{E}[|X_k|^p] \leq p!\sigma^2/2$ we recover the second inequality of Lemma 3.4, i.e. Bernstein's inequality for $\lambda \in [0, 1/2)$

$$\log \mathbb{E}\left[e^{\lambda Z}\right] \leq n\sigma^2\lambda^2/2(1-\lambda).$$

Concerning the variance factors we obtain, $v = \mathbb{E}[Z] + n\sigma^2$ in the first case and $v = 2\mathbb{E}[Z] + n\sigma^2$ in the second case, they are also sub-optimal. Indeed, under the assumptions of Theorem 6.2 we have

$$\mathsf{Var}[Z] \leq \sum_{k=1}^{n} \mathbb{E}\left[(Z - \mathbb{E}_k[Z])^2\right] \leq n\sigma^2.$$

Finally we present a refinement of the Theorem 5.2 which applies when the increments are bounded on both sides.

Theorem 6.3. *Assume Z is sub-additive. If $0 \leq Z - Z_k \leq 1$ a.s., and $\mathbb{E}[Z] \leq n/2$, for all $\lambda \geq 0$ such that*

$$\frac{\mathbb{E}[Z]}{n} \leq \frac{1 - e^{\lambda} + \lambda e^{\lambda}}{(e^{\lambda} - 1)^2},$$

we have

$$n^{-1} \log \mathbb{E}\left[e^{\lambda Z}\right] \leq \log\left(1 + \frac{\mathbb{E}[Z]}{n}(e^{\lambda} - 1)\right).$$

As a result, for all $0 \leq t \leq 3(1 - 2\mathbb{E}[Z]/n)/4$

$$\mathbb{P}[Z \geq \mathbb{E}[Z] + nt] \leq \left(\frac{1-p}{1-t-p}\right)^{n(1-p-t)} \left(\frac{p}{p+t}\right)^{n(p+t)},$$

with $p = \mathbb{E}[Z]/n$.

Compared to Rio's result (second inequality of Theorem 5.2), we are able to replace the Poisson type bound by an optimal Bernoulli one. However, this improvement is valid only for a limited range of values of λ which translates into a limited range of values of t in the tail bound.

Next result weakens further the conditions on Z. This allows to obtain upper bounds on the Laplace transform of Z of Bernstein type, provided one controls the Laplace transform of two quantities: the first being the sum of increments and the second being the sum of the squares of these increments. This result can be considered as a partial refinement of Theorem 5.4.

Theorem 6.4. *Let $Z, Z_1, \dots, Z_n$ be random variables defined as in (2.1). Let V and W be two $\mathcal{A}$-measurable random variables such that*

$$\sum_{k=1}^{n}(Z - Z_k) \leq V \ a.s.\,,$$

and

$$\sum_{k=1}^{n} \mathbb{E}_k \left[Y_k^2\right] \leq W \ a.s.\,.$$

Assume moreover that $Y_k \leq Z - Z_k \leq 1$ a.s., $\mathbb{E}_k\left[Y_k\right] \geq 0$ and $Y_k \leq a$ a.s.

Then for all $\theta > 0$, and all $\lambda \in [0, (1+a)/\theta)$ we have

$$\log \mathbb{E}\left[e^{\lambda(Z-\mathbb{E}[Z])}\right] \leq \frac{\lambda}{1-\lambda\theta/(1+a)}\left(\log \mathbb{E}\left[e^{\lambda V}\right] + \frac{\theta}{1+a}\log \mathbb{E}\left[e^{\lambda W/\theta}\right]\right).$$

7. Applications

7.1. Previous Results for Empirical Processes

Concentration inequalities for suprema of empirical processes has been a hot topic ever since the first result obtained by Talagrand in 1994. Suprema of empirical processes are the objects that prompted the deepest advances of Talagrand in the induction method [18]. They also are at the origin of the introduction by Ledoux of the entropy method [12].

Later, the entropy method was further refined, still in connection to the suprema of empirical processes. In particular, Massart first obtained small constants for such concentration inequalities, as next theorem shows.

Theorem 7.1 (Massart [14]). *Consider n independent random variables $X_1, \dots, X_n$ taking their values in a space $\mathcal{X}$ and a countably infinite set $\mathcal{F}$ of functions mapping from $\mathcal{X}$ into $\mathbb{R}$ such that $\|f\|_\infty \leq 1$ for all $f \in \mathcal{F}$. Let Z denote either*

$$\sup_{f\in\mathcal{F}}\left|\sum_{i=1}^{n} f(X_i)\right| \quad or \quad \sup_{f\in\mathcal{F}}\left|\sum_{i=1}^{n} f(X_i) - \mathbb{E}\left[f(X_i)\right]\right|.$$

Let $V = \mathbb{E}\left[\sup_{f\in\mathcal{F}}\sum_{i=1}^{n} f^2(X_i)\right]$, then for $x > 0$ we have

$$\mathbb{P}\left[Z \geq \mathbb{E}\left[Z\right] + x\right] \leq 2\exp\left(-cx\log(1+\frac{cx}{V})\right),$$

where $c \leq 2/25$.

Notice that the variance factor V obtained above can be bounded (see [14]) by

$$V \leq \sup_{f\in\mathcal{F}}\sum_{k=1}^{n} \mathbb{E}_k\left[f^2(X_k)\right] + 16\mathbb{E}\left[Z\right].$$

More recently, Rio [17] obtained improved constants for the concentration of those objects, still using the entropy method.

Theorem 7.2 (Rio [17]). *Assume the X_i are identically distributed according to P. Let $\mathcal{F}$ be a countable set of functions from $\mathcal{X}$ to $\mathbb{R}$ and assume that all functions f in $\mathcal{F}$ are P-measurable, square-integrable and satisfy $\mathbb{E}[f] = 0$. Assume $\sup_{f\in\mathcal{F}} \operatorname{ess\,sup} f \leq 1$ and denote*

$$Z = \sup_{f\in\mathcal{F}} \sum_{i=1}^{n} f(X_i) .$$

Let σ be a positive real number such that $\sigma^2 \geq \sup_{f\in\mathcal{F}} \mathsf{Var}[f(X_1)]$ almost surely, then for all $x \geq 0$, we have

$$\mathbb{P}[Z \geq \mathbb{E}[Z] + x] \leq \exp\left(-\frac{x}{2}\log(1+\frac{x}{v})\right) ,$$

with $v = n\sigma^2 + 2\mathbb{E}[Z]$ and also

$$\mathbb{P}\left[Z \geq \mathbb{E}[Z] + \sqrt{2xv} + \frac{x}{2}\right] \leq e^{-x} .$$

7.2. Improved Results

The next result is an application of Theorem 6.1 which gives a function generalization of Bennett's inequality. More precisely, it gives a bound on the deviation above its mean of the supremum of an empirical process indexed by a class of upper bounded or bounded functions. The bound we obtain reduces to the classical Bennett's inequality for sums of i.i.d. random variables when the index set is a singleton.

This result improves the main result in [17] and provides a positive answer to the question raised in [14] about the possibility of obtaining a functional version of Bennett's inequality with optimal constants.

Theorem 7.3. *Let $\mathcal{F}$ be a countable set of functions from $\mathcal{X}$ to $\mathbb{R}$ and assume that all functions f in $\mathcal{F}$ are measurable, square-integrable and satisfy $\mathbb{E}[f(X_k)] = 0$. Assume $\sup_{f\in\mathcal{F}} \operatorname{ess\,sup} f \leq 1$ and denote*

$$Z = \sup_{f\in\mathcal{F}} \sum_{k=1}^{n} f(X_k) .$$

Let σ be a positive real number such that $n\sigma^2 \geq \sum_{k=1}^{n} \sup_{f\in\mathcal{F}} \mathbb{E}f^2(X_k)$, then for all $x \geq 0$, we have

$$\mathbb{P}[Z \geq \mathbb{E}[Z] + x] \leq \exp\left(-vh_1\left(\frac{x}{v}\right)\right) ,$$

with $v = n\sigma^2 + 2\mathbb{E}[Z]$ and h_1 defined in (2.4) and also

$$\mathbb{P}\left[Z \geq \mathbb{E}[Z] + \sqrt{2xv} + \frac{x}{3}\right] \leq e^{-x} .$$

Proof. We denote by f_k a function such that

$$\sum_{i\neq k} f_k(X_i) = \sup_{f\in\mathcal{F}} \sum_{i\neq k} f(X_i) .$$

We consider Z_k as in Lemma C.1 and introduce the following auxiliary random variables for $k = 1, \dots, n$,

$$Y_k = f_k(X_k)\,,$$

Denoting by f_0 the function achieving the maximum in Z, we have

$$Y_k \leq Z - Z_k \leq f_0(X_k) \leq 1 \text{ a.s. },$$

and

$$\mathbb{E}_k\left[Y_k\right] = 0\,.$$

Also we have

$$\sum_{k=1}^{n} \mathbb{E}_k\left[Y_k^2\right] = \sum_{k=1}^{n} \mathbb{E}_k\left[f_k^2(X_k)\right] \leq \sum_{k=1}^{n} \sup_{f\in\mathcal{F}} \mathbb{E} f^2(X_k)\,.$$

The result follows from applying Theorem 6.1. □

Notice that the variance factor we obtain is not sharp when $\mathbb{E}\left[Z\right]/n$ is large as shows Lemma 4.1.

7.3. Randomized Processes

We now consider an application of Theorem 6.2. The result we obtain is useful in the context of regression with unbounded errors [1].

Theorem 7.4. *Let $\mathcal{F}$ be a countable subset of $\mathbb{R}^n$ such that $\sup_{f\in\mathcal{F}} \|f\|_\infty \leq 1$. Let*

$$Z := \sup_{f\in\mathcal{F}} \sum_{k=1}^{n} X_k f_k\,,$$

where X_k are random variables that satisfy $\mathbb{E}\left[X_k\right] = 0$, and $\mathbb{E}_k\left[|X_k|^p\right] \leq p!\sigma^2/2$ for $p \geq 2$. We have for all $\lambda \in (0, 1/2]$,

$$\log \mathbb{E}\left[e^{\lambda(Z-\mathbb{E}[Z])}\right] \leq \frac{v\lambda^2}{2(1-\lambda)}\,,$$

with $v = n\sigma^2 + 2\mathbb{E}\left[Z\right]$. Moreover, for all $\lambda \in (\sqrt{2}/4 - 1/2, 0]$,

$$\log \mathbb{E}\left[e^{\lambda(Z-\mathbb{E}[Z])}\right] \leq \frac{v\lambda^2}{2(1+2\lambda)}\,,$$

with $v = n\sigma^2 + 4\mathbb{E}\left[Z\right]$.

Proof. Sub-additivity follows from Lemma C.1 since we have an empirical process indexed by functions of the form $f : X_k \mapsto X_k f_k$. We define as before,

$$Z_k := \sup_{f\in\mathcal{F}} \sum_{i\neq k} X_i f_i\,.$$

We have

$$\mathbb{E}_k\left[Z\right] \geq \sup_{f\in\mathcal{F}} \mathbb{E}_k\left[\sum_{i=1}^{n} X_i f_i\right] = Z_k\,.$$

Moreover, denoting by g the vector which achieves the supremum in Z, and by g' the vector which achieves the supremum in Z_k we obtain

$$X_k g'_k \leq Z - Z_k \leq X_k g_k \,,$$

so that

$$|Z - Z_k|^q \leq |X_k|^q \sup_{f \in \mathcal{F}} \max_k |f_k|^q \,.$$

We then have to apply Theorem 6.2 to get the upper tail result. For the lower tail, we use $\mathbb{E}_k \left[|Z - \mathbb{E}_k\left[Z\right]|^q\right] \leq 2^q \mathbb{E}_k\left[|Z - Z_k|^q\right]$ and Theorem 6.2. □

We notice that in the above theorem, the variance factor is suboptimal since it involves n times the individual variance, where n is a crude upper bound on $\|f\|_2^2$ (see Lemma 4.2).

8. Proofs

We first briefly explain the techniques we use to obtain the main results. Our starting point is Lemma 2.1. We then write

$$\psi(\lambda(Z - Z_k))e^{\lambda Z} = e^{\lambda Z_k}\nu(-\lambda(Z - Z_k)) \,.$$

The second step is to upper bound the functions of x, $\psi(\lambda x)$ or $\nu(-\lambda x)$ using results in Lemma A.3. Then we may use decoupling techniques that allow for example to upper bound the expectation of the product of two random variables by the expectations of the random variables themselves.

We will thus obtain bounds on the entropy of the increments $Z - Z_k$ which will lead to bounds on the entropy of Z via Lemma 2.1. The obtained inequalities will be differential inequalities involving the function $\mathbb{E}\left[e^{\lambda Z}\right]$. Solving these inequalities will give bounds on this function.

8.1. Decoupling Lemmas

We first give a simple but powerful result on the entropy which immediately gives a simple decoupling inequality.

Lemma 8.1 (Variational Formulation for Entropy).

$$\mathbb{H}\left[e^{\lambda Z}\right] = \sup_T \mathbb{E}\left[e^{\lambda Z} \log \frac{T}{\mathbb{E}\left[T\right]}\right] ,$$

where the supremum is extended to all non-negative random variables T such that the right-hand side is well-defined. As a consequence, if V and Z are two $\mathcal{A}$ measurable random variables, we have for any λ and any $\theta > 0$,

$$\lambda \mathbb{E}\left[V e^{\lambda Z}\right] \leq \theta \mathbb{H}\left[e^{\lambda Z}\right] + \theta \mathbb{E}\left[e^{\lambda Z}\right] \log \mathbb{E}\left[e^{\lambda V/\theta}\right] .$$

Note that the tensorization lemma can be obtained as a consequence of the variational formulation (see e.g. [13, 16]).

Following an idea of Rio [17], we extend the above decoupling inequality by combining it with Lemma 2.1.

Lemma 8.2. *Assume $\lambda \geq 0$ and $Z, Z_1, \ldots, Z_n$ are defined as in (2.1). If $\sum_{k=1}^n (Z - Z_k) \leq V$, then*

$$\mathbb{E}\left[\sum_{k=1}^n e^{\lambda Z} - e^{\lambda Z_k}\right] \leq \mathbb{E}\left[e^{\lambda Z}\right] \log \mathbb{E}\left[e^{\lambda V}\right],$$

and

$$\mathbb{E}\left[\sum_{k=1}^n e^{\lambda Z} - e^{\lambda Z_k} + \lambda U_k e^{\lambda Z}\right] \leq \mathbb{E}\left[e^{\lambda Z}\right] \log \mathbb{E}\left[e^{\lambda \sum_{k=1}^n U_k}\right] + \mathbb{E}\left[\lambda V e^{\lambda Z}\right].$$

Also

$$\mathbb{E}\left[\sum_{k=1}^n e^{\lambda Z} - e^{\lambda Z_k} \left(\mathbb{E}\left[e^{\lambda V}\right]\right)^{1/n}\right] \leq 0,$$

and

$$\mathbb{E}\left[\sum_{k=1}^n e^{\lambda Z} - e^{\lambda \mathbb{E}_k[Z]}\right] \leq \mathbb{E}\left[e^{\lambda Z}\right] \log \mathbb{E}\left[e^{\lambda V}\right] - \mathbb{E}\left[\sum_{k=1}^n \lambda \mathbb{E}_k\left[Z - Z_k\right] e^{\lambda Z}\right].$$

Proof. By the variational formulation and the tensorization

$$\begin{aligned}
\mathbb{E}\left[e^{\lambda Z}\right] \log \mathbb{E}\left[e^{\lambda V}\right] - \mathbb{E}\left[\lambda V e^{\lambda Z}\right] &\geq \mathbb{E}\left[e^{\lambda Z}\right] \log \mathbb{E}\left[e^{\lambda Z}\right] - \mathbb{E}\left[\lambda Z e^{\lambda Z}\right] \\
&\geq \mathbb{E}\left[\sum_{k=1}^n e^{\lambda Z} - e^{\lambda Z_k} - \lambda (Z - Z_k) e^{\lambda Z}\right],
\end{aligned}$$

and also by hypothesis

$$\mathbb{E}\left[\lambda V e^{\lambda Z}\right] \geq \mathbb{E}\left[\sum_{k=1}^n \lambda (Z - Z_k) e^{\lambda Z}\right],$$

which gives the first result. Similarly, one can get

$$\mathbb{E}\left[e^{\lambda Z}\right] \log \mathbb{E}\left[e^{\lambda \sum_{k=1}^n U_k}\right] \geq \mathbb{E}\left[\sum_{k=1}^n e^{\lambda Z} - e^{\lambda Z_k} - \lambda (Z - Z_k) e^{\lambda Z} + \lambda U_k e^{\lambda Z}\right],$$

so that

$$\mathbb{E}\left[\sum_{k=1}^n e^{\lambda Z} - e^{\lambda Z_k} + \lambda U_k e^{\lambda Z}\right] \leq \mathbb{E}\left[e^{\lambda Z}\right] \log \mathbb{E}\left[e^{\lambda \sum_{k=1}^n U_k}\right] + \mathbb{E}\left[\sum_{k=1}^n \lambda (Z - Z_k) e^{\lambda Z}\right].$$

□

8.2. Bounds on the Entropy for Single Random Variables

We now give upper bounds on the entropy for single random variables.

Lemma 8.3. *If $X \in [0,1]$ a.s., then for all $\lambda \in \mathbb{R}$,*

$$\begin{aligned} \mathbb{H}\left[e^{\lambda X}\right] &\leq \inf_{p\in\mathbb{R}} \mathbb{E}\left[(e^{\lambda(p-1)} - e^{\lambda p} + \lambda)Xe^{\lambda X} + (e^{\lambda p} - 1 - \lambda p)e^{\lambda X}\right] \\ &= \mathbb{H}\left[e^{\lambda X}\right] + \mathbb{E}\left[e^{\lambda X}\right]\log\left((e^{-\lambda}-1)\mathbb{E}\left[Xe^{\lambda X}\right] + \mathbb{E}\left[e^{\lambda X}\right]\right) \\ &\leq \psi(\lambda)\mathbb{E}\left[Xe^{\lambda X}\right] . \end{aligned}$$

If $X \leq 1$ a.s., for $\lambda \geq 0$,

$$\begin{aligned} \mathbb{H}\left[e^{\lambda X}\right] &\leq \inf_{p\in\mathbb{R}} \mathbb{E}\left[e^{\lambda p} - e^{\lambda X} - \lambda p e^{\lambda X} + \frac{\lambda e^{\lambda}}{e^{\lambda}-1}(e^{\lambda X}-1)\right] \\ &= -\mathbb{E}\left[e^{\lambda X}\right]\log\mathbb{E}\left[e^{\lambda X}\right] + \frac{\lambda e^{\lambda}}{e^{\lambda}-1}\left(\mathbb{E}\left[e^{\lambda X}\right]-1\right) \\ &\leq \frac{\nu(-\lambda)}{e^{\lambda}-1}\left(\mathbb{E}\left[e^{\lambda X}\right]-1\right) . \end{aligned}$$

If $X \leq 1$ a.s. then for $\lambda \geq 0$ and $\alpha \in [1/2, 1]$,

$$\mathbb{H}\left[e^{\lambda X}\right] \leq \frac{\nu(-\lambda)}{e^{\lambda}+\alpha-1}\mathbb{E}\left[Xe^{\lambda X} + \alpha X^2 - X\right] , \tag{8.1}$$

and

$$\mathbb{H}\left[e^{\lambda X}\right] \leq \frac{\nu(-\lambda)}{e^{\lambda}-1-\lambda+\lambda\alpha}\mathbb{E}\left[e^{\lambda X} - 1 + \lambda(\alpha X^2 - X)\right] . \tag{8.2}$$

Proof. Consider the case $X \in [0,1]$. The first result follows from the convexity of $x \mapsto e^{\lambda x} - 1 - \lambda x$. The second equality follows from minimization in p. The third inequality follows from the first one with $p = 0$.

Now in the other cases, one simply uses the fact that

$$\mathbb{H}\left[e^{\lambda X}\right] = \inf_{p\in\mathbb{R}} \mathbb{E}\left[\psi(\lambda(X-p))e^{\lambda X}\right]$$

and the inequalities of Lemma A.3 to obtain the results. □

8.3. Bounds on the Entropy of Sub-Additive Functions

Combining the bounds on the entropy of individual random variables and the tensorization of entropy (Lemma 2.1) we obtain results for sub-additive functions.

The next two theorems consider conditions on the differences $Z - \mathbb{E}_k[Z]$ which will, in next section, be related to the increments.

Theorem 8.4. *Assume Z is sub-additive and defined as in (2.1). If $Z - \mathbb{E}_k[Z] \leq 1$ almost surely, then for all $\alpha \in [1/2, 1]$ and all $\lambda \geq 0$,*

$$\mathbb{H}\left[e^{\lambda Z}\right] \leq \frac{\nu(-\lambda)}{e^{\lambda}+\alpha-1}\mathbb{E}\left[Ze^{\lambda Z} + \sum_{k=1}^{n}(Z_k - \mathbb{E}_k[Z] + \alpha\mathsf{Var}_k[Z])e^{\lambda Z}\right] .$$

and

$$\mathbb{H}\left[e^{\lambda Z}\right] \leq \frac{\nu(-\lambda)}{\psi(-\lambda)+\alpha\lambda}\mathbb{E}\left[e^{\lambda Z}\log\mathbb{E}\left[e^{\lambda Z}\right] + \sum_{k=1}^{n}\lambda(Z_k - \mathbb{E}_k\left[Z\right] + \alpha\mathsf{Var}_k\left[Z\right])e^{\lambda Z}\right].$$

Proof. We apply respectively Inequalities (8.1) and (8.2) of Lemma 8.3 to the quantity $\mathbb{H}_k\left[e^{\lambda(Z-\mathbb{E}_k[Z])}\right]$. We then use Lemma 2.1 to get

$$\mathbb{H}\left[e^{\lambda Z}\right] \leq \mathbb{E}\left[\sum_{k=1}^{n} e^{\lambda \mathbb{E}_k[Z]}\mathbb{H}_k\left[e^{\lambda(Z-\mathbb{E}_k[Z])}\right]\right].$$

We then use Jensen's inequality to get

$$e^{\lambda\mathbb{E}_k[Z]} \leq \mathbb{E}_k\left[e^{\lambda Z}\right].$$

Using the sub-additivity condition then gives the first inequality. For the second one, we first obtain

$$\mathbb{H}\left[e^{\lambda Z}\right] \leq \frac{\nu(-\lambda)}{\psi(-\lambda)+\alpha\lambda}\mathbb{E}\left[\sum_{k=1}^{n} e^{\lambda Z} - e^{\lambda Z_k} + \lambda\alpha\mathsf{Var}_k\left[Z\right]e^{\lambda Z}\right]. \tag{8.3}$$

Applying the last inequality of Lemma 8.2 gives the result. □

Theorem 8.5. *Assume Z is sub-additive and defined as in (2.1). If for all $k = 1, \ldots, n$, $\mathbb{E}_k\left[(Z-\mathbb{E}_k\left[Z\right])^2\right] \leq \sigma_k$ and $\|Z-\mathbb{E}_k\left[Z\right]\|_\infty \leq 1$ then for $\lambda \in (-\log 2, 1]$,*

$$\mathbb{H}\left[e^{\lambda Z}\right] \leq g(\lambda)\mathbb{E}\left[Ze^{\lambda Z} + \sum_{k=1}^{n}\left(Z_k - \mathbb{E}_k\left[Z\right] + \sigma_k^2\right)e^{\lambda Z}\right],$$

with $g(\lambda) = \nu(-|\lambda|)/(1-\mathrm{sgn}(\lambda)+\mathrm{sgn}(\lambda)e^{|\lambda|})$. If for all $p \geq 2$, $\mathbb{E}_k\left[|Z-\mathbb{E}_k\left[Z\right]|^p\right] \leq p!\sigma_k^2/2$ then for $\lambda \in (\sqrt{2}/2-1, 1/2]$

$$\mathbb{H}\left[e^{\lambda Z}\right] \leq \lambda^2\mathbb{E}\left[Ze^{\lambda Z} + \sum_{k=1}^{n}\left(Z_k - \mathbb{E}_k\left[Z\right] + \frac{\sigma_k^2}{2}\right)e^{\lambda Z}\right],$$

with $g(\lambda) = \lambda^2/(\mathrm{sgn}(\lambda) + (1-\mathrm{sgn}(\lambda))(1-|\lambda|)^2)$.

Proof. Given a non-negative function $g(\lambda)$ to be specified later, we use the decomposition

$$\begin{aligned} &\mathbb{E}\left[\psi(\lambda(Z-\mathbb{E}_k\left[Z\right]))e^{\lambda Z}\right] \\ &= \mathbb{E}\left[\mathbb{E}_k\left[1-e^{\lambda(Z-\mathbb{E}_k[Z])} + \lambda(Z-\mathbb{E}_k\left[Z\right])e^{\lambda(Z-\mathbb{E}_k[Z])}\right]e^{\lambda\mathbb{E}_k[Z]}\right] \\ &= \mathbb{E}\left[g(\lambda)(Z-\mathbb{E}_k\left[Z\right])e^{\lambda Z} + \mathbb{E}_k\left[f(Z-\mathbb{E}_k\left[Z\right])\right]e^{\lambda\mathbb{E}_k[Z]}\right], \end{aligned}$$

where $f(x) = 1 - e^{\lambda x} + (\lambda - g(\lambda))xe^{\lambda x}$. We thus have

$$f(x) = -g(\lambda)x + \sum_{q=2}^{\infty}(1-\frac{g(\lambda)}{\lambda})\frac{\lambda^q x^q}{(q-1)!} - \frac{\lambda^q x^q}{q!}.$$

Note that in the sum above, the coefficient of x^q is $\lambda^q(q-1-qg(\lambda)/\lambda)/q!$. So that if λ is such that $q-1-qg(\lambda)/\lambda \geq 0$, we can upper bound

$$f(x) \leq -g(\lambda)x + \sum_{q=2}^{\infty} |x|^q|\lambda|^q(q-1-q\frac{g(\lambda)}{\lambda})/q!\,,$$

so that

$$\mathbb{E}_k\left[f(Z-\mathbb{E}_k\left[Z\right])\right] \leq \sum_{q=2}^{\infty} \mathbb{E}_k\left[|Z-\mathbb{E}_k\left[Z\right]|^q\right]|\lambda|^q(q-1-q\frac{g(\lambda)}{\lambda})/q!\,.$$

Now, when $\mathbb{E}_k\left[|Z-\mathbb{E}_k\left[Z\right]|^q\right] \leq \sigma^2$, we obtain

$$\mathbb{E}_k\left[f(Z-\mathbb{E}_k\left[Z\right])\right] \leq (1-e^{|\lambda|})g(\lambda)\operatorname{sgn}(\lambda) + \nu(-|\lambda|)\,,$$

so that choosing $g(\lambda) = \nu(-|\lambda|)/(1-\operatorname{sgn}(\lambda)+\operatorname{sgn}(\lambda)e^{|\lambda|})$ gives

$$\mathbb{E}_k\left[f(Z-\mathbb{E}_k\left[Z\right])\right] \leq g(\lambda)\,,$$

for $\lambda \in (-\log 2, 1]$. When $\mathbb{E}_k\left[|Z-\mathbb{E}_k\left[Z\right]|^q\right] \leq q!\sigma^2/2$, we obtain

$$2\mathbb{E}_k\left[f(Z-\mathbb{E}_k\left[Z\right])\right] \leq g(\lambda)\operatorname{sgn}(\lambda)(1-\frac{1}{(1-|\lambda|)^2}) + \frac{\lambda^2}{(1-|\lambda|)^2}\,,$$

so that choosing $g(\lambda) = \lambda^2/(\operatorname{sgn}(\lambda)+(1-\operatorname{sgn}(\lambda))(1-|\lambda|)^2)$ gives

$$\mathbb{E}_k\left[f(Z-\mathbb{E}_k\left[Z\right])\right] \leq \frac{g(\lambda)}{2}\,,$$

for $\lambda \in (\sqrt{2}/2-1, 1/2]$.

Using Lemma 2.1 and Jensen's inequality gives the results. □

8.4. Consequences

We now consider conditions on the increments and apply the two theorems of previous section.

Corollary 8.6. *Assume Z is sub-additive. If $Y_k \leq Z - Z_k \leq 1$, $\mathbb{E}_k\left[Y_k\right] \geq 0$ and $Y_k \leq a$ then*

$$\mathbb{H}\left[e^{\lambda Z}\right] \leq \frac{\nu(-\lambda)}{e^{\lambda}+\alpha-1}\mathbb{E}\left[Ze^{\lambda Z} + \alpha\sum_{k=1}^{n}\mathbb{E}_k\left[Y_k^2\right]e^{\lambda Z}\right],$$

and

$$\mathbb{H}\left[e^{\lambda Z}\right] \leq \frac{\nu(-\lambda)}{\psi(-\lambda)+\alpha\lambda}\mathbb{E}\left[e^{\lambda Z}\log\mathbb{E}\left[e^{\lambda Z}\right] + \alpha\lambda\sum_{k=1}^{n}\mathbb{E}_k\left[Y_k^2\right]e^{\lambda Z}\right],$$

with $\alpha = 1/(1+a)$.

Proof. These inequalities follow from first two inequalities of Theorem 8.4. The only extra step to be taken can be written as follows

$$\begin{aligned} Z_k - \mathbb{E}_k[Z] + \alpha \mathsf{Var}_k[Z] &\leq \mathbb{E}_k\left[\alpha(Z-Z_k)^2 - (Z-Z_k)\right] \\ &\leq \mathbb{E}_k\left[\alpha Y_k^2 - Y_k\right] \\ &\leq \alpha \mathbb{E}_k\left[Y_k^2\right], \end{aligned}$$

where the second line follows from the fact that $\alpha x^2 - x$ is non-increasing on $(-\infty, a]$ and its value on $[a, 1]$ is less than its value in a (for $\alpha = 1/(1+a)$). □

Corollary 8.7. *Assume Z is sub-additive. If $Y_k \leq Z - Z_k \leq 1$, $\mathbb{E}_k[Y_k] \geq 0$ and $Y_k \leq a$ then for $\alpha = 1/(1+a)$, $\lambda \geq 0$ and $\theta > 0$,*

$$\left(1 - \frac{\alpha\theta f(\lambda)}{\lambda}\right)\mathbb{H}\left[e^{\lambda Z}\right] \leq f(\lambda)\mathbb{E}\left[Ze^{\lambda Z} + \frac{\alpha\theta}{\lambda}e^{\lambda Z}\log\mathbb{E}\left[e^{\lambda\sum_{k=1}^n \mathbb{E}_k[Y_k^2]/\theta}\right]\right],$$

with $f(\lambda) = \nu(-\lambda)(e^\lambda + \alpha - 1)^{-1}$. And

$$(1 - \alpha\theta f(\lambda))\,\mathbb{H}\left[e^{\lambda Z}\right] \leq f(\lambda)\mathbb{E}\left[e^{\lambda Z}\log\mathbb{E}\left[e^{\lambda Z}\right] + \alpha\theta e^{\lambda Z}\log\mathbb{E}\left[e^{\lambda\sum_{k=1}^n \mathbb{E}_k[Y_k^2]/\theta}\right]\right],$$

with $f(\lambda) = \nu(-\lambda)(\psi(-\lambda) + \alpha\lambda)^{-1}$.

Proof. Both inequalities follow from Corollary 8.6 and Lemma 8.1. □

8.5. Solving the Differential Inequalities

In this section, we explain how to solve the differential inequalities that arise from previous results. The first lemma explains that if one has a function which satisfies the reverse differential inequality, then all solution of the initial inequality are upper bounded by this function. The consequence is that to obtain an upper bound on the solution, we simply have to 'guess' a function which satisfies the reverse inequality (or the corresponding equality).

Lemma 8.8. *Let f, g, h be continuous functions such that f and g are non-negative. Let L be a solution of*

$$f(x)L'(x) - g(x)L(x) \leq h(x), \tag{8.4}$$

and L_0 be a function that satisfies

$$f(x)L'(x) - g(x)L(x) \geq h(x), \tag{8.5}$$

then if $L(0) = L_0(0)$ and $L'(0) = L_0'(0)$ then $L \leq L_0$.

Proof. Consider first the differential inequality

$$f(x)L'(x) - g(x)L(x) \leq 0 \tag{8.6}$$

We look for solutions of this inequality that are defined on $[0, +\infty)$ and satisfying $F(0) = F'(0) = 0$.

There are three possible cases

- F is equal to 0 on $[0, a]$ with $a > 0$.
- F is (strictly) negative on $(0, a)$ with $a > 0$.

- F is (strictly) positive on $(0, a)$ with $a > 0$.

In the first case, translation allows to end up in either the second or the third case.

In the second case, we have because of (8.6), F' non-positive on $(0, a)$ and thus F is non-increasing so that it stays strictly negative. This implies (by continuity of F) that F is non-positive wherever it is defined.

In the third case, we can rewrite (8.6) as

$$f(x)\frac{F'}{F} \leq g(x)$$

which can be integrated on $[c, x]$ to give

$$F(x) \leq F(c) \exp \int_c^x \frac{g(u)}{f(u)} du\,.$$

Making c tend to 0, by continuity, we obtain $F(x) \leq 0$. Thus in all cases, we obtain that if F satisfies (8.6) with $F(0) = F'(0) = 0$ then $F(x) \leq 0$.

Let us consider the differential equation

$$f(x)L'(x) - g(x)L(x) = h(x)\,. \tag{8.7}$$

Now if L is a solution of (8.4) and L_1 is a solution of (8.7) with $L_1(0) = L(0)$ and $L'(0) = L_1'(0)$, then $F = L - L_1$ is a solution of (8.6) with $F(0) = F'(0) = 0$ which gives $L \leq L_1$.

Now notice that (8.5) is equivalent to

$$f(x)(-L'(x)) - g(x)(-L(x)) \leq -h(x)\,,$$

so that the same argument as before shows that $-L_0 \leq -L_1$ which concludes the proof. □

Now we can give the results in different situations.

Lemma 8.9. *Let F be any non-negative continuous function with continuous derivative defined on an interval I containing 0 and such that $F(0) = 1$ and $F'(0) = 0$. If F satisfies on I, respectively*

$$\lambda F' - F \log F \leq \lambda^2 (F' + \gamma F)$$

$$\lambda F' - F \log F \leq \frac{\lambda^2}{1 + 4\lambda + 2\lambda^2}(F' + \gamma F)\,,$$

$$\lambda F' - F \log F \leq \frac{1 - e^\lambda + \lambda e^\lambda}{e^\lambda + \alpha - 1}(F' + \gamma F)$$

$$\lambda F' - F \log F \leq \frac{1 - e^\lambda + \lambda e^\lambda}{e^\lambda - 1 - \lambda + \alpha\lambda}(F \log F + \gamma\lambda F)$$

$$\lambda F' - F \log F \leq \frac{1 - e^{-\lambda} - \lambda e^{-\lambda}}{1 + \alpha - e^{-\lambda}}(F' + \gamma F)\,,$$

then F satisfies on I, respectively

$$\log F(\lambda) \leq \frac{\gamma\lambda^2}{1-\lambda}$$
$$\log F(\lambda) \leq \frac{\gamma\lambda^2}{1+\lambda}$$
$$\log F(\lambda) \leq \frac{\gamma}{\alpha}(e^{\lambda}-1-\lambda)$$
$$\log F(\lambda) \leq \frac{\gamma}{\alpha}(e^{\lambda}-1-\lambda)$$
$$\log F(\lambda) \leq \frac{\gamma}{\alpha}(e^{-\lambda}-1+\lambda)$$

Proof. Consider the differential inequality satisfied by $L(\lambda) = \log F(\lambda)$ (with $L(0) = L'(0) = 0$), check that the proposed upper bound satisfies this inequality with an equality and use Lemma 8.8. □

Lemma 8.10. *For $\lambda \in [0, z)$, any solution F of*

$$\lambda F' - \frac{n\lambda}{1-e^{-\lambda}} F(1 - F^{-1/n}) \leq 0\,,$$

with $F(0) = 1$ and $F'(0) = np$ satisfies

$$F(\lambda) \leq (1 + p(e^{\lambda} - 1))^n\,.$$

Proof. Notice that the function $G(\lambda) = F^{1/n}(\lambda)$ satisfies $G(0) = 1$, $G'(0) = 0$ and

$$\lambda G' - \frac{\lambda}{1-e^{-\lambda}}(G-1) \leq 0\,.$$

Notice that the proposed upper bound satisfies the above with an equality and use Lemma 8.8 to conclude. □

8.6. Proofs of Main Theorems

The main theorems now follow from combining all the previous steps.

Proof of Theorem 6.1. We use Corollary 8.6 and Lemma 8.9. □

Proof of Theorem 6.2. We use Theorem 8.5 and the fact that $\mathbb{E}_k[Z] \geq Z_k$. The result can be obtained from Lemma 8.9. For the case when the moments of $Z - Z_k$ are controlled, one can use the same reasoning as in Theorem 8.5 replacing $\mathbb{E}_k[Z]$ by Z_k. Using the fact that $\mathbb{E}_k[Z] \geq Z_k$ gives the same result (since $e^{\lambda Z_k} \leq e^{\lambda \mathbb{E}_k[Z]} \leq \mathbb{E}_k\left[e^{\lambda Z}\right]$). □

Proof of Theorem 6.3.

$$\begin{aligned}
&\mathbb{H}\left[e^{\lambda Z}\right] \\
&\quad \le \mathbb{E}\left[\sum_{k=1}^{n} e^{\lambda Z_k}\mathbb{H}\left[e^{\lambda(Z-Z_k)}\right]\right] \\
&\qquad \text{(by Lemma 2.1)} \\
&\quad \le \inf_{p\in\mathbb{R}} \mathbb{E}\left[\sum_{k=1}^{n} e^{\lambda(p+Z_k)} - e^{\lambda Z} - \lambda p e^{\lambda Z} + \frac{\lambda}{1-e^{-\lambda}}(e^{\lambda Z} - e^{\lambda Z_k})\right] \\
&\qquad \text{(by Lemma 8.3)} \\
&\quad \le \mathbb{E}\left[\sum_{k=1}^{n} e^{\lambda Z_k}\mathbb{E}\left[e^{\lambda Z}\right]^{1/n} - e^{\lambda Z} - \frac{1}{n}e^{\lambda Z}\log \mathbb{E}\left[e^{\lambda Z}\right] + \frac{\lambda}{1-e^{-\lambda}}(e^{\lambda Z} - e^{\lambda Z_k})\right] \\
&\qquad \text{(choosing } p = \tfrac{1}{n\lambda}\log \mathbb{E}\left[e^{\lambda Z}\right]) \\
&\quad = \left(\mathbb{E}\left[\sum_{k=1}^{n} e^{\lambda Z_k}\right] - nF^{1-1/n}\right)\left(F^{1/n} - \frac{\lambda}{1-e^{-\lambda}}\right) - F\log F \\
&\qquad + \frac{n\lambda}{1-e^{-\lambda}}F\left(1 - F^{-1/n}\right)
\end{aligned}$$

where in the last step we denoted $F(\lambda) = \mathbb{E}\left[e^{\lambda Z}\right]$. By Lemma 8.2 we know that the first factor of the first term is non-negative so that we have to study the term of the second factor. Now assume $F^{1/n} \le \lambda/(1-e^{-\lambda})$, then we obtain

$$\mathbb{H}\left[e^{\lambda Z}\right] \le -F\log F + \frac{n\lambda}{1-e^{-\lambda}}F(1-F^{-1/n}),$$

which implies

$$\lambda F' - \frac{n\lambda}{1-e^{-\lambda}}F(1-F^{-1/n}) \le 0,$$

so that by Lemma 8.10

$$F(\lambda) \le (1 + \mathbb{E}[Z](e^{\lambda}-1)/n)^n.$$

Moreover, if λ_0 is the solution of $\mathbb{E}[Z]/n = (e^{\lambda}-1)^{-2}(1-e^{\lambda}+\lambda e^{\lambda})$, then for all $\lambda \in [0,\lambda_0)$, we have $1 + \mathbb{E}[Z](e^{\lambda}-1)/n \le \lambda/(1-e^{-\lambda})$. □

Proof of Theorem 6.4. We start from the second inequality of Corollary 8.7. Recall that

$$f(\lambda) = \frac{\nu(-\lambda)}{\psi(-\lambda)+\lambda\alpha},$$

with $\alpha = 1/(1+a)$. Now notice that for all $\lambda \ge 0$ and $a \in [0,1]$, we have

$$(1 - \frac{1}{1+a})\lambda^2 \le \frac{\lambda^2}{2} \le e^{\lambda} - 1 - \lambda,$$

so that we obtain $f(\lambda) \leq \lambda$. Denoting $F(\lambda) = \mathbb{E}\left[e^{\lambda Z}\right]$, we have

$$\left(1 - \frac{f(\lambda)\theta}{1+a}\right)(\lambda F'(\lambda) - F(\lambda)\log F(\lambda)) \leq f(\lambda)F(\lambda)\left(\log \mathbb{E}\left[e^{\lambda V}\right] + \frac{\theta}{1+a}\log \mathbb{E}\left[e^{\lambda W/\theta}\right]\right),$$

for all $\theta \geq 0$.

Since without loss of generality we can assume that $F(\lambda) > 0$, we can rewrite the above for $0 < f(\lambda) < (1+a)/\theta$ as

$$\frac{1}{\lambda}\frac{F'(\lambda)}{F(\lambda)} - \frac{1}{\lambda^2}\log F(\lambda) \leq \frac{f(\lambda)}{\lambda^2(1 - f(\lambda)\theta/(1+a))}\left(G(\lambda) + \frac{\theta}{1+a}H(\lambda/\theta)\right),$$

where $G(\lambda) = \log \mathbb{E}\left[e^{\lambda V}\right]$ and $H(\lambda) = \log \mathbb{E}\left[e^{\lambda W}\right]$.

Since $f(\lambda) \leq \lambda$, the condition $0 < f(\lambda) < (1+a)/\theta$ is satisfied when $0 < \lambda < (1+a)/\theta$. For any such λ, we can integrate the above differential inequality since the left hand side is simply the derivative of $\lambda \mapsto \frac{1}{\lambda}\log F(\lambda)$ (which is equal to $\mathbb{E}[Z]$ for $\lambda = 0$). We thus get

$$\frac{1}{\lambda}\log F(\lambda) - \mathbb{E}[Z] \leq \int_0^\lambda \frac{f(s)}{s^2(1 - f(s)\theta/(1+a))}\left(G(s) + \frac{\theta}{1+a}H(s/\theta)\right) ds,$$

Now notice that since G is convex and $G(0) = 0$, $G(\lambda)/\lambda$ is non-decreasing for $\lambda > 0$. The same is true for H. We can thus upper bound the right hand side by

$$\frac{1}{\lambda}\left(G(\lambda) + \frac{\theta}{1+a}H(\lambda/\theta)\right)\int_0^\lambda \frac{f(s)}{s(1 - f(s)\theta/(1+a))}ds.$$

Now using $f(s) \leq s$ we get

$$\int_0^\lambda \frac{f(s)}{s(1 - f(s)\theta/(1+a))}ds \leq \int_0^\lambda \frac{1}{(1 - s\theta/(1+a))}ds \leq \frac{\lambda}{1 - \lambda\theta/(1+a)},$$

where we used $\log x \leq x - 1$.

Finally we obtain

$$\log \mathbb{E}\left[e^{\lambda(Z - \mathbb{E}[Z])}\right] \leq \frac{\lambda}{1 - \lambda\theta/(1+a)}\left(G(\lambda) + \frac{\theta}{1+a}H(\lambda/\theta)\right),$$

which completes the proof. □

Acknowledgment

We would like to thank Emmanuel Rio for suggestions on an earlier version of this work, Pascal Massart for many inspiring discussions and an anonymous reviewer for helpful comments, especially concerning improvements of Lemma 4.1.

Appendix A. Additional Material

In this section we provide additional technical results that are useful in the above proofs. We first recall a simple consequence of L'Hospital rule which is useful for studying the behavior of functions formed by ratios.

Lemma A.1. *Let f and g two functions such that $f(0) = g(0) = 0$, and there exists some $A > 0$ such that for all $x \in [0, A)$, $g'(x) > 0$ and $f'(x)/g'(x)$ is non-decreasing on $[0, A)$. Then $f(x)/g(x)$ is non decreasing on $[0, A)$.*

Lemma A.2. *The following facts hold true.*

(i) $\psi \geq 0$, $\nu \geq 0$, $\psi(0) = \nu(0) = 0$.

(ii) ψ *is convex.*

(iii) *For all* $x \in [0,1]$ *and all* λ,

$$\psi(\lambda x) \leq x\psi(\lambda)\,.$$

(iv) *For* $x \geq 0$,

$$\psi(x) \leq \frac{x^2}{2(1+x/3)} \leq \frac{x^2}{2} \quad \text{and} \quad \nu(x) \leq \frac{x^2}{2}\,.$$

(v) *For* $x \leq 0$,

$$\psi(x) \geq \frac{x^2}{2} \quad \text{and} \quad \nu(x) \geq \frac{x^2}{2}\,.$$

(vi) *For* $x \leq 1$ *and* $\lambda \leq 0$,

$$\psi(\lambda x) \leq \psi(\lambda)x_+^2 + \frac{\lambda^2}{2}x_-^2 \leq \psi(\lambda)x^2\,.$$

Proof. Recall that if f is convex, then

$$f(x) - f(y) \leq f'(x)(x-y)\,.$$

We thus have $e^x - 1 \leq xe^x$ and also $1 - e^{-x} \leq x$ for all x which proves (i). (ii) is easily verified. (iii) follows from (ii).

In order to prove (iv), consider $f(x) = (1 + x/3)\psi(x) - x^2/2$. For $x \geq 0$, we have $f'''(x) = -xe^{-x}/3 \leq 0$ so that $f''(x) \leq f''(0) = 0$ and $f'(x) \leq f'(0) = 0$ thus $f(x) \leq f(0) = 0$. Similarly, consider $g(x) = \nu(x) - x^2$. For $x \geq 0$, we have $g''(x) = (1-x)e^{-x} - 1 \leq e^{-x} - 1 \leq 0$ and again, since $g'(0) = g(0) = 0$ we have $g(x) \leq 0$.

The same computations can be used to prove (v) since for $x \leq 0$, $f''(x) \geq 0$ and $g''(x) \geq 0$.

In order to prove (vi), let $\lambda \leq 0$ and $x \leq 1$. Let us introduce $h(x) = \psi(\lambda x)/x^2$. Let us first consider $x > 0$. We have

$$h'(x) = \frac{(2x - \lambda x^2)(e^{-\lambda x} - 1)}{x^4} \geq 0\,,$$

so that h is non-decreasing on $(0, 1]$ and thus $h(x) \leq h(1) = \psi(\lambda)$. When $x \leq 0$, we use (iv). The second part of the inequality in (vi) follows from (v). □

Next we give some simple properties of the functions ψ and ν that are used in the proofs.

Lemma A.3. *For all* $\lambda \geq 0$,

$$x \mapsto \frac{e^{\lambda x} - 1 - \lambda x}{x^2}$$

is non-decreasing on $\mathbb{R}$, *and it is non-increasing on* $\mathbb{R}$ *when* $\lambda \leq 0$.

For all $\lambda \geq 0$, *and* $\alpha \in [1/2, 1]$,

$$x \mapsto \frac{1 - e^{\lambda x} + \lambda x e^{\lambda x}}{x e^{\lambda x} + \alpha x^2 - x}$$

is non-decreasing for $x \leq 1$.

For all $\alpha \in (0, 1]$ *and* $\lambda \in (-\log(1 + \alpha), 0]$,

$$x \mapsto \frac{1 - e^{\lambda x} + \lambda x e^{\lambda x}}{x e^{\lambda x} + \alpha x^2 - x}$$

is non-increasing for $x \geq -1$.

For all $\lambda \geq 0$,

$$x \mapsto \frac{\lambda x e^{\lambda x}}{e^{\lambda x} - 1}$$

is non-decreasing and non-negative on $\mathbb{R}$.

For all $\lambda \leq 0$,

$$x \mapsto \frac{\lambda x e^{\lambda x}}{e^{\lambda x} - 1}$$

is non-increasing and non-negative on $\mathbb{R}$.

For all $\lambda \geq 0$ *and all* $\alpha \in [1/2, 1]$,

$$x \mapsto \frac{1 - e^{\lambda x} + \lambda x e^{\lambda x}}{e^{\lambda x} - 1 + \lambda(\alpha x^2 - x)}$$

is non-decreasing for $x \leq 1$.

Proof. We only prove the second result. All the other ones can be derived in a similar way.

Notice that since $\lambda \geq 0$, we have $x e^{\lambda x} \geq x$ so that $x e^{\lambda x} + \alpha x^2 - x \geq \alpha x^2$ so that this quantity is > 0 for $x \neq 0$. Moreover the inequality is trivially satisfied for $x = 0$. We thus have to prove that

$$\frac{1 - e^{\lambda x} + \lambda x e^{\lambda x}}{x e^{\lambda x} + \alpha x^2 - x} \leq f(\lambda)\,.$$

Since $f(\lambda)$ is equal to the value of the left hand side when $x = 1$, it is enough to prove that this function of x is non-decreasing for $x \leq 1$. For this, we will use Lemma A.1. Consider $h(x) = 1 - e^{\lambda x} + \lambda x e^{\lambda x}$ and $j(x) = x e^{\lambda x} + \alpha x^2 - x$. We have $h'(x) = \lambda^2 x e^{\lambda x}$ and $j'(x) = (1 + \lambda x)e^{\lambda x} + 2\alpha x - 1$. Let $m(x) = h'(x)/j'(x)$, i.e.

$$m(x) = \frac{\lambda^2 x}{1 + \lambda x + (2\alpha x - 1)e^{-\lambda x}}\,.$$

The sign of $m'(x)$ is the same as the sign of

$$1 + \lambda x + (2\alpha x - 1)e^{-\lambda x} - x(\lambda + (2\alpha + \lambda - 2\lambda\alpha x)e^{-\lambda x}) = \nu(\lambda x) + 2\lambda\alpha x^2 e^{-\lambda x},$$

which is non-negative. As a consequence, Lemma A.1 can be applied to the ratio $h(x)/j(x)$ which yields the desired result. □

Appendix B. From Laplace Transforms to Deviation Bounds

In this section we show how to get bounds on the probability of deviation from a bound on the moment generating function. These results are quite standard but we give them for the sake of completeness.

Lemma B.1. *Let Z be a random variable. Assume that for any $\lambda > 0$ we have for some constant A*

$$\log \mathbb{E}\left[e^{\lambda(Z-\mathbb{E}[Z])}\right] \leq A\lambda^2,$$

then for all $t > 0$,

$$\mathbb{P}\left[Z \geq \mathbb{E}\left[Z\right] + t\right] \leq e^{-\frac{t^2}{4A}},$$

which can be written, for all $x > 0$,

$$\mathbb{P}\left[Z \geq \mathbb{E}\left[Z\right] + \sqrt{2Ax}\right] \leq e^{-x}.$$

Assume that for any $\lambda < 0$ we have for some constant A

$$\log \mathbb{E}\left[e^{\lambda(Z-\mathbb{E}[Z])}\right] \leq A\lambda^2,$$

then for all $t > 0$,

$$\mathbb{P}\left[Z \leq \mathbb{E}\left[Z\right] - t\right] \leq e^{-\frac{t^2}{4A}},$$

which can be written, for all $x > 0$,

$$\mathbb{P}\left[Z \leq \mathbb{E}\left[Z\right] - \sqrt{2Ax}\right] \leq e^{-x}.$$

Proof. The proof follows from a straightforward application of Markov's inequality. We have, for all $\lambda > 0$,

$$\mathbb{P}\left[Z \geq \mathbb{E}\left[Z\right] + t\right] \leq \mathbb{E}\left[e^{\lambda(Z-\mathbb{E}[Z])-\lambda t}\right] \leq e^{A\lambda^2 - \lambda t},$$

and the right-hand side is minimized for $\lambda = \frac{2A}{t}$. The same reasoning applies for the other inequality. □

The next lemmas are also standard results that can be obtained by Chernoff's method.

Lemma B.2. *Let Z be a random variable. Assume that for some constants $A, B > 0$ and any $0 \leq \lambda < 1/B$ we have*

$$\log \mathbb{E}\left[e^{\lambda(Z-\mathbb{E}[Z])}\right] \leq \frac{A\lambda^2}{2(1 - B\lambda)},$$

then for all $t > 0$,

$$\mathbb{P}\left[Z \geq \mathbb{E}\left[Z\right] + t\right] \leq \exp\left(-\frac{A}{2B^2}\left(1 - \sqrt{1 + \frac{2Bt}{A}}\right)^2\right) \leq \exp\left(-\frac{t^2}{2A + 2Bt}\right),$$

which can be written, for all $x \geq 0$,

$$\mathbb{P}\left[Z \geq \mathbb{E}\left[Z\right] + \sqrt{2Ax} + Bx\right] \leq e^{-x}.$$

Assume that for some constants $A, B > 0$ *and any* $\lambda < 0$ *we have*

$$\log \mathbb{E}\left[e^{\lambda(Z - \mathbb{E}[Z])}\right] \leq \frac{A\lambda^2}{2(1 - B\lambda)},$$

then for all $0 \leq t \leq A/2B$,

$$\mathbb{P}\left[Z \leq \mathbb{E}\left[Z\right] - t\right] \leq \exp\left(-\frac{A}{2B^2}\left(1 - \sqrt{1 - \frac{2Bt}{A}}\right)^2\right) \leq \exp\left(-\frac{t^2}{2A - 2Bt}\right),$$

which can be written, for all $0 \leq x \leq A/2B^2$,

$$\mathbb{P}\left[Z \leq \mathbb{E}\left[Z\right] - \sqrt{2Ax} + Bx\right] \leq e^{-x}.$$

Lemma B.3 (See e.g. [14]). *Let* Z *be a random variable. Assume that for some constant* $A > 0$ *and any* $\lambda \geq 0$ *we have*

$$\log \mathbb{E}\left[e^{\lambda(Z - \mathbb{E}[Z])}\right] \leq A\psi(-\lambda),$$

then for all $t > 0$,

$$\mathbb{P}\left[Z \geq \mathbb{E}\left[Z\right] + t\right] \leq \exp\left(-(t + A)\log(1 + \frac{t}{A}) + t\right)$$
$$\leq \exp\left(-\frac{t}{2}\log(1 + \frac{t}{A})\right) \leq \exp\left(-\frac{t^2}{2A + 2t/3}\right),$$

which can be written, for all $x \geq 0$,

$$\mathbb{P}\left[Z \geq \mathbb{E}\left[Z\right] + \sqrt{2Ax} + \frac{x}{3}\right] \leq e^{-x}.$$

Assume that for some constant $A > 0$ *and any* $\lambda < 0$ *we have*

$$\log \mathbb{E}\left[e^{\lambda(Z - \mathbb{E}[Z])}\right] \leq A\psi(-\lambda),$$

then for all $0 \leq t \leq A$,

$$\mathbb{P}\left[Z \geq \mathbb{E}\left[Z\right] + t\right] \leq \exp\left(-(t + A)\log(1 + \frac{t}{A}) + t\right) \leq \exp\left(-\frac{t^2}{2A + 2t/3}\right),$$

and, for all $x \geq 0$,

$$\mathbb{P}\left[Z \leq \mathbb{E}\left[Z\right] - \sqrt{2Ax}\right] \leq e^{-x}.$$

Lemma B.4. *Let Z be a random variable. Assume that for any $\lambda \in \mathbb{R}$ we have*

$$\log \mathbb{E}\left[e^{\lambda Z}\right] \leq \mathbb{E}[Z]\left(e^{\lambda}-1\right),$$

then for all $x \geq 0$,

$$\mathbb{P}\left[Z \geq \mathbb{E}[Z]+\sqrt{2x\mathbb{E}[Z]}+\frac{x}{3}\right] \leq e^{-x},$$

and

$$\mathbb{P}\left[Z \leq \mathbb{E}[Z]-\sqrt{2x\mathbb{E}[Z]}\right] \leq e^{-x}.$$

Now assume that for $\lambda \geq 0$,

$$\log \mathbb{E}\left[e^{\lambda Z}\right] \leq \frac{\mathbb{E}[Z]}{1-\mathbb{E}[Z]/n}\left(e^{\lambda(1-\mathbb{E}[Z]/n)}-1\right),$$

then for $t > 0$,

$$\mathbb{P}[Z \geq \mathbb{E}[Z]+t] \leq e^{-\frac{\mathbb{E}[Z]}{1-\mathbb{E}[Z]/n} h_1\left(\frac{t}{\mathbb{E}[Z]}\right)},$$

and if for $\lambda \leq 0$,

$$\log \mathbb{E}\left[e^{\lambda Z}\right] \leq n \log\left(1+\frac{\mathbb{E}[Z]}{n}\left(e^{\lambda}-1\right)\right),$$

then for $t \geq 0$,

$$\mathbb{P}[Z \leq \mathbb{E}[Z]-t] \leq \left(\frac{n-\mathbb{E}[Z]}{n-\mathbb{E}[Z]+t}\right)^{n-\mathbb{E}[Z]+t}\left(\frac{\mathbb{E}[Z]}{\mathbb{E}[Z]-t}\right)^{\mathbb{E}[Z]-t}$$

Appendix C. Other Lemmas and Proofs

The following lemma shows that suprema of empirical processes are indeed sub-additive functions.

Lemma C.1. *Let $\mathcal{F}$ be a set of functions and let*

$$Z=\sup_{f \in \mathcal{F}} \sum_{k=1}^{n} f(X_k).$$

Then, defining $Z_k=\sup_{f \in \mathcal{F}} \sum_{i \neq k} f(X_i)$, Z is sub-additive, i.e.

$$\sum_{k=1}^{n}(Z-Z_k) \leq Z \text{ almost surely.}$$

The same is true if Z is defined as

$$Z=\sup_{f \in \mathcal{F}}\left|\sum_{k=1}^{n} f(X_k)\right|,$$

and $Z_k=\sup_{f \in \mathcal{F}}\left|\sum_{i \neq k} f(X_i)\right|$.

Proof. We have, denoting by f_0 the function where the supremum is attained in Z,

$$(n-1)Z = \sum_{k=1}^{n} \sum_{i \neq k} f_0(X_i) = \sum_{k=1}^{n} \sum_{i \neq k} f_0(X_i) \leq \sum_{k=1}^{n} Z_k \,.$$

This extends to the case where the supremum is not reached in $\mathcal{F}$ (by taking limits). Considering the set $\mathcal{F} \cup -\mathcal{F}$ gives the second result. □

Lemma C.2. *If $p/n \in [0, 1/2]$ and $t/n \in [0, \frac{3}{4}(1-2(p/n))]$, then the infimum of $n\log(1+p/n(e^\lambda - 1)) + \lambda(p+t)$ satisfies*

$$p/n \leq \frac{1 - e^\lambda + \lambda e^\lambda}{(1-e^\lambda)^2} \,.$$

Lemma C.3 (Bennett [2]). *For all $\lambda \geq 0$, $\alpha \geq 0$ and $x \leq 1$,*

$$\begin{aligned}(1+\alpha)^2 e^{\lambda x} &\leq x^2\left(e^\lambda - (\lambda(1+\alpha)+1)e^{-\lambda\alpha}\right) \\ &\quad + x\left(2\alpha e^\lambda - (2\alpha + \lambda(\alpha^2-1))e^{-\lambda\alpha}\right) \\ &\quad + \alpha^2 e^\lambda + (1 + 2\alpha + \lambda\alpha(1+\alpha))e^{-\lambda\alpha} \,.\end{aligned}$$

Proof of Lemma 3.3.

$$\begin{aligned}\mathbb{E}\left[(X - \mathbb{E}X)^2\right] &= \mathbb{E}X^2 - (a+b)\mathbb{E}X - (\mathbb{E}X)^2 + (a+b)\mathbb{E}X \\ &\leq \mathbb{E}Y^2 - (a+b)\mathbb{E}Y - (\mathbb{E}X)^2 + (a+b)\mathbb{E}X\end{aligned}$$

$$\begin{aligned}\log \mathbb{E}\left[e^{\lambda X}\right] &\leq \mathbb{E}\left[e^{\lambda X} - 1\right] \\ &= \mathbb{E}\left[e^{\lambda X} - 1 - \lambda e^\lambda X + \lambda e^\lambda X\right] \\ &\leq \mathbb{E}\left[e^{\lambda Y} - 1 - \lambda e^\lambda Y + \lambda e^\lambda X\right] \\ &= \mathbb{E}\left[e^{\lambda Y} - 1 - \lambda Y\right] + \lambda(1 - e^\lambda)\mathbb{E}Y + \lambda e^\lambda \mathbb{E}X \\ &\leq \sigma^2(e^{-\lambda b} - 1 + \lambda b) + \lambda e^\lambda \mathbb{E}X\end{aligned}$$

$$\begin{aligned}\log \mathbb{E}\left[e^{\lambda X}\right] &\leq \mathbb{E}\left[e^{\lambda X} - 1\right] \\ &= \mathbb{E}\left[e^{\lambda X} - 1 - \lambda e^\lambda X + \lambda e^\lambda X\right] \\ &\leq \mathbb{E}\left[e^{\lambda Y} - 1 - \lambda e^\lambda Y + \lambda e^\lambda X\right] \\ &= \mathbb{E}\left[e^{\lambda Y} - 1 - \lambda Y\right] + \lambda(1 - e^\lambda)\mathbb{E}Y + \lambda e^\lambda \mathbb{E}X \\ &\leq \sigma^2 \frac{\lambda^2}{2(1+b\lambda)} + \lambda e^\lambda \mathbb{E}X\end{aligned}$$

□

References

[1] Y. Baraud. Model selection for regression on a fixed design. *Probab. Theory Related Fields*, **117**:467–493, 2000.

[2] G. Bennett. Probability inequalities for the sum of independent random variables. *Journal of the American Statistical Association*, **57**(297):33–45, 1962.

[3] S. Bernstein. Sur une modification de l'inégalité de tchebichef. *Annals Science Institute Sav. Ukraine, Sect. Math. I*, 1924.

[4] S. Bobkov. Some extremal properties of the Bernoulli distribution. *Theory of Probability and its Applications*, **41**(4):748–755, 1996.

[5] S. Boucheron, O. Bousquet, G. Lugosi, and P. Massart. Moment inequalities for functions of independent random variables. Preprint, 2003.

[6] S. Boucheron, G. Lugosi, and P. Massart. A sharp concentration inequality with applications. *Random Structures and Algorithms*, **16**(3):277–292, 2000.

[7] S. Boucheron, G. Lugosi, and P. Massart. Concentration inequalities using the entropy method. *Annals of Probability*, 2002. To appear.

[8] D. Chafaï. On ϕ-entropies and ϕ-Sobolev inequalities. arXiv.math.PR/0211103, 2002. Preprint.

[9] H. Cramér. Problems in probability theory. *Annals of Mathematical Statistics*, **18**:165–193, 1947.

[10] W. Hoeffding. Probability inequalities for sums of bounded random variables. *Journal of the American Statistical Association*, **58**:13–30, 1963.

[11] R. Latala and C. Oleszkiewicz. Between Sobolev and Poincaré. In *GAFA, Israel Seminar, Lecture Notes in Mathematics*, 1745, pages 147–168. Springer, 2000.

[12] M. Ledoux. On Talagrand's deviation inequalities for product measures. *ESAIM Probability and Statistics*, **1**:63–87, 1996.

[13] M. Ledoux. *The concentration of measure phenomenon.* Mathematical Surveys and Monographs. American Mathematical Society, 2001.

[14] P. Massart. About the constants in Talagrand's inequality for empirical processes. *The Annals of Probability*, **29**(2):863–884, 2000.

[15] C. McDiarmid. *Surveys in Combinatorics*, chapter On the method of bounded differences, pages 148–188. Cambridge University Press, Cambridge, 1989.

[16] E. Rio. Inégalités de concentration pour les processus empiriques de classes de parties. *Probability Theory and Related Fields*, **119**:163–175, 2000.

[17] E. Rio. Une inegalité de Bennett pour les maxima de processus empiriques. In *Colloque en l'honneur de J. Bretagnolle, D. Dacunha-Castelle et I. Ibragimov*, 2001. To appear.

[18] M. Talagrand. A new look at independence. *Annals of Probability*, **24**:1–34, 1996.

[19] A. van der Vaart and J. Wellner. *Weak convergence and empirical processes with applications to statistics.* John Wiley & Sons, New York, 1996.

Max Planck Institute for Biological Cybernetics, Spemannstr. 38,
D-72076 Tübingen, Germany
E-mail address: olivier.bousquet@tuebingen.mpg.de

Progress in Probability, Vol. 56, 249–278

Ratio Limit Theorems for Empirical Processes

Evarist Giné[1], Vladimir Koltchinskii[2], and Jon A. Wellner[3]

Abstract. Concentration inequalities are used to derive some new inequalities for ratio-type suprema of empirical processes. These general inequalities are used to prove several new limit theorems for ratio-type suprema and to recover a number of the results from [1] and [2]. As a statistical application, an oracle inequality for nonparametric regression is obtained via ratio bounds.

1. Introduction

Let $\mathcal{F}$ be a uniformly bounded class of real valued measurable functions on a probability space $(S, \mathcal{A}, P)$. To be specific, we assume most often that $\mathcal{F}$ takes values in $[0,1]$ (although, in some places below, the class will be scaled differently). Let X, X_i, $i \in \mathbb{N}$, be i.i.d. random variables in $(S, \mathcal{A})$ with distribution P. We denote by P_n the empirical measure based on the sample $(X_1, \ldots, X_n)$, $P_n = n^{-1}\sum_{i=1}^n \delta_{X_i}$. Suppose that $\sigma_P(f)$ is defined in such a way that

$$\sigma_P^2(f) \geq \mathrm{Var}_P(f) := Pf^2 - (Pf)^2, \ f \in \mathcal{F}.$$

In particular, $\sigma_P(f)$ can be the standard deviation itself, or in can be equal to $\sqrt{Pf}$ (recall that f takes values in $[0,1]$). In this note we present a simple technique to study the asymptotic behavior of the supremum of the standardized empirical process,

$$\sup_{f\in\mathcal{F}, \sigma_P(f)>\delta_n} \frac{\sqrt{n}|P_n f - Pf|}{\sigma_P(f)},$$

as well as some of its variations such as

$$\sup_{f\in\mathcal{F}, Pf>\delta_n} \left|\frac{P_n f}{Pf} - 1\right| \quad \text{and} \quad \sup_{f\in\mathcal{F}, \sigma_P(f)>\delta_n} \frac{\sqrt{n}|P_n f - Pf|}{\omega(\sigma_P(f))}$$

for suitable 'moduli' ω and properly chosen 'cutoffs' δ_n depending on the complexity of the class $\mathcal{F}$. These questions for Vapnik-Červonenkis (VC) classes of sets

---

Key words and phrases. empirical processes, concentration inequalities, ratio limit theorems, nonparametric regression, oracle inequalities.

[1]Research partially supported by NSF Grant No. DMS-0070382.

[2]Research partially supported by NSA Grant No. MDA904-02-1-0075.

[3]Research partially supported by NSF Grants DMS-9532039 and DMS-0203320.

were studied by Alexander in [2], and his proofs were technically rather sophisticated. Our results apply to very general classes of functions and, particularly in the form they take when we specialize them to VC classes of functions, they may be considered as analogues of some of Alexander's results for VC classes of sets. The need for this kind of results, in the generality given here, is illustrated by an example. Indeed, as an application of our general theorems we obtain an 'oracle inequality' in a simple but quite general non-parametric regression setting (cf., [16], [19]). So, the type of inequalities proved in this article may turn out to be useful for bounding errors of prediction in Statistics and in Machine Learning.

The main advance on empirical process theory since 1987, when Alexander proved his results, has been Talagrand's discovery of concentration inequalities ([30],[31]). This tool allows us to handle ratios very easily by proving several simple exponential bounds expressed in terms of expectations of localized sup norms of empirical processes. These bounds are obtained by stratifying the class $\mathcal{F}$ according to variance size, applying Talagrand's inequality to each stratum and then collecting terms. This approach, as carried out here, originated in the more specialized setting of statistical learning theory and was developed by several authors (see, e.g., Koltchinskii and Panchenko [20], Koltchinskii [21], Panchenko [26, 27] Bousquet, Koltchinskii and Panchenko [10], Bartlett, Bousquet and Mendelson [5] and, especially, the Ph. D. dissertations of Panchenko [25] and Bousquet [7]). A very close approach has been developed in some other statistical applications even earlier (see [23] and references therein), and in one form or other is also present in [2]. The exponential bounds for ratios together with some new bounds on expectations of suprema of empirical processes over VC classes of functions ([30], [12], [11], [24]) allow one to obtain Alexander type theorems without any effort. The present approach may open a possibility to understand much better, and for much more general classes than VC, this important class of limit theorems for empirical processes, and in particular, to widen the scope of their applicability.

There is an extensive literature on ratio limit theorems for classical empirical processes (see e.g. Wellner [35]). For general empirical processes indexed by sets or functions some important references are [1], [2], [3], [4], [15], [19], [28], [32], [34].

In order to avoid measurability problems, in what follows, we will assume that the supremum over the class $\mathcal{F}$ or over any of the subclasses we consider is in fact a countable supremum. In this case we say that the class $\mathcal{F}$ is measurable.

With some abuse of notation we will write $\log m$ for $1 \vee \log m$ and $\log\log m$ for $1 \vee \log\log m$.

2. Ratio limit theorems: normalization with $\sigma_P(f)$

We introduce some notations used in what follows. We set

$$\mathcal{F}(r) := \{f \in \mathcal{F} : \sigma_P(f) \leq r\}$$

and, for $r < s$,

$$\mathcal{F}(r, s] := \mathcal{F}(s) \setminus \mathcal{F}(r).$$

We denote

$$\xi_n(r,s] := \sup_{f\in\mathcal{F}(r,s]} \frac{|P_n f - Pf|}{\sigma_P(f)}.$$

Let now (given $r, s,\ r < s$) $q > 1$ and suppose that $s := rq^l$, for some $l \in \mathbb{N}$, so that

$$l = \log_q \frac{s}{r}.$$

[This will not be a loss of generality since the choice of q will be in our hands.] Let

$$\rho_j := rq^j,\ j = 0, \ldots, l$$

(with $\rho_0 = r,\ \rho_l = s$). Then we define a function $\psi_{n,q}$ from $(r, s]$ into the real line by setting

$$\psi_{n,q}(u) := \mathbb{E}\|P_n - P\|_{\mathcal{F}(\rho_{j-1},\rho_j]},\ u \in (\rho_{j-1}, \rho_j],\ j = 1, \ldots, l,$$

and we also set

$$\beta_{n,q}(r,s] := \sup_{u\in(r,s]} \frac{\psi_{n,q}(u)}{u}.$$

Given two sequences $\{r_n\},\ \{s_n\}$ of positive numbers such that $r_n < s_n$ we set

$$\xi_n := \xi_n(r_n, s_n],$$

and if $q_n \downarrow 1$ (so that $s_n = r_n q_n^{l_n}$ for an integer l_n) we define

$$\beta_n := \beta_{n,q_n}(r_n, s_n].$$

Our first goal is to prove the following general theorems. The only assumption on $\mathcal{F}$ is that it is a measurable class of functions taking values on $[0, 1]$.

Theorem 1. *Suppose that*

$$\sqrt{\frac{\log\log_{q_n}\frac{s_n}{r_n}}{n}} \bigvee \frac{1}{nr_n} = o(\beta_n).$$

Then

$$\frac{\xi_n}{\beta_n} \to 1 \text{ in Pr} \quad \text{as } n \to \infty.$$

The following a.s. version holds under slightly stronger assumptions. For simplicity, we consider only the case of $s_n \equiv 1$.

Theorem 2. *Suppose that*

$$\sqrt{\frac{\log\log_{q_n}\frac{1}{r_n} + \log\log n}{n}} \bigvee \frac{\log\log n}{nr_n} = o(\beta_n).$$

In addition, suppose that

$$r_n \searrow \text{ and } \frac{\beta_n}{\sqrt{n}} \searrow.$$

Then

$$\limsup_{n\to\infty} \frac{\xi_n}{\beta_n} = 1 \text{ a.s.}$$

Somewhat stronger assumptions lead to a.s. convergence to 1:

Theorem 3. *Suppose that*

$$\sqrt{\frac{\log\log_{q_n}\frac{s_n}{r_n}+\log n}{n}} \bigvee \frac{\log n}{nr_n} = o(\beta_n).$$

Then

$$\lim_{n\to\infty}\frac{\xi_n}{\beta_n} = 1 \text{ a.s.}$$

Remarks. 1. It is easy to see that the conditions of Theorem 1 are satisfied if

$$\sqrt{\frac{\log\log n}{n}} = o(\beta_n)$$

and

$$r_n\sqrt{n\log\log n}\to\infty.$$

2. Only formal modifications in the proof given below for Theorem 1 also show that if

$$\sqrt{\frac{\log\log_{q_n}\frac{s_n}{r_n}}{n}} = O(\beta_n)$$

and the sequence $\{nr_n\beta_n\}$ is bounded away from 0, then

$$\frac{\xi_n}{\beta_n} = O_p(1).$$

3. Likewise, it can also be shown that if

$$\limsup\frac{1}{\beta_n}\sqrt{\frac{\log\log_{q_n}\frac{s_n}{r_n}}{n}} < 1/2$$

and

$$\frac{1}{nr_n} = o(\beta_n),$$

then the sequence $\left\{\frac{\xi_n}{\beta_n}\right\}$ is both stochastically bounded and stochastically bounded away from 0.

4. If in Theorem 2 we replace $q_n\downarrow 1$ by $q>1$, and take $\beta_n\geq\beta_{n,q}(r_n,1]$, then the conditions

$$\sqrt{\frac{\log\log_{q_n}\frac{1}{r_n}+\log\log n}{n}} \bigvee \frac{\log\log n}{nr_n} = O(\beta_n),$$

$$r_n\searrow \text{ and } \frac{\beta_n}{\sqrt{n}}\searrow$$

imply

$$\limsup_{n\to\infty}\frac{\xi_n}{\beta_n} = R \text{ a.s.}$$

for some $R < \infty$. (This follows from obvious modifications in the proof of Theorem 2 given below together with the Kolmogorov 0-1 law as $nr_n\beta_n \to \infty$.)

5. For a general $\{s_n\}$, the extra conditions needed in Theorem 2 are $s_n \searrow$, $ns_n \nearrow$ and, most importantly,

$$\lim_{\varepsilon\to 0} \frac{\beta_{n,q_n}(r_n, s_n(1+\varepsilon)]}{\beta_n} = 1.$$

These assumptions (as well as the additional monotonicity assumptions in Theorem 2) come from a version of Lemma 7.2 in [2] that we use in the current proof. These assumptions might just be of a technical nature, and thus, perhaps superfluous.

The proofs are based on the following lemma.

Lemma 1. *For $t > 0$, define*

$$\Delta^{q,+}_{(r,s]}(t) := \sqrt{2q^2\frac{t+2\log\log_q \frac{s}{r}}{n} + 4\frac{t+c_q}{nr}\beta_{n,q}(r,s]} + \frac{t+c_q}{3nr}$$

and

$$\Delta^{q,-}_{(r,s]}(t) := \sqrt{2q^2\frac{t+2\log\log_q \frac{s}{r}}{n} + 4\frac{t+c_q}{nr}\beta_{n,q}(r,s]} + \frac{8}{3}\frac{t+c_q}{nr},$$

where $c_q := 2q\sup_{1\le j\le l} q^{-j}\log j$. Then

$$\mathbb{P}\left\{\sup_{f\in\mathcal{F}(r,s]}\frac{|P_nf-Pf|}{\sigma_P(f)} \ge \beta_{n,q}(r,s] + \Delta^{q,+}_{(r,s]}(t)\right\} \le 2e^{-t}$$

and

$$\mathbb{P}\left\{\sup_{f\in\mathcal{F}(r,s]}\frac{|P_nf-Pf|}{\sigma_P(f)} \le \frac{1}{q}\left[\beta_{n,q}(r,s] - \Delta^{q,-}_{(r,s]}(t)\right]\right\} \le 2e^{-t}.$$

Proof. For $q > 1$ and $0 < u \le 1$ define

$$\mathcal{F}_q(u) := \mathcal{F}(u/q, u]$$

and consider the following events

$$E^+_{q,u}(t) := \left\{\|P_n - P\|_{\mathcal{F}_q(u)} \le \mathbb{E}\|P_n - P\|_{\mathcal{F}_q(u)} + \sqrt{2\frac{t}{n}(u^2 + 2\psi_{n,q}(u))} + \frac{t}{3n}\right\}$$

and

$$E^-_{q,u}(t) := \left\{\|P_n - P\|_{\mathcal{F}_q(u)} \ge \mathbb{E}\|P_n - P\|_{\mathcal{F}_q(u)} - \sqrt{2\frac{t}{n}(u^2 + 2\psi_{n,q}(u))} - \frac{8t}{3n}\right\}.$$

By Talagrand's concentration inequalities for empirical processes (see [17], Theorem 1.1 and [9], Theorem 7.4), we have

$$\mathbb{P}(E^+_{q,u}(t)) \ge 1 - e^{-t} \text{ and } \mathbb{P}(E^-_{q,u}(t)) \ge 1 - e^{-t}$$

Let $\mathcal{F}_j := \mathcal{F}_{(\rho_{j-1},\rho_j]} = \mathcal{F}_q(\rho_j)$ and

$$E^+_j := E^+_{q,\rho_j}(t + 2\log j),\ E^-_j := E^-_{q,\rho_j}(t + 2\log j).$$

Then

$$\mathbb{P}\Big(\bigcap_{j=1}^{\infty} E_j^+\Big) \geq 1 - \sum_{j=1}^{\infty} e^{-t-2\log j} \geq 1 - e^{-t}\sum_{j=1}^{\infty} j^{-2} \geq 1 - 2e^{-t}$$

and similarly

$$\mathbb{P}\Big(\bigcap_{j=1}^{\infty} E_j^-\Big) \geq 1 - 2e^{-t}.$$

On the event $\bigcap E_j^+$, we have

$\forall j \; \forall f \in \mathcal{F}_j,$

$$\frac{|P_n f - Pf|}{\sigma_P(f)} \leq \frac{\psi_{n,q}(\sigma_P(f))}{\sigma_P(f)} + \sqrt{2\frac{t+2\log j}{n}\left(q^2 + 2\frac{\psi_{n,q}(\sigma_P(f))}{\sigma_P^2(f)}\right)} + \frac{t+2\log j}{3n\sigma_P(f)}.$$

Note that, for $f \in \mathcal{F}_j$,

$$\frac{t+2\log j}{n\sigma_P(f)} \leq \frac{t+2\log j}{n\rho_{j-1}} \leq \frac{t}{nr} + 2\sup_j \frac{2\log j}{q^{j-1}}\frac{1}{nr} = \frac{t+c_q}{nr}.$$

Since also

$$j \leq l = \log_q \frac{s}{r},$$

we get

$$\begin{aligned}
&\frac{\psi_{n,q}(\sigma_P(f))}{\sigma_P(f)} + \sqrt{2\frac{t+2\log j}{n}\left(q^2 + 2\frac{\psi_{n,q}(\sigma_P(f))}{\sigma_P^2(f)}\right)} + \frac{t+2\log j}{3n\sigma_P(f)} \\
\leq\;& \beta_{n,q}(r,s] + \sqrt{2q^2\frac{t+2\log\log_q\frac{s}{r}}{n} + 4\frac{t+c_q}{nr}\beta_{n,q}(r,s]} + \frac{t+c_q}{3nr} \\
=\;& \beta_{n,q}(r,s] + \Delta^{q,+}_{(r,s]}(t).
\end{aligned}$$

Thus, on the event $\bigcap E_j^+$,

$$\forall\, f \in \mathcal{F}(r,s], \quad \frac{|P_n f - Pf|}{\sigma_P(f)} \leq \beta_{n,q}(r,s] + \Delta^{q,+}_{(r,s]}(t),$$

and the first bound follows.

Similarly, on the event $\bigcap_j E_j^-$, we have

$$\begin{aligned}
\sup_{f\in\mathcal{F}(r,s]} \frac{|P_n f - Pf|}{\sigma_P(f)} &= \sup_j \sup_{f\in\mathcal{F}_j} \frac{|P_n f - Pf|}{\sigma_P(f)} \geq \sup_j \frac{\|P_n - P\|_{\mathcal{F}_j}}{\rho_j} \\
&\geq \sup_j \frac{\psi_{n,q}(\rho_j)}{\rho_j} - \sqrt{2\frac{t+2\log j}{n}\left(1 + 2\frac{\psi_{n,q}(\rho_j)}{\rho_j^2}\right)} - \frac{8}{3}\frac{t+2\log j}{n\rho_j} \\
&\geq \frac{1}{q}\sup_j \sup_{\rho\in(\rho_{j-1},\rho_j]}\left[\frac{\psi_{n,q}(\rho)}{\rho} - \sqrt{2\frac{t+2\log j}{n}\left(1 + 2\frac{\psi_{n,q}(\rho)}{\rho^2}\right)} - \frac{8}{3}\frac{t+2\log j}{n\rho}\right]
\end{aligned}$$

and, exactly as in the case of the upper bound, this can be shown to be

$$\geq \frac{1}{q}\Big[\beta_{n,q}(r,s] - \Delta^{q,-}_{(r,s]}(t)\Big],$$

which yields the second inequality. □

Proof of Theorem 1. The condition of the theorem means that

$$\log\log_{q_n}\frac{s_n}{r_n} = o(n\beta_n^2)$$

and

$$n\beta_n r_n \to \infty \text{ as } n \to \infty.$$

It follows from Lemma 1 that with probability at least $1 - 4e^{-t}$,

$$q_n^{-1}\left(1 - \frac{1}{\beta_n}\Delta^{q_n,-}_{(r_n,s_n]}(t)\right) \leq \frac{\xi_n}{\beta_n} \leq 1 + \frac{1}{\beta_n}\Delta^{q_n,+}_{(r_n,s_n]}(t).$$

Then, the conditions of the theorem immediately imply that for all $t > 0$

$$\frac{1}{\beta_n}\Delta^{q_n,+}_{(r_n,s_n]}(t) \to 0 \text{ as } n \to \infty,$$

and the same is true for $\Delta^{q_n,-}_{(r_n,s_n]}(t)$, which is enough to complete the proof. □

Proof of Theorem 2. Passing from a probability upper bound to an a.s. upper bound, given Lemma 1, is standard, and the details within the present framework are worked out in [2], Lemmas 7.1 and 7.2. Alexander's Lemma 7.2 is stated for classes of sets, but its proof applies as well for classes of functions taking values in $[0,1]$. His lemma, adapted to our case, is as follows:

If $\beta_n/n^{1/2} \searrow$, $r_n \searrow$, then the condition

$$\mathbb{P}\left\{\frac{\xi_n}{\beta_n} > 1 + \varepsilon\right\} = O\left((\log n)^{-1-\theta}\right) \tag{A}$$

for some $\varepsilon, \theta > 0$ implies

$$\mathbb{P}\left\{\frac{\xi_n}{\beta_n} > 1 + 2\varepsilon \text{ i.o.}\right\} = 0. \tag{B}$$

To establish (A), we take in Lemma 1 $t = 2\log\log n$, $q = q_n$, $s = 1$, $r = r_n$. The first bound of Lemma 1 then gives

$$\mathbb{P}\left\{\frac{\xi_n}{\beta_n} \geq 1 + \frac{1}{\beta_n}\Delta^{q_n,+}_{(r_n,1]}(2\log\log n)\right\} \leq 2\exp\{-2\log\log n\} = 2(\log n)^{-2}.$$

It follows from the conditions of the theorem that

$$\log\log_{q_n}\frac{1}{r_n} + \log\log n = o(n\beta_n^2)$$

and

$$\frac{n\beta_n r_n}{\log\log n} \to \infty \text{ as } n \to \infty.$$

This yields, by a straightforward computation, that

$$\frac{1}{\beta_n}\Delta^{q_n,+}_{(r_n,1]}(2\log\log n) \to 0 \text{ as } n \to \infty,$$

which establishes (A). Alexander's lemma now gives (B) for any $\varepsilon > 0$, so that

$$\limsup_{n\to\infty} \frac{\xi_n}{\beta_n} \leq 1.$$

Hence, it remains to prove that the lim sup is ≥ 1. The second bound of Lemma 1 (with the same t, q, r, s as before) gives

$$\mathbb{P}\left\{\frac{\xi_n}{\beta_n} \leq q_n^{-1}\left(1 - \frac{1}{\beta_n}\Delta^{q_n,-}_{(r_n,1]}(2\log\log n)\right)\right\} \leq 2(\log n)^{-2}.$$

Take $n = n_k = e^k$. Since again, for all large enough n,

$$\frac{1}{\beta_n}\Delta^{q_n,-}_{(r_n,1]}(2\log\log n) < \varepsilon$$

we get

$$\mathbb{P}\left\{\frac{\xi_{n_k}}{\beta_{n_k}} \leq 1 - \varepsilon\right\} = O(k^{-2}),$$

which by Borel–Cantelli Lemma implies that for all $\varepsilon > 0$

$$\mathbb{P}\left\{\frac{\xi_{n_k}}{\beta_{n_k}} \leq 1 - \varepsilon \text{ i.o.}\right\} = 0,$$

and the result follows. □

Proof of Theorem 3. It is a straightforward application of Lemma 1 along with Borel–Cantelli Lemma. One should take $t := 2\log n$ this time. □

3. Continuity moduli of empirical processes

Our goal in this section is to study the asymptotic behavior of

$$\sup_{f\in\mathcal{F}(r_n,s_n]} \frac{n^{1/2}|P_n f - Pf|}{\omega(\sigma_P(f))}$$

with a properly chosen "continuity modulus" ω. This will provide a piece of information about the local continuity modulus of the empirical process $n^{1/2}(P_n - P)$ at $f = 0$. The global continuity modulus can be studied quite similarly. (See [2] for definitions and motivation.) Also, we concentrate on "in probability" results (their almost sure versions can be also obtained with a little extra work).

We use the notations of the previous section and define

$$\omega_n(u) := n^{1/2}\psi_{n,q_n}(u),\ u \in (r_n, s_n].$$

Theorem 4. *Let ω be a nonnegative nondecreasing bounded function on $[0,1]$, satisfying the conditions $\frac{\omega(u)}{u} \searrow$ and*

$$\sup_{u\in(0,1]} \frac{u}{\omega(u)}\sqrt{\log\log\frac{1}{u}} < \infty.$$

Suppose that

$$\omega_n(u) \le \omega(u),\ u \in [r_n, s_n].$$

If

$$\sup_n \frac{\log\log_{q_n}\frac{s_n}{r_n}}{\omega(r_n)\sqrt{n}} < \infty,$$

then

$$\sup_{f\in\mathcal{F}(r_n,s_n]} \frac{n^{1/2}|P_n f - Pf|}{\omega(\sigma_P(f))}$$

is stochastically bounded and uniformly (in n) bounded in L_1:

$$\sup_n \mathbb{E}\left\{\sup_{f\in\mathcal{F}(r_n,s_n]} \frac{n^{1/2}|P_n f - Pf|}{\omega(\sigma_P(f))}\right\} < \infty. \tag{3.1}$$

In Theorem 4, $q_n \downarrow 1$ can be replaced by $q > 1$ or by $1 < q_n < C < \infty$.

Theorem 5. *Let ω be a nonnegative nondecreasing bounded function on $[0,1]$, satisfying the conditions $\frac{\omega(u)}{u} \searrow$ and*

$$\frac{u}{\omega(u)}\sqrt{\log\log\frac{1}{u}} \to 0 \text{ as } u \to 0.$$

Suppose that

$$\sup_{u\in(r_n,s_n]}\left|\frac{\omega_n(u)}{\omega(u)} - 1\right| \to 0 \text{ as } n \to \infty.$$

If also

$$\frac{\omega(r_n)\sqrt{n}}{\log\log_{q_n}\frac{s_n}{r_n}} \to \infty,$$

then

$$\sup_{f\in\mathcal{F}(r_n,s_n]} \frac{n^{1/2}|P_n f - Pf|}{\omega(\sigma_P(f))} \to 1 \text{ as } n \to \infty \text{ in Pr.}$$

Proof of Theorem 4. We follow the proof of Lemma 1 with $r = r_n, s = s_n, q = q_n, l = l_n$. The definition of the events E_j^+, E_j^- is slightly different:

$$E_j^+ := E_{q,\rho_j}^+(t + 2\log(l-j+1)), E_j^- := E_{q,\rho_j}^-(t + 2\log(l-j+1)),$$

but we still have

$$\mathbb{P}\Big(\bigcap_{j=1}^{l} E_j^+\Big) \ge 1 - 2e^{-t},\ \mathbb{P}\Big(\bigcap_{j=1}^{l} E_j^-\Big) \ge 1 - 2e^{-t}.$$

On the event $\bigcap_{j=1}^{l} E_j^+$, we have

$$\forall j\ \forall\ f \in \mathcal{F}_j,\ n^{1/2}|P_n f - Pf| \le \omega_n(\sigma_P(f))$$
$$+\sqrt{2\,(t+2\log(l-j+1))\left(\sigma_P^2(f)+\frac{2\omega_n(\sigma_P(f))}{\sqrt{n}}\right)}+\frac{t+2\log(l-j+1)}{3\sqrt{n}},$$

which under the assumptions about ω implies that

$$\forall j\ \forall\ f \in \mathcal{F}_j,\ \ \frac{n^{1/2}|P_n f - Pf|}{\omega(\sigma_P(f))}$$
$$\le 1+\sqrt{2(t+2\log(l_n-j+1))\left(\frac{\rho_j^2}{\omega^2(\rho_j)}+\frac{2}{\sqrt{n}\omega(\rho_{j-1})}\right)}+\frac{t+2\log(l-j+1)}{3\sqrt{n}\omega(\sigma_P(f))}.$$

We have

$$\max_{1\le j\le l}\frac{\rho_j}{\omega(\rho_j)}\sqrt{\log(l-j+1)} \le \max_{1\le j\le l}\frac{\rho_j}{\omega(\rho_j)}\sqrt{\log\log_q\frac{s}{\rho_{j-1}}}$$
$$\le C\sup_{u\in(0,1]}\frac{u}{\omega(u)}\sqrt{\log\log\frac{1}{u}} =: K < +\infty$$

(for some constants C, K). Also, for all $j = 1, \dots, l$

$$\log(l-j+1)\frac{1}{\sqrt{n}\omega(\rho_{j-1})} \le \log\log_{q_n}\frac{s_n}{r_n}\frac{1}{\sqrt{n}\omega(r_n)},$$

which is bounded by the conditions. This allows us to easily conclude that on the event $\bigcap_{j=1}^{l} E_j^+$,

$$\forall j\ \forall\ f\in\mathcal{F}_j\ \ \frac{n^{1/2}|P_n f - Pf|}{\omega(\sigma_P(f))} \le K_1 t + K_2$$

with some constants K_1, K_2. Thus

$$\mathbb{P}\left\{\sup_{f\in\mathcal{F}(r_n,s_n]}\frac{n^{1/2}|P_n f - Pf|}{\omega(\sigma_P(f))} \ge K_1 t + K_2\right\} \le 2e^{-t},$$

implying the stochastic boundedness of the sequence in question. This bound also implies the boundedness in L_1 by simply integrating the tail bound. □

The proof of Theorem 5 requires to work out just several more details (including the lower bounds on the supremum) and it will not be given here.

4. Ratios $\frac{P_n f}{Pf}$: uniform LLN

We now turn to the study of

$$\sup_{Pf>r_n^2}\left|\frac{P_n f}{Pf}-1\right|.$$

Assuming that $nr_n^2 \to \infty$, we concentrate on determining necessary and sufficient conditions for the above suprema to converge to 0 in probability. Other types of ratio limit theorems can be studied as well using the methods of the previous sections.

In this section we set $\sigma_P(f) := \sqrt{Pf}$ and use all the notations of Section 2 (such as $\mathcal{F}(r,s]$, for instance). In particular, we need the functions $\psi_{n,q}$ to define the quantity

$$E_{n,q}(r,s] := \sup_{u\in(r,s]} \frac{\psi_{n,q}(u)}{u^2}.$$

Let the sequences r_n, s_n, q_n be as in Section 2 and let

$$E_n := E_{n,q_n}(r_n, s_n].$$

Theorem 6. *Suppose that $nr_n^2 \to \infty$ as $n \to \infty$. Choose $q_n \downarrow 1$ so that*

$$\log \frac{1}{q_n - 1} = o(nr_n^2) \text{ as } n \to \infty.$$

Then the condition $E_n \to 0$ as $n \to \infty$ is necessary and sufficient for

$$\sup_{f\in\mathcal{F}(r_n,s_n]} \left|\frac{P_n f}{Pf} - 1\right| \to 0 \text{ in Pr.}$$

The proof is based on the following lemma.

Lemma 2. *For $t > 0$,*

$$\mathbb{P}\left\{\sup_{f\in\mathcal{F}(r,s]} \left|\frac{P_n f}{Pf} - 1\right| \geq E_{n,q}(r,s] + \sqrt{2\frac{t}{nr^2}(q^2 + 2E_{n,q}(r,s])} + \frac{t}{3nr^2}\right\} \leq \frac{q^2}{q^2-1}\frac{q}{t}e^{-t/q}$$

and

$$\mathbb{P}\left\{\sup_{f\in\mathcal{F}(r,s]} \left|\frac{P_n f}{Pf} - 1\right| \leq q^{-1}\left(E_{n,q}(r,s] - \sqrt{2\frac{t}{nr^2}(1 + 2E_{n,q}(r,s])} - \frac{8t}{3nr^2}\right)\right\} \leq \frac{q^2}{q^2-1}\frac{q}{t}e^{-t/q}.$$

Proof. It is similar to that of Lemma 1 and we use the notations introduced in that proof. The sets E_j^+, E_j^- are now defined as follows:

$$E_j^+ := E_{q,\rho_j}^+(tq^{2(j-1)}),\ E_j^- := E_{q,\rho_j}^-(tq^{2(j-1)}).$$

With this definition, we have

$$\mathbb{P}\Big(\bigcup_{j=1}^{\infty}(E_j^+)^c\Big) \leq \sum_{j=0}^{\infty} e^{-tq^{2j}} = \frac{q^2}{q^2-1}\sum_{j=0}^{\infty} q^{-2j}\exp\{-tq^{2j}\}(q^{2j} - q^{2(j-1)})$$

$$\leq \frac{q^2}{q^2-1}\int_{1/q}^{\infty} x^{-1}\exp\{-tx\}dx \leq \frac{q^2}{q^2-1}\int_{t/q}^{\infty} y^{-1}e^{-y}dy \leq \frac{q^2}{q^2-1}\frac{q}{t}e^{-t/q},$$

and similarly

$$\mathbb{P}\Big(\bigcup_{j=1}^{\infty}(E_j^-)^c\Big) \le \frac{q^2}{q^2-1}\frac{q}{t}e^{-t/q}.$$

On the event $\bigcap_{j=1}^{\infty} E_j^+$, we have

$$\forall j\ \forall f \in \mathcal{F}_j,\ \frac{|P_n f - Pf|}{Pf}$$

$$\le \frac{\psi_{n,q}(\sqrt{Pf})}{Pf} + \sqrt{2\frac{tq^{2(j-1)}}{nPf}\left(q^2 + 2\frac{\psi_{n,q}(\sqrt{Pf})}{Pf}\right)} + \frac{tq^{2(j-1)}}{3nPf}.$$

Since for $f \in \mathcal{F}_j$,

$$q^{2(j-1)}r^2 < Pf \le q^{2j}r^2,$$

we get

$$\forall j\ \forall f \in \mathcal{F}_j,\quad \frac{|P_n f - Pf|}{Pf} \le E_{n,q}(r,s] + \sqrt{2\frac{t}{nr^2}\left(q^2 + 2E_{n,q}(r,s]\right)} + \frac{t}{3nr^2},$$

which proves the first bound. The second one can be proved similarly. □

Proof of Theorem 6. Choose $t_n \to \infty$ so that $t_n = o(nr_n^2)$ and $\log(q_n-1)^{-1} = o(t_n)$. Then

$$\frac{q_n^2}{q_n^2-1}\frac{1}{t_n}e^{-t_n} \to 0$$

and we have

$$q_n^{-1}\left(E_n - \sqrt{2\frac{t_n}{nr_n^2}(1+2E_n)} - \frac{8t_n}{3nr_n^2}\right) \le \sup_{f\in\mathcal{F}(r_n,s_n]}\left|\frac{P_n f}{Pf} - 1\right|$$

$$\le E_n + \sqrt{2\frac{t_n}{nr_n^2}(q_n^2 + 2E_n)} + \frac{t_n}{3nr_n^2}$$

with probability $1 - o(1)$. This immediately implies the result. □

Here are two useful corollaries of Lemma 2 which we will use in section 7.

Corollary 1. *Suppose that $\mathcal{F}$ is a measurable class of functions with values in $[0,1]$ satisfying a.s.*

$$\log N(\mathcal{F}, L_2(P_n), \tau) \le A\tau^{-\alpha}$$

for all $\tau > 0$ and some finite A and $\alpha \in (0,2)$. Then for $n \in \mathbb{N}$, $0 < \varepsilon \le 1$, $q > 1$, $1 \ge \delta \ge \delta_n = (n\varepsilon)^{-2/(\alpha+2)}$, and a constant $C = C(A,\alpha,q)$ depending only on A, α and q,

$$\mathbb{P}\left\{\sup_{f\in\mathcal{F}:\, Pf>\delta}\left|\frac{P_n f}{Pf} - 1\right| \ge C\sqrt{\varepsilon}\right\} \le \frac{q^2}{q^2-1}\frac{2q}{n\varepsilon\delta}e^{-n\varepsilon\delta/(2q)} = \frac{D}{n\varepsilon\delta}e^{-n\varepsilon\delta/(2q)}.$$

Proof. Choosing $r^2 = \delta$ and $t = n\varepsilon\delta/2$ in the first inequality of Lemma 2 gives

$$\mathbb{P}\left\{ \sup_{f\in\mathcal{F}:Pf>\delta} \left| \frac{P_n f}{Pf} - 1 \right| \geq E_{n,q}(\sqrt{\delta},1] + \sqrt{\varepsilon(q^2 + 2E_{n,q}(\sqrt{\delta},1]) + \varepsilon/6} \right\} \leq \frac{q^2}{q^2-1}\frac{2q}{n\varepsilon\delta}e^{-n\varepsilon\delta/(2q)} .$$

To bound $E_{n,q}(\sqrt{\delta},1]$ we argue as follows: By a standard symmetrization inequality, Dudley's entropy bound for Rademacher processes and our random entropy hypothesis

$$\begin{aligned}
\mathbb{E}\|P_n - P\|_{\mathcal{F}(r)} &\leq \frac{K}{\sqrt{n}}\mathbb{E}\left\{ \int_0^{\sqrt{\sup_{f\in\mathcal{F}(r)} P_n(f^2)}} \sqrt{\log N(\mathcal{F}(r), L_2(P_n), \tau)}\, d\tau \right\} \\
&\leq \frac{K}{\sqrt{n}} \int_0^{\sqrt{\mathbb{E}(\sup_{f\in\mathcal{F}(r)} P_n(f^2))}} \sqrt{A\tau^{-\alpha}}\, d\tau \\
&\leq \frac{K}{\sqrt{n}} \int_0^{\sqrt{\mathbb{E}(\sup_{f\in\mathcal{F}(r)} P_n(f))}} \sqrt{A\tau^{-\alpha}}\, d\tau \\
&\leq \frac{K}{\sqrt{n}} \int_0^{\sqrt{\mathbb{E}(\sup_{f\in\mathcal{F}(r)} |(P_n-P)(f)|)+r^2}} \sqrt{A\tau^{-\alpha}} d\tau \\
&\leq \frac{2}{2-\alpha}\frac{K\sqrt{A}}{\sqrt{n}}\left(E\|P_n - P\|_{\mathcal{F}(r)} + r^2\right)^{1/2-\alpha/4} \\
&\leq \frac{B}{\sqrt{n}}\left\{ (E\|P_n - P\|_{\mathcal{F}(r)})^{1/2-\alpha/4} \vee (r^2)^{1/2-\alpha/4} \right\}
\end{aligned}$$

with $B = 2^{3/2-\alpha/4}K\sqrt{A}/(2-\alpha)$, where the second inequality follows from concavity of the integral $\int_0^x h(t)dt$ when h is non-increasing. Thus if $\mathbb{E}\|P_n - P\|_{\mathcal{F}(r)} \leq r^2$, then

$$\mathbb{E}\|P_n - P\|_{\mathcal{F}(r)} \leq \frac{B}{\sqrt{n}} r^{1-\alpha/2} ,$$

while if $\mathbb{E}\|P_n - P\|_{\mathcal{F}(r)} > r^2$, then

$$\mathbb{E}\|P_n - P\|_{\mathcal{F}(r)} \leq \left(\frac{B}{\sqrt{n}}\right)^{4/(2+\alpha)} = \tilde{B}n^{-2/(\alpha+2)} .$$

Combining these bounds yields

$$\mathbb{E}\|P_n - P\|_{\mathcal{F}(r)} \leq \frac{B}{\sqrt{n}} r^{1-\alpha/2} \vee \tilde{B}n^{-2/(\alpha+2)} .$$

By taking $r = \rho_j = \sqrt{\delta_n} q^{2j}$ it follows that

$$\begin{aligned}
\mathbb{E}\|P_n - P\|_{\mathcal{F}(\rho_j)} &\leq \frac{B}{\sqrt{n}}\rho_j^{1-\alpha/2} \vee \tilde{B} n^{-2/(\alpha+2)} \\
&= \left\{B\sqrt{\varepsilon}(n\varepsilon)^{-2/(\alpha+2)} q^{j(2-\alpha)}\right\} \bigvee \left\{\tilde{B}\varepsilon^{2/(\alpha+2)}(n\varepsilon)^{-2/(\alpha+2)}\right\} \\
&\leq M\sqrt{\varepsilon}\delta_n q^{j(2-\alpha)}
\end{aligned}$$

with $M = B \vee \tilde{B}$. Hence it follows that

$$\begin{aligned}
E_{n,q}(\sqrt{\delta}, 1] &= \sup_{\sqrt{\delta}<u\leq 1} u^{-2}\psi_{n,q}(u) \leq \max_{j\geq 1} \rho_{j-1}^{-2}\mathbb{E}\|P_n - P\|_{\mathcal{F}(\rho_j)} \\
&\leq \max_{j\geq 1}\left\{M\sqrt{\varepsilon}q^4 q^{-j(2+\alpha)}\right\} = M\sqrt{\varepsilon}q^{2-\alpha} \leq Mq^{2-\alpha}.
\end{aligned}$$

Combining this bound with the first display of the proof yields the claimed inequality with

$$C \equiv C(A, \alpha, q) = Mq^{2-\alpha} + \sqrt{q^2 + 2Mq^{2-\alpha}} + 1/6.$$

Similarly we can prove the following, using Lemma 2 and either direct computation or the bound in Corollary 3 below (or in Proposition 2.1 of [12]).

Corollary 2. *Suppose that $\mathcal{F}$ is a measurable class of functions with values in $[0,1]$ satisfying a.s.*

$$N(\mathcal{F}, L_2(P_n), \tau) \leq \left(\frac{A}{\tau}\right)^v, \quad 0 < \tau \leq 1,$$

for some $v \geq 1$ and $A \geq 2e\sqrt{v}$ (in particular, this holds if $\mathcal{F}$ is a VC class). Let $q > 1$ and let n, ε, δ and δ_n satisfy

$$\frac{1}{n} \leq \varepsilon \leq 1, \quad \text{and} \quad 1 \geq \delta \geq \delta_n := \frac{v \log \frac{A\sqrt{n\varepsilon}}{\sqrt{v}}}{n\varepsilon}.$$

Then, there exists a universal constant C such that

$$\mathbb{P}\left\{\sup_{f\in\mathcal{F}:\, Pf>\delta}\left|\frac{P_n f}{Pf} - 1\right| \geq Cq^4\sqrt{\varepsilon}\right\} \leq \frac{q^2}{q^2-1}\frac{2q}{n\varepsilon\delta}e^{-n\varepsilon\delta/(2q)} = \frac{D}{n\varepsilon\delta}e^{-n\varepsilon\delta/(2q)}.$$

The constant C can be taken to be five times the constant in Corollary 3 below.

5. An inequality for expected values of empirical processes indexed by VC classes of functions

In the previous sections, either the results themselves or the conditions for their application are in terms of $\psi_{n,q}(u)$, that is, of $\mathbb{E}\|P_n - P\|_{\mathcal{F}(r,s]}$, and therefore require, for their application, of good estimates of the expectation of suprema of empirical processes indexed by general classes of functions. These are often available if the L_2 covering numbers of the classes are under control, as in the case of Vapnik-Červonenkis classes or the classes considered in Corollary 1. Next we give estimates for VC classes that improve on some of the estimates in the literature in that, instead of being in terms of σ and $\|F\|_\infty$, where F is a measurable envelope of the class, they are in terms of σ and $\|F\|_{L_2(P)}$ (when the functions in the class take values between -1 and 1). These estimates will be used in Section 6, and we think they can be useful elsewhere as well.

Let $\mathcal{F}$ be a uniformly bounded class of real valued measurable functions on a probability space $(S, \mathcal{A}, P)$. To be specific, assume the functions in $\mathcal{F}$ take values in $[-1, 1]$ and are centered. Assume also that the class $\mathcal{F}$ is adequately measurable (as described in the introduction) and VC, in particular,

$$N(\mathcal{F}, L_2(Q), \tau) \le \left(\frac{A\|F\|_{L_2(Q)}}{\tau}\right)^v \tag{5.1}$$

for all $0 < \tau < \|F\|_{L_2(Q)}$ and some finite A and v, that we assume $A \ge 2$ and $v \ge 1$ without loss of generality. Here, $1 \ge F \ge \sup_{f\in\mathcal{F}} |f|$ is a measurable envelope of the class $\mathcal{F}$. Let X, X_i, $i \in \mathbb{N}$, be i.i.d. (P) random variables (coordinates on a product probability space), and let P_n be the empirical measure corresponding to the variables X_i, as in previous sections. Let σ^2 be any number such that $\sup_f \mathbb{E}f^2(X) \le \sigma^2 \le \mathbb{E}F^2(X)$. The norm signs without specification will denote sup over the class $\mathcal{F}$. Here is the bound:

Theorem 7. *Under the assumptions in the above paragraph we have that for all* $n \in \mathbb{N}$,

$$\mathbb{E}\left\|\sum_{i=1}^n f(X_i)\right\|_{\mathcal{F}} \le C\left[\sqrt{v}\sqrt{n}\sigma\sqrt{\log\frac{A\|F\|_{L_2(P)}}{\sigma}} \vee v\log\frac{A\|F\|_{L_2(P)}}{\sigma}\right.$$
$$\left.\vee\sqrt{v}\sqrt{n}A\|F\|_{L_2(P)}\exp\left(-\frac{9}{8}n\|F\|^2_{L_2(P)}\right)\right].$$

Proof. The square root trick for probabilities in [14], Lemma 3.3 and its remark - that misses a factor of 8–, give that for all $t \ge 47n\sigma^2$,

$$\mathbb{P}\left\{\left\|\sum_{i=1}^n f^2(X_i)\right\| \ge t\right\} \le \mathbb{E}\left[1 \wedge \left(8\left(\frac{A\|F\|_{L_2(P_n)}}{\sigma}\right)^v e^{-t/16}\right)\right].$$

By concavity of the function $1 \wedge x$ on $[0, \infty)$ and Hölder, we have

$$\mathbb{E}\left[1 \wedge \left(8\left(\frac{A\|F\|_{L_2(P_n)}}{\sigma}\right)^v e^{-t/16}\right)\right]$$
$$= \mathbb{E}\left[1 \wedge \left(8^{1/v}\left(\frac{A\|F\|_{L_2(P_n)}}{\sigma}\right) e^{-t/(16v)}\right)\right]^v$$
$$\le \mathbb{E}\left[1 \wedge \left(8^{1/v}\left(\frac{A\|F\|_{L_2(P_n)}}{\sigma}\right) e^{-t/(16v)}\right)\right]$$

$$\leq 1 \wedge \left(\frac{8^{1/v} A \|F\|_{L_2(P)}}{\sigma} e^{-t/(16v)} \right).$$

Integrating this tail estimate one readily obtains:

Lemma 3. *Let $\mathcal{F}$ be a measurable VC class of P-centered functions taking values between -1 and 1, with $A \geq 2$ and $v \geq 1$ in (5.1). Let $F \geq \sup_{f \in \mathcal{F}} |f|$ be a measurable envelope of the class $\mathcal{F}$ and let $\sup_f \mathbb{E} f^2(X) \leq \sigma^2 \leq \mathbb{E} F^2(X)$. Then, for all $n \in \mathbb{N}$,*

$$\mathbb{E} \left\| \sum_{i=1}^n f^2(X_i) \right\| \leq 135 \left[n\sigma^2 \vee v \log \left(\frac{A \|F\|_{L_2(P)}}{\sigma} \right) \right].$$

Let ε_i, $i \in \mathbb{N}$, be independent Rademacher variables independent from the variables X_j, and let $\mathbb{E}_\varepsilon$ denote conditional expectation given the sequence $\{X_i\}$. The subgaussian entropy bound gives that

$$\mathbb{E}_\varepsilon \left\| \sum_{i=1}^n \varepsilon_i f(X_i)/\sqrt{n} \right\|_{\mathcal{F}} \leq C \int_0^{\left(\| \sum_{i=1}^n f^2(X_i) \| / n \right)^{1/2}} \sqrt{v \log \left(\frac{A \|F\|_{L_2(P_n)}}{\varepsilon} \right)} d\varepsilon$$

for some universal constant C. Since

$$\int_0^{\left(\| \sum_{i=1}^n f^2(X_i) \| / n \right)^{1/2}} \sqrt{v \log \left(\frac{A \|F\|_{L_2(P_n)}}{\varepsilon} \right)} d\varepsilon \leq D \sqrt{v} A \|F\|_{L_2(P_n)},$$

where $D = \int_0^1 \sqrt{\log u^{-1}} du$, the above integral is dominated by

$$\int_0^{\left(\| \sum_{i=1}^n f^2(X_i) \| / n \right)^{1/2}} \sqrt{v \log \left(\frac{2A \|F\|_{L_2(P)}}{\varepsilon} \right)} d\varepsilon$$

$$+ D \sqrt{v} A \|F\|_{L_2(P_n)} I_{[\|F\|_{L_2(P_n)} > 2\|F\|_{L_2(P)}]}.$$

Regarding the second summand, Hölder's inequality followed by Bernstein's exponential inequality give

$$\mathbb{E} \left(\|F\|_{L_2(P_n)} I_{[\|F\|_{L_2(P_n)} > 2\|F\|_{L_2(P)}]} \right) \leq \|F\|_{L_2(P)} \exp \left(-\frac{9}{8} n \|F\|^2_{L_2(P)} \right).$$

For the first summand, we note that, by concavity of the integral of a decreasing function (as in Corollary 1), we have

$$\mathbb{E} \left[\int_0^{\left(\| \sum_{i=1}^n f^2(X_i) \| / n \right)^{1/2}} \sqrt{v \log \left(\frac{2A \|F\|_{L_2(P)}}{\varepsilon} \right)} d\varepsilon \right]$$

$$\leq \int_0^{\left(\mathbb{E} \| \sum_{i=1}^n f^2(X_i) \| / n \right)^{1/2}} \sqrt{v \log \left(\frac{2A \|F\|_{L_2(P)}}{\varepsilon} \right)} d\varepsilon.$$

Now, by regular variation, this integral is dominated by a constant times

$$\sqrt{v}\frac{1}{\sqrt{n}}\left(\mathbb{E}\left\|\sum_{i=1}^n f^2(X_i)\right\|\right)^{1/2}\left(\log\frac{2A\|F\|_{L_2(P)}}{\left(\mathbb{E}\left\|\sum_{i=1}^n f^2(X_i)\right\|/n\right)^{1/2}}\right)^{1/2},$$

which, by the lemma, is in turn dominated by a constant times

$$\frac{1}{\sqrt{n}}\left[\sqrt{n}\sigma\sqrt{v}\sqrt{\log\frac{A\|F\|_{L_2(P)}}{\sigma}}\vee v\log\frac{A\|F\|_{L_2(P)}}{\sigma}\right].$$

Collecting the above bounds and applying a desymmetrization inequality we obtain the desired bound. □

Corollary 3. *If in the previous theorem we also have $n\sigma^2 \geq A$, then there exists a universal constant C such that, for all $n \in \mathbb{N}$,*

$$\mathbb{E}\left\|\sum_{i=1}^n f(X_i)\right\|_{\mathcal{F}} \leq C\left[\sqrt{v}\sqrt{n}\sigma\sqrt{\log\frac{A\|F\|_{L_2(P)}}{\sigma}}\vee v\log\frac{A\|F\|_{L_2(P)}}{\sigma}\right].$$

Proof. It follows from the previous theorem and the inequality

$$\sqrt{\log x} \geq x\exp\left(-\frac{9}{8}x^2\right), \quad x \geq 2,$$

where we take $x = A\|F\|_{L_2(P)}/\sigma \geq A \geq 2$. □

The proof of Theorem 7 substantially modifies the proof of a similar bound (simpler, but with $U = \|F\|_\infty$ instead of $\|F\|_{L_2(P)}$) in [12]. In that proof, an abstract version of the square root trick (due to Ledoux and Talagrand [22]) was used, whereas here, as in [11], we use the Giné and Zinn [13] version of Le Cam's square root trick.

Remark. Bounds on expectations of empirical processes that take into account the norm of the envelope of the class can be obtained also under different assumptions on the entropy (in particular, for instance, in the setting of Corollary 1 of the previous section). We are not presenting these bounds here.

6. Ratio limit theorems for VC classes of functions

In this section we combine the main results from Sections 2–4 with the moment bound in Section 5 in order to obtain analogues for VC classes of functions of some of the results in [2] for classes of sets. In what follows the class $\mathcal{F}$ is assumed to be a measurable VC class of functions (as defined in Section 5) taking values between 0 and 1, and otherwise, we resume the notation set up in Sections 1–4.

Let us fix $q > 1$. For $0 < r < 1$ we define

$$\mathcal{F}(r) := \{f \in \mathcal{F} : \sigma_P(f) \leq r\}, \quad \mathcal{F}_q(r) := \mathcal{F}(r) \setminus \mathcal{F}(r/q),$$

and let $F_{q,r} \le 1$ ($F_r \ge 1$) be a measurable envelope of the class $\mathcal{F}_q(r)$ (resp. $\mathcal{F}(r)$). These localized envelopes will play an important role in ratio limit theorems. The analogue for functions of the 'capacity function' in [2] is precisely

$$g(r) := \frac{q\|F_r\|_{L_2(P)}}{r} \vee 1,$$

however it is more convenient to localize a little more and define $g_q(r)$ as any function on $(0,1]$ such that

$$\frac{q\|F_{q,r}\|_{L_2(P)}}{r} \le g_q(r) \le \frac{q}{r}.$$

The following result is a version of Theorem 3.1, case (ii), in [2] for functions.

Theorem 8. *Let $\mathcal{F}$ be a measurable VC class of functions taking values on $[0,1]$ and let q be any number larger than 1. Define*

$$r_n := \sup\left\{r > 0 :\ r \le \sqrt{\frac{\log g_q(r) \vee \log\log n}{n}}\right\}$$

and

$$b_n := \sqrt{\log g_q(r_n) \vee \log\log n}.$$

Then, the sequence

$$\sup_{f\in\mathcal{F},\sigma_P(f)>r_n} \frac{n^{1/2}|P_n f - Pf|}{b_n\sigma_P(f)}, \quad n \in \mathbb{N},$$

is stochastically bounded. If moreover the sequence b_n/n is nonincreasing, then there is $R < \infty$ such that

$$\limsup_{n\to\infty} \sup_{f\in\mathcal{F},\sigma_P(f)>r_n} \frac{n^{1/2}|P_n f - Pf|}{b_n\sigma_P(f)} = R \quad a.s.$$

Proof. It suffices to show that r_n and $\beta_n := Kb_n/\sqrt{n}$ for $K < \infty$ to be specified later, satisfy the conditions of the second remark following Theorem 3 for stochastic boundedness, and of the fourth for a.s. boundedness. The definitions readily imply that

$$r_n \ge \sqrt{\frac{\log\log n}{n}},$$

which immediately gives

$$\sqrt{\frac{\log\log r_n^{-1} \vee \log\log n}{n}} \bigvee \frac{\log\log n}{nr_n} = O\left(b_n/\sqrt{n}\right).$$

So, it remains to verify that

$$Kb_n/\sqrt{n} \ge \beta_{n,q}(r_n, 1]$$

for some $K < \infty$ and from some n on. If $r > r_n$, then for some $K < \infty$ and for all $n > \exp(e^{q^2 A^2})$ we have

$$r > \sqrt{\frac{\log g_q(r) \vee \log\log n}{n}} \geq \frac{1}{\sqrt{2}}\sqrt{\frac{\log(Ag_q(r)) \vee \log\log n}{n}} \geq \frac{1}{\sqrt{2}}\sqrt{\frac{\log(Ag_q(r))}{n}},$$

and also $n\sigma^2 \geq nr^2/q^2 > A^2$. Thus, the corollary to Theorem 7 shows that there is a constant K depending only on A and v such that

$$\mathbb{E}\,\|P_n - P\|_{\mathcal{F}_q(r)} \leq Kr\sqrt{\frac{\log(g_q(r))}{n}} \leq \frac{Krb_n}{\sqrt{n}},$$

which gives

$$Kb_n/\sqrt{n} \geq \beta_{n,q}(r_n, 1].$$

The next result is an analogue for VC classes of functions of Theorem 4.4 in [2].

Theorem 9. *Let $\mathcal{F}$ be a measurable VC class of functions taking values on $[0,1]$ and let q be any number larger than 1. Define*

$$r_n := \sup\left\{r > 0 : \ r \leq \sqrt{\frac{\log g_q(r) \vee \log\log r^{-1}}{n}}\right\}$$

and

$$\omega(r) := r\sqrt{\log g_q(r) \vee \log\log_q r^{-1}}.$$

Assume $\omega(r) \nearrow$ and $\omega(r)/r \searrow$. Then, the sequence

$$\sup_{f\in\mathcal{F}.\sigma_P(f)>r_n} \frac{n^{1/2}|P_n f - Pf|}{\omega(\sigma_P(f))}, \quad n \in \mathbb{N},$$

is stochastically bounded.

Proof. We will check the conditions of Theorem 4 for $\bar{\omega} := K\omega$, for $K < \infty$ conveniently chosen, and for $q_n = q > 1$. If $r > r_n$ then, by the definition of r_n,

$$r > \sqrt{\frac{\log g_q(r) \vee \log\log r^{-1}}{n}}$$

and in particular, $r > \sqrt{(\log\log r^{-1})/n}$, which implies that $nr^2 > A^2$ from some n on, n depending only on A. These two observations imply, by the corollary to Theorem 7, that

$$\|P_n - P\|_{\mathcal{F}_q(r)} \leq \frac{K\omega(r)}{\sqrt{n}},$$

that is,

$$\omega_n(r) \leq \bar{\omega}(r), \quad r \in (r_n, 1]$$

if we take $\bar{\omega} = K\omega$. The definition of ω already implies that

$$\frac{u\sqrt{\log\log u^{-1}}}{\omega(u)} \leq 1$$

for all $u \in (0,1)$. The definition of r_n also implies that $r_n \geq \sqrt{(\log\log r_n^{-1})/n}$, and this immediately gives

$$\frac{\log\log r_n^{-1}}{\sqrt{n}\omega(r_n)} \leq 1.$$

Now, the result follows from Theorem 4. □

We conclude with an analogue for functions of part of Theorem 5.1 in [2].

Theorem 10. *Let $\mathcal{F}$ be a measurable VC class of functions taking values on $[0,1]$. Let $r_n \to 0$ and $nr_n^2 \to \infty$, and let $q_n \downarrow 1$ be such that*

$$\log\frac{1}{q_n - 1} = o(nr_n^2).$$

Assume $g_{q_n}(r)$ is nonincreasing. Then, if

$$\frac{\log g_{q_n}(r_n)}{nr_n^2} \to 0,$$

we have

$$\sup_{f\in\mathcal{F}, Pf > r_n^2}\left|\frac{P_n f}{Pf} - 1\right| \to 0 \text{ in Pr.}$$

Proof. We check the conditions of Theorem 6 with $s_n = 1$. We will apply the corollary to Theorem 7 with $\sigma = \sup_f \sqrt{Pf}$, and envelope the square root of the natural envelope, which we can since the functions in $\mathcal{F}$ take values on $[0,1]$. Obviously $nr_n^2 > A^2$ from some $n = n(A)$ on. Then, for these values of n, by the corollary to Theorem 7,

$$\begin{aligned} E_{n,q_n}(r_n, 1] &\leq \sup_{r>r_n}\frac{1}{r^2}\mathbb{E}\left\|P_n - P\right\|_{\mathcal{F}(r/q_n, r)} \\ &\leq K\sup_{r>r_n}\left(\sqrt{\frac{\log Ag_{q_n}(r)}{nr^2}} + \frac{\log Ag_{q_n}(r)}{nr^2}\right) \end{aligned}$$

for some constant K that depends only on A and v. But the monotonicity of g_{q_n} implies that this sup is attained at $r = r_n$, which implies that $E_{n,q_n}(r_n, 1] \to 0$. Now, Theorem 6 implies the result. □

When specialized to VC classes of sets, the last three theorems completely recover the results of [2] mentioned in this section, up to constants. In particular then, one gets the classical results for the empirical distribution function and the empirical measure of intervals when P is uniform on $[0,1]^d$. For example, if $\mathcal{F}_1 = \{I_{[0,a]} : 0 \leq a \leq 1, \prod_{i=1}^d a_i \leq 1/2\}$, (here $a = (a_1, \ldots, a_d)$ and $[0,a] = \{(x_1, \ldots, x_d) : 0 \leq x_i \leq a_i, i = 1, \ldots, d\}$) then we take $\sigma_P^2[0,a] = P[0,a]$, and we find that $F_{q,r}(x) = 1\{x \in [0,1]^d : \prod_{j=1}^d x_j \leq r^2\}$ so, with $X = (X_1, \ldots, X_d)$ and

$X_i \sim$ Uniform$[0,1]$,

$$\begin{aligned}
\|F_{q,r}\|^2_{L_2(P)} &= P(X_1 \cdots X_d \leq r^2) \\
&= P(-\log X_1 - \cdots - \log X_d > -\log(r^2)) \\
&= P(\text{Gamma}(d,1) > -\log(r^2)) \\
&= P(\text{Poisson}(-\log(r^2)) < d) = \sum_{j=0}^{d-1} r^2 \frac{(-\log(r^2))^j}{j!} \\
&\sim r^2 \frac{(2\log(1/r))^{d-1}}{(d-1)!} \qquad \text{as } r \downarrow 0\,.
\end{aligned}$$

So, we have $g_q(r) \simeq (2\log r^{-1})^{(d-1)/2}/\sqrt{(d-1)!}$, $r_n \simeq \sqrt{(\log\log n)/n}$ in Theorems 8 and 9, $b_n \simeq \sqrt{\log\log n}$ and $\omega(r) \simeq r\sqrt{\log\log r^{-1}}$. For Theorem 10, we can take any r_n such that $nr_n^2 \to \infty$ for $d = 1$ (thus recovering a result of [35]), and such that $nr_n^2/\log\log n \to \infty$ for $d > 1$. Likewise, if $\mathcal{F}_2 = \{I_{[a,b]} : 0 \leq a_i \leq b_i \leq 1, \prod_{i=1}^d (b_i - a_i) \leq 1/2\}$, then $\|F_{q,r}\|_{L_2(P)} = 1$, $g_q(r) \simeq 1/r$, $r_n \simeq \sqrt{(\log n)/n}$ in Theorems 8 and 9, $b_n \simeq \sqrt{\log n}$ and $\omega(r) \simeq r\sqrt{\log r^{-1}}$; and for Theorem 10, we can take any r_n such that $nr_n^2/\log n \to \infty$.

The results in this section and the previous examples illustrate one of the main points of this article, namely, that very general theorems, that apply to classes of functions that may not even be VC and which have very simple proofs, are sharp (at least up to constants) when specialized to VC classes of sets and functions and, in particular, to the classical settings of distribution functions and the empirical measure of intervals.

7. An Oracle Inequality for Regression via Ratio Bounds

Here we give an application of the ratio bounds in Section 6 to a statistical problem in the setting of nonparametric regression. The type of inequality we prove in this section provides an 'in probability' type of 'oracle inequality' for a simple version of this type of problem. For a nice introduction to oracle inequalities more generally, see [16]. The flavor of our result here is somewhat akin to the results of [18]. For an example of some L_2−type oracle inequalities see e.g. [19]. Massart in [23] develops a very general framework for oracle inequalities in many statistical problems including regression.

Consider the following regression model:

$$Y_i = f_0(X_i) + \xi_i$$

where the variables X_i are i.i.d. with law P, f_0 is a bounded measurable function and the variables ξ_i's are i.i.d. $N(0,1)$ (other distributions are possible), independent from the variables X_j. For a class of functions $\mathcal{F}$ define

$$\widehat{f}_n \equiv \operatorname{argmin}_{f\in\mathcal{F}} n^{-1}\sum_{j=1}^{n}(Y_j - f(X_j))^2 ,$$

and

$$\overline{f} \equiv \operatorname{argmin}_{f\in\mathcal{F}} P(f-f_0)^2 ,$$

(assuming, for simplicity, the existence of the argmins). Since the only norms occurring in this section are $L_2(P)$ norms, we set, from here on, $\|\cdot\| = \|\cdot\|_{L_2(P)}$.

Theorem 11. *Suppose that $\mathcal{F}$ is a measurable class of functions taking values in $[0,1]$, and with $L_2(P_n)$ metric entropies bounded a.s. by $A\tau^{-\alpha}$ with $0<\alpha<2$, as in Corollary 1, and let $C = C(2^\alpha A, \alpha, 3/2)$ be as defined in this corollary. Then there exist constants $C_i = C_i(A,\alpha)$, $i=1,2,3$, depending only on A and α, such that for all $n\in\mathbb{N}$ and $\varepsilon\in(0,1/3]$ satisfying $\log\log_{3/2}(n\varepsilon^2/C^2)^{1/(2+\alpha)} \le (n\varepsilon^2/C^2)^{2\alpha/(2+\alpha)}$, the bound*

$$\|\widehat{f}_n - f_0\|^2 \le \frac{1+2\varepsilon}{1-2\varepsilon}\|\overline{f}-f_0\|^2 + \frac{C_1(A,\alpha)}{\varepsilon(n\varepsilon^2)^{2/(2+\alpha)}}$$

holds with probability at least $1-\tau_n$, where

$$\tau_n = C_2\exp\left\{-C_3(n\varepsilon^2)^{\alpha/(2+\alpha)}\right\}.$$

Proof. Let $C = C(2^\alpha A,\alpha,3/2)$ and $D = D(3/2) = 27/5$ be constants as defined in Corollary 1 and, given $0<\varepsilon\le 1/3$, set $\delta_n := (n\varepsilon^2/C^2)^{-2/(2+\alpha)}$. On the event

$$L_n := \left\{\|\widehat{f}_n - \overline{f}\|^2 < \delta_n\right\}$$

we have

$$\|\widehat{f}_n - f_0\| \le \sqrt{\delta_n} + \|\overline{f}-f_0\| .$$

This yields, using $ab\le(a^2+b^2)/2$,

$$\begin{aligned}\|\widehat{f}_n - f_0\|^2 &\le \|\overline{f}-f_0\|^2 + 2\sqrt{\delta_n}\|\overline{f}-f_0\| + \delta_n\\ &= \|\overline{f}-f_0\|^2 + 2\sqrt{\frac{\delta_n}{\varepsilon}}\sqrt{\varepsilon}\|\overline{f}-f_0\| + \delta_n\\ &\le (1+\varepsilon)\|\overline{f}-f_0\|^2 + \left(1+\frac{1}{\varepsilon}\right)\delta_n . \end{aligned} \tag{7.1}$$

So, if $\mathbb{P}(L_n)\ge 1-\tau_n$, the theorem is proved. Otherwise, we must look at $\|\widehat{f}_n - f_0\|$ on L_n^c. To this end, we first note that, letting P and P_n denote respectively the law of (X,ξ) and the empirical measure of the variables (X_i,ξ_i), $i=1,\dots,n$, when both types of variables occur (and only the law of X and the empirical measure of the X_i when ξ or ξ_i are not present),

$$\begin{aligned}\|\widehat{f}_n - f_0\|^2 &= \|\overline{f} - f_0\|^2 + P\left[(\widehat{f}_n - Y)^2 - (\overline{f} - Y)^2\right] \\ &\le \|\overline{f} - f_0\|^2 + (P - P_n)\left[(\widehat{f}_n - Y)^2 - (\overline{f} - Y)^2\right],\end{aligned}$$

where the identity follows from the orthogonality with respect to P between ξ and any square integrable function of X, and the inequality holds because $\widehat{f}_n$ minimizes the empirical squared error. The idea is to apply the ratio bound in Corollary 1 to the last term, but we cannot apply it directly (since the class of functions involved is not bounded because of the Gaussian noise) and some modifications are needed. Note that

$$(\widehat{f}_n - Y)^2 - (\overline{f} - Y)^2 = (\widehat{f}_n - f_0)^2 - (\overline{f} - f_0)^2 - 2(\widehat{f}_n - \overline{f})\xi,$$

and therefore, that

$$\begin{aligned}(P - P_n)\left[(\widehat{f}_n - Y)^2 - (\overline{f} - Y)^2\right] &\le (P - P_n)\left[(\widehat{f}_n - f_0)^2 \vee \delta_n - (\overline{f} - f_0)^2 \vee \delta_n\right] \\ &\quad +2P_n\left[(\widehat{f}_n(x) - \overline{f}(x))y\right] + 2\delta_n.\end{aligned}$$

If we apply Corollary 1 for the class

$$\mathcal{G} \equiv \{(f - f_0)^2 \vee \delta_n : \ f \in \mathcal{F}\}$$

to the first term and with $q = 3/2$ (note that the $L_2(Q)$ entropies of $\mathcal{G}$ are dominated by $\overline{A}\tau^{-\alpha} = 2^\alpha A\tau^{-\alpha}$), we obtain that the two bounds

$$(P - P_n)\left((\widehat{f}_n - f_0)^2 \vee \delta_n\right) \le \varepsilon\|\widehat{f}_n - f_0\|^2 + \varepsilon\delta_n$$

and

$$(P_n - P)\left((\overline{f}_n - f_0)^2 \vee \delta_n\right) \le \varepsilon\|\overline{f}_n - f_0\|^2 + \varepsilon\delta_n$$

hold together with probability at least

$$1 - \frac{C^2 D}{n\varepsilon^2\delta_n}\exp\left\{-\frac{n\varepsilon^2\delta_n}{3C^2}\right\}.$$

We thus conclude that the inequality

$$(1 - \varepsilon)\|\widehat{f}_n - f_0\|^2 \le (1 + \varepsilon)\|\overline{f} - f_0\|^2 + \frac{2}{n}\sum_{i=1}^n \xi_i(\widehat{f}_n - \overline{f})(X_i) + 2(1 + \varepsilon)\delta_n \quad (7.2)$$

holds with probability at least

$$1 - \frac{C^2 D}{n\varepsilon^2\delta_n}\exp\left\{-\frac{n\varepsilon^2\delta_n}{3C^2}\right\}.$$

Now we need to bound the second term on L_n^c. First we consider

$$A_n := \frac{2}{n}\sum_{j=1}^n \xi_j(\widehat{f}_n(X_j) - \overline{f}(X_j))I\left(P_n(\widehat{f}_n - \overline{f})^2 \le \delta_n\right).$$

We have

$$|A_n| \leq \frac{2}{\sqrt{n}} \sup_{f,g\in\mathcal{F},P_n(f-g)^2\leq\delta_n} \left| \frac{1}{\sqrt{n}} \sum_{i=1}^n \xi_i(f(X_i) - g(X_i)) \right|,$$

and by the Gaussian entropy bound, if we let $\mathbb{E}_\xi$ denote conditional expectation given the variables X_i,

$$\begin{aligned}
\mathbb{E}_\xi|A_n| &\leq \frac{2K}{\sqrt{n}} \int_0^{\sqrt{\delta_n}} \sqrt{\log N(\mathcal{F} - \mathcal{F}, L_2(P_n), \tau)}d\tau \\
&\leq \frac{2K}{\sqrt{n}} \int_0^{\sqrt{\delta_n}} \sqrt{\log N^2(\mathcal{F}, L_2(P_n), \tau/2)}d\tau \\
&\leq \frac{\kappa'(A,\alpha)}{\sqrt{n}} \int_0^{\sqrt{\delta_n}} \sqrt{\tau^{-\alpha}}\, d\tau \\
&\leq \frac{\kappa(A,\alpha)\delta_n^{(1-\alpha/2)/2}}{\sqrt{n}},
\end{aligned}$$

where K is a universal constant and κ and κ' are constants that depend only on A and α, and we can assume $\kappa(A,\alpha) \geq 1$. Then, the Borell–Sudakov–Tsirel'son inequality (cf. [22], page 57, comments following (3.2) or [33], Proposition A.2.1 and its proof) gives

$$\mathbb{P}_\xi \left\{ |A_n| \geq \frac{\kappa(A,\alpha)\sqrt{\delta_n^{1-\alpha/2}}}{\sqrt{n}} + \sqrt{\frac{8\delta_n t}{n}} \right\} \leq e^{-t},$$

so that, taking $t = \kappa^2(A,\alpha)\delta_n^{-\alpha/2}$, we obtain that

$$|A_n| \leq 4\kappa(A,\alpha)\sqrt{\frac{\delta_n^{1-\alpha/2}}{n}} \tag{7.3}$$

with probability at least

$$1 - \exp\left\{-\kappa^2(A,\alpha)\delta_n^{-\alpha/2}\right\} = 1 - \exp\left\{-\kappa^2(A,\alpha)(n\varepsilon^2/C^2)^{\alpha/(2+\alpha)}\right\}.$$

Let now

$$B_n := \frac{2}{n}\sum_{j=1}^n \xi_j(\widehat{f}_n(X_j) - \overline{f}(X_j))I\left(P_n(\widehat{f}_n - \overline{f})^2 > \delta_n\right),$$

which we decompose as

$$B_n = \left\{P_n(\widehat{f}_n - \overline{f})^2\right\}^{1/2} \frac{\frac{2}{n}\sum_{i=1}^n \xi_i(\widehat{f}_n - \overline{f})(X_i)}{\left\{P_n(\widehat{f}_n - \overline{f})^2\right\}^{1/2}} I\left(P_n(\widehat{f}_n - \overline{f})^2 > \delta_n\right).$$

By using the ratio bound again (Corollary 1), it follows that the intersection of L_n^c with the event that

$$\left\{P_n(\widehat{f}_n - \overline{f})^2\right\}^{1/2} \leq (1+\varepsilon)^{1/2}\|\widehat{f}_n - \overline{f}\|$$

has probability at least

$$\mathbb{P}(L_n^c) - (C^2D/(n\varepsilon^2\delta_n))\exp\left\{-n\varepsilon^2\delta_n/(3C^2)\right\}.$$

So, the intersection of L_n^c with the event that

$$B_n \leq \frac{2(1+\varepsilon)^{1/2}\|\widehat{f}_n - \overline{f}\|}{\sqrt{n}} \cdot \sup_{\substack{f,g\in\mathcal{F} \\ P_n(f-g)^2>\delta_n}} \frac{\sum_{i=1}^n \xi_i(f-g)(X_i)/\sqrt{n}}{\{P_n(f-g)^2\}^{1/2}}$$

also has at least this probability. To estimate

$$C_n := \sup_{\substack{f,g\in\mathcal{F} \\ P_n(f-g)^2>\delta_n}} \frac{\sum_{i=1}^n \xi_i(f-g)(X_i)/\sqrt{n}}{\{P_n(f-g)^2\}^{1/2}}$$

we proceed as in Lemmas 1 and 2, and we isolate the computation:

Lemma 4. *Let $\xi := (\xi_1, \ldots, \xi_n)$ be $N(0, I)$ and let*

$$\mathcal{A}_n := \left\{\left(\frac{(f-g)(X_i)/\sqrt{n}}{\{P_n(f-g)^2\}^{1/2}} : i = 1, \ldots, n\right) : f, g \in \mathcal{F}, P_n(f-g)^2 > \delta_n\right\}.$$

Then, we have that, for the constant $\kappa(A,\alpha)$ in the bound (7.3) for A_n and for any $q > 1$,

$$\sup_{a\in\mathcal{A}_n} |\langle a, \xi\rangle| \leq 2\kappa(A,\alpha)q\delta_n^{-\alpha/4}$$

with probability at least $1 - 2\exp\left\{-\kappa^2(A,\alpha)q^2\delta_n^{-\alpha/2} + 2\log\log_q \delta_n^{-1/2}\right\}$.

Proof of the Lemma. Set $r_n = \sqrt{\delta_n}$, let $q > 1$ be such that $r_n q^{l_n} = 1$, l_n a positive integer, and, for $r \geq r_n$, let

$$\mathcal{A}_n(r) := \left\{a \in \mathcal{A}_n : \sqrt{\mathbb{E}_\xi\langle a, \xi\rangle^2} \in (r/q, r]\right\}.$$

Then, by the Gaussian entropy bound we have, as above,

$$\begin{aligned} \mathbb{E}_\xi \|\langle a, \xi\rangle\|_{\mathcal{A}_n(r)} &\leq \frac{q}{r}\mathbb{E}_\xi\left(\left|\sum_{i=1}^n \xi_i(f-g)(X_i)/\sqrt{n}\right| I\left(P_n(f-g)^2 \leq r^2\right)\right) \\ &\leq \kappa(A,\alpha)qr^{-\alpha/2}/2, \end{aligned}$$

Hence, again by Borell–Sudakov–Tsirel'son,

$$\mathbb{P}_\xi\left\{\|\langle a, \xi\rangle\|_{\mathcal{A}_n(r)} > \kappa(A,\alpha)qr^{-\alpha/2}/2 + \sqrt{2t}\right\} \leq e^{-t}.$$

So, if

$$E_j^+(t) = \left\{\|\langle a, \xi\rangle\|_{\mathcal{A}_n(r_n q^j)} \le \kappa(A,\alpha) q (r_n q^j)^{-\alpha/2}/2 + \sqrt{2(t + 2\log j)}\right\},$$

we have

$$\mathbb{P}\left(\bigcap_{j=1}^{\infty} E_j^+(t)\right) \ge 1 - 2e^{-t},$$

and, on the event $\bigcap_{j=1}^{\infty} E_j^+(t)$,

$$\forall j, \forall a \in \mathcal{A}_n(r_n q^j), \quad |\langle a, \xi\rangle| \le C(A,\alpha) q r_n^{-\alpha/2}/2 + \sqrt{2\left(t + 2\log\log_q \frac{1}{r_n}\right)}$$

(see the proof of Lemma 1). If we now take

$$t = \kappa^2(A,\alpha) q^2 r_n^{-\alpha} - 2\log\log_q r_n^{-1},$$

the lemma follows (we are making the tacit assumption that this quantity is positive; if it is not positive, the lemma is true but meaningless). □

Thus, we conclude that, at least with the large probability prescribed by the lemma, we have (with $q = 3/2$)

$$C_n \le 3\kappa(A,\alpha)\delta_n^{-\alpha/4}.$$

Hence, the probability of the intersection of L_n^c with the event

$$B_n \le \frac{6\kappa(A,\alpha)(1+\varepsilon)^{1/2}}{n^{1/2}\delta_n^{\alpha/4}} \|\widehat{f}_n - \overline{f}\| \tag{7.4}$$

has probability at least

$$\mathbb{P}(L_n^c) \; - \; (D/(n\varepsilon^2/C^2)^{\alpha/(2+\alpha)}) \exp\left\{-(n\varepsilon^2/C^2)^{\alpha/(2+\alpha)}/3\right\}$$
$$-2\exp\left\{-\frac{9}{4}\kappa^2(A,\alpha)(n\varepsilon^2/C^2)^{\alpha/(2+\alpha)} + 2\log\log_{3/2}(n\varepsilon^2/C^2)^{1/(2+\alpha)}\right\},$$

where we have replaced δ_n by its value $(n\varepsilon^2/C^2)^{-2/(2+\alpha)}$.

Collecting the bounds (7.2)–(7.4) together with their probabilities, and using $\|\widehat{f}_n - \overline{f}\| \le \|\widehat{f}_n - f_0\| + \|\overline{f} - f_0\|$, we obtain that the intersection of L_n^c with the event that

$$\begin{aligned}
&(1-\varepsilon)\|\widehat{f}_n - f_0\|^2 \\
&\quad \le (1+\varepsilon)\|\overline{f} - f_0\|^2 + \frac{2(1+\varepsilon)}{(n\varepsilon^2/C^2)^{2/(2+\alpha)}} + 4\kappa(A,\alpha)\frac{\varepsilon/C}{(n\varepsilon^2/C^2)^{2/(2+\alpha)}} \\
&\qquad + 6\kappa(A,\alpha)\frac{\varepsilon(1+\varepsilon)^{1/2}/C}{(n\varepsilon^2/C^2)^{1/(2+\alpha)}}\left(\|\widehat{f}_n - f_0\| + \|\overline{f} - f_0\|\right)
\end{aligned}$$

has probability at least

$$\mathbb{P}(L_n^c) - 2(D/(n\varepsilon^2/C^2)^{\alpha/(2+\alpha)}) \exp\left\{-(n\varepsilon^2/C^2)^{\alpha/(2+\alpha)}/3\right\}$$
$$-3\exp\left\{-\kappa^2(A,\alpha)(n\varepsilon^2/C^2)^{\alpha/(2+\alpha)} + 2\log\log_{3/2}(n\varepsilon^2/C^2)^{1/(2+\alpha)}\right\}.$$

Since D is a constant and C depends only on A and α and can be taken to be at least 1, it is clear that, under the assumption in the theorem about n and ε, we can find $C_2(A,\alpha)$ and $C_3(A,\alpha)$ such that the above probability is at least $\mathbb{P}(L_n^c) - \tau_n$, with τ_n as in the statement of the theorem. Combining with the bound (7.10), that holds on L_n, we obtain that

$$\begin{aligned}
&(1-\varepsilon)\|\widehat{f}_n - f_0\|^2 \\
&\le (1+\varepsilon)\|\overline{f} - f_0\|^2 + \frac{1+1/\varepsilon}{(n\varepsilon^2/C^2)^{2/(2+\alpha)}} + 4\kappa(A,\alpha)\frac{\varepsilon/C}{(n\varepsilon^2/C^2)^{2/(2+\alpha)}} \\
&\quad + 6\kappa(A,\alpha)\frac{\varepsilon(1+\varepsilon)^{1/2}/C}{(n\varepsilon^2/C^2)^{1/(2+\alpha)}}\left(\|\widehat{f}_n - f_0\| + \|\overline{f} - f_0\|\right)
\end{aligned}$$

holds with probability at least $1-\tau_n$. Using $ab \le (a^2+b^2)/2$, $0 < \varepsilon \le 1/3$, and collecting terms, the above inequality implies the following one:

$$\begin{aligned}
(1-2\varepsilon)\|\widehat{f}_n - f_0\|^2 &\le (1+2\varepsilon)\|\overline{f} - f_0\|^2 \\
&\quad + \left(\varepsilon\lambda(A,\alpha) + \frac{2C^{4/(2+\alpha)}}{\varepsilon}\right)\frac{1}{(n\varepsilon)^{2/(2+\alpha)}},
\end{aligned}$$

where

$$\lambda(A,\alpha) := 4\kappa(A,\alpha)C^{(2-\alpha)/(2+\alpha)} + 48\kappa^2(A,\alpha)C^{-2\alpha/(2+\alpha)}.$$

Dividing both sides by $1-2\varepsilon \ge 1/3$, the bound in the theorem follows e.g. for $C_1(A,\alpha) = \lambda(A,\alpha)/9 + 2C^{4/(2+\alpha)}/3$. □

The rate prescribed in Theorem 11 obviously depends on the complexity of the class, in particular, a better rate obtains for VC type classes. Using Corollary 2 instead of Corollary 1 in the above proof, and taking

$$\delta_n = \frac{\log\frac{(2/3)^4 A\sqrt{n}\varepsilon}{C\sqrt{v}}}{(2/3)^8 n\varepsilon^2/(vC^2)}$$

gives the following:

Theorem 12. *Let $\mathcal{F}$ be a measurable class of functions taking values in $[0,1]$ satisfying the entropy condition of Corollary 2 with $v \ge 1$ and $A \ge 2\sqrt{ev}$. Let n and $C^2(3/2)^8/n < \varepsilon^2 \le 1/9$ be such that $\delta_n \le 1/(2A^2)$, and assume*

$\log \delta_n^{-1} > \log\log_{3/2} \delta_n^{-1/2}$. *Then, there exist constants* $C_i = C_i(A, v) > 1$, $i = 1, 2, 3$, *depending only on* A *and* v, *such that*

$$\|\widehat{f}_n - f_0\|^2 \le \frac{1+2\varepsilon}{1-2\varepsilon}\|\overline{f} - f_0\|^2 + C_1 \frac{\log(n\varepsilon^2)}{n\varepsilon^3},$$

with probability at least $1 - C_2(n\varepsilon^2)^{-C_3}$.

Acknowledgement. We are grateful to Olivier Bousquet for several interesting exchanges on different aspects of this article and for making his manuscript [9] available to us.

References

[1] Alexander, K.S., *Rates of growth for weighted empirical processes.* In: Proc. of the Berkeley Conference in Honor of Jerzy Neyman and Jack Kiefer, Volume II, L. Le Cam and R. Olshen (Eds.) (1985), Wadsworth, Inc., 475–493.

[2] Alexander, K.S., *Rates of Growth and Sample Moduli for Weighted Empirical Processes Indexed by Sets.* Probability Theory and Related Fields **75** (1987), 379–423.

[3] Alexander, K.S., *The Central Limit Theorem for Weighted Empirical Processes Indexed by Sets.* J. Multivariate Analysis **22** (1987), 313–339.

[4] Bartlett, P. and Lugosi, G., *An inequality for uniform deviations of sample averages from their means.* Statistics and Probability Letters **44** (1999), 55–62.

[5] Bartlett, P., Bousquet, O. and Mendelson, S. *Localized Rademacher Complexities.* In: Computational Learning Theory, Lecture Notes in Artificial Intelligence (2002), Springer, 44–58.

[6] Bercu, B., Gassiat, E., and Rio, E. *Concentration inequalities, large and moderate deviations for self-normalized empirical processes.* Annals of Probability **30**, 1576–1604.

[7] Bousquet, O. *Concentration Inequalities and Empirical Processes Theory Applied to the Analysis of Learning Algorithms.* Ph.D. Thesis, Ecole Polytechnique, Paris (2002).

[8] Bousquet, O. *A Bennett concentration inequality and its applications to empirical processes.* C.R. Acad. Sci. Paris, Ser. I **334** (2002), 495–500.

[9] Bousquet, O. Concentration Inequalities for Sub-Additive Functions Using the Entropy Method (2003). This volume, 217–252.

[10] Bousquet, O., Koltchinskii, V. and Panchenko, D. *Some Local Measures of Complexity of Convex Hulls and Generalization Bounds.* In: Computational Learning Theory, Lecture Notes in Artificial Intelligence (2002), Springer, 59–73.

[11] Einmahl, U. and Mason, D., *An empirical process approach to the uniform consistency of kernel type function estimators.* J. Theor. Probab. **13** (2000), 1–37.

[12] Giné, E. and Guillou, A., *On consistency of kernel density estimators for randomly censored data: rates holding uniformly over adaptive intervals.* Ann. I. H. Poincaré **4** (2001), 503–522.

[13] Giné, E. and Zinn, J., *Some limit theorems for empirical processes.* Ann. Probab. **12** (1984), 929–989.

[14] Giné, E. and Zinn, J., *Lectures on the central limit theorem for empirical processes.* Probability and Banach Spaces, Zaragoza, Spain. Lecture Notes in Math., **1221** (1986), 50–113.

[15] Haussler, D., *Decision theoretic generalizations of the PAC model for neural nets and other learning applications.* Information and Computation **100** (1992), 78–150.

[16] Johnstone, I. M. *Oracle inequalities an nonparametric function estimation.* Proceedings of the International Congress of Mathematicians. Vol. III, Berlin (1998). Doc. Math. **1998**, Extra Vol. III, 267–278.

[17] Klein, T. *Une inégalité de concentration á gauche pour les processus empiriques.* C. R. Acad. Sci. Paris, Ser. I, (2002), **334**:500–505.

[18] Kneip, A. *Ordered linear smoothers.* Ann. Statist. **22**, 835–866.

[19] Kohler, M. *Inequalities for uniform deviations of averages from expectations with applications to nonparametric regression.* J. of Statistical Planning and Inference **89** (2000), 1–23.

[20] Koltchinskii, V. and Panchenko, D. *Empirical Margin Distributions and Bounding the Generalization Error of Combined Classifiers.* Ann. Statist. **30** (2002), 1–50.

[21] Koltchinskii, V. *Bounds on Margin Distributions in Learning Problems.* Ann. Inst. H. Poincaré, (2003) to appear.

[22] Ledoux, M. and Talagrand, M., *Probability in Banach spaces.* Springer, New York, 1991.

[23] Massart, P. *Some applications of concentration inequalities in statistics.* Annales de la Faculté des Sciences de Tolouse IX (2000), 245–303.

[24] Mendelson, S. *Rademacher averages and phase transitions in Glivenko–Cantelli classes,* IEEE Transactions on Information Theory (2002), to appear.

[25] Panchenko, D. *Concentration inequalities in product spaces and applications to statistical learning theory,* Ph.D. Thesis, University of New Mexico, Albuquerque (2002).

[26] Panchenko, D. *Some extensions of an inequality of Vapnik and Chervonenkis.* Electronic Communic. in Probab. **7** (2002).

[27] Panchenko, D. *Symmetrization Approach to Concentration Inequalities for Empirical Processes.* Ann. Probab., (2003) to appear.

[28] Pollard, D., *Uniform ratio limit theorems for empirical processes* Scandinavian J. Statistics **22** (1995), 271–278.

[29] Rio, E., *Une inégalité de Bennett pour les maxima de processus empiriques* Colloque en l'honneur de J. Bretagnolle, D. Dacunha-Castelle et I. Ibragimov (2001).

[30] Talagrand, M., *Sharper bounds for Gaussian and empirical processes.* Ann. Probab. **22** (1994), 28–76.

[31] Talagrand, M., *New concentration inequalities in product spaces.* Invent. Math. **126** (1996), 505–563.

[32] van de Geer, S.A. *Applications of Empirical Processes Theory.* Cambridge University Press, (2000).

[33] van der Vaart, A. W. and Wellner, J. A. *Weak Convergence and Empirical Processes.* Springer

[34] Vapnik, V.N. *Statistical Learning Theory.* John Wiley & Sons, (1998).

[35] Wellner, J.A. *Limit theorems for the ratio of the empirical distribution function to the true distribution function.* Z. Wahrscheinlichkeitstheor. Verw. Geb. **45** (1978), 108–123.

Departments of Mathematics and Statistics, University of Connecticut, Storrs, CT 06269, USA
E-mail address: gine@uconnvm.uconn.edu

Department of Mathematics and Statistics, University of New Mexico, Albuquerque, NM 87131-1141, USA
E-mail address: vlad@math.unm.edu

University of Washington, Department of Statistics, Box 354322, Seattle, Washington 98195-4322, USA
E-mail address: jaw@stat.washington.edu

Progress in Probability, Vol. 56, 279–298

Asymptotic Distributions of Trimmed Wasserstein Distances Between the True and the Empirical Distribution Function

Erich Haeusler[1] and David M. Mason[2]

Abstract. If the distribution function F has a finite mean, then the Wasserstein distance $d(F_n, F) = \int_{-\infty}^{\infty} |F_n(x) - F(x)|\, dx$ between F and the corresponding empirical distribution function F_n based on a sample of size n converges almost surely to zero as $n \to \infty$. In [6] del Barrio, Giné and Matrán have provided an exhaustive study of the distributional limit theorems associated with this law of large numbers. Nothing can be said about $d(F_n, F) = \infty$ almost surely for all $n \geq 1$ if F has no finite mean. In the present paper we modify $d(F_n, F)$ into a finite quantity for all F by an adaptation of the notion of trimming from statistics, and study the asymptotic distributions of these trimmed Wasserstein distances for appropriate classes of distribution functions F via weighted approximation results for uniform empirical processes.

1. Introduction

The Wasserstein distance between two probability measures P_1 and P_2 on the real line $\mathbb{R}$ with finite means is given by

$$\begin{aligned} d(P_1, P_2) &= \inf\Big\{ \int |x-y|\, d\mu(x,y) : \mu \text{ is a probability measure} \\ &\qquad\qquad \text{on } \mathbb{R}^2 \text{ with marginals } P_1 \text{ and } P_2 \Big\} \\ &= \int_{-\infty}^{\infty} |F_1(x) - F_2(x)|\, dx\,, \end{aligned}$$

where F_1, F_2 are the (right-continuous) distribution functions of P_1, P_2, respectively. The interest in this distance in probabilistic limit theory stems from the fact that it metrizes convergence in distribution together with convergence of the first absolute moments.

2000 *Mathematics Subject Classification.* Primary 60F05; Secondary 60F07, 62G30.
Key words and phrases. Wasserstein distance, empirical distribution functions, trimming, asymptotic distribution.
[1]This work was completed with the support of a NATO Grant. [2] This work was completed with the support of a NATO Grant, NSA Grant MDA904–02–1–0034 and NSF Grant DMS–0203865.

Let $X_i, i \geq 1$, be a sequence of independent and identically distributed random variables with common distribution function F. For every integer $n \geq 1$, let

$$F_n(x) = \frac{1}{n} \sum_{i=1}^{n} 1\{X_i \leq x\}, \quad -\infty < x < \infty,$$

denote the empirical distribution function based on $X_1, \ldots, X_n$. If F has a finite mean, then by the Glivenko-Cantelli theorem and the strong law of large numbers almost surely

$$d(F_n, F) = \int_{-\infty}^{\infty} |F_n(x) - F(x)|\, dx \to 0 \quad \text{as } n \to \infty. \tag{1.1}$$

On the other hand, $d(F_n, F) = \infty$ almost surely for all $n \geq 1$, whenever F does not have finite mean. Recently, in [6] del Barrio, Giné and Matrán have provided an exhaustive study of the distributional limit theorems associated to the law of large numbers (1.1). As it turns out, the natural settings for these limit theorems are the domains of attraction $D(\alpha)$ of α-stable laws with $1 \leq \alpha \leq 2$, assuming a finite mean in the case $\alpha = 1$.

Of course, nothing can be said about $d(F_n, F) = \infty$ when F does not have a finite mean. In this case we may modify $d(F_n, F)$ into a finite quantity by adaptating the notion of trimming from statistics, that is, by cropping off a little bit from the upper and lower tails of F by considering

$$d_T(F_n, F) = \int_{x_{n,-}}^{x_{n,+}} |F_n(x) - F(x)|\, dx \tag{1.2}$$

for appropriate trimming levels $x_{n,-} \to -\infty$ and $x_{n,+} \to \infty$ as $n \to \infty$. We shall show that distributional limit theorems for these *trimmed Wasserstein distances* between F_n and F are available for certain classes of distribution functions F with or *without* finite means. Thus, trimming has the same effect on $d(F_n, F)$ as it has on sums of independent and identically distributed random variables: It allows for distributional limit theorems which are otherwise unavailable. For a comprehensive account of the probabilistic and statistical effects of trimming on the behavior of sums of independent and identically distributed random variables the reader is referred to [7].

A trimmed version of the L_2–Wasserstein distance, where $x_{n,-}$ and $x_{n,+}$ converge to finite or infinite values as $n \to \infty$, has recently received some attention in the statistical literature. Consult, in particular, Munk and Czado [10].

To define the trimming levels in (1.2), introduce the quantile function Q of F , defined for $0 < u < 1$, to be $Q(u) = \inf\{x : F(x) \geq u\}$, and let $k_n, n \geq 1$, be a sequence of positive constants each less than n, satisfying

$$k_n/n \to 0 \quad \text{at the rate} \quad k_n \to \infty. \tag{1.3}$$

Set $x_{n,-} = Q(k_n/n)$ and $x_{n,+} = Q(1 - k_n/n)$. Then $x_{n,-} \to -\infty$ and $x_{n,+} \to \infty$ as $n \to \infty$, as required, provided that Q is unbounded in both tails, which we will assume henceforth to exclude trivial cases. For these trimming levels, we shall rewrite the random variables $d_T(F_n, F)$, appropriately centered and normalized, in terms of a sequence $U_1, U_2, \ldots$ of independent Uniform (0,1) random variables. For every integer $n \geq 1$, let G_n denote the uniform empirical distribution function based on $U_1, \ldots, U_n$, defined to be

$$G_n(t) = n^{-1} \sum_{i=1}^{n} 1\{U_i \leq t\}, \quad 0 \leq t \leq 1,$$

and let u_n be the corresponding uniform empirical process given by

$$u_n(t) = \sqrt{n}(G_n(t) - t), \quad 0 \leq t \leq 1.$$

Then by the usual quantile function representation $X_i =_d Q(U_i)$ and a change of variables for every integer $n \geq 1$ we have

$$\sqrt{n}\{d_T(F_n, F) - E(d_T(F_n, F))\} =_d \int_{k_n/n}^{1-k_n/n} \{|u_n(t)| - E\,|u_n(t)|\}\, dQ\,(t)\,. \quad (1.4)$$

We will use weighted approximation methodology to transfer the problem of deriving the asymptotic distribution of this sequence of integral functionals of the empirical process to finding that of the sequence of integral functionals of the Brownian bridge

$$I_n := \int_{k_n/n}^{1-k_n/n} \{|B_n(t)| - E\,|B_n(t)|\}\, dQ\,(t)\,, \quad (1.5)$$

where for each $n \geq 1$, B_n is a Brownian bridge. This method has found numerous applications in probability and statistics to derive asymptotic distributions; see e.g. [1], [2], [3], [4], [5], [7], [11] and [12]. It will prove its versatility also in the present setting.

The result that will allow us to replace u_n by B_n will be the following weighted approximation result which is contained in Theorem 1.2 of [9]. Before we can state this result we first need to introduce some notation. Assume that on the same probability space on which the sequence of independent Uniform (0,1) random variables sit there exists a sequence $B_n, n \geq 1$, of Brownian bridges. For any $0 \leq \nu < 1/2$, $n \geq 2$, let

$$\Delta_{n,\nu} = \sup_{1/n \leq t \leq 1-1/n} \frac{n^{\nu}|u_n(t) - B_n(t)|}{(t(1-t))^{1/2-\nu}}\,. \quad (1.6)$$

Theorem 1.2 of [9] implies the following result:

Theorem. *There exists a probability space $(\Omega, \mathcal{A}, P)$ with independent Uniform $(0,1)$ random variables $U_1, U_2, \ldots$, and a sequence of Brownian bridges $B_1, B_2, \ldots$, such that for every $0 \leq \nu < 1/2$ there exists a $\gamma > 0$ satisfying*

$$\sup_{n>1} E \exp(\gamma \Delta_{n,\nu}) < \infty\,. \quad (1.7)$$

To formulate our basic replacement result, let $\theta(u)$ be any nonincreasing function defined on $(0,1)$ that is positive for small enough values of $u \in (0,1)$. Assume that for a given quantile function Q

$$\limsup_{u \searrow 0} \frac{u\left\{Q^2(u)+Q^2(1-u)\right\}}{\theta^2(u)} < \infty . \tag{1.8}$$

Then we have

Proposition 1. *Let θ and Q satisfy (1.8), and let $k_n, n \geq 1$, be any sequence of positive constants each less than n, satisfying $k_n/n \to 0$ at the rate $k_n \to \infty$. Then for every $0<\nu<1/2$ there exists a constant $M_\nu > 0$, such that for all large enough $n \geq 1$,*

$$\frac{1}{\theta(k_n/n)} \int_{k_n/n}^{1-k_n/n} |u_n(t) - B_n(t)|\, dQ(t) \leq M_\nu k_n^{-\nu} \Delta_{n,\nu} . \tag{1.9}$$

Proof. We see that for any $0<\nu<1/2$

$$\begin{aligned} &\frac{1}{\theta(k_n/n)} \int_{k_n/n}^{1-k_n/n} |u_n(t) - B_n(t)|\, dQ(t) \\ &\leq \frac{\Delta_{n,\nu}}{n^\nu \theta(k_n/n)} \int_{k_n/n}^{1-k_n/n} (t(1-t))^{1/2-\nu}\, dQ(t) \\ &\leq \frac{\Delta_{n,\nu}}{n^\nu \theta(k_n/n)} \left[\int_{k_n/n}^{1/2} t^{1/2-\nu}\, dQ(t) + \int_{1/2}^{1-k_n/n} (1-t)^{1/2-\nu}\, dQ(t)\right]. \end{aligned}$$

Integrating by parts we get

$$\frac{1}{\theta(k_n/n)} \int_{k_n/n}^{1/2} t^{1/2-\nu}\, dQ(t)$$

$$= \frac{(1/2)^{1/2-\nu} Q(1/2) - (k_n/n)^{1/2-\nu} Q(k_n/n)}{\theta(k_n/n)} - \frac{1/2-\nu}{\theta(k_n/n)} \int_{k_n/n}^{1/2} Q(t) t^{-1/2-\nu}\, dt ,$$

which by using (1.8) is readily shown to be

$$\leq M_\nu^{(1)} (k_n/n)^{-\nu}$$

for some constant $M_\nu^{(1)} > 0$ and all large enough $n \geq 1$. Similarly one can show that

$$\frac{1}{\theta(k_n/n)} \int_{1/2}^{1-k_n/n} (1-t)^{1/2-\nu}\, dQ(t) \leq M_\nu^{(2)} (k_n/n)^{-\nu}$$

for some constant $M_\nu^{(2)} > 0$ and all large enough $n \geq 1$. Setting $M_\nu = M_\nu^{(1)} + M_\nu^{(2)}$ finishes the proof of the proposition. □

The first step in any application of Proposition 1 is to find a normalizing function $\theta(u)$ such that (1.8) is satisfied. One possible choice is provided by Lemma 2.1 in

[3] which was crucial for the proofs of all the trimmed sum central limit theorems in that paper: For any non-degenerate quantile function Q, we have

$$\limsup_{u\searrow 0} \frac{u\left\{Q^2(u)+Q^2(1-u)\right\}}{\sigma^2(u)} < \infty, \tag{1.10}$$

where

$$\sigma^2(u) = \int_u^{1-u}\int_u^{1-u} (s\wedge t - st)\, dQ(s)\, dQ(t).$$

Therefore, $\theta^2 = \sigma^2$ works in Proposition 1 for any non-degenerate quantile function, and we will use this normalizing function in this and the next section. To include also the centering in the replacement, we set

$$\Lambda_n(k_n/n) = \frac{1}{\sigma(k_n/n)} \int_{k_n/n}^{1-k_n/n} \{|u_n(t)| - E|u_n(t)|\}\, dQ(t) \tag{1.11}$$

and

$$\Gamma_n(k_n/n) = \frac{1}{\sigma(k_n/n)} \int_{k_n/n}^{1-k_n/n} \{|B_n(t)| - E|B_n(t)|\}\, dQ(t). \tag{1.12}$$

Now, applying Proposition 1, we immediately get the following corollary:

Corollary 1. *Let Q be any non-degenerate quantile function, and let $k_n, n \geq 1$, be any sequence of positive constants each less than n, satisfying $k_n/n \to 0$ at the rate $k_n \to \infty$. Then for every $0 < \nu < 1/2$ there exists a constant $M_\nu > 0$, such that for all large enough $n \geq 1$,*

$$|\Lambda_n(k_n/n) - \Gamma_n(k_n/n)| \leq M_\nu k_n^{-\nu}\Delta_{n,\nu} + M_\nu k_n^{-\nu} E\Delta_{n,\nu}. \tag{1.13}$$

Note that on the probability space of the Theorem we have $\Delta_{n,\nu} = O_P(1)$ and $E\Delta_{n,\nu} = O(1)$ so that the right hand side of (1.13) converges to zero in probability as $n \to \infty$. Consequently, under the assumptions of Corollary 1, the random variables $\Lambda_n(k_n/n)$ and $\Gamma_n(k_n/n)$ necessarily have the same asymptotic distributions, if any. In Section 2 we shall derive these asymptotic distributions for certain classes of underlying distribution functions F by determining the asymptotic distributions of $\Gamma(k_n/n)$, which together with (1.13) is the core of the weighted approximation methodology. Because of (1.4) these limit theorems are in fact limit theorems for the centered trimmed Wasserstein distances $d_T(F_n, F) - E(d_T(F_n, F))$, properly normalized.

Moreover, note that (1.7) and (1.13) imply that for $0 \leq \nu < 1/2$, a sufficiently large n_0 and some $\gamma > 0$,

$$\sup_{n\geq n_0} E\exp\left(\gamma k_n^\nu |\Lambda_n(k_n/n) - \Gamma_n(k_n/n)|\right) < \infty. \tag{1.14}$$

This fact can be exploited as follows. By recopying steps from the proof of Theorem 5.1 of [6] (also see Proposition 6.2 in [6]), based on Borell's inequality, one gets the following exponential inequality: For all $z > 0$ and any $0 < u < 1/2$ such that $\sigma(u) > 0$,

$$P\left\{\frac{1}{\sigma(u)}\left|\int_u^{1-u}\{|B(t)| - E|B(t)|\}\,dQ(t)\right| > z\right\} \le 2\exp\left(-\frac{2z^2}{\pi^2}\right).$$

Notice that absolutely no assumptions are required on the underlying F. Applying this exponential inequality we get whenever $k_n/n \to 0$ that for some $n_0 \ge 1$ and $\gamma > 0$

$$\sup_{n \ge n_0} E\exp(\gamma|\Gamma_n(k_n/n)|) < \infty, \tag{1.15}$$

and whenever $k_n/n \to 0$ and $k_n \to \infty$, (1.15), in combination with (1.14), gives

$$\sup_{n \ge n_0} E\exp(\gamma|\Lambda_n(k_n/n)|) < \infty. \tag{1.16}$$

Clearly, the bounds (1.15) and (1.16) yield convergence of all positive moments whenever $\Gamma_n(k_n/n)$ and $\Lambda_n(k_n/n)$ converge in distribution.

2. Asymptotic distributions

In this section we shall derive the asymptotic distributions of the random variables $\Gamma_n(k_n/n)$ from (1.12) and $\Lambda_n(k_n/n)$ from (1.11) and hence of $d_T(F_n, F)$, properly centered and normalized, for appropriate classes of underlying distribution functions F. In Subsection 2.1 we shall consider a parametric family of distributions with infinite variances in the domain of attraction of a normal law. For these distributions it has been shown in [6] that $d(F_n, F)$, properly centered and normalized, is asymptotically normal. Relying heavily on [6] we shall verify asymptotic normality also for the trimmed Wasserstein distances $d_T(F_n, F)$, properly centered and normalized. In Subsection 2.2 we shall derive the asymptotic distribution of $d_T(F_n, F)$ for all distributions in the domain of attraction of a non-normal stable law, and in Subsection 2.3 for symmetric distributions with slowly varying tails. In both of these cases the asymptotic distributions turn out to be non-normal. Thus, the trimmed Wasserstein distances exhibit a significantly different behavior than the integrals

$$\int_{k_n/n}^{1-k_n/n} u_n(t)\,dQ(t),$$

which are asymptotically normal for *all* non-degenerate quantile functions, when appropriately normalized; see Lemma 2.2 in [3]. These integrals represent the 'middle portions' of sums of independent and identically distributed random variables in the weighted approximation methodology approach to the asymptotic distribution of sums of i.i.d. random variables. In fact, the asymptotic normality of these middle portions for all non-degenerate quantile functions was a key factor in the investigations in [3] and [4].

2.1. Domain of normal attraction case

Whenever the quantile function Q of a distribution function F satisfies

$$\limsup_{s\searrow 0} sQ^2(s) < \infty \quad \text{or} \quad Q(s) = L_1(s)s^{-1/2} \tag{2.1}$$

and

$$\limsup_{s\searrow 0} sQ^2(1-s) < \infty \quad \text{or} \quad Q(1-s) = L_2(s)s^{-1/2}\,, \tag{2.2}$$

where L_1 and L_2 are slowly varying functions at zero, then F is easily shown to be in the domain of attraction of a normal law. A special case of such quantile functions are those which have a derivative of the form

$$Q'(t) = \frac{\left|\frac{1}{2}\log\left(\frac{t}{1-t}\right)\right|^{\beta/2}}{\left(t(1-t)\right)^{3/2}}, \quad \text{for } 0 < t < 1\,, \tag{2.3}$$

for some β. If $\beta \geq -1$, then Q has infinite variance. We will confine our attention to this case.

We shall prove

Theorem 1. *Whenever Q has a derivative of the form (2.3) with $\beta \geq -1$ and $k_n, n \geq 1$, is a sequence of positive constants each less than n satisfying $k_n/n \to 0$, then*

$$\Gamma_n(k_n/n) \to_d N\left(0, 1 - \frac{2}{\pi}(2 - \log 2)\right), \tag{2.4}$$

and if, in addition, $k_n \to \infty$, then

$$\Lambda_n(k_n/n) \to_d N\left(0, 1 - \frac{2}{\pi}(2 - \log 2)\right). \tag{2.5}$$

Moreover, all the positive moments of $\Gamma_n(k_n/n)$ and $\Lambda_n(k_n/n)$ converge to those of a $N\left(0, 1 - \frac{2}{\pi}(2 - \log 2)\right)$ random variable.

Proof. We repeat here many steps from [6], with some significant differences. Applying the well-known transformation

$$B(t) = \left(t(1-t)\right)^{1/2} V\left(\frac{1}{2}\log\left(\frac{t}{1-t}\right)\right)$$

between the Brownian bridge B and the stationary Ornstein-Uhlenbeck process V, we get that the I_n in (1.5) is equal in distribution to

$$\int_{k_n/n}^{1-k_n/n} \left\{\left|V\left(\frac{1}{2}\log\left(\frac{t}{1-t}\right)\right)\right| - E\left|V\left(\frac{1}{2}\log\left(\frac{t}{1-t}\right)\right)\right|\right\}\left(t(1-t)\right)^{1/2} dQ(t)$$

$$= \int_{-s_n/2}^{s_n/2} \{|V(u)| - E|V(u)|\} \frac{e^u}{1+e^{2u}}\, dQ\left(\frac{e^{2u}}{1+e^{2u}}\right) =: J_n\,,$$

with $s_n = \log\left(\frac{1-k_n/n}{k_n/n}\right)$, where the last equality is based on the transformation $u = \frac{1}{2}\log\left(\frac{t}{1-t}\right)$ and $t = \frac{e^{2u}}{1+e^{2u}}$. Since we assume that Q has derivative Q' as in (2.3), we obtain

$$J_n = 2\int_{-s_n/2}^{s_n/2} \{|V(u)| - E|V(u)|\} |u|^{\beta/2}\, du.$$

Observing that $s_n \to \infty$, we can now apply Theorem 6.8 in [6] to state the asymptotic distribution of I_n. For $\beta > -1$, let

$$\sigma_\beta^2 = \frac{2^{3-\beta}}{1+\beta}\left(1 - \frac{2}{\pi}(2 - \log 2)\right) \tag{2.6}$$

and for $\beta = -1$, set

$$\sigma_{-1}^2 = 16\left(1 - \frac{2}{\pi}(2 - \log 2)\right). \tag{2.7}$$

For $\beta > -1$, we get

$$s_n^{-(\beta+1)/2} J_n \to_d N\left(0, \sigma_\beta^2\right), \tag{2.8}$$

and for $\beta = -1$,

$$(\log s_n)^{-1/2} J_n \to_d N\left(0, \sigma_{-1}^2\right). \tag{2.9}$$

We shall use (2.8) and (2.9) to derive the asymptotic distribution of $\Gamma(k_n/n) = \sigma(k_n/n)^{-1} I_n =_d \sigma(k_n/n)^{-1} J_n$.

Since $Q \in D(2)$ has infinite variance, from (3.4) in [5], we get

$$\sigma^2(u) \sim \int_u^{1-u} Q^2(t)\, dt, \quad \text{as } u \searrow 0.$$

Now routine analysis shows that for our Q,

$$-Q(1-u) \sim Q(u) \sim -2^{1-\beta/2} |\log u|^{\beta/2} u^{-1/2}, \quad \text{as } u \searrow 0,$$

which gives for $\beta > -1$,

$$\sigma(u) \sim \left(\frac{2^{3-\beta}}{1+\beta}\right)^{1/2} |\log u|^{\frac{\beta+1}{2}},$$

and for $\beta = -1$,

$$\sigma(u) \sim 4(\log|\log u|)^{1/2}.$$

Thus by (2.8), for $\beta > -1$, and by (2.9), for $\beta = -1$,

$$\sigma(k_n/n)^{-1} I_n \to_d N\left(0, 1 - \frac{2}{\pi}(2 - \log 2)\right). \tag{2.10}$$

This proves (2.4). (The limiting distributional result (2.10) was proved in [6] for the case $k_n/n = 1/n$. We have repeated their argument in order to make our presentation complete.) Assertion (2.5) follows from (2.4) by applying Corollary 1. Finally by applying (1.15) and (1.16), we see that we have convergence of all positive moments of $\Gamma(k_n/n)$ and $\Lambda_n(k_n/n)$ to those of a $N\left(0, 1 - \frac{2}{\pi}(2 - \log 2)\right)$ random variable. □

Remark. When $\beta = 0$, statement (2.4) is a special case of Lemma 3.4 of Chapter 5 in [2]. Likewise, when $\beta = 0$, assertion (2.5) follows from Theorem 3.3 of Chapter 5 in [2].

2.2. Non-normal domain of attraction case

In this subsection, let the distribution function F be in the domain of attraction of a non-normal stable law. Set $G(x) = P\{|X_1| \le x\}$, for $-\infty < x < \infty$, and let $K(s) = \inf\{x : G(x) \ge s\}$, for $0 < s < 1$. In [1] it is shown that the classical conditions for $F \in D(\alpha)$, with $0 < \alpha < 2$, are equivalent to the following three conditions:

(S.i) $K(1-s) = s^{-1/\alpha} L(s)$ with L slowly varying at zero;

(S.ii) $\lim_{s \searrow 0} Q(s) / K(1-s) = -q^{1/\alpha}$;

(S.iii) $\lim_{s \searrow 0} Q(1-s) / K(1-s) = p^{1/\alpha}$,

with $0 \le p \le 1$ and $q = 1 - p$.

In this case by (3.1) in [5]

$$\lim_{s \searrow 0} \sigma^2(s) / s^{1-2/\alpha} L^2(s) = 2\left(p^{2/\alpha} + q^{2/\alpha}\right) / (2-\alpha) =: c^2(\alpha). \tag{2.11}$$

Define

$$\Upsilon(\alpha) = c_1(\alpha) \int_0^1 \left\{\left|W^{(1)}(t)\right| - E|W(t)|\right\} t^{1/\alpha - 2}\, dt$$
$$+ c_2(\alpha) \int_0^1 \left\{\left|W^{(2)}(t)\right| - E|W(t)|\right\} t^{1/\alpha - 2}\, dt,$$

where W, $W^{(1)}$ and $W^{(2)}$ are independent standard Wiener processes on $[0,1]$ and

$$c_1(\alpha) = \alpha^{-1} c^{-1}(\alpha) q^{1/\alpha} \quad \text{and} \quad c_2(\alpha) = \alpha^{-1} c^{-1}(\alpha) p^{1/\alpha}.$$

Notice that

$$c_1^2(\alpha) + c_2^2(\alpha) = \frac{2-\alpha}{2\alpha^2}.$$

We shall prove

Theorem 2. *Whenever $F \in D(\alpha)$, with $0 < \alpha < 2$, and k_n, $n \ge 1$, is a sequence of positive constants each less than n satisfying $k_n/n \to 0$, then*

$$\Gamma_n(k_n/n) \to_d \Upsilon(\alpha), \tag{2.12}$$

and if, in addition, $k_n \to \infty$, then

$$\Lambda_n(k_n/n) \to_d \Upsilon(\alpha). \tag{2.13}$$

Moreover, all the positive moments of $\Gamma_n(k_n/n)$ and $\Lambda_n(k_n/n)$ converge to those of $\Upsilon(\alpha)$.

Proof. We shall write each Brownian bridge B_n as

$$B_n(s) = W_n(s) - sW_n(1),$$

where W_n is a standard Wiener process on $[0,1]$. Let $W_{n,1}(s) = W_n(s)$ and $W_{n,2}(s) = W_n(s+1/2) - W_n(1/2)$ for $0 \le s \le 1/2$. Clearly $W_{n,1}$ and $W_{n,2}$ are independent standard Wiener processes on $[0,1/2]$. We shall further write each W_n as

$$W_n(s) = \begin{cases} W_{n,1}(s), & \text{for } 0 \le s \le 1/2, \\ W_{n,2}(s-1/2) + W_{n,1}(1/2), & \text{for } 1/2 < s \le 1. \end{cases}$$

For any $\lambda > 1$, set

$$\begin{aligned}(\Upsilon_{n,1}(\lambda), \Upsilon_{n,2}(\lambda)) &= \left(\frac{1}{\sigma(k_n/n)} \int_{k_n/n}^{\lambda k_n/n} \{|B_n(t)| - E|B_n(t)|\}\, dQ(t), \right. \\ &\qquad \left. \frac{1}{\sigma(k_n/n)} \int_{1-\lambda k_n/n}^{1-k_n/n} \{|B_n(t)| - E|B_n(t)|\}\, dQ(t) \right).\end{aligned}$$

The following lemma is the crucial step in the proof.

Lemma 1. *Under the assumptions of Theorem* 2 *for any* $\lambda > 1$,

$$(\Upsilon_{n,1}(\lambda), \Upsilon_{n,2}(\lambda)) \to_d (\Upsilon_1(\alpha,\lambda), \Upsilon_2(\alpha,\lambda)),$$

where for $i = 1, 2$,

$$\Upsilon_i(\alpha,\lambda) = c_i(\alpha) \int_{1/\lambda}^{1} \left\{ \left| W^{(i)}(t) \right| - E|W(t)| \right\} t^{1/\alpha - 2}\, dt.$$

Proof. Set

$$\overline{\Upsilon}_{n,1}(\lambda) = \frac{1}{\sigma(k_n/n)} \int_{k_n/n}^{\lambda k_n/n} \{|W_{n,1}(t)| - E|W_{n,1}(t)|\}\, dQ(t)$$

and

$$\overline{\Upsilon}_{n,2}(\lambda) = \frac{1}{\sigma(k_n/n)} \int_{1-\lambda k_n/n}^{1-k_n/n} \{|W_{n,2}(t)| - E|W_{n,2}(t)|\}\, dQ(t).$$

Note that

$$\left| \Upsilon_{n,1}(\lambda) - \overline{\Upsilon}_{n,1}(\lambda) \right| \le [|W_n(1)| + E|W_n(1)|] \frac{\int_{k_n/n}^{\lambda k_n/n} t\, dQ(t)}{\sigma(k_n/n)}.$$

By (S.i), (S.ii) and (2.11) we have $\int_{k_n/n}^{\lambda k_n/n} t\, dQ(t) / \sigma(k_n/n) \to 0$ as $n \to \infty$. Thus

$$\Upsilon_{n,1}(\lambda) - \overline{\Upsilon}_{n,1}(\lambda) \to_P 0$$

and similarly

$$\Upsilon_{n,2}(\lambda) - \overline{\Upsilon}_{n,2}(\lambda) \to_P 0.$$

Since for each $n \ge 1$, $\overline{\Upsilon}_{n,1}(\lambda)$ and $\overline{\Upsilon}_{n,2}(\lambda)$ are independent, to finish the proof of the lemma it suffices to show that for each $i = 1, 2$,

$$\overline{\Upsilon}_{n,i}(\lambda) \to_d \Upsilon_i(\alpha,\lambda).$$

Note that by a change of variables each

$$\overline{\Upsilon}_{n,1}(\lambda) = \int_1^\lambda \left\{ \left| \sqrt{\frac{n}{k_n}} W_{n,1}(sk_n/n) \right| - E \left| \sqrt{\frac{n}{k_n}} W_{n,1}(sk_n/n) \right| \right\} d\mu_n(s),$$

where

$$d\mu_n(s) = \frac{\sqrt{\frac{k_n}{n}}\, dQ(sk_n/n)}{\sigma(k_n/n)}, \tag{2.14}$$

which in turn is equal in distribution to

$$\int_1^\lambda \{|W(s)| - E|W(s)|\}\, d\mu_n(s).$$

Using elementary properties of regularly varying functions, in combination with (S.ii) and (2.11), it is routine to verify that the measure defined by $d\mu_n$ on $[1,\lambda]$ converges weakly to the measure μ on $[1,\lambda]$ given by

$$d\mu(s) = \alpha^{-1} c^{-1}(\alpha)\, q^{1/\alpha} s^{-1-1/\alpha}\, ds = c_1(\alpha)\, s^{-1-1/\alpha}\, ds.$$

Thus

$$\overline{\Upsilon}_{n,1}(\lambda) \to_d c_1(\alpha) \int_1^\lambda \{|W(s)| - E|W(s)|\}\, s^{-1-1/\alpha}\, ds.$$

For $t \in [0,1]$ set

$$W^{(1)}(t) = tW(1/t).$$

We see by the change of variables $s = t^{-1}$ that

$$c_1(\alpha) \int_{1/\lambda}^1 \left\{ \left| W^{(1)}(t) \right| - E \left| W^{(1)}(t) \right| \right\} t^{1/\alpha - 2}\, dt = \Upsilon_1(\alpha, \lambda).$$

Similarly

$$\overline{\Upsilon}_{n,2}(\lambda) \to_d \Upsilon_2(\alpha, \lambda),$$

where necessarily $\Upsilon_1(\alpha,\lambda)$ and $\Upsilon_2(\alpha,\lambda)$ are independent, finishing the proof of the lemma. □

To complete the proof of Theorem 2, note that for any $\lambda > 1$

$$\operatorname{Var}\left(\frac{1}{\sigma(k_n/n)} \int_{\lambda k_n/n}^{1-\lambda k_n/n} \{|B_n(t)| - E|B_n(t)|\}\, dQ(t) \right) =: \frac{\kappa^2(\lambda k_n/n)}{\sigma^2(k_n/n)},$$

which by the fact that $0 \le \varphi(\rho) \le |\rho|$ for all $\rho \in [-1,1]$, is

$$\le \frac{\sigma^2(\lambda k_n/n)}{\sigma^2(k_n/n)}.$$

(See (3.1), (3.2) and (3.9) in Section 3 for the definitions of $\kappa^2(u)$ and $\varphi(\rho)$, along with their properties.) Now by (2.11)

$$\lim_{n\to\infty} \frac{\sigma^2(\lambda k_n/n)}{\sigma^2(k_n/n)} = \lambda^{1-2/\alpha}.$$

Since $\lambda > 1$ can be chosen arbitrarily large, (2.12) now follows readily from Lemma 1. Assertion (2.13) follows from (2.12) by applying Corollary 1. Finally by applying (1.15) and (1.16), we see that we have convergence of all positive moments of $\Gamma_n(k_n/n)$ and $\Lambda_n(k_n/n)$ to those of $\Upsilon(\alpha)$. □

Remark. Since $\int_0^1 |W^{(i)}(t)|t^{1/\alpha-2}\,dt$ for $i = 1,2$ are positive random variables, $\Upsilon(\alpha)$ is clearly non-normal. Also Theorem 2 improves upon Theorem 3.3 (iii) of Chapter 5 of [2] for $p = 1$ by giving convergence of the moments.

2.3. Slowly varying tails case

We now assume that F is symmetric about zero and has slowly varying tails. In this case its quantile function Q is rapidly varying at zero, meaning that for all $\lambda > 1$, both

$$Q(s)/Q(\lambda s) \to \infty \quad \text{and} \quad Q(1-s)/Q(1-\lambda s) \to \infty, \qquad \text{as } s \searrow 0 \tag{2.15}$$

and

$$\sigma^2(s) \sim 2sQ^2(s), \qquad \text{as } s \searrow 0. \tag{2.16}$$

(For this last fact refer to formula (3.5) in [3].) Set

$$\Upsilon(0) = \frac{1}{\sqrt{2}}\left\{\left|Z^{(1)}\right| - E\left|Z^{(1)}\right|\right\} + \frac{1}{\sqrt{2}}\left\{\left|Z^{(2)}\right| - E\left|Z^{(2)}\right|\right\},$$

where $Z^{(1)}$ and $Z^{(2)}$ are independent standard normal random variables. We shall prove the following result:

Theorem 3. *Whenever F is symmetric about zero and has slowly varying tails and $k_n, n \geq 1$, is a sequence of positive constants each less than n satisfying $k_n/n \to 0$, then*

$$\Gamma_n(k_n/n) \to_d \Upsilon(0), \tag{2.17}$$

and if, in addition, $k_n \to \infty$, then

$$\Lambda_n(k_n/n) \to_d \Upsilon(0). \tag{2.18}$$

Moreover, all the positive moments of $\Gamma_n(k_n/n)$ and $\Lambda_n(k_n/n)$ converge to those of $\Upsilon(0)$.

Proof. The proof follows the same lines as that of Theorem 2. For any $\lambda > 1$, let $(\Upsilon_{n,1}(\lambda), \Upsilon_{n,2}(\lambda))$ and $\left(\overline{\Upsilon}_{n,1}(\lambda), \overline{\Upsilon}_{n,2}(\lambda)\right)$ be defined as in the proof of Theorem 2. The following lemma corresponding to Lemma 1 is the crucial step.

Lemma 2. *Under the assumptions of Theorem 3 for any $\lambda > 1$,*

$$(\Upsilon_{n,1}(\lambda), \Upsilon_{n,2}(\lambda)) \to_d (\Upsilon_1(0), \Upsilon_2(0)),$$

where for $i = 1,2$,

$$\Upsilon_i(0) = \left\{\left|Z^{(i)}\right| - E\left|Z^{(i)}\right|\right\}/\sqrt{2}.$$

Proof. As before, observe that

$$\left|\Upsilon_{n,1}(\lambda) - \overline{\Upsilon}_{n,1}(\lambda)\right| \leq \left[\left|W_n(1)\right| + E\left|W_n(1)\right|\right] \frac{\int_{k_n/n}^{\lambda k_n/n} t\, dQ(t)}{\sigma(k_n/n)}.$$

Now an application of elementary properties of rapidly varying functions gives $\int_{k_n/n}^{\lambda k_n/n} t\, dQ(t) / \sigma(k_n/n) \to 0$ as $n \to \infty$. Thus

$$\Upsilon_{n,1}(\lambda) - \overline{\Upsilon}_{n,1}(\lambda) \to_P 0$$

and by the same argument

$$\Upsilon_{n,2}(\lambda) - \overline{\Upsilon}_{n,2}(\lambda) \to_P 0.$$

Since for each $n \geq 1$, $\overline{\Upsilon}_{n,1}(\lambda)$ and $\overline{\Upsilon}_{n,2}(\lambda)$ are independent, to complete the proof of the lemma it is sufficient to show that for each $i = 1, 2$,

$$\overline{\Upsilon}_{n,i}(\lambda) \to_d \Upsilon_i(0).$$

As in the proof of Lemma 1, by a change of variables each

$$\overline{\Upsilon}_{n,1}(\lambda) =_d \int_1^{\lambda} \{|W(s)| - E|W(s)|\}\, d\mu_n(s),$$

where $d\mu_n(s)$ is defined exactly as in (2.14). Using (2.15) and (2.16) it is easy to see that the measure defined by $d\mu_n$ on $[1, \lambda]$ converges weakly to the point mass measure μ on $[1, \lambda]$ given by

$$\mu(\{1\}) = 1/\sqrt{2}.$$

Thus

$$\overline{\Upsilon}_{n,1}(\lambda) \to_d \left\{\left|Z^{(1)}\right| - E\left|Z^{(1)}\right|\right\} / \sqrt{2}$$

and similarly

$$\overline{\Upsilon}_{n,2}(\lambda) \to_d \Upsilon_2(0),$$

where necessarily $\Upsilon_1(0)$ and $\Upsilon_2(0)$ are independent, finishing the proof of the lemma. □

To complete the proof of (2.17), notice that as at the end of the proof of Theorem 2 we have for any $\lambda > 1$,

$$\operatorname{Var}\left(\frac{1}{\sigma(k_n/n)} \int_{\lambda k_n/n}^{1-\lambda k_n/n} \{|B_n(t)| - E|B_n(t)|\}\, dQ(t)\right) \leq \frac{\sigma^2(\lambda k_n/n)}{\sigma^2(k_n/n)},$$

which by (2.15) and (2.16) converges to 0 as $n \to \infty$. Assertion (2.18) follows from (2.17) using Corollary 1. Now by applying (1.15) and (1.16), we see that all of the positive moments of $\Gamma_n(k_n/n)$ and $\Lambda_n(k_n/n)$ converge to those of $\Upsilon(0)$. □

Remark. Clearly, $\Upsilon(0)$ is non-normal.

3. Some partial results and open problems

Let B denote any Brownian bridge. It is natural to normalize the random variable

$$\int_{k_n/n}^{1-k_n/n} \{|B_n(t)| - E|B_n(t)|\}\, dQ(t)$$

by its standard deviation $\kappa(k_n/n)$, where for $0 < u < 1/2$

$$\kappa^2(u) = \int_u^{1-u}\int_u^{1-u} \operatorname{cov}(|B(t)|, |B(s)|)\, dQ(s)\, dQ(t), \tag{3.1}$$

which, in turn,

$$= \int_u^{1-u}\int_u^{1-u} \varphi\left(\frac{s\wedge t - st}{\sqrt{s(1-s)}\sqrt{t(1-t)}}\right)\sqrt{s(1-s)}\sqrt{t(1-t)}\, dQ(s)\, dQ(t),$$

where

$$\varphi(\rho) = \frac{2}{\pi}\left(\rho \arcsin\rho + \sqrt{1-\rho^2} - 1\right), \quad \rho \in [-1,1]; \tag{3.2}$$

see [13] and references therein. For future reference we note that $\varphi(\rho)$ has the Taylor series expansion

$$\varphi(\rho) = \frac{1}{\pi}\sum_{n=0}^{\infty}\binom{-1/2}{n}\frac{(-1)^n}{2n+1}\frac{1}{n+1}\rho^{2n+2}, \quad \rho \in [-1,1]. \tag{3.3}$$

In order to replace u_n by B_n in the random variable

$$\frac{1}{\kappa(k_n/n)}\int_{k_n/n}^{1-k_n/n} \{|u_n(t)| - E|u_n(t)|\}\, dQ(t)$$

using weighted approximation methods, by Proposition 1 we need to verify that

$$\limsup_{u\searrow 0} \frac{u\{Q^2(u) + Q^2(1-u)\}}{\kappa^2(u)} < \infty. \tag{3.4}$$

It is an open problem whether (3.4) is true in general. Here are some sufficient conditions under which (3.4) holds.

Proposition 2. *Whenever*

$$\limsup_{u\searrow 0} uQ^2(u) < \infty \tag{3.5}$$

or for some $0 < \lambda_1 < 1$

$$\limsup_{s\searrow 0} \frac{Q(2s)}{Q(s)} =: \lambda_1 < 1, \tag{3.6}$$

and

$$\limsup_{u \searrow 0} uQ^2(1-u) < \infty \tag{3.7}$$

or for some $0 < \lambda_2 < 1$

$$\limsup_{s \searrow 0} \frac{Q(1-2s)}{Q(1-s)} =: \lambda_2 < 1, \tag{3.8}$$

then (3.4) *holds.*

Proof. First by using the Taylor expansion of $\varphi(\rho)$ given in (3.3) it can be readily shown that

$$\frac{1}{\pi}\rho^2 \le \varphi(\rho) \le \frac{2}{\pi}\rho^2, \quad \rho \in [-1,1]. \tag{3.9}$$

Therefore to establish (3.4), it suffices to prove that

$$\limsup_{u \searrow 0} \frac{u\left\{Q^2(u) + Q^2(1-u)\right\}}{\tau^2(u)} < \infty, \tag{3.10}$$

where

$$\tau^2(u) = \int_u^{1-u}\int_u^{1-u} \frac{(s \wedge t - st)^2}{\sqrt{s(1-s)}\sqrt{t(1-t)}}\, dQ(s)\, dQ(t).$$

It is readily checked that $\sigma^2(u) \ge \tau^2(u)$. However, this is of no help in establishing (3.10).

We will first show that

$$\limsup_{u \searrow 0} \frac{uQ^2(u)}{\tau^2(u)} < \infty. \tag{3.11}$$

Here we can assume that (3.6) holds, because otherwise, under (3.5), statement (3.11) is obvious. Moreover, we can assume without loss of generality that

$$Q(s) \to -\infty \quad \text{as } s \searrow 0. \tag{3.12}$$

This of course implies that $Q(s) < 0$ for $0 < s \le \gamma$, for some $0 < \gamma < 1/2$. Furthermore, we will assume that $Q(s)$ is not slowly varying at zero. Otherwise, we have automatically that

$$sQ^2(s) \to 0 \quad \text{as } s \searrow 0,$$

and there is nothing to prove. Note that in general when $Q(s)$ is not slowly varying at zero, we necessarily have

$$\liminf_{s \searrow 0} \frac{Q(2s)}{Q(s)} < 1.$$

Notice that for $0 < u \leq \gamma < 1/2$

$$\begin{aligned}\tau^2(u) &\geq \int_u^\gamma \int_u^\gamma \frac{(s \wedge t - st)^2}{\sqrt{s(1-s)}\sqrt{t(1-t)}} \, dQ\,(s)\, dQ\,(t) \\ &\geq \int_u^\gamma \int_u^\gamma \frac{1\,\{s \leq t\}\,(s(1-t))^2}{\sqrt{s(1-s)}\sqrt{t(1-t)}} \, dQ\,(s)\, dQ\,(t) \\ &\geq (1/2)^{3/2} \int_u^\gamma s^{3/2} \int_s^\gamma t^{-1/2} \, dQ\,(t)\, dQ\,(s)\,.\end{aligned}$$

Now

$$\begin{aligned}&\int_s^\gamma t^{-1/2}\, dQ\,(t) = \gamma^{-1/2} Q(\gamma) - s^{-1/2} Q(s) + \frac{1}{2}\int_s^\gamma t^{-3/2} Q\,(t)\, dt \\ &\geq \gamma^{-1/2} Q(\gamma) - s^{-1/2} Q(s) + \frac{1}{2} Q(s) \int_s^{2s} t^{-3/2}\, dt + \frac{1}{2}\int_{2s}^\gamma t^{-3/2} Q\,(t)\, dt \\ &\geq \gamma^{-1/2} Q(\gamma) - (2s)^{-1/2}\, Q(s) + \frac{1}{2} Q(2s) \int_{2s}^\infty t^{-3/2}\, dt \\ &= \gamma^{-1/2} Q(\gamma) + (2s)^{-1/2}\, \{Q(2s) - Q(s)\}\,,\end{aligned}$$

which by the assumptions (3.12) and (3.6), is for some $\tau > 0$ and all $s > 0$ small enough

$$> -\tau s^{-1/2} Q(s)\,.$$

Therefore choosing $0 < \gamma' < \gamma$ sufficiently small, we have

$$(1/2)^{3/2} \int_u^\gamma s^{3/2} \int_s^\gamma t^{-1/2}\, dQ\,(t)\, dQ\,(s) \tag{3.13}$$

$$\begin{aligned}&\geq (1/2)^{3/2} \int_u^{\gamma'} s^{3/2} \int_s^\gamma t^{-1/2}\, dQ\,(t)\, dQ\,(s) \\ &\geq -\frac{u\tau}{4} \int_u^{\gamma'} Q(s)\, dQ\,(s) \geq \frac{u}{8}\tau \left(Q^2(u) - Q^2(\gamma')\right),\end{aligned} \tag{3.14}$$

which because of (3.12) for all $u > 0$ small is

$$\geq \frac{u}{16}\tau Q^2(u)\,.$$

For inequality (3.14) we use the integration by parts formula on page 419 of [8]. Thus for all $u > 0$ small

$$\tau^2(u) \geq \frac{u}{16}\tau Q^2(u)\,.$$

Similarly, assuming (3.8), there exists a $\tau' > 0$ such that for all $u > 0$ small,

$$\tau^2(u) \geq \frac{u}{16}\tau' Q^2(1-u). \qquad \square$$

Remark. Two additional sufficient conditions for (3.4) are that (3.6) holds and for all $s > 0$ small enough

$$Q^2\,(s) \geq Q^2\,(1-s)\,, \tag{3.15}$$

and vice versa.

Presently we do not know how to treat the cases when

$$\limsup_{s\searrow 0}\frac{Q(2s)}{Q(s)}=1 \quad\text{and}\quad \liminf_{s\searrow 0}\frac{Q(2s)}{Q(s)}<1$$

or

$$\limsup_{s\searrow 0}\frac{Q(1-2s)}{Q(1-s)}=1 \quad\text{and}\quad \liminf_{s\searrow 0}\frac{Q(1-2s)}{Q(1-s)}<1\,.$$

Notice that whenever

$$\int_0^{1/2} Q^2(t)\,dt<\infty \quad\text{or}\quad Q(s)=L_1(s)s^{-\beta_1} \tag{3.16}$$

and

$$\int_{1/2}^{1} Q^2(t)\,dt<\infty \quad\text{or}\quad Q(1-s)=L_2(s)s^{-\beta_2}, \tag{3.17}$$

where L_1 and L_2 are slowly varying functions at zero and $0<\beta_1,\beta_2<\infty$, then the conditions of Proposition 2 hold.

Next, for the three classes of distribution functions considered in Section 2 we will show that $\kappa^2(u)$ and $\sigma^2(u)$ are exactly of the same order and compute the value of

$$\lim_{u\searrow 0}\kappa^2(u)/\sigma^2(u)\,. \tag{3.18}$$

In particular, we will prove that for $F\in D(\alpha)\,,\ 0<\alpha<2\,,$

$$\lim_{u\searrow 0}\kappa^2(u)/\sigma^2(u)=1-\frac{2}{\pi}\left[\frac{\alpha\left(\frac{1}{\alpha}-\frac{1}{2}\right)B\left(\frac{1}{\alpha}-\frac{1}{2},\frac{3}{2}\right)-2}{\alpha-2}\right]=:\rho(\alpha)\,. \tag{3.19}$$

Towards this end, define for $0<\alpha<2$

$$V_\alpha=\int_0^1\left\{\left|W^{(1)}(t)\right|-E\left|W(t)\right|\right\}t^{1/\alpha-2}\,ds$$

and let

$$\psi(\alpha)=\alpha\left(\frac{1}{\alpha}-\frac{1}{2}\right)B\left(\frac{1}{\alpha}-\frac{1}{2},\frac{3}{2}\right)=\frac{\alpha\Gamma\left(\frac{1}{\alpha}+\frac{1}{2}\right)\Gamma\left(\frac{3}{2}\right)}{\Gamma\left(\frac{1}{\alpha}+1\right)}\,.$$

We shall need the following proposition in order to establish (3.19).

Proposition 3. *For all* $0<\alpha<2\,,$

$$\operatorname{Var}(V_\alpha)=\frac{2\alpha^2}{2-\alpha}\left(1-\frac{2}{\pi}\left[\frac{\psi(\alpha)-2}{\alpha-2}\right]\right).$$

Proof. We see by the covariance formulas in [13] that

$$\begin{aligned}\operatorname{Var}(V_\alpha) &= \int_0^1\int_0^1 \operatorname{cov}\left(\left|\frac{W(u)}{\sqrt{u}}\right|, \left|\frac{W(v)}{\sqrt{v}}\right|\right) u^{1/\alpha-3/2}v^{1/\alpha-3/2}\,du\,dv \\ &= \int_0^1\int_0^1 \varphi\left(\frac{u\wedge v}{\sqrt{uv}}\right) u^{1/\alpha-3/2}v^{1/\alpha-3/2}\,du\,dv \\ &= 2\int_0^1\int_0^v \varphi\left(\sqrt{\frac{u}{v}}\right) u^{1/\alpha-3/2}\,du\,v^{1/\alpha-3/2}\,dv \\ &= \frac{2\alpha}{2-\alpha}\int_0^1 \varphi\left(\sqrt{t}\right) t^{1/\alpha-3/2}\,dt,\end{aligned}$$

and, recalling the definition of φ, we get

$$= \frac{4\alpha}{\pi(2-\alpha)}\left[\int_0^1 \arcsin\left(\sqrt{t}\right) t^{1/\alpha-1}dt + B\left(\frac{1}{\alpha}-\frac{1}{2},\frac{3}{2}\right) - \frac{2\alpha}{2-\alpha}\right].$$

Moreover, after integrating by parts we have

$$\int_0^1 \arcsin\left(\sqrt{t}\right) t^{1/\alpha-1}\,dt = \frac{\alpha\pi}{2} - \frac{2-\alpha}{2}B\left(\frac{1}{\alpha}-\frac{1}{2},\frac{3}{2}\right).$$

The proof is completed by inserting this last expression into the previous one and doing some rearranging. □

We are now ready to prove (3.19). Theorem 2 implies that whenever $F \in D(\alpha)$, with $0<\alpha<2$, and $k_n/n \to 0$ as $n\to\infty$,

$$\operatorname{Var}(\Gamma_n(k_n/n)) = \frac{\kappa^2(k_n/n)}{\sigma^2(k_n/n)} \to \left(c_1^2(\alpha)+c_2^2(\alpha)\right)\operatorname{Var}\left(\int_0^1 |W(t)|\,t^{1/\alpha-2}\,dt\right),$$

which by Proposition 3

$$= 1 - \frac{2}{\pi}\left[\frac{\alpha\left(\frac{1}{\alpha}-\frac{1}{2}\right)B\left(\frac{1}{\alpha}-\frac{1}{2},\frac{3}{2}\right)-2}{\alpha-2}\right].$$

This says that whenever $F\in D(\alpha)$, with $0<\alpha<2$, then (3.19) holds.

For the distributions in the domain of attraction of a normal law given by (2.3) of Subsection 2.1, by Theorem 1, whenever $k_n/n\to 0$ as $n\to\infty$,

$$\operatorname{Var}(\Gamma_n(k_n/n)) = \frac{\kappa^2(k_n/n)}{\sigma^2(k_n/n)} \to 1-\frac{2}{\pi}(2-\log 2),$$

whence

$$\lim_{u\searrow 0}\kappa^2(u)/\sigma^2(u) = 1-\frac{2}{\pi}(2-\log 2). \tag{3.20}$$

This, of course, implies that $\kappa^{-1}(k_n/n)\,I_n \to_d N(0,1)$.

Finally, when F is symmetric and has slowly varying tails we get by arguing exactly as above via Theorem 3, that whenever $k_n/n\to 0$ as $n\to\infty$,

$$\frac{\kappa^2(k_n/n)}{\sigma^2(k_n/n)} \to 1-\frac{2}{\pi},$$

which implies that

$$\lim_{u \searrow 0} \kappa^2(u)/\sigma^2(u) = 1 - \frac{2}{\pi}. \tag{3.21}$$

In all three cases we considered we see then that we could have normalized I_n by $\kappa(k_n/n)$.

Remark. One finds that $\rho(\alpha)$ as in (3.19) is a decreasing function on $(0,2)$ satisfying

$$\lim_{\alpha \searrow 0} \rho(\alpha) = 1 - \frac{2}{\pi} \quad \text{and} \quad \lim_{\alpha \nearrow 2} \rho(\alpha) = 1 - \frac{2}{\pi}(2 - \log 2). \tag{3.22}$$

For a proof, differentiating $\psi(\alpha)$, we get

$$\psi'(\alpha) = \frac{\Gamma\left(\frac{1}{\alpha} + \frac{1}{2}\right)\Gamma\left(\frac{3}{2}\right)}{\Gamma\left(\frac{1}{\alpha} + 1\right)} - \frac{\alpha^{-1}\Gamma\left(\frac{3}{2}\right)\Gamma'\left(\frac{1}{\alpha} + \frac{1}{2}\right)}{\Gamma\left(\frac{1}{\alpha} + 1\right)} + \frac{\alpha^{-1}\Gamma\left(\frac{3}{2}\right)\Gamma\left(\frac{1}{\alpha} + \frac{1}{2}\right)\Gamma'\left(\frac{1}{\alpha} + 1\right)}{\Gamma^2\left(\frac{1}{\alpha} + 1\right)}.$$

This gives

$$\psi'(2) = 1 - \frac{1}{2}\Gamma'(1) + \frac{1}{2}\frac{\Gamma'\left(\frac{3}{2}\right)}{\Gamma\left(\frac{3}{2}\right)} = 1 - \frac{1}{2}\left(\Gamma'(1) - \frac{\Gamma'\left(\frac{3}{2}\right)}{\Gamma\left(\frac{3}{2}\right)}\right) = 2 - \log 2.$$

Thus

$$\lim_{\alpha \nearrow 2} \rho(\alpha) = 1 - \frac{2}{\pi}(2 - \log 2).$$

Also one finds by using Stirling's formula that

$$\lim_{\alpha \searrow 0} \rho(\alpha) = 1 - \frac{2}{\pi}.$$

We see then there is a smooth transition in the limit from the domain of attraction of a normal case to the slowly varying tail case.

Acknowledgement. The authors would like to thank the referee for a careful reading of the manuscript and pointing them to the Munk and Czado (1998) paper.

References

[1] M. Csörgő, S. Csörgő, L. Horváth and D.M. Mason, *Normal and Stable Convergence of Integral Functions of the Empirical Distribution Function.* Ann. Probab. **14** (1986), 86–118.

[2] M. Csörgő and L. Horváth, *Weighted Approximations in Probability and Statistics.* Wiley, 1993.

[3] S. Csörgő, E. Haeusler and D.M. Mason, *The Asymptotic Distribution of Trimmed Sums.* Ann. Probab. **16** (1988), 672–699.

[4] S. Csörgő, E. Haeusler and D.M. Mason, *A Probabilistic Approach to the Asymptotic Distribution of Sums of Independent, Identically Distributed Random Variables.* Adv. Appl. Math. **9** (1988), 259–333.

[5] S. Csörgő, L. Horváth and D.M. Mason, *What Portion of the Sample Makes a Partial Sum Asymptotically Stable or Normal?* Probab. Theory Related Fields **72** (1986), 1–16.

[6] E. del Barrio, E. Giné and C. Matrán, *Central Limit Theorems for the Wasserstein Distance Between the Empirical and the True Distributions.* Ann. Probab. **27** (1999), 1009–1071. Correction note, Ann. Probab. **31** (2003).

[7] M.G. Hahn, D.M. Mason and D.C Weiner (Eds.), *Sums, Trimmed Sums and Extremes.* Birkhäuser, 1991.

[8] E. Hewitt and K. Stromberg, *Real and Abstract Analysis.* Springer-Verlag, 1969.

[9] D.M. Mason, An exponential inequality for a weighted approximation to the uniform empirical process with applications. In: *State of the Art in Statistics and Probability Theory, Festschrift for Willem R. Van Zwet.* (M. de Gunst, C. Klaassen, A. van der Vaart, Eds.), I.M.S. Lect. Notes in Math. Statist. **36**, pp. 477–498, 2001.

[10] A. Munk and C. Czado, *Nonparametric validation of similar distributions and assessment of goodness of fit.* J. R. Stat. Soc., Ser. B, Stat. Methodol. **60** (1998), 223–241.

[11] G.R. Shorack, *Probability for Statisticians.* Springer-Verlag, 2000.

[12] G.R. Shorack and J.A. Wellner, *Empirical Processes with Applications to Statistics.* Wiley, 1986.

[13] J.A. Wellner and R.T. Smythe, *Computing the Covariance of Two Brownian Area Integrals.* Statist. Neerl. **56** (2002), 101–109.

Mathematical Institute, University of Giessen,
Arndtstr. 2, D-35392 Giessen, Germany
E-mail address: erich.haeusler@math.uni-giessen.de

Department of Food & Resource Economics, University of Delaware,
206 Townsend Hall, Newark, DE 19716, USA
E-mail address: davidm@udel.edu

Part IV

Stochastic Differential Equations

Progress in Probability, Vol. 56, 301–321

On the Rate of Convergence of Splitting-up Approximations for SPDEs

István Gyöngy and Nicolai Krylov

Abstract. We consider the stochastic PDE

$$du(t,x) = (Lu(t,x) + f(t,x))\,dt + (M_k u(t) + g_k(t,x))\circ dW^k,$$

where L and M^k are second and first order partial differential operators, W is a multi-dimensional Wiener process, and '$\circ$' stands for the Stratonovich differential. We approximate the solution by the splitting-up method (method of alternative directions) and we estimate the error in terms of Sobolev's norm $W_p^m(\mathbb{R}^d)$, and hence also in the supremum norm. We also show that our rate of convergence estimate is sharp.

1. Introduction

Stochastic partial differential equations (SPDEs) arise in many areas of science and engineering, from population genetics to signal processing. Therefore it is important to find effective methods to solve SPDEs numerically. Various numerical methods have been used for SPDEs: finite difference method, Galerkin's approximation, finite element method, particle method, Wiener chaos decomposition, splitting-up method, etc (see, for example [1], [2], [3], [5], [7], [14], [15], [18] and the references therein.) A new approach to the splitting-up method is presented in [9]. This approach not only allows us in [9] to improve the results of [1], [2], [4], and [10] in the direction of convergence in sup norm but also to put forth the splitting-up method for general filtering equations. A typical result from [9] reads as follows. For each $T > 0$ and $q > 0$ and $X := W_2^m(\mathbb{R}^d)$ there is a constant N such that

$$E \max_{t\in T_n} \|u^{(n)}(t) - u(t)\|_X^q \le N/n^q \tag{1.1}$$

for all integers $n \ge 1$, where $u^{(n)}$ are constructed by splitting-up method with time step $1/n$. In [9] we assumed as little smoothness on the coefficients of the differential

---

2000 *Mathematics Subject Classification.* 60H15, 65M12, 65M15, 93E11.
Key words and phrases. Stochastic partial differential equations, Splitting-up.
The work of the second author was partially supported by NSF Grant DMS-0140405.

operators as we could, at the expense of complicated calculations, which we were not able to adjust to estimate the rate of convergence in W_p^m in place of W_2^m.

In the present paper we estimate the rate of convergence in $X := W_p^m$ for any integer $m \geq 0$ and real number $p \geq 2$ by simple calculations, but at the expense of somewhat stronger smoothness assumptions. By using Sobolev's embedding theorem for W_p^m with large p we can get the same estimate of the rate of convergence in $X := C(\mathbb{R}^d)$, with the supremum norm, under substantially weaker smoothness assumptions than in [9]. As in [9], we present our result for the multistage splitting-up method. We also prove that, generally speaking, the rate of convergence cannot be higher than $1/n$ (see Sec. 6). It is also worth noting that we prove estimate (1.1) for $X = W_p^m$ with $q \in (0, p]$. This is done on the basis of stochastic inequalities from Lemma 3.2. As in [9], we could give an application of our results to filtering problems. We prefer not to do this for shortness of presentation and because this would be just a repetition, almost word for word, of what is written in [9].

The paper is organized as follows. In Sec. 2 we introduce our general setting but state the results only for the case of time independent data for the sake of simplicity of presentation. Theorem 2.5 is proved in Sec. 2 on the basis of Theorem 2.4, which in turn is proved in Sec. 4, after we prepare some auxiliary facts in Sec. 3. In Sec. 5 we generalize Theorem 2.3 for time dependent and random coefficients. Sec. 6 is devoted to the discussion of the sharpness of the rate $1/n$.

In conclusion we introduce some notation used everywhere below. Throughout the paper d, d_0, d_1 are fixed integers, K, T are fixed finite positive constants, and

$$D_i = \partial/\partial x^i, \quad D_{ij} = \partial^2/\partial x^i \partial x^j .$$

Let $(\Omega, \mathcal{F}, P)$ be a complete probability space, $\mathcal{F}_t$, $t \geq 0$, be an increasing filtration of sub σ-fields of $\mathcal{F}$, such that $\mathcal{F}_0$ is complete with respect to $(\mathcal{F}, P)$. By $\mathcal{P}$ we denote the σ-field of predictable subsets of $\Omega \times (0, \infty)$ generated by $\mathcal{F}_t$. We assume that on Ω we are given a continuous $\mathcal{F}_t$-martingale $Y_t = (Y_t^1, ..., Y_t^{d_0})$.

We always assume the summation convention over repeated integer-valued indices.

2. The case of time independent coefficients

For $\varepsilon = 0, 1$ and $r = 0, 1, ..., d_1$ (notice r can be zero), let $V_{t,\varepsilon}^r$ be continuous increasing processes defined for $t \in [0, T]$. Consider the following equation

$$du(t,x) = (L_r u(t,x) + f_r(t,x))\, dV_{t,\varepsilon}^r + (M_k u(t,x) + g_k(t,x))\, dY_t^k \tag{2.1}$$

for $t \in (0, T]$, $x \in \mathbb{R}^d$ with initial condition $u(0,x) = u_{0\varepsilon}(x)$, where the operators L_r and M_k are written as

$$L_r = a_r^{ij}(t,x) D_{ij} + a_r^i(t,x) D_i + a_r(t,x), \quad M_k = b_k^i(t,x) D_i + b_k(t,x). \tag{2.2}$$

In order to formulate our assumptions we fix an integer $m \geq 0$.

Assumption 2.1 (smoothness of the coefficients). All the coefficients $a_r^{ij}(t,x)$, $a_r^i(t,x)$, $a_r(t,x)$, $b_k^i(t,x)$, $b_k(t,x)$ are predictable for any $x \in \mathbb{R}^d$ and, for any $(\omega,t) \in \Omega \times (0,\infty)$, their derivatives up to order $m+4$ exist, are continuous, and by magnitude are bounded by K.

Assumption 2.2. The processes $V_{t,\varepsilon}^r$ are predictable $V_{0,\varepsilon}^r = 0$, $V_{t,\varepsilon}^0 =: V_t^0$ is independent of ε, and there is a predictable increasing process V_t such that

$$V_0 = 0, \quad V_T \leq K, \quad \sum_{r,\varepsilon} dV_{t,\varepsilon}^r + d\langle Y\rangle_t \leq dV_t \tag{2.3}$$

in the sense of measures on $[0,T]$.

Remark 2.1. Actually (2.3) is always satisfied with $V_t = \sum_{r,\varepsilon} V_{t,\varepsilon}^r + \langle Y\rangle_t$, provided that this process is bounded by K on $[0,T]$. Also notice that we single out one of $V_{t,\varepsilon}^r$ with $r=0$, because it plays a special role and is designed to go with Y_t. The latter processes are independent of ε and so is V^0. This is also the reason why later we do not impose Assumption 5.1 on L_0.

Equation (2.1) is supposed to be parabolic in the usual stochastic sense.

Assumption 2.3. For any $\omega \in \Omega$, $\varepsilon = 0,1$, $x, \lambda \in \mathbb{R}^d$, we have

$$2a_r^{ij}(t,x)\lambda^i\lambda^j\, dV_{t,\varepsilon}^r - b_k^i(t,x)b_l^j(t,x)\lambda^i\lambda^j\, d\langle Y^k, Y^l\rangle_t \geq 0$$

in the sense of measures on $[0,T]$. (Recall that summation convention is used over repeated integer valued indices, and that r runs through $0,1,...,d_1$.)

We investigate the convergence of not only functions themselves but also of their derivatives in L_p. Therefore we need the spaces W_p^m of L_p functions whose generalized derivatives up to order m are also in L_p. We define the norm $\|u\|_{m,p}$ in W_p^m by

$$\|u\|_{m,p}^p := \sum_{|\alpha|\leq m} \int_{\mathbb{R}^d} |D^\alpha u(x)|^p\, dx,$$

where $\alpha = (\alpha_1, \ldots, \alpha_d)$ are multi-indices,

$$|\alpha| := \alpha_1 + \cdots + \alpha_d, \quad D^\alpha := D_1^{\alpha_1} \cdots D_d^{\alpha_d}.$$

We always assume that $p \geq 2$ and we take and fix a number $q \in (0,p]$.

Assumption 2.4. For each $\omega \in \Omega$, the functions $f_r(t) = f_r(t,\cdot)$ are weakly continuous as W_p^{m+4}-valued functions, $g_k(t) = g_k(t,\cdot)$ are weakly continuous as W_p^{m+5}-valued functions, and the initial condition $u_{0\varepsilon}$ is an $\mathcal{F}_0$-measurable W_p^{m+4}-valued function for each ε. Furthermore, f_r and g_k are predictable and

$$E \sup_{t\in[0,T]} \|f(t)\|_{m+4,p}^q + E \sup_{t\in[0,T]} \|g(t)\|_{m+5,p}^q + E\|u_0\|_{m+4,p}^q \leq K,$$

where $E\|u_0\|_{n,p}^q = E\|u_{01}\|_{n,p}^q + E\|u_{00}\|_{n,p}^q$, $\|f(t)\|_{n,p}^q = \sum_r \|f_r(t)\|_{n,p}^q$ and $\|g(t)\|_{n,p}^q = \sum_k \|g_k(t)\|_{n,p}^q$.

Definition 2.2. By a solution of (2.1) with initial data u_0 we mean a W^1_p-valued function $u(t) = u(t,\cdot)$, defined on $\Omega \times [0,T]$, which is $\mathcal{F}_t$-adapted and weakly continuous in t, and such that for any $\phi \in C_0^\infty$ the equation

$$(u(t,\cdot),\phi) = (u_0,\phi) + \int_0^t \big[-(a_r^{ij} D_i u(s), D_j\phi) + \big((a_r^i - a_{rx^j}^{ij}) D_i u(s) + a_r u(s) + f_r(s),\phi\big)\big]\, dV^r_{s,\varepsilon} + \int_0^t (b_k^i D_i u(s) + b_k u(s) + g_k(s),\phi)\, dY_s^k \tag{2.4}$$

holds for all $t \in [0,T]$ at once with probability 1, where $(\xi,\eta) = \int_{\mathbb{R}^d} \xi\eta\, dx$.

Theorem 2.3. *Under Assumptions* 2.1, 2.2, 2.3, *and* 2.4, *for* $\varepsilon = 0,1$, *equation* (2.1) *with initial condition* $u_{0\varepsilon}$ *has a unique solution* $u_\varepsilon(t)$. *Furthermore,* $u_\varepsilon(t)$ *is weakly continuous in* W_p^{m+4} *for each* ω.

This theorem is a particular case of Theorem 3.1. The following is the basic tool of proving our estimate of convergence for the splitting-up method.

Theorem 2.4. *Let* $a_r^{ij}, a_r^i, a_r, b_k^i, b_k$, f_r *and* g_k *be independent of* t. *Then under Assumptions* 2.1, 2.2, 2.3, *and* 2.4, *there is a constant* N *depending only on* d, d_0, d_1, K, p, q, m, *and* T, *such that*

$$E \sup_{t\in[0,T]} \|u_1(t) - u_0(t)\|^q_{m,p} \le N(E\|u_{01} - u_{00}\|^q_{m,p} + A^q),$$

where

$$A = \sup_{\omega\in\Omega} \max_{t\in[0,T]} \max_r |V^r_{t,1} - V^r_{t,0}|.$$

Theorem 2.4 is proved in Sec. 4. Now we give its application to the splitting-up method along the lines introduced in [9]. In $(0,T]\times\mathbb{R}^d$ we consider the following equation

$$du(t,x) = \sum_{r=1}^{d_1} (L_r u(t,x) + f_r(t,x))\, dt + (L_0 u(t,x) + f_0(t,x))\, dV_t^0 + (M_k u(t,x) + g_k(t,x))\, dY_t^k, \tag{2.5}$$

with the same operators L_r and M_k as above and initial condition $u(0,x) = u_0(x)$.

Assumption 2.5. Assumptions 2.1 and 2.4 are satisfied with $u_{0\varepsilon} = u_0$. The process V_t^0 is predictable continuous increasing and starting at zero. We have $V_T^0 + \langle Y\rangle_T \le K$. The matrices (a_r^{ij}) are non-negative and, for any $\omega \in \Omega$, $x, \lambda \in \mathbb{R}^d$, we have

$$2a_0^{ij}(t,x)\lambda^i\lambda^j\, dV_t^0 - b_k^i(t,x) b_r^j(t,x) \lambda^i\lambda^j\, d\langle Y^k, Y^r\rangle_t \ge 0$$

in the sense of measures on $[0,T]$.

By $u(t)$ we denote the unique solution of (2.5) with initial condition $u(0,x) = u_0(x)$, which exists owing to Theorem 2.3.

Next set $T_n := \{t_i := iT/n : i = 0,1,2,\dots,n\}$, $\delta := T/n$ for an integer $n \geq 1$, and define the approximation $u^{(n)}$ according to $u^{(n)}(0) := u_0$,

$$u^{(n)}(t_{i+1}) := \mathbf{P}_\delta^{(d_1)} \cdots \mathbf{P}_\delta^{(2)} \mathbf{P}_\delta^{(1)} \mathbf{Q}_{t_i t_{i+1}} u^{(n)}(t_i), \quad i = 0,1,2,\dots,n-1, \tag{2.6}$$

where $\mathbf{P}_t^{(\gamma)}\psi := v(t)$, $\gamma = 1,2,\dots,d_1$, and $\mathbf{Q}_{st}\psi := \tilde{v}(t)$ denote the solutions of the equations

$$dv(t,x) = (L_\gamma v(t,x) + f_\gamma(x))\,dt, \quad t \geq 0,$$

$$d\tilde{v}(t,x) = (L_0\tilde{v}(t,x) + f_0(x))\,dV_t^0 + (M_k\tilde{v}(t,x) + g_k(x))\,dY_t^k, \quad t \geq s,$$

respectively, with initial conditions $v(0,x) = \psi(x)$ and $\tilde{v}(s,x) = \psi(x)$, respectively. The functions $u^{(n)}$ are called the splitting-up approximations since, while constructing $u^{(n)}$, we split equation (2.5) into several subequations.

Theorem 2.5. *Let $a_r^{ij}, a_r^i, a_r, b_k^i, b_k$, f_r and g_k be independent of t. Then under Assumption 2.5 there is a constant N depending only on d, d_0, d_1, K, p, q, m and T, such that*

$$E \max_{t\in T_n} \|u^{(n)}(t) - u(t)\|_{m,p}^q \leq N n^{-q}$$

for all $n \geq 1$.

Proof. Set $d' := d_1 + 1$, fix an integer $n \geq 1$ and let $\delta := T/n$. According to our idea from [9] we change time by using the following function

$$\kappa(t) := \begin{cases} t - kd_1\delta & \text{for} \quad t \in [kd'\delta, (kd'+1)\delta], k = 0,1,\dots, \\ (k+1)\delta & \text{for} \quad t \in [(kd'+1)\delta, (k+1)d'\delta], k = 0,1,\dots, \\ 0 & \text{for} \quad t \leq 0. \end{cases} \tag{2.7}$$

Define

$$\bar{Y}^k(t) := Y_{\kappa(t)}^k, \quad \bar{\mathcal{F}}_t = \mathcal{F}_{\kappa(t)}, \quad \bar{V}_{t,0}^0 = \bar{V}_{t,1}^0 := V_{\kappa(t)}^0,$$

$$\bar{V}_{t,0}^r := \kappa(t), \quad \bar{V}_{t,1}^r := \kappa(t - r\delta) \quad \text{for} \quad r = 1,2,\dots,d_1.$$

Consider the equations

$$du_\varepsilon(t) = (L_r u_\varepsilon(t) + f_r)\,d\bar{V}_{t,\varepsilon}^r + (M_k u_\varepsilon(t) + g_k)\,d\bar{Y}_t^k, \quad \varepsilon = 0,1, \tag{2.8}$$

with $u_0(0,x) = u_1(0,x) = u_0(x)$. It is easy to see that Assumptions 2.2 and 2.3 hold also with $\bar{Y}^k$ and $\bar{V}_\varepsilon^r$ ($\varepsilon = 0,1$) in place of Y^k and V_ε^r, respectively. Thus by Theorem 2.3 the solutions u_0 and u_1 exist, and by virtue of Theorem 2.4 there is a constant N depending only on d, d_0, d_1, p, q, m, K, and T, such that

$$E \sup_{t\in[0,Td']} \|u_1(t) - u_0(t)\|_{m,p}^q \leq N \sup_{t\in[0,Td']} \sup_{r\leq d_1} |\kappa(t+r\delta) - \kappa(t)|^q = NT^q n^{-q},$$

which implies the theorem, since clearly $u_0(d't) = u(t)$ and $u_1(d't) = u^{(n)}(t)$ for $t \in T_n$. □

Remark 2.6. We can define the approximation $u^{(n)}$ by splitting-up in any order, i.e., we can define $u^{(n)}$ by

$$u^{(n)}(t_{i+1}) := \mathbf{P}_\delta^{(d_1)} \cdots \mathbf{P}_\delta^{(l+1)} \mathbf{Q}_{t_i t_{i+1}} \mathbf{P}_\delta^{(l)} \cdots \mathbf{P}_\delta^{(2)} \mathbf{P}_\delta^{(1)} u^{(n)}(t_i),$$

in place of (2.6). Then one can easily see from its proof that Theorem 2.5 remains valid.

3. An auxiliary result

We consider the equation

$$du(t,x) = (Lu(t,x) + f(t,x))\, dV_t + (M_k u(t,x) + g_k(t,x))\, dY_t^k \tag{3.1}$$

for $t \in (0,T], x \in \mathbb{R}^d$ with initial condition $u(0,x) = u_0(x)$, where $T \in (0,\infty)$ is a fixed number and the operators L and M_k are written as

$$L = a^{ij}(t,x)D_{ij} + a^i(t,x)D_i + a(t,x), \quad M_k = b_k^i(t,x)D_i + b_k(t,x).$$

For convenience we enumerate some further assumptions regarding (3.1). Fix an integer $m = 1, 2, \ldots$, real numbers $p \geq 2$, $q \in (0,p]$, $K \in [0,\infty)$.

Assumption 3.1 (smoothness of the coefficients). All the coefficients $a^{ij}(t,x)$, $a^i(t,x)$, $a(t,x)$, $b_k^i(t,x)$, $b_k(t,x)$ are predictable for any $x \in \mathbb{R}^d$ and, for any $(\omega,t) \in \Omega \times (0,\infty)$, their derivatives up to order m and for a^{ij} up to order $2 \vee m$ exist, are continuous, and by magnitude are bounded by K.

Assumption 3.2 (stochastic parabolicity of (3.1)). The process V_t is increasing, continuous, predictable, $V_0 = 0$, and $V_T \leq K$. We have $d\langle Y \rangle_t \leq dV_t$ and for any $x, \lambda \in \mathbb{R}^d$, in the sense of measures on $[0,T]$

$$2a^{ij}\lambda^i\lambda^j\, dV_t - b_k^i b_r^j \lambda^i \lambda^j\, d\langle Y^k, Y^r\rangle_t \geq 0.$$

Assumption 3.3. In equation (3.1) the function $f(t) = f(t,\cdot)$ is predictable W_p^m-valued, $g_k(t) = g_k(t,\cdot)$ are predictable W_p^{m+1}-valued, u_0 is W_p^m-valued and $\mathcal{F}_0$-measurable. Furthermore,

$$E\|u_0\|_{m,p}^q + EK_{m,p}^q(T) < \infty,$$

where

$$K_{l,p}^p(t) := \int_0^t \{\|f(s)\|_{l,p}^p + \|g(s)\|_{l+1,p}^p\}\, dV_s.$$

Solutions of equation (3.1) are always understood according to Definition 2.2.

Theorem 3.1. *Under Assumptions* 3.1, 3.2, *and* 3.3 *there exists a unique solution of* (3.1) *with initial condition* u_0. *In addition,* $u(t)$ *is weakly continuous with values in* W_p^m *for each* ω, *and for any integer* $l \in [0,m]$

$$E \sup_{t\in[0,T]} \|u(t)\|_{l,p}^q \leq NE\|u_0\|_{l,p}^q + NEK_{l,p}^q(T), \tag{3.2}$$

where N *depends only on* $d, d_0, K, m, p, q,$ *and* T.

Proof. First we prove the existence. Assume temporarily that, in addition to our assumptions, we have that $q = p$ and

$$E\|u_0\|_{m,2}^2 + EK_{m,2}^2(T) < \infty. \tag{3.3}$$

In this situation the theorem is only a slight modification of Theorem 3.1 of [13] and can be proved by the same method. The only difference is that $V_t = t$ and Y_t is a d_1 dimensional Wiener process in [13]. In fact, under the additional assumption, one can easily obtain our result directly from Theorem 3.1 of [13]. Indeed, replacing V_t with $V_t + t$ (and multiplying the corresponding coefficients by $dV_t/(dV_t + dt)$) allows us to assume that V_t is strictly increasing. After that a time change reduces the whole situation to the one with $V_t = t$. To deal with Y_t one uses the fact that any continuous martingale can be written as a stochastic integral against a Wiener process. Anyhow in this way we obtain the existence and *uniqueness* of a solution $u(t)$ in the class of adapted processes which are weakly continuous with values in W_p^m and W_2^m on $[0, T]$ and we also get that, for any stopping time $\tau \le T$,

$$E \sup_{t\in[0,\tau]} \|u(t)\|_{l,p}^p \le NE\|u_0\|_{l,p}^p + NEK_{l,p}^p(\tau), \tag{3.4}$$

where N depends only on d, d_0, K, m, p, and T. This is (3.2) if $\tau = T$ and $q = p$.

To abandon the assumption that $q = p$ we use the following variant of well-known transformations of stochastic inequalities.

Lemma 3.2. *Let $T \in [0, \infty]$ and let f_t and g_t be nonnegative continuous $\mathcal{F}_t$-adapted processes such that, for any constant $c > 0$,*

$$Ef_\tau I_{g_0 \le c} \le Eg_\tau I_{g_0 \le c} \tag{3.5}$$

for any bounded stopping time $\tau \le T$. Then, for any bounded stopping time $\tau \le T$, for $\gamma \in (0, 1)$

$$E \sup_{t\le\tau} f_t^\gamma \le \tfrac{2-\gamma}{1-\gamma} E \sup_{t\le\tau} g_t^\gamma. \tag{3.6}$$

Proof. By replacing f_t and g_t with $f_{t\wedge T}$ and $g_{t\wedge T}$, respectively, we see that we may assume that $T = \infty$. Then we replace g_t with $\max_{s\le t} g_s$ and see that without losing generality we may assume that g_t is non-decreasing. In that case fix a constant $c > 0$ and let $\theta_f = \inf\{t \ge 0 : f_t \ge c\}$, $\theta_g = \inf\{t \ge 0 : g_t \ge c\}$. Then

$$\begin{aligned} P(\sup_{t\le\tau} f_t > c) &\le P(\theta_f \le \tau) \le P(\theta_g \le \tau) + P(\theta_f \le \tau \wedge \theta_g, \theta_g > \tau) \\ &\le P(g_\tau \ge c) + P(g_0 \le c, f_{\tau\wedge\theta_g\wedge\theta_f} \ge c) \\ &\le P(g_\tau \ge c) + \frac{1}{c} EI_{g_0\le c} f_{\tau\wedge\theta_g\wedge\theta_f}. \end{aligned}$$

In the light of (3.5) we replace the expectation with

$$\begin{aligned} EI_{g_0\le c} g_{\tau\wedge\theta_g\wedge\theta_f} &\le EI_{g_0\le c} g_{\tau\wedge\theta_g} = EI_{g_0\le c}(g_\tau \wedge g_{\theta_g}) \\ &\le EI_{g_0\le c}(g_\tau \wedge c) \le E(g_\tau \wedge c). \end{aligned}$$

Hence

$$P(\sup_{t\leq\tau} f_t > c) \leq P(g_\tau \geq c) + \frac{1}{c}E(c\wedge g_\tau)$$

Now it only remains to substitute $c^{1/\gamma}$ in place of c and integrate with respect to c over $(0,\infty)$. The lemma is proved. □

Remark 3.3. (i) Condition (3.5) is obviously satisfied if g_0 is non-random and

$$Ef_\tau \leq Eg_\tau \tag{3.7}$$

for any bounded stopping time $\tau \leq T$. By the way, generally (3.7) implies (3.6) only if g_0 is non-random which is easily seen if one takes $f_t \equiv f_0$ and $g_t \equiv g_0$.

(ii) Another situation in which condition (3.5) is satisfied presents if $E(f_\tau|G) \leq E(g_\tau|G)$ (a.s.), where G is a σ-field such that g_0 is G-measurable.

Now we continue the proof of the theorem. The argument showing (3.4) also proves that for any $\mathcal{F}_0$-measurable event A we have

$$EI_A \sup_{t\in[0,\tau]} \|u(t)\|_{l,p}^p \leq NEI_A\|u_0\|_{l,p}^p + NEI_AK_{l,p}^p(\tau), \tag{3.8}$$

with the same constant N as in (3.4). Therefore, owing to Lemma 3.2 we could conclude that (3.2) holds for any $q\in(0,p]$ if we knew that $\|u(t)\|_{l,p}^p$ were continuous in t. Generally, we do not know if it is true, but for any $\phi_1,\dots,\phi_n \in C_0^\infty(\mathbb{R}^d)$, the process $\eta(t) := \max_{i\leq n}(\phi_i, u(t))$ is continuous. If the norms of ϕ_i in $W_{p'}^{-l}$, with $p' = p/(p-1)$, are equal to 1, then we can substitute $\eta(t)$ in place of $\|u(t)\|_{l,p}$ on the left in (3.8). Then we get a corresponding modification of (3.2). After that it suffices to use the monotone convergence theorem upon taking ϕ_i to be dense on the unit sphere in $W_{p'}^{-l}$.

In order to show that condition (3.3) is not needed, we take a nonnegative $\zeta \in C_0^\infty(\mathbb{R}^d)$ such that $\zeta(x) = 1$ for $|x|\leq 1$ and $\zeta(x) = 0$ for $|x|\geq 2$ and, for $R>0$, define $\zeta_R(x) = \zeta(x/R)$. Also define $u_R(t)$ as weakly W_p^m-continuous and weakly W_2^m-continuous solutions of (3.1) with $\zeta_R f$ and $\zeta_R g$ in place of f and g, respectively, and with initial condition $\zeta_R u_0$.

By the above result and, in particular, by the uniqueness emphasized before (3.4) we have

$$\begin{aligned} &E\sup_{t\in[0,T]}\|u_R(t)-u_S(t)\|_{l,p}^q \\ &\quad\leq NE\|(\zeta_R-\zeta_S)u_0\|_{l,p}^q \\ &\qquad + NE\left(\int_0^T \{\|(\zeta_R-\zeta_S)f(s)\|_{l,p}^p + \|(\zeta_R-\zeta_S)g(s)\|_{l+1,p}^p\}\,dV_s\right)^{q/p}. \end{aligned}$$

By the dominated convergence theorem the left-hand side tends to zero as $R,S\to\infty$, and the rest is standard.

Now we pass to proving the uniqueness assuming without losing generality that $f=0$, $g=0$, $u_0=0$. Also owing to the first part of the proof we may assume that $p > 2$. Observe that, for each $R > 0$, the function $\zeta_R u(t)$ satisfies (3.1) with zero initial condition and with f_R and g_R in place of f and g respectively, where

$$f_R = -2a^{ij} D_i\zeta_R D_j u - u(a^{ij} D_{ij}\zeta_R + a^i D_i\zeta_R), \quad g_{Rk} = ub^i_k D_i\zeta_R.$$

The process $\zeta_R u(t)$ is W^1_2-valued and weakly continuous. Therefore, by (3.4) we have

$$E \sup_{t\in[0,\tau]} \|\zeta_R u(t)\|^2_{0,2} \le NE \int_0^\tau \{\|f_R(s)\|^2_{0,2} + \|g_R(s)\|^2_{1,2}\} \, dV_s \tag{3.9}$$

for any stopping time $\tau \le T$.

Introduce

$$\tau_n := T \wedge \inf\{t \ge 0 : \|u(t)\|_{1,p} > n\}, \quad n = 1, 2, \dots,$$

and assume that τ_n are stopping times. Then (3.9) holds with $\tau = \tau_n$. Observe that $\|u(t)\|_{1,p} \le n$ for $t < \tau_n$, so that Hölder's inequality and the fact that $p > 2$ easily imply that, for each n, for $s < \tau_n$, $\|f_R(s)\|^2_{0,2} + \|g_R(s)\|^2_{1,2}$ is bounded by a constant and tends to zero as $R \to \infty$. Then (3.9) shows that $u(t) = 0$ for $t \le \tau_n$, Furthermore, since $u(t)$ is weakly continuous in W^1_p, for each ω, its norm $\|u(t)\|_{1,p}$ is a bounded function on $[0, T]$ for each ω. It follows that $\tau_n \uparrow T$ as $n \to \infty$, so that $u(t) = 0$ for all $t \le T$.

Thus, it only remains to show that τ_n is a stopping time. Observe that while proving uniqueness from the very beginning we could take $\mathcal{F}_{t+}$ instead of $\mathcal{F}_t$ with the goal to prove the uniqueness in a larger class of functions. Therefore, without loss of generality, we assume that the family of σ-fields $(\mathcal{F}_t)_{t\ge 0}$ is right continuous.

Next notice that, since $u(t)$ is weakly continuous in W^1_p, for each ω, its norm $\|u(t)\|_{1,p}$ is lower semicontinuous. It follows that, for any $s \in [0, T)$ and ρ being the set of rationals on $[0, \infty)$, we have

$$\{\omega : \tau_n > s\}$$
$$= \bigcap_{m=1}^{\infty} \bigcup_{\substack{t\in\rho \\ s+1/m>t>s}} \bigcap_{r\in\rho\cap[0,t]} \{\omega : \|u(r)\|_{1,p} \le n\} \in \bigcap_{m=1}^{\infty} \mathcal{F}_{s+1/m} = \mathcal{F}_{s+} = \mathcal{F}_s.$$

The theorem is proved. □

4. Proof of Theorem 2.4

Clearly $v(t) := u_1(t) - u_0(t)$ satisfies the equation

$$dv(t) = L_r v(t) \, dV^r_{t,1} + M_k v(t) \, dY^k_t + (L_r u_0(t) + f_r) \, dA^r_t$$

with $v(0) = u_{10} - u_{00}$, where $A^r_t := V^r_{t,1} - V^r_{t,0}$. By Itô's formula from [8]

$$\begin{aligned}(L_r u_0(t) + f_r) \, dA^r_t =& d((L_r u_0(t) + f_r) A^r_t) \\ &- A^r_t (L_r L_j u_0(t) + L_r f_j) \, dV^j_{t,0} \\ &- A^r_t (L_r M_k u_0(t) + L_r g_k) \, dY^k_t.\end{aligned} \tag{4.1}$$

Hence for $\varphi(t) := v(t) - (L_r u_0(t) + f_r)A_t^r$ we have the equation

$$d\varphi(t) = (L_r\varphi(t) + F_r^1(t))\,dV_{t,1}^r + F_r^0(t)\,dV_{t,0}^r + (M_k\varphi(t) + G_k(t))\,dY_t^k, \quad (4.2)$$

with

$$F_r^1(t) := L_r(L_j u_0(t) + f_j)A_t^j, \quad F_r^0(t) := -L_j(L_r u_0(t) + f_r)A_t^j,$$
$$G_k(t) := [M_k, L_j]u_0(t)A_t^j + (M_k f_j - L_j g_k)A_t^j.$$

By Theorem 3.1 we have that $E\sup_{t\le t}\|u_0(t)\|_{m+4}^q \le N$. Thus by Assumptions 2.1, 2.2, and 2.4

$$E\sup_{t\le T}\|F_r^\varepsilon(t)\|_{m,p}^q \le NA^q \qquad \varepsilon = 0, 1,$$
$$E\sup_{t\le T}\|G_k(t)\|_{m+1,p}^q \le NA^q,$$
$$E\sup_{t\le T}\|\bar V_t^r(L_r u_0(t) + f_r)\|_{m,p}^q \le NA^q.$$

Hence by virtue of Theorem 3.1 we get

$$E\sup_{t\le T}\|\varphi(t)\|_{m,p}^q \le N\Big(E\|v(0)\|_{m,p}^q + A^q\Big),$$

which implies

$$E\sup_{t\le T}\|u_1(t) - u_0(t)\|_{m,p}^q \le N\Big(E\|u_{10} - u_{00}\|_{m,p}^q + A^q\Big)$$

by the triangle inequality. The proof of Theorem 2.4 is complete. □

5. The case of time dependent coefficients

Here we consider equation (2.1) keeping Assumptions 2.1, 2.2, 2.3 and 2.4, and assuming that the following condition also holds, in which

$$h(t,x) = (a_\gamma^{ij}(t,x), a_\gamma^i(t,x), a_\gamma(t,x), f_\gamma(t,x) : \gamma = 1, 2, \ldots, d_1, i, j = 1, \ldots, d).$$

Notice that g_k and the coefficients of the operators L_0 and M_k are not included in $h(t,x)$. In this section we stipulate that Greek integer-valued indices run through $1, 2, \ldots, d_1$.

Assumption 5.1. There exists a continuous $\mathcal{F}_t$-martingale

$$Z_t = (Z_t^1, \ldots, Z_t^{d_2})$$

and for any $x \in \mathbb{R}^d$ there exist bounded predictable functions

$$h_r(t,x) = (a_{\gamma r}^{ij}(t,x), a_{\gamma r}^i(t,x), a_{\gamma r}(t,x), f_{\gamma r}(t,x))$$

defined on $\Omega \times (0,T]$ for $r = 0, 1, \ldots, d_2$, such that

(i) $d\langle Z\rangle_t \le dV_t$,

(ii) $h(t,x) = h(0,x) + \int_0^t h_0(s,x)\,dV_s + \int_0^t h_r(s,x)\,dZ_s^r$ for all ω and t, where, as usual, the summation in r is carried over all *possible* values, which in this case are $1, 2, \ldots, d_2$. Furthermore, h_r are continuously differentiable with respect to x up to order $m+1$ and $|D^\beta h_r| \le K$ for $|\beta| \le m+1$.

Theorem 5.1. *Under Assumptions* 2.1, 2.2, 2.3, 2.4, *and* 5.1 *there is a constant* N *depending only on* $d, d_0, d_1, d_2, K, p, q, m,$ *and* T, *such that*

$$E \sup_{t\in[0,T]} \|u_1(t) - u_0(t)\|_{m,p}^q \le N(E\|u_{01} - u_{00}\|_{m,p}^q + A^q).$$

Proof. We can repeat the proof of Theorem 2.4 with straightforward changes. Define

$$L_{\gamma r} = a_{\gamma r}^{ij} D_{ij} + a_{\gamma r}^{i} D_i + a_{\gamma r}, \quad A_t^r := V_{t,1}^r - V_{t,0}^r,$$

and observe that, since $A_t^0 \equiv 0$ and now the coefficients of L_γ and f_γ depend on time, there is the following additional term on the right-hand side of formula (4.1):

$$-A_t^\gamma (L_{\gamma 0} u_0(t) + f_{\gamma 0})\, dV_t - A_t^\gamma L_{\gamma r}(M_k u_0(t) + g_k(t))\, d\langle Y^k, Z^r\rangle_t$$
$$-A_t^\gamma (L_{\gamma r} u_0(t) + f_{\gamma r}(t))\, dZ_t^r.$$

Thus equation (4.2) now has the additional term $H(t)\,dV_t + J_r(t)\,dZ_t^r$ on the right-hand side, where

$$H(t) := -A_t^\gamma (L_{\gamma 0} u_0(t) + f_{\gamma 0}(t)) - A_t^\gamma C_{kr} L_{\gamma r}(M_k u_0(t) + g_k(t)),$$
$$J_r(t) := -A_t^\gamma (L_{\gamma r} u_0(t) + f_{\gamma r}(t)),$$
$$C_{kr}(t) := d\langle Y^k, Z^r\rangle_t / dV(t).$$

Clearly,

$$E \sup_{t\le T} \|H(t)\|_{m,p}^q \le NA^q, \qquad E \sup_{t\le T} \|J_r(t)\|_{m+1,p}^q \le NA^q,$$

where N is a constant depending only on K, m, p, q, d_1, d_2, T. Thus with these modifications the proof of Theorem 5.1 is exactly the same as that of Theorem 2.4. □

By using the above theorem we can extend our result on splitting-up approximations, Theorem 2.5, to SPDEs with time dependent coefficients. Let us consider the solution $u(t)$ of equation (2.5) in $(0, T] \times \mathbb{R}^d$ with initial condition $u(0, x) = u_0(x)$, and remember that $T_n := \{t_i = iT/n : i = 0, 1, 2, \ldots, n\}$.

Since now L_r, f_r, M_k, g_k may depend on t, it is convenient to exhibit their dependence on t following the example $L_r(t)$. For $\gamma = 1, 2, \ldots, d_1$ and $s \in [0, T]$, let $\mathbf{P}_t^\gamma(s)\varphi$ denote the solution of the equation

$$dv(t) = (L_\gamma(s) v(t) + f_\gamma(s))\, dt, \quad t \ge 0, \quad v(0) = \varphi. \tag{5.1}$$

Notice that the coefficients of L_γ and f_γ are "frozen" at time s. Then $u^{(n)}(t)$ for $t \in T_n$ is defined recursively as follows: $u^{(n)}(0) = u_0$,

$$u^{(n)}(t_{i+1}) := \mathbf{P}_\delta^{d_1}(t_{i+1}) \cdots \mathbf{P}_\delta^2(t_{i+1}) \mathbf{P}_\delta^1(t_{i+1}) \mathbf{Q}_{t_i t_{i+1}}(u^{(n)}(t_i)) \tag{5.2}$$

for $i = 0, 1, 2, \ldots, n-1$, where $\delta = T/n$ and $\mathbf{Q}_{st}\varphi$ denotes the solution of the equation

$$d\tilde{v}(t) = (L_0(t)\tilde{v}(t) + f_0(t))\, dV_t^0 + (M_k(t)\tilde{v}(t) + g_k(t))\, dY_t^k, \quad t \ge s,$$
$$\tilde{v}(s) = \varphi.$$

Theorem 5.2. *Under Assumptions* 2.5 *and* 5.1 *there is a constant* N *depending only on* $d, d_0, d_1, d_2, K, p, q, m,$ *and* T*, such that*

$$E \max_{t \in T_n} \|u^{(n)}(t) - u(t)\|_{m,p}^q \leq N n^{-q}$$

for all integers $n \geq 1$.

Proof. The proof is almost exactly the same as that of the corresponding statement, Theorem 2.5, in the time independent case. We define $d' := d_1 + 1$, $\kappa(t)$, $\bar{V}^r_{t,\varepsilon}$, and $\bar{Y}^k(t)$ in the same way. Consider the counterparts of equations (2.8)

$$du_\varepsilon(t) = \big(L_r(\kappa(t))u_\varepsilon(t) + f_r(\kappa(t))\big)\, d\bar{V}^r_{t,\varepsilon} + \big(M_k(\kappa(t))u_\varepsilon(t) + g_k(\kappa(t))\big)\, d\bar{Y}^k_t$$

for $\varepsilon = 0$, 1, with initial data $u_\varepsilon(0) = u_0$.

Then it is almost obvious that the assumptions of Theorem 5.1 are satisfied with the same constant K and with $d'T$ in place of T. We apply this theorem and after that, as in the proof of Theorem 2.5, it only remains to observe that $u_0(d't) = u(t)$ and $u_1(d't) = u^{(n)}(t)$ for $t \in T_n$. The theorem is proved. □

Remark 5.3. We can define the approximation $u^{(n)}$ by

$$u^{(n)}(t_{i+1}) := \mathbf{P}^{d_1}_\delta(t_{i+1}) \cdots \mathbf{P}^{l+1}_\delta(t_{i+1})\mathbf{Q}_{t_i t_{i+1}}\mathbf{P}^l_\delta(t_i) \cdots \mathbf{P}^2_\delta(t_i)\mathbf{P}^1_\delta(t_i)u^{(n)}(t_i)$$

in place of (5.2), where $1 \leq l \leq d_1$ is a fixed integer. By obvious modifications of the above proof one can show that Theorem 5.2 holds also for this approximation.

Remark 5.4. One can define a splitting-up approximation for the solution of equation (2.5) also by

$$u^{(n)}(t_{i+1}) := \mathbf{P}^{d_1}_{t_i t_{i+1}} \cdots \mathbf{P}^2_{t_i t_{i+1}} \mathbf{P}^1_{t_i t_{i+1}} \mathbf{Q}_{t_i t_{i+1}}(u^{(n)}(t_i))$$

in place of (5.2), where $v(t) := \mathbf{P}^\gamma_{st}\varphi$ denotes the solution of the equation

$$dv(t) = (L_\gamma v(t) + f_\gamma(t))\, dt \quad t \geq s, \qquad v(s) = \varphi. \tag{5.3}$$

By a straightforward modification of the proof of Theorem 5.2 one can see that it remains true also for this approximation. We prefer the splitting-up approximation defined by (5.2), because in practice it is usually more convenient to solve the time-independent PDE (5.1) than the time-dependent PDE (5.3).

Let $C^l = C^l(\mathbb{R}^d)$ denote the Banach space of functions $f = f(x)$, $x \in \mathbb{R}^d$, having continuous derivatives up to the order l, such that

$$\|f\|_{C^l} := \sup_{x \in \mathbb{R}^d} \sum_{|\beta| \leq l} |D^\beta f(x)| < \infty.$$

We get the following corollary from the previous theorem by Sobolev's theorem on embedding of W^m_p into C^l.

Corollary 5.5. *If Assumptions* 2.5 *and* 5.1 *hold with* $m > l + d/p$ *and nonnegative integer* l, *then for some* $N = N(d, d_0, d_1, d_2, K, p, q, m, T)$

$$E \max_{t \in T_n} \|u^{(n)}(t) - u(t)\|^q_{C^l} \leq N n^{-q}$$

for all $n \geq 1$.

6. Sharpness of the rate of convergence estimate

Here we show that, generally speaking, the rate $1/n$ is sharp for equation (2.5) with $d_1 = 1$. The main point of the argument below is based on a quite surprising fact that, for deterministic equations, the splitting-up method provide almost the true solution in the middle points of the partition. For simplicity we only show this for time independent coefficients and free terms. Including time dependence is straightforward and would only lead to somewhat longer formulas.

First we investigate the case of deterministic PDEs. We consider the PDE

$$\frac{\partial}{\partial t} u(t, x) = (L_0 + L_1) u(t, x) + f(x) \tag{6.1}$$

with initial condition $u(0, x) = u_0(x)$, where f is a function on $\mathbb{R}^d$, and L_r, $r = 0, 1$, are operators as in (2.2) with coefficients a_r^{ij}, a_r^i, a_r, depending *only* on $x \in \mathbb{R}^d$. We split this PDE into two:

$$\frac{\partial}{\partial t} u(t, x) = L_0 u(t, x) + f_0(x), \tag{6.2}$$

$$\frac{\partial}{\partial t} u(t, x) = L_1 u(t, x) + f_1(x), \tag{6.3}$$

such that $f = f_0 + f_1$. We fix an integer $m \geq 0$, a real number $p \geq 2$, and make the following assumptions.

Assumption 6.1. The matrices (a_r^{ij}) are non-negative definite for all $x \in \mathbb{R}^d$, $r = 0, 1$.

Assumption 6.2. The partial derivatives of the coefficients a_r^i, a_r up to order m and for a^{ij} up to order $2 \vee m$ exist, are continuous, and by magnitude are bounded by K.

Assumption 6.3. The data u_0, f_0, f_1 are in W_p^m,

$$\|u_0\|_{m,p} + \|f_0\|_{m,p} + \|f_1\|_{m,p} \leq K.$$

If the above assumptions hold with $m \geq 1$, then each of equations (6.1), (6.2), (6.3), with initial condition u_0, admits a unique generalized solution (in the sense of Definition 2.2), by virtue of Theorem 3.1. Let u denote the solution of equation (6.1) with initial condition u_0, and let $\mathbf{P}_t^{(0)}$ and $\mathbf{P}_t^{(1)}$ denote the solution operators corresponding to (6.2) and (6.3), respectively. By virtue of Theorem

2.5, if Assumptions 6.1–6.3, with $m+4$ in place of m, hold then the splitting-up approximations

$$u^{(n)}(t_0)=u_0, \qquad u^{(n)}(t_{k+1})=\mathbf{P}_\delta^{(1)}\mathbf{P}_\delta^{(0)}u^{(n)}(t_k), \quad k=0,1,2,\dots,n-1 \quad (6.4)$$

converge in W_p^m to the true solution u with rate at least $1/n$. Below we show that by choosing appropriately the initial values of the splitting-up equations or choosing appropriately the splitting $f=f_0+f_1$ of the free term f, we can achieve at least $1/n^2$ as the rate of convergence for the splitting-up approximations $u^{(n)}(t_{k+\frac{1}{2}})=\mathbf{P}_{\delta/2}^{(0)}u^{(n)}(t_k)$ in the middle points

$$t_{k+1/2}:=(k+\frac{1}{2})\delta, \qquad k=0,1,2,\dots,n-1$$

of the partition $T_n:=\{t_k:=k\delta:k=0,1,2\dots,n\}$ with mesh size $\delta:=T/n$.

Introduce the following auxiliary splitting-up approximations u_n. For each $\delta=T/n$ take a $u_0^\delta\in W_p^{m+6}$ and let

$$u_n(t_{1/2}):=\mathbf{P}_{\delta/2}^{(0)}u_0^\delta,$$

$$u_n(t_k):=\mathbf{P}_{\delta/2}^{(1)}\mathbf{P}_{\delta/2}^{(0)}u_n(t_{k-\frac{1}{2}}), \quad u_n(t_{k+\frac{1}{2}}):=\mathbf{P}_{\delta/2}^{(0)}\mathbf{P}_{\delta/2}^{(1)}u_n(t_k), \quad k\geq 1.$$

Observe that, for $k\geq 1$,

$$u_n(t_{k-\frac{1}{2}})=\mathbf{P}_{\delta/2}^{(0)}(\mathbf{P}_\delta^{(1)}\mathbf{P}_\delta^{(0)})^{k-1}u_0^\delta, \quad u_n(t_k)=\mathbf{P}_{\delta/2}^{(1)}\mathbf{P}_\delta^{(0)}(\mathbf{P}_\delta^{(1)}\mathbf{P}_\delta^{(0)})^{k-1}u_0^\delta. \quad (6.5)$$

A few times below, for functions h on $\mathbb{R}^d$, we write $h\not\equiv 0$ to mean that $h\neq 0$ on a set of positive Lebesgue measure.

Theorem 6.1. *Let Assumptions* 6.1–6.3 *be satisfied with* $m+6$ *in place of* m. *Then there exists a constant* N, *depending only on* d, K, T, m, p, *such that the following statements hold:*

(i) We have

$$\max_{1\leq k\leq 2n}\|u(t_{k/2})-u_n(t_{k/2})\|_{m,p}\leq Nn^{-2}$$
$$+N\|u_0+(\delta/2)(L_1u_0+f_1)-u_0^\delta\|_{m,p}, \quad (6.6)$$

and, if $u_0^\delta=u_0$, *then*

$$\|u(t_{1/2})-u_n(t_{1/2})\|_{m,p}\geq(\delta/2)\|L_1u_0+f_1\|_{m,p}-Nn^{-2}. \quad (6.7)$$

(ii) Assume that $L_1u_0(x)+f_1(x)=0$ *(a.e.). Then, for the splitting-up approximations* $u^{(n)}$ *defined in* (6.4), *we have*

$$\max_{0\leq k\leq n-1}\|u(t_{k+1/2})-\mathbf{P}_{\delta/2}^{(0)}u^{(n)}(t_k)\|_{m,p}\leq Nn^{-2}.$$

(iii) Assume that $L_1u_0(x)+f_1(x)=0$ *(a.e.) and take* $0\leq r_1\leq r_2\leq T$. *Then*

$$\lim_{n\to\infty}n\max_{t\in T_n\cap[r_1,r_2]}\|u(t)-u^{(n)}(t)\|_{m,p}=(T/2)\sup_{t\in[r_1,r_2]}\|L_1u(t)+f_1\|_{m,p}. \quad (6.8)$$

In particular, if

$$\exists t\in[0,T]: \quad L_1u(t)+f_1\not\equiv 0, \quad (6.9)$$

then

$$\lim_{n\to\infty} n \max_{t\in T_n} \|u(t) - u^{(n)}(t)\|_{m,p} > 0.$$

Remark 6.2. If we take $u_0^\delta = \mathbf{P}^{(1)}_{\delta/2}u_0$, so that $u_n(t_k) = (\mathbf{P}^{(1)}_{\delta/2}\mathbf{P}^{(0)}_{\delta}\mathbf{P}^{(1)}_{\delta/2})^k u_0$ for $k = 1, \dots, n$, then the second term on the right in (6.6) is of the same order n^{-2} as the first one.

Indeed, for $h(t) := \mathbf{P}^{(1)}_t u_0$ we have

$$\begin{aligned} h(t) = h(0) + \int_0^t (L_1 h(s) + f_1)\, ds &= u_0 + t(L_1 u_0 + f_1) \\ &+ \int_0^t (t-s)L_1(L_1 h(s) + f_1)\, ds, \end{aligned}$$

which implies

$$\|u_0 + (\delta/2)(L_1 u_0 + f_1) - u_0^\delta\|_{m,p} \le N\delta^2 \sup_{t\le\delta}[\|h(t)\|_{m+4,p} + \|f_1\|_{m+2,p}].$$

In this case (6.6) is related to the so-called Strang's splitting formula (or the symmetric Trotter's formula, see [6], [17]).

Remark 6.3. If we take $u_0^{T/n} = u_0^\delta = u_0$, then

$$\begin{aligned} (T/2)\|L_1 u_0 + f_1\|_{m,p} &\le \varliminf_{n\to\infty} n \max_{1\le k\le 2n} \|u(kT/(2n)) - u_n(kT/(2n))\|_{m,p} \\ &\le \varlimsup_{n\to\infty} n \max_{1\le k\le 2n} \|u(kT/(2n)) - u_n(kT/(2n))\|_{m,p} \\ &\le N\|L_1 u_0 + f_1\|_{m,p}. \end{aligned}$$

Remark 6.4. By introducing $f_1 := -L_1 u_0$ and $f_0 = f - f_1$, one always achieves that $L_1 u_0 + f_1 = 0$. This way of defining f_i looks particularly innocent if $u_0 = 0$, so that one lets f go together with L_0.

Remark 6.5. Assume that u_0 is a *unique* solution of $L_1 u_0 + f_1 = 0$ in class W_p^{m+6}. Then, obviously, condition (6.9) is fulfilled if and only if $u(t) \not\equiv u_0$ on $[0, T]$. Also notice that the uniqueness holds if $-a_1$ is greater than a sufficiently big constant. This condition on a_1 can be satisfied on the account of changing a_0 in such a way that the sum $a_0 + a_1$ remains unchanged. Actually both a_0 and a_1 can be made as large negative as one likes by replacing u in equation (6.1) with $ue^{\lambda t}$, where λ is an appropriate constant.

Remark 6.6. Assume that $L_1 u_0 + f_1 = 0$. Then $u(t) \not\equiv u_0$ on $[0, T]$ if and only if $L_0 u_0 + f_0 \ne 0$.

Indeed, if $L_0 u_0 + f_0 \ne 0$, the assumption that $u(t) \equiv u_0$ leads to $\partial u/\partial t = 0$ and $(L_0 + L_1)u_0 + f = 0$, which contradicts $L_0 u_0 + f_0 \ne 0$. On the other hand, if $L_0 u_0 + f_0 = 0$, then $v(t) := u_0$ obviously is a solution of (6.1) with initial data u_0. By uniqueness $v = u$ and hence $u(t) \equiv u_0$.

Remark 6.7. Remarks 6.5 and 6.6 imply that if u_0 is a unique solution of $L_1 u_0 + f_1 = 0$ in class W_p^{m+6}, then condition (6.9) is fulfilled if and only if $L_0 u_0 + f_0 \ne 0$.

To prove Theorem 6.1, we first present a lemma about the PDE

$$du(t,x) = Lu(t,x)\,dt + G_r(t,x)\,dH_t^r, \qquad (t,x) \in (0,T] \times \mathbb{R}^d, \tag{6.10}$$

where G_r is a real-valued function of $(t,x) \in [0,T] \times \mathbb{R}^d$, H^r is a continuous real function of bounded variation on $[0,T]$ for every $r = 0,1,\dots,d_1$ (here we allow d_1 to be any integer), and

$$L = a^{ij}(t,x)D_{ij} + a^i(t,x)D_i + a(t,x)$$

with non-random coefficients $(a^{ij}) \geq 0$, a^i, a.

Lemma 6.8. *Assume that $G_r \in L_p([0,T], W_p^{m+2})$ and*

$$d(G_r(t),\phi)_0 = (G_r'(t),\phi)\,dt, \qquad r = 0,1,\dots,d_1$$

for all $\phi \in C_0(\mathbb{R}^d)$, with some functions $G_r' \in L_p([0,T], W_p^m)$. Suppose that the derivatives in x of a, a^i up to order m and for a^{ij} up to order $m \vee 2$ exist and are bounded by the constant K. Then equation (6.10) *with initial condition $\psi \in W_p^m$ has a unique generalized solution u and*

$$\sup_{t\in[0,T]} \|u(t)\|_{m,p}^p \leq N\|\psi\|_{m,p}^p + N \sup_{t\in[0,T]} |H_t^r|^p \|G_r(t)\|_{m,p}^p$$

$$+ N \sup_{t\in[0,T]} |H_t^r|^p \int_0^T (\|G_r(t)\|_{m+2,p}^p + \|G_r'(t)\|_{m,p}^p)\,dt,$$

where N is a constant depending only on p, d, d_1, K, m, T.

Proof. Clearly

$$d((G_r(t),\phi)H_t^r) = (G_r(t),\phi)\,dH_t^r + H_t^r(G_r'(t),\phi)\,dt,$$

for all $\phi \in C_0(\mathbb{R}^d)$. Hence, u is a generalized solution of equation (6.10), if and only if $w(t,x) := u(t,x) - H_t^r G_r(t,x)$ is a generalized solution of

$$dw(t,x) = \{Lw(t,x) + H_t^r(LG_r(t,x) - G_r'(t,x))\}\,dt$$

with initial condition $w(0,x) = \psi - H_0^r G_r(0,x)$. Hence the lemma follows immediately from Theorem 3.1. □

Proof of Theorem 6.1. For each fixed integer $n \geq 1$ we change the time by using the time scale $\kappa(t)$ with $d_1 = 1$ defined by (2.7). We consider the equations

$$dv_0(t) = (L_1 v_0(t) + f_1)\,dA_t^0 + (L_0 v_0(t) + f_0)\,dA_t^0, \tag{6.11}$$

$$dv_1(t) = (L_1 v_1(t) + f_1)\,dA_t^1 + (L_0 v_1(t) + f_0)\,dA_t^0, \tag{6.12}$$

with initial conditions $v_0(0) = u_0$, $v_1(0) = u_0^\delta$, where $A_t^0 := \kappa(t)$ and $A_t^1 := \kappa(t-\delta)$. They admit unique generalized solutions v_0 and v_1, respectively, which are weakly continuous W_p^{m+6}-valued functions on $[0,T]$.

Notice that $u(A_t^0) = v_0(t)$, so that $u(t_{k/2}) = v_0(k\delta - \delta/2)$ for $k = 1,\dots,2n$. Furthermore, formulas (6.5) easily imply that $u_n(t_{k/2}) = v_1(k\delta - \delta/2)$ for $k = 1,\dots,2n$. Hence,

$$\max_{k=1,\dots,2n} \|u(t_{k/2}) - u_n(t_{k/2})\|_{m,p} = \max_{k=1,\dots,2n} \|(v_0 - v_1)(k\delta - \delta/2)\|_{m,p}.$$

Thus, to prove (6.6), it suffices to prove that

$$\max_{k=1,\dots,2n} \|(v_0 - v_1)(k\delta - \delta/2)\|_{m,p} \le Nn^{-2} + N\|u_0 + (\delta/2)(L_1 u_0 + f_1) - u_0^\delta\|_{m,p}. \tag{6.13}$$

To this end we set

$$\bar{A}_t := A_t^0 - A_t^1 - \delta/2, \quad w(t) := v_0(t) - v_1(t) - \bar{A}_t(L_1 v_0(t) + f_1). \tag{6.14}$$

Then by taking into account that

$$d\big((L_1 v_0(t) + f_1, \phi)\bar{A}_t\big) = (L_1(Lv_0(t) + f), \phi)\bar{A}_t\, dA_t^0 + (L_1 v_0(t) + f_1, \phi)\, d\bar{A}_t \tag{6.15}$$

for all $\phi \in C_0(\mathbb{R}^d)$, from equations (6.11), (6.12) we easily get that

$$d(v_0 - v_1)(t) = L_r(v_0 - v_1)(t)\, dA_t^r + (L_1 v_0(t) + f_1)\, d\bar{A}_t,$$

and that w is a generalized solution of the equation

$$dw(t) = L_r w(t)\, dA_t^r + L_r(L_1 v_0(t) + f_1)\bar{A}_t\, dA_t^r - L_1(Lv_0(t) + f)\bar{A}_t\, dA_t^0$$

with initial condition $w(0) = u_0 - u_0^\delta + \delta(L_1 u_0 + f_1)/2$. We rewrite this equation into

$$dw(t) = \hat{L}w(t)\, dt + G_r(t)\, dH_t^r$$

with

$$\hat{L} := \dot{A}_t^r L_r, \qquad \dot{A}_t^r := \frac{d}{dt}A_t^r, \qquad H_t^r := \int_0^t \bar{A}_s\, dA_s^r,$$

and obvious G_r. It is easy to see that G_r are differentiable in t and $G_r(t)$ and $G_r'(t)$ satisfy

$$\sup_{t\in[0,2T]} \|G_r(t)\|_{m+2,p} \le N\big(\sup_{t\in[0,2T]} \|v_0(t)\|_{m+6,p} + \sum_r \|f_r\|_{m+4,p}\big),$$

$$\sup_{t\in[0,2T]} \|G_r'(t)\|_{m,p} \le N\big(\sup_{t\in[0,2T]} \|v_0(t)\|_{m+6,p} + \sum_r \|f_r\|_{m+4,p}\big).$$

By Theorem 3.1, the latter quantities are bounded above by a constant depending only on d, K, p, m, T. Furthermore, for $r = 0, 1$, we have

$$\sup_{t\ge 0} |H_t^r| \le \delta^2/8 = T^2 n^{-2}/8.$$

Hence, by Lemma 6.8,

$$\sup_{t\in[0,2T]} \|w(t)\|_{m,p} \le Nn^{-2} + N\|u_0 + (\delta/2)(L_1 u_0 + f_1) - u_0^\delta\|_{m,p}. \tag{6.16}$$

By combining this with (6.14) and the equality $\bar{A}_{(k-1/2)\delta} = 0$, $k = 1, 2, \dots, 2n$, we get (6.13). This proves (6.6).

As for (6.7), we define $h(t) = v_0(t) - v_1(t)$, observe that $h(0) = 0$, and, by computing the time derivative, for $t \in [0, \delta]$ we obtain

$$h(t) = \int_0^t (Lv_0(s) + f - L_0 v_1(s) - f_0)\,ds$$
$$= t(L_1 u_0 + f_1) + \int_0^t (t-s)[L(Lv_0(s) + f) - L_0(L_0 v_1(s) + f_0)]\,ds.$$

Hence,

$$\|u(t_{1/2}) - u_n(t_{1/2})\|_{m,p} = \|v_0(t_{1/2}) - v_1(t_{1/2})\|_{m,p} = \|h(t_{1/2})\|_{m,p}$$
$$\geq (\delta/2)\|L_1 u_0 + f_1\|_{m,p} - \delta^2 N \sup_{t \leq \delta/2} [\|v_0(t)\|_{m+4,p} + \|v_1(t)\|_{m+4,p}]$$
$$- \delta^2 N[\|f\|_{m+2,p} + \|f_0\|_{m+2,p}],$$

which proves (6.7).

Assertion (ii) follows from (i) since $\mathbf{P}^{(0)}_{\delta/2} u^{(n)}(t_k) = u_n(t_{k+\frac{1}{2}})$ (cf. (6.5)).

To prove (iii), take $u_0^\delta = u_0$. Then it is easy to see that

$$\max_{t \in T_n \cap [r_1, r_2]} \|u(t) - u^{(n)}(t)\|_{m,p} = \max_{t \in T_n \cap [r_1, r_2]} \|v_0(2t) - v_1(2t)\|_{m,p}.$$

Due to (6.14) we further get

$$(\delta/2) \max_{t \in T_n \cap [r_1, r_2]} \|L_1 v_0(2t) + f_1\|_{m,p} + \sup_{t \in [0, 2T]} \|w(t)\|_{m,p}$$
$$\geq \max_{t \in T_n \cap [r_1, r_2]} \|u(t) - u^{(n)}(t)\|_{m,p}$$
$$\geq (\delta/2) \max_{t \in T_n \cap [r_1, r_2]} \|L_1 v_0(2t) + f_1\|_{m,p} - \sup_{t \in [0, 2T]} \|w(t)\|_{m,p}.$$

Next, we use (6.16) along with the observation that the second term on the right in (6.16) is zero by our assumptions. We also use the fact that $v_0(2k\delta) = u(k\delta)$ and that u is a weakly continuous function of t with values in W_p^{m+2}, and therefore $\|L_1 u(t) + f_1\|_{m,p}$ is lower continuous in $t \in [0, T]$. Then we arrive at (6.8). The theorem is proved. □

Now we consider stochastic PDE (2.5) with $d_1 = 1$, $V_t^0 = t$ and with a d_0-dimensional Wiener process W in place of the martingale Y:

$$du(t,x) = (L_1 u(t,x) + f_1(x))\,dt$$
$$+ (L_0 u(t,x) + f_0(x))\,dt + (M_k u(t,x) + g_k(t,x))\,dW_t^k, \quad (6.17)$$

where the operators L_r, M_k are of the form

$$L_r = a_r^{ij}(x) D_{ij} + a_r^i(x) D_i + a_r(x), \quad M_k = b_k^i(x) D_i + b_k(x).$$

We assume that all data are deterministic and that the coefficients a_r^{ij}, a_r^i, a_r and the free terms f_r do not depend on t. Take $p \geq 2$, an integer $m \geq 0$, and $q \in [1, p]$.

Theorem 6.9. *Let Assumption* 2.5 *be satisfied with* $m+2$ *in place of* m. *Suppose that* u_0 *is a unique (nonrandom) solution of the equation* $L_1 u_0 + f_1 = 0$ *in* W_p^{m+6} *(see Remark* 6.5*). Also assume that* $L_0 u_0 + f_0 \neq 0$ *and define the splitting-up approximation* $u^{(n)}$ *for equation* (6.17), *as before (see* (2.6)*), by* $u^{(n)}(0) := u_0$,

$$u^{(n)}(t_{i+1}) := \mathbf{P}_\delta^{(1)} \mathbf{Q}_{t_i,t_{i+1}} u^{(n)}(t_i), \quad i = 0, 1, \ldots, n-1.$$

Then

$$\underline{\lim}_{n\to\infty} n^q \max_{t\in T_n} E\|u^{(n)}(t) - u(t)\|_{m,p}^q > 0.$$

Proof. Since by Jensen's inequality, Minkowski's inequality, and the triangle inequality

$$\begin{aligned}(E\|u^{(n)}(t) - u(t)\|_{m,p}^q)^{1/q} &\geq E\|u^{(n)}(t) - u(t)\|_{m,p} \\ &\geq \|Eu^{(n)}(t) - Eu(t)\|_{m,p} \\ &\geq |\|Eu^{(n)}(t)\|_{m,p} - \|Eu(t)\|_{m,p}|,\end{aligned}$$

it suffices to prove that

$$\underline{\lim}_{n\to\infty} n \max_{t\in T_n} |\, \|Eu^{(n)}(t)\|_{m,p} - \|Eu(t)\|_{m,p}\,| > 0.$$

To this end we notice that $\bar{u}(t) := Eu(t)$ is a generalized solution of the problem

$$d\bar{u}(t) = (L_1\bar{u}(t) + f_1)\,dt + (L_0\bar{u}(t) + f_0)\,dt, \quad \bar{u}(0) = u_0. \tag{6.18}$$

Indeed, for all $\phi \in L_{p'}([0,T], \mathbb{R}^d)$, $p' = p/(p-1)$, we have

$$E\Big(\int_0^t (b_k^i D_i u(s) + b_k u(s) + g_k(s), \phi)\, dW_s^k\Big)^2$$

$$\leq N\|\phi\|_{p'}^2 E\int_0^t (\|u(s)\|_{1,p}^2 + \sum_k \|g_k(s)\|_{0,p}^2)\, dt < \infty$$

by virtue of Theorem 3.1, which implies that the expectation of the stochastic integral in the definition of the generalized solution u (see Definition 2.2) vanishes. Thus taking the expectation in both sides of equation (2.4) and using that all data are deterministic, by interchanging the derivation and expectation we get that $\bar{u}$ is a generalized solution of (6.18). We get similarly that $\bar{u}^{(n)} := Eu^{(n)}$ is the splitting-up approximation

$$\bar{u}^{(n)}(t_0) = u_0, \qquad \bar{u}^{(n)}(t_{k+1}) = \mathbf{P}_\delta^{(1)}\mathbf{P}_\delta^{(0)}\bar{u}^{(n)}(t_k)$$

for $\bar{u}$, where $\mathbf{P}_t^{(r)}$ is the solution operator of equation

$$du(t) = (L_r u(t) + f_r)\,dt.$$

Hence the theorem follows from statement (iii) of Theorem 6.1 and Remarks 6.7. The theorem is proved. □

Remark 6.10. From the above proof it is clear that Theorem 6.9 holds also for the splitting-up approximations

$$u^{(n)}(0) := u_0, \quad u^{(n)}(t_{i+1}) := \mathbf{Q}_{t_i,t_{i+1}}\mathbf{P}_\delta^{(1)}u^{(n)}(t_i), \quad i = 0, 1, \ldots, n-1,$$

if we interchange L_1 with L_0 and f_1 with f_0 in the formulation of the theorem.

References

[1] A. Bensoussan, R. Glowinski and A. Rascanu, *Approximation of Zakai equation by the splitting-up method.* In: *Stochastic Systems and Optimization, Warsaw, 1988.* Lect. Notes in Control and Information Sciences **136**, 257–265, Springer, 1989.

[2] A. Bensoussan, R. Glowinski, and A. Rascanu, *Approximation of some stochastic differential equations by the splitting-up method.* Appl. Math. Optim. **25**, 81–106.

[3] D. Crisan and T. Lyons, *A particle approximation of the solution of the Kushner-Stratonovitch equation.* Probab. Theory Related Fields **115** (1999), 549–578.

[4] P. Florchinger and F. Le Gland, *Time-discretization of the Zakai equation for diffusion processes observed in correlated noise.* Stochastics and Stochastics Reports **35** (1991), 233–256.

[5] A. Germani and M. Piccioni, *Semi-discretization of stochastic partial differential equations on $\mathbb{R}^d$ by a finite element technique.* Stochastics **23** (1988), 131–148.

[6] D. Gottlieb, *Strang-type difference schemes for multidimensional problems.* SIAM J. Numer. Anal. **9** (1972), 650–661.

[7] W. Grecksch and P.E. Kloeden, *Time-discretized Galerkin approximations of parabolic stochastic PDEs.* Bulletin of the Australian Mathematical Society **54** (1996), 79–85.

[8] I. Gyöngy and N. V. Krylov, *On stochastic equations with respect to semimartingales II, Itô formula in Banach spaces.* Stochastics **6** (1982), 153–174.

[9] I. Gyöngy and N. V. Krylov, *On splitting-up method and stochastic partial differential equations.* The Annals of Probability **31** (2003), 564–591.

[10] K. Ito and B. Rozovskii, *Approximation of the Kushner equation for nonlinear filtering.* Control Optim. **38** (2000), 893–915.

[11] N. V. Krylov, *Introduction to the theory of diffusion processes.* Amer. Math. Soc., RI, 1995.

[12] N. V. Krylov and B. L. Rozovskii, *Stochastic evolution equations.* J. of Soviet Mathematics **16** (1981), 1233–1277.

[13] N. V. Krylov and B.L. Rozovskii, *On the characteristics of degenerate second order parabolic Itô equations.* Trudy seminara imeni Petrovskogo **8** (1982), 153–168 (in Russian); English translation in *J. Soviet Math.* **32**, (1986), 336–348.

[14] S. V. Lototsky, *Problems in statistics of stochastic differential equations.* Thesis, University of Southern California, 1996.

[15] S. V. Lototsky, R. Mikulevicius and B. Rozovskii, *Nonlinear filtering revisited: a spectral approach.* SIAM J. Control Optim. **35** (1997), 435–461.

[16] B. Rozovskii, *Stochastic evolution systems. Linear theory and applications to nonlinear filtering.* Kluwer, Dordrecht, (1990).

[17] G. Strang, *On the construction and comparison of difference schemes.* SIAM J. Numer. Anal. **5** (1968), 506–517.

[18] H. Yoo, *An analytic approach to stochastic differential equations and its applications.* Ph.D. thesis, University of Minnesota, 1998.

Acknowledgment

The first author gratefully acknowledges the financial support provided by a grant of Marta Sanz-Solé, and by CRM, Institut d'Estudis Catalans, towards his participation in the 'Stochastic Inequalities' conference in Barcelona in June 2002, where this paper was presented. The work of the second author was partially supported by NSF Grant DMS-0140405.

School of Mathematics, University of Edinburgh, King's Buildings, Edinburgh, EH9 3JZ, United Kingdom
E-mail address: `gyongy@maths.ed.ac.uk`

127 Vincent Hall, University of Minnesota, Minneapolis, MN, 55455, USA
E-mail address: `krylov@math.umn.edu`

Progress in Probability, Vol. 56, 323–338

Lower Bounds for Densities of Uniformly Elliptic Non-homogeneous Diffusions

Arturo Kohatsu-Higa

Abstract. In this article we interpret heuristically the conditions of the definition of a uniformly elliptic random variable on Wiener space which allow to obtain Aronson type estimates for the density of this random variable. As an example we apply this concept to uniformly elliptic non-homogeneous diffusions.

1. Introduction

In a recent article (see [10]), we have described some minimal conditions that ensure that a random variable F defined on Wiener space has a density with a lower bound of Gaussian type. We called a random variable satisfying such conditions a uniformly elliptic random variable (see Definition 1). This class of random variables is broad enough to be applied to various stochastic equations. In particular, it includes the solution to the uniformly elliptic stochastic heat equation and the uniformly elliptic hyperbolic stochastic partial differential equation, also known as the biparametric diffusion (see [4]). A natural environment for the study of densities is Malliavin Calculus. We refer the reader to any of the well known books available on the matter (see for example, [15], [19], [20]).

The purpose of this article is to clarify the role of each requirement in the definition of uniformly elliptic random variable and show how it can be verified in a non-trivial simple example. Still, the result we obtain seems to be new.

We consider the case of the non-homogeneous uniformly elliptic diffusion. That is, let X be the solution to the following stochastic differential equation

$$X_t = x + \int_0^t b(s, X_s)ds + \int_0^t \sigma_j(s, X_s)dW_s^j$$

where $b : [0,T] \times \mathbb{R}^q \to \mathbb{R}^q$, $\sigma : [0,T] \times \mathbb{R}^q \to \mathbb{R}^q \times \mathbb{R}^k$ and W is a k-dimensional Wiener process. Assume that the coefficients $b(t,\cdot)$, $\sigma(t,\cdot) \in C_b^\infty(\mathbb{R}^q)$ uniformly in time and that the functions are jointly measurable. Under these conditions the

---

2000 *Mathematics Subject Classification.* Primary 60H07; Secondary 35B65 (2000MSC).
Key words and phrases. Malliavin Calculus, density estimates, Aronson estimates, non-homogeneous diffusions.
This research was partially supported with grants BFM 2000-807 and BFM 2000-0598.

solution to the above equation exists and is unique. Furthermore assume that the solution process is uniformly elliptic. That is, for any vectors x, $\xi \in \mathbb{R}^q$ and $t \in [0,T]$, there exists a strictly positive constant c so that

$$\xi'\sigma(t,x)\sigma(t,x)'\xi \geq c\,\|\xi\|^2 .$$

It is natural to expect under this condition that the random variable $F = X_t$ should behave as a Gaussian random variable. In fact, one expects to have that X_t should have a density for $t > 0$, denoted by $p_t(x,\cdot)$, and that there exists two positive constants m and M such that

$$m^{-1}t^{-q/2}\exp\left(-\frac{m\,\|y-x\|^2}{t}\right) \leq p_t(x,y) \leq Mt^{-q/2}\exp\left(-\frac{\|y-x\|^2}{Mt}\right) \tag{1.1}$$

The fact that the density exists and that it has a Gaussian upper bound can be solved with well known techniques of Malliavin Calculus (see [23]). Also there are very well known established analytical techniques to solve the problem of finding upper and lower bounds for fundamental solutions of homogeneous or non-homogeneous partial differential equations. See for example, [1], [5], [6], [17], [21], [22] and [14].

In the stochastic framework using Malliavin Calculus, the problem of finding a detailed global lower bound for the density of X_t was first studied for the hypoelliptic diffusion case in [13]. Furthermore, the metric in the exponent is more explicit than the one in (1.1) and is closely related to the Riemmanian metric defined through the coefficients of the stochastic differential equation.

Azencott (see [2]) obtained a Gaussian lower bound estimate for the uniformly elliptic non-homogeneous diffusion in the case the coefficients are smooth in time and space. We generalize this result, via a different approach, weakening the restrictions on the time variable. Furthermore, we want to stress the generality of the definition of uniformly elliptic random variables. In fact, the result presented here for non-homogeneous diffusions can be easily generalized to the case of solutions of uniformly elliptic stochastic differential equations with random coefficients under appropriate continuity and differentiability conditions on the coefficients.

$C_b^\infty(\mathbb{R}^d)$ denotes the space of real bounded functions on $\mathbb{R}^d$ such that they are infinitely differentiable with bounded derivatives. $C_p^\infty(\mathbb{R}^d)$ stands for a similar space but the functions and their derivatives have polynomial growth instead. C, c, m and M denote constants in general that may change from one line to another unless stated otherwise. We also use the double index summation convention. $\|\cdot\|$ without any subindices denotes the usual Euclidean norm in $\mathbb{R}^l$. The dimension l should be clear from the context. For stochastic processes we use indistinctively $X(t)$ or X_t.

2. Preliminaries

Let W be a k-dimensional Wiener process indexed in $[0,T]$. Our base space is a sample space $(\Omega, \mathcal{F}, P)$ where the Wiener process is defined (for details see [19], Section 1.1 and [20]). The associated filtration will be defined as $\{\mathcal{F}_t; 0 \le t \le T\}$, where $\mathcal{F}_t$ is the σ-field generated by the random variables $\{W(s),\ s \in [0,t]\}$. On the sample space $(\Omega, \mathcal{F}, P)$ one can define a derivative operator D, associated domains $(\mathbb{D}^{n,p}, \|\cdot\|_{n,p})$ where n denotes the order of differentiation and p denotes the $L^p(\Omega)$ space where the derivatives lie. The high order stochastic derivative is denoted by D^v_α for $v \in \{1,\dots,k\}^n$ and $\alpha \in [0,T]^n$. We say that F is smooth if $F \in \mathbb{D}^\infty = \cap_{n\in\mathbf{N},p>1}\mathbb{D}^{n,p}$.

For a q-dimensional random variable $F \in \mathbb{D}^{1,2}$, we denote by ψ_F the Malliavin covariance matrix associated with F. That is, $\psi_F^{i,j} = \langle DF^i, DF^j\rangle_{L^2[0,T]}$. One says that the random variable is non-degenerate if $F \in \mathbb{D}^\infty$ and the matrix ψ_F is invertible a.s. and $(\det \psi_F)^{-1} \in \bigcap_{p\ge1} L^p(\Omega)$. In such a case expressions of the type $E(\delta_y(F))$, where δ_y denotes the Dirac delta function, have a well defined meaning through the integration by parts formula.

The integration by parts formula of Malliavin Calculus can be briefly described as follows. Suppose that F is a non-degenerate random variable and $G \in \mathbb{D}^\infty$. Then for any function $g \in C_p^\infty(\mathbb{R}^q)$ and a finite sequence of multi-indexes $\beta \in \cup_{l\ge1}\{1,\dots,q\}^l$, we have that there exists a random variable $H^\beta(F,G) \in \mathbb{D}^\infty$ so that

$$E(g^\beta(F)G) = E(g(F)H^\beta(F,G))$$

Here g^β denotes the high order derivative of order $l(\beta)$ and whose partial derivatives are taken according the index vector β. This inequality can be obtained following the calculations in Lemma 12 of [16]. In some cases we will consider the above norms and definitions on a conditional form. That is, we will use partial Malliavin Calculus. We will denote this by adding a further time sub-index in the norms. For example, if one completes the space of smooth functionals with the norm

$$\begin{aligned} \|F\|_{2,s} &= (E(\|F\|^2/\mathcal{F}_s))^{1/2} \\ \|F\|^2_{1,2,s} &= \|F\|^2_{2,s} + E(\int_s^T \|D_uF\|^2\,du/\mathcal{F}_s), \end{aligned}$$

we obtain the space $\mathbb{D}^{1,2}_s$.

To simplify the notation we will sometimes denote $E_s(\cdot) = E(\cdot/\mathcal{F}_s)$. Analogously we will write H^β_s and $\psi_F(s)$ when considering integration by parts formula and the Malliavin covariance matrix conditioned on $\mathcal{F}_s$. That is, $\psi_F^{i,j}(s) = \langle DF^i, DF^j\rangle_{L^2[s,T]}$. Also we say that $F \in \overline{\mathbb{D}}^{1,2}_s$ when $F \in \mathbb{D}^{1,2}_s$ and $\|F\|_{1,2,s} \in \bigcap_{p\ge1} L^p(\Omega)$. Similarly, we say that F is s-conditionally non-degenerate if $F \in \overline{\mathbb{D}}^{\infty}_s$ and $(\det\psi_F(s))^{-1} \in \cap_{p>1} L^p_s(\Omega)$. In such a case, as before, expressions like $E(\delta_y(F)/\mathcal{F}_s)$ have a well defined meaning through the partial integration by parts formula or via an approximation of the delta function.

We will also have to deal with similar situations for sequences F_i that are $\mathcal{F}_{t_i}$-measurable random variables, $i = 1, \ldots, N$ for a partition $0 = t_0 < t_1 < \cdots < t_N$. In this case we say that $\{F_i; i = 1, \ldots, N\} \subseteq \overline{\mathbb{D}}^{\infty}$ uniformly if $F_i \in \overline{\mathbb{D}}^{\infty}_{t_{i-1}}$ for all $i = 1, \ldots, N$ and for any $l > 1$ one has that there exists a finite positive constant $C(n, p, l)$ such that

$$\sup_{N} \sup_{i=1,..,N} E \|F_i\|^l_{n,p,t_{i-1}} \leq C(n, p, l).$$

In what follows we will sometimes expand our basic sample space to include further increments of another independent Wiener process, $\overline{W}$ (usually these increments are denoted by $Z_i = \overline{W}(i+1) - \overline{W}(i) \sim N(0,1)$) independent of W in such a case we denote the expanded filtration by $\overline{\mathcal{F}}_t = \mathcal{F}_t \vee \sigma(\{\overline{W}(s); s \leq i+1, t_i \leq t\})$. We do this without further mentioning and suppose that all norms and expectations are considered in the extended space. Sometimes we will write $F \in \mathcal{F}_t$ which stands for F is a $\mathcal{F}_t$-measurable random variable.

We use the notation $I^i_j(h) = \int_{t_{i-1}}^{t_i} h(s) dW^j(s)$ for $j = 1, \ldots, k$ and $h : \Omega \to L^2([t_{i-1}, t_i]; \mathbb{R}^q)$ a $\mathcal{F}_{t_{i-1}}$-measurable smooth random processes.

3. Some heuristics

In order to motivate the definition of uniformly elliptic random variable we give a brief idea of how to obtain a lower Gaussian estimate for the density of this r.v. (for a complete proof, see [10]). We use the case of non-homogeneous diffusion in order to draw a parallel with some well known concepts. This will also help us guide in the application of the Definition 1 in Section 4. All references to the definition of uniformly elliptic random variable are in Definition 1. The idea for the proof which also appears graphically in Figure 1 can be explained as follows: We say that an estimate for a density $p(x)$ is global if the estimate is valid for all x. In contrast, we say that an estimate is local if it is only valid for points close to x. In order to obtain a global Gaussian type lower bound for a random variable $F = X(t)$ generated by a Wiener process, we first identify a time component in the random variable that will assure adaptedness. In our example this is t. Next, given any sequence of partitions of the time interval $[0, t]$ we assume there exists a sequence of adapted approximations along the time axis. In our case, let $0 = t_0 < t_1 < \cdots < t_N = t$ be any partition of $[0, t]$. Then we define the sequence of approximations through $F_i = X(t_i)$. Denote by $p(t_i, x_i; t_{i+1}, x_{i+1})$ the conditional density of F_{i+1} at x_{i+1} with respect to $F_i = x_i$.

Idea of the proof. In order to obtain the global Gaussian type lower bound one uses the Chapman–Kolmogorov formula. That is, if one can have a good local lower Gaussian estimate of the density of $F_i = X(t_i)$ conditioned to $\mathcal{F}_{t_{i-1}}$ then one can hope that the global lower estimate for the density of $X(t)$ should be satisfied. In order to obtain this local estimate one needs to consider partitions of small size.

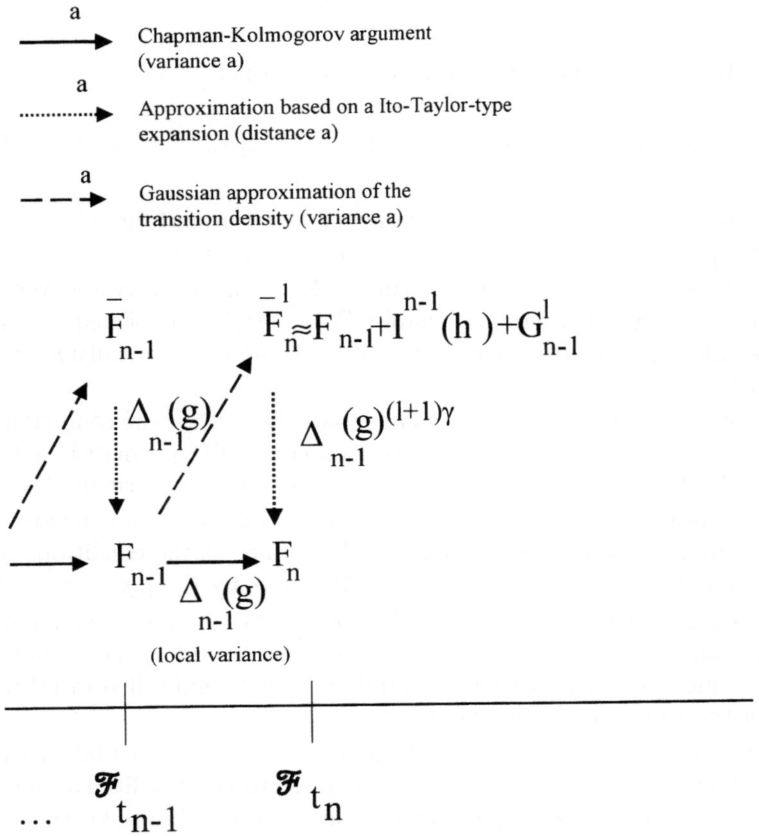

FIGURE 1. Idea of the proof to obtain Gaussian lower bounds for F_n. Here $I^{n-1}(h)$ is conditionally Gaussian and its local variance is of the order $\Delta_{n-1}(g)$. G^l_{n-1} is a residue in the sense that is of smaller order than $I^{n-1}(h)$. As l becomes bigger $\overline{F}^l_n$ approaches F_n.

The main idea of the argument is to use Chapman-Kolmogorov formula to obtain that

$$p(0,x;t,y) = \int_{\mathbb{R}^q} \cdots \int_{\mathbb{R}^q} p(0,x;t_1,x_1)\cdots p(t_{n-1},x_{n-1};t_n,y)dx_1\cdots dx_{n-1}.$$

Here $\pi = \{t_i;\ i=0,\ldots,N\}$ is a partition of $[0,t]$ with $0=t_0<\cdots<t_n=t$. This is noted in Figure 1 by the solid arrows. Next we localize the previous estimation by considering balls $B_i = B(x_i,r_i)$ for some appropiate radius r_i and $x_n = y$.

Then the above can be bounded by below giving that

$$p(0,x;t,y) \geq \int_{B_n} \cdots \int_{B_2} p(0,x;t_1,x_1) \dots p(t_{n-1},x_{n-1};t_n,y)dx_1 \dots dx_{n-1}.$$

hence, one needs to find a local Gaussian type lower bound for $p(t_i,x_i;t_{i+1},x_{i+1})$. Nevertheless, still, there are various points to this argument:

1. It is difficult to see how to obtain the local Gaussian type lower bound for the density of a general random variable $F_i = X(t_i)$ conditioned to $\mathcal{F}_{t_{i-1}}$. First one needs that the conditional density exists. This is the condition in **(H2b)** in Definition 1.

2. In order to obtain a Gaussian type lower bound for the conditional density of $F_i = X(t_i)$ with respect to $\mathcal{F}_{t_{i-1}}$, one expands $X(t_i)$, conditioned on $\mathcal{F}_{t_{i-1}}$, using a Itô-Taylor expansion (this is (3.1)) so that the main term in this expansion, $\sum_{j=1}^{k} I_j^i(h_j)$, is a Gaussian random variable. This Gaussian random variable should be non-degenerated and have the same local variance as the conditional density of $F_i = X(t_i)$ conditioned to $\mathcal{F}_{t_{i-1}}$ which we will denote by $\Delta_{i-1}(g) = \int_{t_{i-1}}^{t_i} \|g(s)\|^2 ds$. Note that g can change with time therefore it effectively measures a local variance. In order for the estimate to lead to a global lower bound we need that this local variance should not vanish or be unbounded. This is embodied in **(H2c)**. This is the core of the uniform elliptic condition.

3. The argument in 2. is not sufficient as we also need that the density of the newly defined random variable has to be close to the conditional density of F_i. In conclusion, instead of finding directly a local lower estimate for the conditional density of $F_i = X(t_i)$ with respect to $\mathcal{F}_{t_{i-1}}$ we find a local lower estimate for the conditional density of the further approximation $\overline{F}_i^l \equiv \overline{F}_i$ with respect to $\mathcal{F}_{t_{i-1}}$. We call this new approximation the truncated approximation, the order of the approximation being determined by the parameter l. In order to find the local estimate for the conditional density of the truncated approximation one needs that the Gaussian term $\sum_{j=1}^{k} I_j^i(h_j)$ has to be the dominant one in the series expansion. That is, one needs to prove that the higher order terms, $\Delta_{i-1}(g)^{(l+1)\gamma} Z_i + G_i^l$ do not contribute significantly to the Gaussian estimate (this is condition **(H2d)**). That is, the variance of the higher order terms is smaller than the variance of the Gaussian term. This step is noted by the dashed arrows in Figure 1.

4. Next one needs to have that the conditional density of the truncated approximation, $\overline{F}_i$, with respect to $\mathcal{F}_{t_{i-1}}$ has to be close to the conditional density of $F_i = X(t_i)$ with respect to $\mathcal{F}_{t_{i-1}}$. For this, one needs to use a high number of terms in the Itô-Taylor series expansion so as to obtain that the conditional density of the truncated series and the conditional density of $X(t_{i+1})$ are close enough. This is embodied in condition **(H2a)**. This step is noted by the dotted arrows in Figure 1.

5. Finally the estimates have to be uniform in the sense that the constants appearing in the estimate should not depend "too much" on the partition or the

sigma field $\mathcal{F}_{t_{i-1}}$. For this reason we need to require that various estimates are uniform such as **(H1)** and that all constants appearing do not depend on $\omega \in \Omega$.

The definition of a uniformly elliptic random variable is as follows

Definition 1. *Let $F \in \mathcal{F}_t$. Suppose that there exists:*
1. $\epsilon > 0$ such that for any sequence of partitions $\pi_N = \{0 = t_0 < t_1 < \cdots < t_N = t\}$ whose norm is smaller than ϵ and $|\pi_N| = \max\{|t_{i+1} - t_i|; i = 0, \ldots, N-1\} \to 0$ as $N \to \infty$ there exists a sequence $F_i \in L^2(\Omega; \mathbb{R}^q)$, $i = 1, \ldots, N$ such that $F_N = F$. F_i is a $\mathcal{F}_{t_i}$-measurable random variable and is a t_{i-1}-conditionally non-degenerate random variable.
2. a function $g : [0, T] \to \mathbb{R}_{>0}$ (the local variance function) and a positive constant $C(T)$ such that $\|g\|_{L^2([0,T])} \leq C(T)$. We denote the local variance $\Delta_{i-1}(g) = \int_{t_{i-1}}^{t_i} \|g(t)\|^2 dt$. This quantity measures approximately the variance in the local Gaussian approximation to the density of F_i conditioned to $\mathcal{F}_{t_{i-1}}$.
3. Suppose that for each F_i and each $l \in \mathbb{N}$ there exists a sequence $\overline{F}_i \equiv \overline{F}_i^l$ such that

$$\overline{F}_i = \Delta_{i-1}(g)^{(l+1)\gamma} Z_i + F_{i-1} + \sum_{j=1}^{k} I_j^i(h_j) + G_i^l. \tag{3.1}$$

Here G_i^l are $\mathcal{F}_{t_i} \cap \overline{\mathbb{D}}_{t_{i-1}}^{\infty}$ random variables and $h_j \equiv h_j|_{[t_{i-1}, t_i]} : \Omega \to L^2([t_{i-1}, t_i]; \mathbb{R}^q)$ is a collection of $\mathcal{F}_{t_{i-1}}$-measurable smooth random processes which satisfies for almost all $\omega \in \Omega$:
(H1) *There exists a constant $C(n, p, T)$ such that*

$$\|F_i\|_{n,p} + \sup_{\omega \in \Omega} \|h_j\|_{L^2([t_{i-1}, t_i])}(\omega) \leq C(n, p, T)$$

for any $j = 1, \ldots, k$, $i = 0, \ldots, N$ and n, $p \in \mathbb{N}$.
Furthermore the following four conditions are satisfied for the approximation sequence $\overline{F}_i$ and any $i = 1, \ldots, N$ and almost all $\omega \in \Omega$:
(H2a) *There exists a constant $\gamma > 0$, such that for any n, p, $l \in \mathbb{N}$, $\left\|F_i - \overline{F}_i\right\|_{n,p,t_{i-1}} \leq C(n, p, T)\Delta_{i-1}(g)^{(l+1)\gamma}$.*
(H2b) *There exists a constant $C(p, T) > 0$ such that for any $p > 1$*

$$\left\|\det \psi_{F_i}^{-1}(t_{i-1})\right\|_{p,t_{i-1}} \leq C(p, T)\Delta_{i-1}(g)^{-q}.$$

(H2c) *Define*

$$A = \Delta_{i-1}(g)^{-1} \begin{pmatrix} \int_{t_{i-1}}^{t_i} \langle h^1(s), h^1(s) \rangle ds & \cdots & \int_{t_{i-1}}^{t_i} \langle h^1(s), h^q(s) \rangle ds \\ \vdots & \cdots & \vdots \\ \int_{t_{i-1}}^{t_i} \langle h^q(s), h^1(s) \rangle ds & \cdots & \int_{t_{i-1}}^{t_i} \langle h^q(s), h^q(s) \rangle ds \end{pmatrix}.$$

We assume that there exists strictly positive constants $C_1(T)$ and $C_2(T)$, such that for all $\xi \in \mathbb{R}^q$,

$$C_1(T)\xi'\xi \geq \xi' A\xi \geq C_2(T)\xi'\xi.$$

(H2d) *There exist constants $\varepsilon > 0$ and $C(n,p,l,T)$ such that*

$$\left\|G_i^l\right\|_{n,p,t_{i-1}} \leq C(n,p,l,T)\Delta_{i-1}(g)^{\frac{1}{2}+\varepsilon}.$$

In the previous definition γ is a constant that may change depending on the characteristics of how the underlying noise appears in the structure of F and the quality of the approximation sequence $\overline{F}_i$. For example in the case of the non-homogeneous diffusion we will use for $\overline{F}_i$ a high order Itô Taylor type of approximation and in such a case $\gamma = 1/2$.

In this setting we try to give conditions for the sequence as close as possible to the general set-up of stochastic differential equations and requiring the least amount of conditions so that the lower bound for the density of the approximative random variable can be obtained. Note that in this definition, $\overline{F}_i$ is measurable with respect to the expanded filtration $\overline{\mathcal{F}}_{t_i}$ as we are adding the variables Z_i to its definition. In particular the norm appearing in condition **(H2a)** is the norm in the extended space.

Also we remark that the random variables $\overline{F}_i$, considered in this Theorem will not necessarily be non-degenerate unless one adds the independent random variable $\Delta_{i-1}(g)^{(l+1)\gamma}Z_i$. Then the main result obtained in [10] is

Theorem 2. *Let F be a uniformly elliptic random variable. Then there exists a constant $M > 0$ that depends on all other constants in Definition 1 such that*

$$p_F(y) \geq \frac{\exp\left(-M\frac{\|y-F_0\|^2}{\|g\|^2_{L^2([0,t])}}\right)}{M\left\|g\right\|_{L^2([0,t])}{}^{q/2}}.$$

4. Lower bound density estimates for non-homogeneous diffusions

Throughout this section we assume that the following hypotheses are satisfied:

(H) The measurable coefficients b and σ satisfy for any $j = 0, 1, ...$

$$\sup_{t\in[0,T]}\left(\sup_{x\in\mathbb{R}^q}\left\|b^{(j)}(t,x)\right\| + \sup_{x\in\mathbb{R}^q}\left\|\sigma^{(j)}(t,\cdot)\right\|\right) < \infty,$$

$$\xi'\sigma(t,x)\sigma(t,x)'\xi \geq c\left\|\xi\right\|^2$$

for some positive constant c and any vectors x, $\xi \in \mathbb{R}^q$ and $t \in [0,T]$.

First we start with a preparatory Lemma that describes the smoothness of the random variables $F_i = X(t_i)$.

Lemma 3. *Assume condition (H). Then $F_i \in \overline{\mathbb{D}}^{\infty}_{t_{i-1}} \cap \mathcal{F}_{t_i}$-measurable r.v. and is a t_{i-1}-conditionally non-degenerate r.v. for all $i = 1, ..., N$*

The proof of this statement is done through the usual techniques of stochastic differentiation. The technique is similar as in the homogeneous case, see for example, [3], [19]. The non-homogeneous case is treated in [11]. Here we only need to do a small modification to their argument in order to incorporate the conditioning.

Proof. We only briefly sketch the main points of this proof. We will prove by induction that for $f \in C_p^\infty(\mathbb{R}^q, \mathbb{R})$, any $n \in \mathbb{N}$, $p > 0$, $T > t > t_{i-1}$, $\alpha \in [t_{i-1}, t]^n$, $\upsilon \in \{1, \ldots, k\}^n$ there exists a positive non-random constant $C(n, p, T)$ such that

$$\sup_{\alpha \in [t_{i-1}, t]^n} E_{t_{i-1}} (D_\alpha^\upsilon f(X(t)))^p \le C(n, p, T). \tag{4.1}$$

Note that the constant $C(n, p, T)$ depends on all the other constants in the problem but is independent of N, ω and the partition. First, let us prove the assertion for $n = 1$, $\alpha \in [t_{i-1}, t]$ and $i \le k$ then we have that by the chain rule for the stochastic derivative

$$D_\alpha^i f(X(t)) = f'(X(t)) D_\alpha^i X(t).$$

Next one obtains using the Picard iteration method that $X(t) \in \mathbb{D}^\infty$ and that

$$D_\alpha^i X(t) = \sigma_i(\alpha, X(\alpha)) + \int_\alpha^t b'(s, X(s)) D_\alpha^i X(s) ds + \int_\alpha^t \sigma_j'(s, X(s)) D_\alpha^i X(s) dW^j(s),$$

then given that the coefficients are bounded with bounded derivatives, it follows that

$$E_{t_{i-1}} \left\| D_\alpha^i X(t) \right\|^p \le C(t) \left(1 + \int_\alpha^t E_{t_{i-1}} \left\| D_\alpha^i X(s) \right\|^p ds \right).$$

for any $p \ge 2$ and $C(t)$ is a positive constant increasing in t and independent of α. Then the conclusion follows applying Gronwall's lemma. Now assume that the assertion is true for $n - 1$. The proof for n follows along the same lines as before. That is, let $\alpha = (s_1, \ldots, s_n)$, $\upsilon = (j_1, \ldots, j_n)$ and denote by $\alpha- = (s_1, \ldots, s_{n-1})$ and $\upsilon- = (j_1, \ldots, j_{n-1})$ then as before using a Picard approximation method, one can prove that $X(t) \in \mathbb{D}_{t_{i-1}}^{n,p}$ for any $n \ge 1$, $p \ge 1$ and $t \ge t_{i-1}$. As a consequence one also obtains the following equation for $D_\alpha^\upsilon X(t)$

$$\begin{aligned} D_{\alpha-}^{\upsilon-} D_{s_n}^{j_n} X(t) &= D_{\alpha-}^{\upsilon-} \sigma_{j_n}(s_n, X(s_n)) + \int_{s_n}^t D_{\alpha-}^{\upsilon-} \left(b'(s, X(s)) D_{s_n}^{j_n} X(s) \right) ds \\ &\quad + D_{\alpha-}^{\upsilon-} \left(\int_{s_n}^t \sigma_j'(s, X(s)) D_{s_n}^{j_n} X(s) dW^j(s) \right), \\ &= D_{\alpha-}^{\upsilon-} \sigma_{j_n}(s_n, X(s_n)) + \int_{s_n}^t \sum_{\upsilon_1} D_{\alpha_1}^{\upsilon_1} b'(s, X(s)) D_{\alpha_2}^{\upsilon_2} D_{s_n}^{j_n} X(s) ds \\ &\quad + \sum_{k=1}^{n-1} D_{\alpha(k)}^{\upsilon(k)} \left(\sigma_{j_k}'(s, X(s)) D_{s_n}^{j_n} X(s) \right) \\ &\quad + \int_{s_n}^t \sum_{\upsilon_1} D_{\alpha_1}^{\upsilon_1} \sigma_j'(s, X(s)) D_{\alpha_2}^{\upsilon_2} D_{s_n}^{j_n} X(s) dW^j(s). \end{aligned}$$

The first sum index is composed of splitting the indexes $\upsilon-$ and $\alpha-$ into two disjoint subsets υ_1 and υ_2 for υ- and $\alpha_1$and α_2 for $\alpha-$. Similarly, $\upsilon(k)$ and $\alpha(k)$ denote the indices without the j_k and s_k component respectively. The final result follows as before by using the chain rule and the Gronwall lemma together with the inductive hypothesis.

The assertion on the conditional non-degeneracy of F_i follows a similar argument as in Theorem 3.5 of [11]. □

In the particular case that σ and b do not depend on t, Kusuoka and Stroock proved in [13] under hypoelliptic conditions that if $b = \sum_{i=1}^{r} a_i \sigma_i$ for some functions $a_i \in C_b^\infty$ then a Gaussian type lower bound is satisfied by the density of X_t.

Here we improve their result in the sense that this extra condition is not required, the diffusion is non-homogeneous and the conditions on the coefficients with respect to the time parameters are minimal. This is obtained by applying the definition of uniformly elliptic random variable. To apply this definition we need to define all its ingredients. That is, define $F = X(t)$ and $g(s) \equiv 1$ for all $s \in [0, T]$. For any partition $0 = t_0 < \ldots < t_n = t$, let $F_i = X(t_i)$. In order to define $\overline{F}_i$ we have to explicitly write an Itô–Taylor expansion for the case of non-homogeneous diffusions. This is obtained considering the difference $F_i - F_{i-1}$. To introduce the Itô–Taylor expansion we will deal with indices $\beta \in \cup_{n \geq 1}\{0, 1, \ldots, k\}^n \cup \{\nu\}$, ν denotes the empty index. In such a case, $l(\beta)$ denotes the length of the multi-index β and $n(\beta)$ the quantity of zeros in β, $l(\nu) = 0$. $\beta(i)$ denotes the i-th component in β, $-\beta$ denotes the index β without its first component. Similarly, one defines $\beta-$.

We also define the following operators for a smooth function $f : [t_{i-1}, t_i]^{l(\beta-)} \times \mathbb{R}^q \to \mathbb{R}^q$, (here we adopt the double index summation notation)

$$
\begin{aligned}
L^r f(u_\beta, x) &= \frac{\partial f}{\partial x_i}(u_{\beta-}, x)\sigma_{ir}(u_{l(\beta)}, x) \\
L^0 f(u_\beta, x) &= \frac{1}{2}\frac{\partial^2 f}{\partial x_i \partial x_j}(u_{\beta-}, x)\sigma_{ir}\sigma_{jr}(u_{l(\beta)}, x) + \frac{\partial f}{\partial x_i}(u_{\beta-}, x) b_i(u_{l(\beta)}, x)
\end{aligned}
$$

where $u_\beta = (u_1, \ldots, u_{l(\beta)})$ and similarly $u_{\beta-} = (u_1, \ldots, u_{l(\beta)-1})$. Then we define inductively

$$
\begin{aligned}
f_\beta(u_\beta, x) &= \left(L^{\beta(l(\beta))} f_{\beta-}\right)(u_\beta, x) \\
f_\nu(u, x) &= f(x).
\end{aligned}
$$

We will frequently use the above formulas for $f(x) = x$ in such a case we note that $\|f_\beta\|_\infty \leq C(\beta)$ for any $\beta \in \cup_{n \geq 1}\{0, 1, \ldots, k\}^n$. I_β denotes the multiple stochastic integral where the indexes of the integral are determined by the set of indices β. That is, for an adapted process Y we define

$$
I_\beta\left[f_\beta(\cdot, Y(\cdot))\right]_{t_{i-1}, t} = \int_{t_{i-1}}^{t} I_{\beta-}\left[f_\beta((\cdot, u_{l(\beta)}), Y(\cdot))\right]_{t_{i-1}, u_{l(\beta)}} dW^{\beta(l(\beta))}(u_{l(\beta)}).
$$

Note that these formulae are not exactly the same as the usual Itô–Taylor expansion (see for example, [9]). In fact, for $\beta = (0,1)$ then for $f(x) = x$, we have

$$I_\beta \left[f_\beta(\cdot, Y(\cdot))\right]_{t_{i-1},t} = \int_{t_{i-1}}^{t} \int_{t_{i-1}}^{u_2} \frac{\partial b}{\partial x_i}(u_1, Y(u_1))\sigma_{i1}(u_2, Y(u_1))du_1 dW^1(u_2).$$

In particular note that the integrand depends on both integrating variables u_1 and u_2 which is not the common in the usual Itô–Taylor formula. In this way one can weaken some restrictive conditions on the coefficients. Then we have the following Itô–Taylor formula for X.

Lemma 4. *Let $f \in C_p^\infty(\mathbb{R}^q)$ and $\mathcal{A}_l = \{\beta \in \cup_{n\geq 1}\{0,1,\dots,k\}^n; 1 \leq l(\beta) + n(\beta) \leq l\}$, $\mathcal{B}_l = \{\beta \in \cup_{n\geq 1}\{0,1,\dots,k\}^n; -\beta \in \mathcal{A}_l, \beta \notin \mathcal{A}_l\}$ then*

$$f(X(t)) - f(X(s)) = \sum_{\beta\in\mathcal{A}_l} I_\beta[f_\beta(\cdot, X_{t_{i-1}})]_{t_{i-1},t_i} + \sum_{\beta\in\mathcal{B}_l} I_\beta[f_\beta(\cdot, X(\cdot))]_{t_{i-1},t_i}. \quad (4.2)$$

In the case that $f(x) = x$ we have the following estimate for $\beta \in \cup_{n\geq 1}\{0,1,\dots,k\}^n$

$$\left\| I_\beta \left[f_\beta(\cdot, X(\cdot))\right]_{t_{i-1},t} \right\|_{n,p,t_{i-1}} \leq C(n,p,T)\,(t - t_{i-1})^{\frac{n \vee l(\beta)}{2}}. \quad (4.3)$$

Proof. First we prove the first statement by induction. Obviously the result is true for $l = 0$. Now suppose that the result is true for l. To prove that the expansion is true for $l+1$, one has to take every term of the type $I_\beta[f_\beta(\cdot, X(\cdot))]_{t_{i-1},t_i}$ for $\beta \in \mathcal{B}_l$ with $n(\beta) + l(\beta) = l+1$ and consider the difference

$$I_\beta[f_\beta(\cdot, X(\cdot))]_{t_{i-1},t_i} - I_\beta[f_\beta(\cdot, X_{t_{i-1}})]_{t_{i-1},t_i} = I_\beta\left[f_\beta(\cdot, X(\cdot)) - f_\beta(\cdot, X_{t_{i-1}})\right]_{t_{i-1},t_i}$$

Then we apply Itô's formula for $f_\beta(u_\beta, X(u_{l(\beta)})) - f_\beta(u_\beta, X_{t_{i-1}})$ fixing the time component to obtain that

$$\begin{aligned} f_\beta(u_\beta, X(u_{l(\beta)})) - f_\beta(u_\beta, X_{t_{i-1}}) &= \int_{t_{i-1}}^{u_{l(\beta)}} L^r f_\beta(u_{(\beta,r)}, X_{u_{l(\beta)+1}}) dW^r(u_{l(\beta)+1}) \\ &\quad + \int_{t_{i-1}}^{u_{l(\beta)}} L^0 f_\beta(u_{(\beta,0)}, X_{u_{l(\beta)+1}}) du_{l(\beta)+1} \end{aligned}$$

Therefore

$$I_\beta[f_\beta(\cdot, X(\cdot))]_{t_{i-1},t_i} = I_\beta[f_\beta(\cdot, X_{t_{i-1}})]_{t_{i-1},t_i} + \sum_{r=0}^{k} I_{(\beta,r)}[f_{(\beta,r)}(\cdot, X(\cdot))]_{t_{i-1},t_i}$$

and $(\beta, r) \in \mathcal{B}_{l+1}$.. To finish the proof of (4.2) one has to prove that:
1. $\mathcal{A}_{l+1} = \mathcal{A}_l \cup \{\beta \in \mathcal{B}_l; n(\beta) + l(\beta) = l+1\}$.
2. $\mathcal{B}_{l+1} = (\mathcal{B}_l - \mathcal{A}_{l+1}) \cup \{(\beta, r); r = 0, \dots, k, \beta \in \mathcal{B}_l, n(\beta) + l(\beta) = l+1\}$.

To prove property 1., take $\beta \in \mathcal{A}_{l+1} - \mathcal{A}_l$ then obviously $n(-\beta) + l(-\beta) \leq l$ and therefore $-\beta \in \mathcal{A}_l$. Therefore the inclusion $\subseteq$ follows. The other inclusion is trivial.

For property 2., the inclusion $\supseteq$ is trivial. For the other, suppose that there exists $\beta' \in \mathcal{B}_{l+1}$ such that $\beta' \notin (\mathcal{B}_l - \mathcal{A}_{l+1}) \cup \{(\beta, r); r = 0, \dots, k, \beta \in \mathcal{B}_l, n(\beta) + l(\beta) = l+1\}$. Then $l(\beta') + n(\beta') = l+2$, $l(-\beta') + n(-\beta') = l$. Therefore $\beta' \in \mathcal{B}_l - \mathcal{A}_{l+1}$ which is a contradiction.

Next we prove the norm estimate (4.3) by induction. For this, we first have that $I_\beta[f_\beta(\cdot, X(\cdot))]_{t_{i-1},t} \in \mathbb{D}^{n,p}_{t_{i-1}}$ for any $n \geq 1$, $p \geq 1$ and any $t \geq t_{i-1}$ due to Lemma 3. Next, suppose that for $\beta \in \bigcup_{n \geq 1}\{0, 1, \dots, k\}^n$ we have $v \subseteq \beta$, then we define $\beta(v)$ as the set of indices that are in β but not in v. We will prove that there exits a positive constant $C(j, a, n, p, T)$ such that for $v = (i_1, \dots, i_a) \in \{1, \dots, k\}^a$ with $v \subseteq \beta$, $v_1 \in \{1, \dots, k\}^n$, $v_2 \in \{1, \dots, k\}^j$, $\alpha = (s_1, \dots, s_a)$, $\alpha_1 \in [t_{i-1}, t]^n$, $\alpha_2 \in [t_{i-1}, t]^j$, $j \geq 0, n \geq 0$ and $a \geq 0$, we have

$$E_{t_{i-1}}\left(\int_{[t_{i-1},t]^{n+j+a}} \left\| D^{v_1}_{\alpha_1} I_{\beta(v)}[\prod_{i=1}^{a} 1(u_i = s_i) D^{v_2}_{\alpha_2} f_\beta(\cdot, X.)]_{t_{i-1},t} \right\|^2 d\alpha_1 d\alpha_2 d\alpha\right)^p$$
$$\leq C(j, a, n, p, T)(t - t_{i-1})^{(n \vee l(\beta(v)) + j + a)p}. \tag{4.4}$$

Here we denote by $u_\beta = (u_1, \dots, u_{l(\beta)})$ the variables of integration in I_β where the first a components correspond to the indices in v. To simplify notation we will denote $\Pi_\alpha = \prod_{i=1}^a 1(u_i = s_i)$. The induction is performed in n. So let us suppose first that $n = 0$. Then we have

$$E_{t_{i-1}}\left(\int_{[t_{i-1},t]^{j+a}} \left\| I_{\beta(v)}[\Pi_\alpha D^{v_2}_{\alpha_2} f_\beta(\cdot, X.)]_{t_{i-1},t} \right\|^2 d\alpha_2 d\alpha\right)^p$$
$$\leq (t - t_{i-1})^{(j+a)(p-1)} \int_{[t_{i-1},t]^{j+a}} E_{t_{i-1}} \left\| I_{\beta(v)}[\Pi_\alpha D^{v_2}_{\alpha_2} f_\beta(\cdot, X.)]_{t_{i-1},t} \right\|^{2p} d\alpha_2 d\alpha$$
$$\leq C(T)(t - t_{i-1})^{(j+a+l(\beta(v)))(p-1)} \int_{[t_{i-1},t]^{j+l(\beta)}} E_{t_{i-1}} \left\| D^{v_2}_{\alpha_2} f_\beta(u_\beta, X_{u_1}) \right\|^{2p} du_\beta d\alpha_2.$$

This step finishes by noting that $E_{t_{i-1}} \left\| D^{v_2}_{\alpha_2} f_\beta(u_\beta, X_{u_1}) \right\|^{2p} \leq C(j, l(\beta), p, T)$ where the positive constant C is independent of α_2, v_2 and β (see (4.1)). Next suppose that the assertion is valid for $n-1$. Then let $v_1 = (j_1, \dots, j_n)$ and $\alpha_1 = (s'_1, \dots, s'_n)$. We consider two cases: First suppose that $j_n \notin \beta(v)$, then we have that

$$D^{v_1}_{\alpha_1} I_{\beta(v)}[\Pi_\alpha D^{v_2}_{\alpha_2} f_\beta(\cdot, X.)]_{t_{i-1},t} = D^{v_1 -}_{\alpha_1 -} I_{\beta(v)}[\Pi_\alpha D^{j_n}_{s'_n} D^{v_2}_{\alpha_2} f_\beta(\cdot, X.)]_{t_{i-1},t}.$$

Then it is clear that the assertion follows from the inductive hypothesis.

Next suppose that $j_n \in \beta(v) = (i_1, \dots, i_{l(\beta(v))})$ and that the indices $i_1, \dots, i_h$, $h \leq l(\beta(v))$, are all the indices in $\beta(v)$ equal to j_n. That is , $j_n \notin \{i_{h+1}, \dots, i_{l(\beta(v))}\}$. Then

$$D^{v_1}_{\alpha_1} I_{\beta(v)}[\Pi_\alpha D^{v_2}_{\alpha_2} f_\beta(\cdot, X.)]_{t_{i-1},t} = D^{v_1 -}_{\alpha_1 -} I_{\beta(v)}[\Pi_\alpha D^{j_n}_{s'_n} D^{v_2}_{\alpha_2} f_\beta(\cdot, X.)]_{t_{i-1},t}$$
$$+ \sum_{r=1}^{h} D^{v_1 -}_{\alpha_1 -} I_{\beta(v, i_r)}[1(u_{l(v)+r} = s'_n)\Pi_\alpha D^{v_2}_{\alpha_2} f_\beta(\cdot, X.)]_{t_{i-1},t}.$$

Again it is clear then that the assertion follows from the inductive hypothesis. From (4.4) one obtains (4.3). □

It should be clear that the estimate (4.4) is not optimal. In fact, the estimate improves as the number of zeros in β increase. Nevertheless this is sufficient for our purposes.

Now we can define the approximation random variables $\overline{F}_i = \overline{X}(t_i)$ as the Itô-Taylor approximation of order l of X based on time t_{i-1}. That is, $\overline{X}$ is defined as follows

$$\begin{aligned}
\overline{X}_{t_i} &= (t_i - t_{i-1})^{\frac{l+1}{2}} Z_i + X_{t_{i-1}} + \sum_{\beta \in \mathcal{A}_1} I_\beta[f_\beta(\cdot, X_{t_{i-1}})]_{t_{i-1},t_i} + G_i^l \\
G_i^l &= \sum_{2 \le l(\beta)+n(\beta) \le l} I_\beta[f_\beta(\cdot, X_{t_{i-1}})]_{t_{i-1},t_i} \\
h_j(u) &= f_{(j)}(u, X_{t_{i-1}}),\ u \in [t_{i-1}, t_i] \\
g(u) &= 1,\ u \in [0,T]
\end{aligned}$$

where $f(x) = x$, $Z_i = \overline{W}(i+1) - \overline{W}(i)$ is a q-dimensional $N(0,I)$ random variable independent of the Wiener process W and Z_j for $j \neq i$.

Theorem 5. *Assume (H) and $t > 0$. Let X be the unique solution of the non-homogeneous diffusion equation. Then $X(t)$ has a smooth density, denoted by $p(t,x,y)$ that satisfies*

$$\frac{m \exp(-\frac{\|x-y\|^2}{mt})}{t^{q/2}} \ge p(t,x,y) \ge \frac{\exp(-M\frac{\|x-y\|^2}{t})}{Mt^{q/2}}$$

for two constants m, $M \in [1, +\infty)$.

Proof. We verify each hypothesis in Definition 1:
Preliminaries: First, $F_i = X(t_i)$ is t_{i-1}-conditionally non-degenerate by Lemma 3, therefore 1. follows. Next, $\|g\|^2_{L^2[0,T]} = T$, therefore 2. follows. $G_i^l \in \mathbb{D}^\infty_{t_{i-1}} \cap \mathcal{F}_{t_i}$ due to Lemma 4, $\|f_\beta\|_\infty \le C(\beta)$ and classical estimates for the $\mathbb{D}^{n,p}$-norms of stochastic integrals. h_j is obviously smooth and adapted to $\mathcal{F}_{t_{i-1}}$. Therefore 3. follows and all the conditions in the preliminaries of the definition of uniformly elliptic random variable are satisfied.

(H1) $\|F_i\|_{n,p} = \|X(t_i)\|_{n,p} \le C(n,p)$ due to Theorem 3.5 in [11].
$\sup_\omega \|h_j\|_{L^2([t_{i-1},t_i])}(\omega) \le C(n,p)$ is satisfied due to the boundeness of h_j.

(H2a) Using (4.2) and (4.3), we obtain that there exists a universal positive constant independent of the initial point,

$$\begin{aligned}\left\|X_{t_i}-\overline{X}_{t_i}\right\|_{n,p,t_{i-1}} &= \left\|(t_i-t_{i-1})^{\frac{l+1}{2}}Z_i+\sum_{\beta\in\mathcal{B}_l}I[f_\beta(\cdot,X_\cdot)]_{t_{i-1},t_i}\right\|_{n,p,t_{i-1}}\\ &\le (t_i-t_{i-1})^{\frac{l+1}{2}}\left\|Z_i\right\|_{n,p,t_{i-1}}+\sum_{\beta\in\mathcal{B}_l}\left\|I[f_\beta(\cdot,X_\cdot)]_{t_{i-1},t_i}\right\|_{n,p,t_{i-1}}\\ &\le C\,(t_i-t_{i-1})^{\frac{l+1}{2}}.\end{aligned}$$

Therefore $\gamma=1/2$ and **(H2a)** follows.

(H2b) It is also known (see Theorem 3.5 in [11]) that

$$\left\|\det\psi^{-1}_{X(t_i)}(t_{i-1})\right\|_{p,t_{i-1}}\le C\,(t_i-t_{i-1})^{-q}$$

for a positive constant C independent of t_{i-1}.

(H2c) In this case we have that

$$A=(t_i-t_{i-1})^{-1}\int_{t_{i-1}}^{t_i}\sigma(s,X(s))\sigma'(s,X(s))ds.$$

Then due to the hypothesis (H), we have uniform ellipticity and uniform boundedness of σ and therefore we have that **(H2c)** is satisfied.

(H2d) $\left\|G_i^l\right\|_{n,p,t_{i-1}}\le\sum_{2\le l(\beta)+n(\beta)\le l}\left\|I_\beta[f_\beta(\cdot,X_{t_{i-1}})]_{t_{i-1},t_i}\right\|_{n,p,t_{i-1}}$
$\le C(n,p,T)(t_i-t_{i-1}).$

To obtain this estimate one can either use known estimates for the $\mathbb{D}^{n,p}$-norms of stochastic integrals (see [19]) or (4.3) for $l(\beta)\ge2$ and compute separately the case $\beta=(0)$. This finishes the proof that the hypothesis in Definition 1 are satisfied and therefore by Theorem 2 the lower bound follows. The upper bound follows using the same technique used for the Itô-Taylor expansion together with classical techniques (see [23], Section 3). □

5. Comments and Applications

Looking at the proof of Theorem 5 one sees that although the definition of uniformly elliptic random variable may seem to be too complicated all its conditions are naturally satisfied. Furthermore most of the properties required in Definition 1 are usually proved when studying the existence and smoothness of densities in the framework of Malliavin Calculus. This result besides giving an explicit Gaussian lower bound estimate for various equations also characterizes the support of an uniformly elliptic random variable as the whole space, $\mathbb{R}^q$. In very loose terms the above estimate also means that the behavior of the non-homogeneous uniformly elliptic diffusion is the same as the Wiener process itself. This has various applications in different areas. Just to show this point let us introduce the following application to potential theory.

Theorem 6. *Assume the same conditions as in the previous theorem with* $d > 2$. *Define* $k(x) = \|x\|^{2-d}$ *and let* $Cap_{d-2}(\cdot)$ *denote the* $d-2$ *Newtonian capacity associated with the kernel* k. *Then one has that for any set* $A \subseteq B(0, R)$ *and any time interval* $[a, b] \subseteq [0, T]$ *there exist two positive constants* $C_1(a, b, R)$ *and* $C_2(a, b, R)$ *such that*

$$C_1 Cap_{d-2}(A) \leq P(X_t \in A, \text{ for some } t \in [a, b]) \leq C_2 Cap_{d-2}(A).$$

Furthermore the Haussdorf dimension of the random set $\{X(t), 0 \leq t\}$ *is 2 if* $d > 2$.

This result follows by combining the result in Theorem 5 in conditional form together with Theorem 2.4 in [4]. One can also use Theorem 5 to improve the results in [24] (quantile estimation) and [7] (statistics of diffusions).

By looking at the proof of Theorem 2 one may be lead to believe that some of the hypotheses are redundant. For example, **(H2b)** and **(H2c)** are obviously related. Similarly, **(H2a)** and **(H2d)** are also related. Nevertheless, for each case, the generality of the truncated approximation sequence in the definition of uniformly elliptic random variable allows for counterexamples. One open problem is how to improve this definition so as to minimize its requirements. However, from the practical point of view, verifying one of these hypotheses is not very different (but not the same) as verifying the other.

References

[1] Aronson, D.G.: Non-negative solutions of linear parabolic equations. *Ann. Scuola Norm. Sup. Pisa* **22** (1968) 607–694.

[2] Azencott, R.: *Densité des diffusions en temps petit: développements asymptotiques*, I. Lecture Notes in Math. **1059**, 402–498. Springer, Berlin 1984.

[3] Cattiaux, P. Mesnager, L.: Hypoelliptic non-homogeneous diffusions. *Probab. Theory Related Fields* **123** (2002) 453–483.

[4] Dalang, R. and Nualart, E. Potential theory for hyperbolic SPDE's. Preprint (2002).

[5] Fabes, E.B. and Stroock, D.: A new proof of Moser's parabolic Harnack inequality using the old ideas of Nash. *Arch. Rational Mech. Anal.* **96**, (1986), 327–338.

[6] Fefferman, C. L. and Sánchez-Calle, A.: Fundamental solutions for second order subelliptic operators. *Ann. of Math.* (2) **124** (1986) 247–272.

[7] Gobet, E.: LAMN property for elliptic diffusion: a Malliavin Calculus approach. *Bernoulli* **7** (2001) 899–912.

[8] Ikeda, N. and Watanabe, S.: *Stochastic Differential Equations and Diffusion Processes.* North-Holland/Kodansha 1989.

[9] Kloeden, P. E. and Platen, E.: *Numerical Solution of Stochastic Differential Equations.* Springer 1992.

[10] Kohatsu-Higa, A.: Lower bounds for densities of uniformly elliptic random variables on Wiener space. Preprint 2002.

[11] Kusuoka, S. and Stroock, D.: *Applications of the Malliavin Calculus, Part I.* In Stochastic Analysis, Proc. Taniguchi Internat. Symp. Katata and Kyoto, 1982, ed. by K. Itô, Kinokuniya, Tokyo, 1984, 271–306.

[12] Kusuoka, S. and Stroock, D.: Applications of the Malliavin Calculus, Part II. *J. Fac. Sci. Univ. Tokyo SectIA Math* **32** (1985) 1–76.

[13] Kusuoka, S. and Stroock, D.: Applications of the Malliavin Calculus, Part III. *J. Fac. Sci. Univ. Tokyo Sect IA Math.* **34** (1987) 391–442.

[14] Liskevich, V. and Semenov, Y.: Estimates for fundamental solutions of second-order parabolic equations. *J. London Math. Soc.* (2) **62** (2000), no. 2, 521–543.

[15] Malliavin, P.: Stochastic Analysis. 1997.

[16] Moret, S. and Nualart, D.: Generalization of Itô's formula for smooth non-degenerate martingales. *Stoch. Processes and Their Appl.* **91** (2001) 115–149.

[17] Nash, J.: Continuity of solutions of parabolic and elliptic equations. *Amer. J. Math.* **80** (1958) 931–954.

[18] Norris, J.R. and Stroock, D.: Estimates on the fundamental solution to heat flows with uniformly elliptic coefficients. *Proc. Math. London Soc.* **62** (1991), 373–402.

[19] Nualart, D.: The Malliavin Calculus and Related Topics. Springer 1995

[20] Nualart, D: *Analysis on Wiener space and anticipating stochastic calculus.* Lecture Notes in Math., **1690**(1998) 123–227.

[21] Sánchez-Calle, A.: Fundamental solutions and geometry of the sum of squares of vector fields. *Invent. Math.* **78** (1984), no. 1, 143–160.

[22] Stroock, D.: Estimates on the heat kernel for the second order divergence form operators. In *Probability Theory,* De Gruyter 1992, 29–44.

[23] Taniguchi, S.: Applications of Malliavin Calculus to time-dependent systems of Heat equations. *Osaka J. Math* **22**(1985), 307–320.

[24] Talay, D. and Zheng, Z.: Quantiles of the Euler scheme for diffusion processes and financial applications. *Math. Finance* **13** (2003), 187–199.

Universitat Pompeu Fabra, Department of Economics and Business, Ramón Trias Fargas 25-27, 08005 Barcelona, Spain
E-mail address: `kohatsu@upf.es`

Progress in Probability, Vol. 56, 339–351

Lyapunov Exponents of Nonlinear Stochastic Differential Equations with Jumps

C. W. Li

Abstract. For a certain kind of nonlinear stochastic differential equations with jumps in $\mathbb{R}^d$, there exists an invariant probability measure μ on $\mathbb{R}^d$. A Lyapunov exponent q_μ can be represented by the Furstenberg–Has'minskii formula as an integral over $\mathbb{R}^d$ with respect to the ergodic invariant measure, so that the almost sure asymptotic stability depends on the sign of q_μ. If the corresponding diffusion is nondegenerate, then μ is unique and has strictly positive invariant density in $C(\mathbb{R}^d)$.

1. Introduction

For linear random dynamical systems in $\mathbb{R}^d$, there exists at most d Lyapunov exponents and the corresponding invariant subspaces as assured by Oseledec's Multiplicative Ergodic Theorem in [15]. If the linear random dynamical system has an infinitesimal generator, then the top Lyapunov exponents can be represented by the Furstenberg–Has'minskii formula as an integral over the unit sphere or the projective space with respect to an ergodic invariant measure, see Arnold [2, Chapter 6]. The linear diffusion case is classical and can be found in Has'minskii [9], Pinsky [16] and Arnold [2], just to name a few. The pure jump case is treated by Li and Blankenship [11] while the nondegenerate jump-diffusion situation is served by Li and et al [13]. The nonlinear diffusions are always linearized at a certain point and are discussed like the linear situation, see Arnold [1], as well as Grorud and Talay [8] by the numerical approximation of the upper Lyapunov exponent. The study of invariant measures can be found in Baxendale [3] for diffusions, also in Menaldi and Robin [14] for jump-diffusions by estimating the solution of the associated integro-differential equations, see also Garroni and Menaldi [6]. The kernel estimation between diffusions and jump-diffusions appears in Garroni and

2000 *Mathematics Subject Classification.* Primary, 28C10, 37H15; Secondary, 60G55, 37A50, 93D20.
Key words and phrases. Stochastic jump-diffusion, invariant measure, Lyapunov exponent.
The work described in this paper was partially supported by a grant from the Research Grants Council of the Hong Kong Special Administrative Region, China (Project No. CityU 1090/97P) and a grant from CityU (Project No. 7001147).

Menaldi [7] as well as Li and Dong [12] while the stability of jump-diffusions via the appropriate Lyapunov functions is considered by Wee [18].

In this article we would like to investigate directly the top Lyapunov exponent of a certain kind of nonlinear jump-diffusions without linearizing the processes locally. As there is insufficient knowledge about the existence and uniqueness of solutions of integro-differential equations, it is much difficult to determine invariant measures for the nonlinear jump-diffusion processes. However due to Skorokhod [17], the gap between jump-diffusions and the corresponding diffusions is bridged.

Let $(\Omega, F, \{F_t\}_{t\geq 0}, I\!P)$ be the underlying filtered complete probability space. Consider the following d-dimensional system of stochastic differential equations with jumps,

$$dX_t = a(X_t)dt + b(X_t)dW_t + \int_Z c(X_{t-}, z)N(dt, dz) \tag{1.1}$$

with initial condition $X_0 = x \in \mathbb{R}^d$. Here $W_t = (W_t^1, \cdots, W_t^m)$ is a standard m-dimensional Brownian motion; $N(dt, dz)$ is a Poisson counting measure on $\mathbb{R}_+ \times Z$ with simultaneous jumps of probability zero, and $\lambda(dz)$ is the deterministic finite characteristic measure on a measurable space Z such that $\widetilde{N}(dt, dz) = N(dt, dz) - \lambda(dz)dt$ is a F_t-martingale measure. We also assume W and N are independent; the coefficients a, c are d-dimensional vector-valued functions, and b is a $d \times m$ matrix-valued function, satisfying the usual conditions such as locally Lipschitz continuity and at most linear growth on $\mathbb{R}^d$, so that the solution of (1.1) is well-defined. Let $f \in C^2(\mathbb{R}^d)$. Then the infinitesimal generator of the process X_t of (1.1) is

$$Lf(x) = \sum_{i=1}^d a_i(x)\partial_i f(x) + \frac{1}{2}\sum_{i,j=1}^d \sigma_{ij}(x)\partial_{ij} f(x) + \int_Z [f(\tilde{c}(x, z)) - f(x)]\lambda(dz), \tag{1.2}$$

where $\sigma_{ij}(x) = \sum_{k=1}^m b_{ik}(x)b_{jk}(x)$ and $\tilde{c}(x, z) = x + c(x, z)$.

This paper is organized as follows. In section 2 we study the existence as well as the uniqueness of invariant measures of a certain kind stochastic differential equations with jumps in $\mathbb{R}^d$ and establish the relation between transition probabilities of jump-diffusions and those of the corresponding diffusions, so that the invariant measures for jump-diffusions and diffusions can be related accordingly. An example is explored for an illustration. In section 3 we investigate the Lyapunov exponents, which can be expressed by the Furstenberg–Has'minskii formula as an integral over $\mathbb{R}^d$ with respect to ergodic invariant measures.

2. Invariant measures

Let us consider the diffusion process X^0 which satisfies the corresponding stochastic differential equation

$$dX_t = a(X_t)dt + b(X_t)dW_t. \tag{2.1}$$

Denote by $P(t,x,A)$ and $P^0(t,x,A)$ the transition probabilities of stochastic differential equations (1.1) and (2.1), respectively. We have the following relation.

Lemma 2.1. *$P(t,x,A)$ and $P^0(t,x,A)$ can be connected by the following relation*

$$\begin{aligned} P(t,x,A) &= e^{-t\lambda(Z)}P^0(t,x,A) \\ &\quad + \int_0^t\int_Z\int_{\mathbb{R}^d} e^{-(t-\tau)\lambda(Z)}P(\tau,x,dy)P^0(t-\tau,\tilde{c}(y,z),A)\lambda(dz)d\tau. \end{aligned} \tag{2.2}$$

or in a series form with $u_0=0$ and $y_0=x$,

$$\begin{aligned} P(t,x,A) &= e^{-t\lambda(Z)}\Big\{P^0(t,x,A)+\sum_{n=1}^{\infty}\int_0^t\int_Z\int_{\mathbb{R}^d}\int_0^{s_n}\int_Z\int_{\mathbb{R}^d}\cdots\int_0^{s_2}\int_Z\int_{\mathbb{R}^d} \\ &\quad \prod_{k=1}^{n}P^0(s_k-s_{k-1},\tilde{c}(y_{k-1},z_{k-1}),dy_k)P^0(t-s_n,\tilde{c}(y_n,z_n),A) \\ &\quad \times\lambda(dz_1)ds_1\cdots\lambda(dz_n)ds_n\Big\}. \end{aligned} \tag{2.3}$$

If transition probability P for the jump-diffusion (1.1) has a density $p(t,x,y)\in C(\mathbb{R}_+\times\mathbb{R}^d\times\mathbb{R}^d)$ if and only if the transition probability P^0 for the diffusion (2.1) has a density $p^0(t,x,y)\in C(\mathbb{R}_+\times\mathbb{R}^d\times\mathbb{R}^d)$ and have the following relation

$$\begin{aligned} p(t,x,y) &= e^{-t\lambda(Z)}p^0(t,x,y) \\ &\quad + \int_0^t\int_Z\int_{\mathbb{R}^d} e^{-\tau\lambda(Z)}p^0(\tau,x,u)p(t-\tau,\tilde{c}(u,z),y)du\lambda(dz)d\tau. \end{aligned} \tag{2.4}$$

Furthermore if $p^0(t,x,y)>0$, then $p(t,x,y)>0$.

We omit the proof as it is similar to that in [13, Lemma 2.3] on the compact sphere and in Skorokhod [17, Theorem 14]. From Lemma 2.1 we can deduce that the smoothness of $p^0(t,x,y)$ together with the integrability of its partial derivatives implies the smoothness of $p(t,x,y)$. On the other hand the smoothness of $p(t,x,y)$ together with the integrability of its partial derivatives also implies the smoothness of $p^0(t,x,y)$. Hence the strong Feller property of X_t^0 in $\mathbb{R}^d$ is equivalent to the strong Feller property of X_t in $\mathbb{R}^d$.

Using (2.2) and the property of invariant measures, we have the following result.

Lemma 2.2. *Let μ be an invariant probability measure of P. Then for $A\in B(\mathbb{R}^d)$, we have*

$$\mu(A)=\int_0^\infty\int_Z\int_{\mathbb{R}^d} e^{-\tau\lambda(Z)}P^0(\tau,\tilde{c}(x,z),A)\mu(dx)\lambda(dz)d\tau. \tag{2.5}$$

Lemma 2.3. *Let μ and μ^0 be invariant probability measures of P and P^0, respectively. Then*

$$\mu^0\ll P^0(t,x,\cdot)\ll P(t,x,\cdot)\ll\mu \tag{2.6}$$

for all $x \in \text{supp } \mu$, $t \geq 0$ *and so* $\text{supp } \mu^0 \subset \text{supp } \mu$. *Indeed if* $\mu(A) = 0$, *then* $P^0(t, y, A) = 0$ *for all* $y \in (\text{supp } \mu) \cup \{\tilde{c}(x, z), x \in \text{supp } \mu\}$, *a.e.* $z \in Z$, $t \geq 0$.

Proof. Suppose $\mu(A)=0$. Then $\int_{\mathbb{R}^d} P(t,x,A)\mu(dx)=\mu(A)=0$ and so $P(t,x,A)=0$, μ-a.e. $x \in \text{supp } \mu$, $t \geq 0$. Suppose $x_n \to x_0$ in $\mathbb{R}^d$ and $P(t, x_n, A) = 0$. If A is closed, let $0 \leq f_k \leq 1$ be continuous functions such that $f_k(x) = 1$ on A and 0 outside a $1/k$-neighborhood of A. Then by Fatou's lemma,

$$\begin{aligned} P(t, x_0, A) &= \inf_{k \geq 1} \int_A f_k(x) P(t, x_0, dy) \\ &\leq \inf_{k \geq 1} \lim_{n \to \infty} \int_A f_k(x) P(t, x_n, dy) = 0 \end{aligned}$$

and so $P(t, x_0, A) = 0$. If A is not closed, then there exist compact sets $A_n \subset A$ such that $P(t, x, A) = \lim_{n\to\infty} P(t, x, A_n)$. As $\mu(A_n) \leq \mu(A) = 0$, $P(t, x, A_n) = 0$ by the above argument and so $P(t, x, A) = 0$ for all $x \in \text{supp } \mu$, $t \geq 0$. By (2.2), we have $P^0(t, x, A) = 0$ for all $x \in \text{supp } \mu$, $t \geq 0$. If μ^0 is an extremal invariant measure, then

$$\mu^0(A) = \lim_{T\to\infty} \frac{1}{T} \int_0^T \chi_A(X_t^0) dt = 0.$$

If μ^0 is not an extremal invariant measure, then μ^0 can be expressed as a convex combination of extremal invariant measures, which are of measure zero on A, so that $\mu^0(A) = 0$. Thus $\mu^0 \ll P^0(t, x, \cdot) \ll P(t, x, \cdot) \ll \mu$ for all $x \in \text{supp } \mu$, $t \geq 0$. Applying a similar argument to (2.5), we have $P^0(t, \tilde{c}(x, z), A) = 0$ for all $x \in \text{supp } \mu$, $t \geq 0$, λ-*a.e.* $z \in Z$ and the remaining assertion follows. On the other hand if μ^0 is an extremal invariant measure and has a density ψ^0, then $P^0(t, u, \cdot)$ has a density $p^0(t, u, \cdot)$ and

$$\psi(x) = \int_0^\infty \int_Z \int_{\mathbb{R}^d} e^{-\tau\lambda(Z)} p^0(\tau, \tilde{c}(u, z), x) \mu(du) \lambda(dz) d\tau.$$

If $\psi^0(x) > 0$, then $p^0(t, u, x) > 0$ for *a.e.* $u \in A$ in $\mathbb{R}^d$ of $\mu^0(A) > 0$ and sufficiently large t, so that $\psi(x) > 0$. The remaining assertion follows as in general μ^0 can be expressed as a convex combination of extremal invariant measures, which have invariant densities. □

In this paper, we assume the coefficients $a(x)$, $b(x)$ and $c(x, z)$ always satisfy the following conditions: There exist some constants $K_a > 0$, $K_b > 0$, $K_c > 0$ and $\eta > 0$ such that for all $x \in \mathbb{R}^d$,

$$\langle x, a(x) \rangle \leq K_a - \eta |x|^\alpha, \quad \alpha > 1, \tag{2.7}$$

$$\|b(x)\|^2 \leq K_b(1 + |x|^\beta), \quad \beta < \alpha, \tag{2.8}$$

$$\int_Z |c(x, z)|^n \lambda(dz) \leq K_{cn}(1 + |x|^{n\gamma}), \quad \gamma < [1 + (\alpha - 2)/n] \wedge (\alpha - 1), \tag{2.9}$$

for some $n \in \mathbb{N}$. If (2.9) holds for some $k = n$, then it holds for all $0 \leq k \leq n$.

Lemma 2.4. *Suppose* $0 \le \upsilon \le \varrho$ *or* $\varrho \le \upsilon \le 0$. *Then for any* $\varepsilon > 0$,

$$y^{\upsilon} \le \varepsilon^{-\upsilon} + \varepsilon^{\varrho-\upsilon} y^{\varrho}, \quad y \ge 0. \tag{2.10}$$

Proof. Let $0 \le \upsilon \le \varrho$ or $\varrho \le \upsilon \le 0$. Then for $\varepsilon > 0$, $(\varepsilon y)^{\upsilon} \le 1 + (\varepsilon y)^{\varrho}$ for all $y \ge 0$, so that (2.10) holds. □

From now on, the generic constant is always denoted by $K(\cdot)$ which depends on certain parameters in the parentheses. The following theorem plays an important role in this paper.

Theorem 2.5. *Let* $X_t(x)$ *be the solution of* (1.1) *with the initial* x. *Suppose the conditions* (2.7)–(2.9) *hold. Then for all* $0 \le p \le n \wedge (n-2+\alpha)$, *there exist* $r_p > 0$ *and* $K_p > 0$ *such that for all* $t > 0$ *and* $x \in \mathbb{R}^d$,

$$\mathbb{E}|X_t(x)|^p \le K_p + |x|^{\alpha_p} e^{-r_p t} \tag{2.11}$$

where $\alpha_p = p$ *if* $\alpha \ge 2$, *and* $\alpha_p = pn/(n-2+\alpha)$ *if* $1 < \alpha < 2$.

Proof. Case (i) $\alpha \ge 2$. For $n \ge 2$, we apply Itô's formula and Lemma 2.4 to obtain

$$\begin{aligned}
&\mathbb{E}|X_t(x)|^n \qquad\qquad (2.12)\\
&= |x|^n + n\int_0^t \mathbb{E}\{|X_s(x)|^{n-2}\langle X_s(x), a(X_s(x))\rangle\}ds \\
&\quad + \frac{n(n-2)}{2}\int_0^t \mathbb{E}\{|X_s(x)|^{n-4}\langle X_s(x), \sigma(X_s(x))X_s(x)\rangle\}ds \\
&\quad + \frac{n}{2}\sum_{i=1}^d \int_0^t \mathbb{E}\{|X_s(x)|^{n-2}\sigma_{ii}(X_s(x))\}ds \\
&\quad + \int_0^t\int_Z \mathbb{E}\{|\tilde{c}(X_s(x),z)|^n - |X_s(x)|^n\}\lambda(dz)ds \\
&\le |x|^n + n\int_0^t \mathbb{E}\{K_a|X_s(x)|^{n-2} - \eta|X_s(x)|^{n-2+\alpha}\}ds \\
&\quad + \frac{n(n-1)}{2}K_b\int_0^t \mathbb{E}\{|X_s(x)|^{n-2}(1+|X_s(x)|^{\beta})\}ds \\
&\quad + \sum_{k=1}^n \binom{n}{k} K_{ck}\int_0^t \mathbb{E}\{|X_s(x)|^{n-k}(1+|X_s(x)|^{k\gamma})\}ds \\
&\le |x|^n + n\int_0^t \mathbb{E}\{K_a[\varepsilon^{-(n-2)} + \varepsilon^{\alpha}|X_s(x)|^{n-2+\alpha}] - \eta|X_s(x)|^{n-2+\alpha}\}ds \\
&\quad + \frac{n(n-1)}{2}K_b\int_0^t \mathbb{E}\{(1+\varepsilon^{-\beta})(\varepsilon^{-(n-2)} + \varepsilon^{\alpha}|X_s(x)|^{n-2+\alpha})\}ds \\
&\quad + \sum_{k=1}^n \binom{n}{k} K_{ck}\int_0^t \mathbb{E}\{(1+\varepsilon^{-k\gamma})(\varepsilon^{-(n-k)} + \varepsilon^{k-2+\alpha}|X_s(x)|^{n-2+\alpha})\}ds
\end{aligned}$$

$$\leq |x|^n + nK(n,\varepsilon,\beta,\gamma)t - n\eta(n,\varepsilon,\alpha,\beta,\gamma)\int_0^t \mathbb{E}|X_s(x)|^{n-2+\alpha}ds,$$

$$\leq |x|^n + nK(n,\varepsilon,\alpha,\beta,\gamma)t - n\eta(n,\varepsilon,\alpha,\beta,\gamma)\int_0^t \mathbb{E}|X_s(x)|^n ds,$$

for some constants $K(n,\varepsilon,\alpha,\beta,\gamma) > 0$ and $\eta(n,\varepsilon,\alpha,\beta,\gamma) > 0$ if $\varepsilon > 0$ is chosen small enough. Let $A_t = K_n + |x|^n e^{-r_n t} - \mathbb{E}|X_t(x)|^n$. Then by (2.12)

$$A_t + r_n \int_0^t A_s ds \tag{2.13}$$

$$\geq K_n + [r_n K_n - nK(n,\varepsilon,\alpha,\beta,\gamma)]t + [n\eta(n,\varepsilon,\alpha,\beta,\gamma) - r_n]\int_0^t \mathbb{E}|X_s(x)|^n ds.$$

If we choose $K_n > nK(n,\varepsilon,\alpha,\beta,\gamma)/r_n$ and $r_n < n\eta(n,\varepsilon,\alpha,\beta,\gamma)$, then $A_t \geq 0$ for all $t \geq 0$; otherwise we may assume $A_t < 0$ for $t \in (0,t_1)$ for some $t_1 > 0$ by the time-invariant of dA_t and the continuity of A_t. We arrive a contradiction as the left hand side of (2.13) is negative while the right hand side of (2.13) is positive for $t \in (0,t_1)$.

As for $0 \leq \varrho \leq 1$, $(1+x)^\varrho \leq 1 + x^\varrho$, $x \geq 0$. We have for all real $0 \leq p \leq n$,

$$\mathbb{E}|X_t(x)|^p \leq (\mathbb{E}|X_t(x)|^n)^{p/n} \leq [K_n + |x|^n e^{-r_n t}]^{p/n} \leq K_n^{p/n} + |x|^p e^{-r_n pt/n}.$$

Case (ii) $1 < \alpha < 2$. Then (2.12) still holds for $n \geq 2$ and by Lemma 2.4, $\mathbb{E}|X_t(x)|^n \geq \varepsilon^{\alpha-2}\mathbb{E}|X_t(x)|^{n-2+\alpha} - \varepsilon^{-n}$. Thus

$$\begin{aligned}\mathbb{E}|X_t(x)|^{n-2+\alpha} \quad &\leq \quad \varepsilon^{-(n-2+\alpha)} + \varepsilon^{2-\alpha}|x|^n + nK(n,\varepsilon,\alpha,\beta,\gamma)t \\ &\quad -n\eta(n,\varepsilon,\alpha,\beta,\gamma)\int_0^t \mathbb{E}|X_s(x)|^{n-2+\alpha}ds.\end{aligned}$$

We can argue as in (2.12) to show

$$\mathbb{E}|X_t(x)|^{n-2+\alpha} \leq K_n + |x|^n e^{-r_n t}. \tag{2.14}$$

By a similar argument as before, we have that (2.11) holds for all real $0 \leq p \leq n - 2 + \alpha$. □

Let $B_b(\mathbb{R}^d)$ be the Banach space of bounded Borel functions defined on $\mathbb{R}^d$ with the sup norm. Let P_t^0 and P_t be the semigroups on $B_b(\mathbb{R}^d)$ of the diffusion process determined by (2.1) and the jump-diffusion process determined by (1.1), respectively, corresponding to the infinitesimal generators L^0 and L as follows:

$$L^0 f(x) = \sum_{i=1}^d a_i(x)\partial_i f(x) + \frac{1}{2}\sum_{i,j=1}^d \sigma_{ij}(x)\partial_{ij} f(x), \tag{2.15}$$

$$L^1 f(x) = \int_Z [f(\tilde{c}(x,z)) - f(x)]\lambda(dz) \tag{2.16}$$

and $L = L^0 + L^1$ as in (1.2). Note that the semigroups P_t^0 and P_t are not continuous in $\mathbb{R}_+$. Then the infinitesimal generators L^0 and L are only defined in the weak sense. Since L^1 is a bounded linear operator on $B_b(\mathbb{R}^d)$, then by [5, Corollary 1.7], P_t^0 and P_t can be related as follows:

$$\begin{aligned} P_t f(x) &= P_t^0 f(x) + \int_0^t P_{t-s}^0 L^1 P_s f(x) ds && (2.17) \\ &= P_t^0 f(x) + \int_0^t P_{t-s} L^1 P_s^0 f(x) ds. && (2.18) \end{aligned}$$

Then by (2.17)–(2.18), the relations between the transition probability functions P and P^0 are as follows:

$$\begin{aligned} P(t,x,A) &= P^0(t,x,A) + \int_0^t \int_Z \int_{\mathbb{R}^d} P^0(t-s,x,du) \\ &\quad \cdot [P(s,\tilde{c}(u,z),A) - P(s,u,A)]\lambda(dz)ds, && (2.19) \end{aligned}$$

$$\begin{aligned} P(t,x,A) &= P^0(t,x,A) + \int_0^t \int_Z \int_{\mathbb{R}^d} P(t-s,x,du) \\ &\quad \cdot [P^0(s,\tilde{c}(u,z),A) - P^0(s,u,A)]\lambda(dz)ds. && (2.20) \end{aligned}$$

Definition 2.1. *A right-continuous process $\{X_t\}_{t\geq 0}$ in $\mathbb{R}^d$ is said to be conservative if $I\!P_x\{X_t \in \mathbb{R}^d\} = 1$ for all $t \geq 0$ and $x \in \mathbb{R}^d$.*

Lemma 2.6. *The jump-diffusion process X_t of* (1.1) *is conservative if and only if the diffusion process X_t^0 of* (2.1) *is conservative.*

Proof. If the diffusion X_t^0 is conservative, then $P^0(t,x,\mathbb{R}^d) = 1$ for all $t \geq 0$ and $x \in \mathbb{R}^d$, and so by (2.17), $P(t,x,\mathbb{R}^d) = P^0(t,x,\mathbb{R}^d) + 0 = 1$ for all $t \geq 0$ and $x \in \mathbb{R}^d$. Thus X_t is conservative. The converse result follows by a similar argument with (2.16). □

Theorem 2.7. *Under the conditions* (2.7)–(2.9)*, there exist invariant probability measures μ and μ^0 for the jump-diffusion process X_t and diffusion process X_t^0, respectively. Indeed μ and μ^0 have finite moments up to order $n \wedge (n-2+\alpha)$ and have the following relations:*

$$\mu(A) = \mu^0(A) + \int_0^\infty \int_Z \int_{\mathbb{R}^d} [P(t,\tilde{c}(x,z),A) - P(t,x,A)]\mu^0(dx)\lambda(dz)dt, \quad (2.21)$$

$$\mu(A) = \mu^0(A) + \int_0^\infty \int_Z \int_{\mathbb{R}^d} [P^0(t,\tilde{c}(x,z),A) - P^0(t,x,A)]\mu(dx)\lambda(dz)dt. \quad (2.22)$$

Furthermore, suppose X^0 is nondegenerate, i.e.,

$$\dim LA\{b_1, \cdots, b_m\}(x) = d, \quad x \in \mathbb{R}^d, \quad (2.23)$$

where b_i denotes the ith column vector of b and LA denotes the Lie algebra generated by the vector fields in the bracket. Then the diffusion (2.1) *and the jump-diffusion (1.1) are strong Feller processes on $\mathbb{R}^d$ and have strictly positive densities $p^0(t,x,y)$, $p(t,x,y) \in C(\mathbb{R}_+ \times \mathbb{R}^d \times \mathbb{R}^d)$, respectively, for sufficiently large t, a.e. $y \in \mathbb{R}^d$, all $x \in \mathbb{R}^d$. In addition the invariant measures μ^0 and μ are unique and have strictly positive invariant densities ψ^0, $\psi \in C(\mathbb{R}^d)$, respectively.*

Proof. Define

$$\mu_T(A) = \frac{1}{T}\int_0^T P(t,x,A)dt.$$

Let $\overline{\mathbb{R}}^d = \mathbb{R}^d \cup \{\infty\}$ be the one-point compactification of $\mathbb{R}^d$. Denote by $P(\overline{\mathbb{R}}^d)$ the space of Borel probability measures on $\overline{\mathbb{R}}^d$ with the weak topology. Then $\overline{\mathbb{R}}^d$ is compact and metrizable, and so is $P(\overline{\mathbb{R}}^d)$. This means that $\{\mu_T\}_{T>0}$ is tight and that there exists $\mu \in P(\overline{\mathbb{R}}^d)$ and a subsequence $T_n \to \infty$ such that $\mu_{T_n} \Rightarrow \mu$ weakly in $P(\overline{\mathbb{R}}^d)$ as $n \to \infty$. Let B_R be the closed ball centered at the origin with the radius R. By the weak convergence of μ_{T_n} and Theorem 2.5, we have

$$\begin{aligned}
\mu\left(B_R^c\right) &\leq \liminf_{n\to\infty} \frac{1}{T_n}\int_0^{T_n} P(t,x,B_R^c)dt \\
&\leq \liminf_{n\to\infty} \frac{1}{T_n}\int_0^{T_n}\int_{B_R^c} \frac{|y|}{R} P(t,x,dy)dt \\
&\leq \liminf_{n\to\infty} \frac{1}{T_n}\int_0^{T_n} \frac{1}{R}\left(K_1 + |x|^{\alpha_1}e^{-r_1 t}\right)dt \\
&\leq \liminf_{n\to\infty} \frac{1}{RT_n}\left[K_1T_n + |x|^{\alpha_1}(1-e^{-r_1T_n})/r_1\right] = \frac{K_1}{R},
\end{aligned}$$

which tends to 0 as $R \to \infty$. Thus $\mu(\infty) = 0$. Without loss of generality, we can regard μ as a probability measure defined on $\mathbb{R}^d$. In addition, for any bounded continuous function f,

$$\begin{aligned}
\mu(P_s f) &= \lim_{n\to\infty}\int_{\mathbb{R}^d} P_s f(y)\mu_{T_n}(dy) \\
&= \lim_{n\to\infty}\frac{1}{T_n}\int_0^{T_n}\int_{\mathbb{R}^d} P_s f(y)P(t,x,dy)dt \\
&= \lim_{n\to\infty}\frac{1}{T_n}\int_0^{T_n}\int_{\mathbb{R}^d} f(y)P(t+s,x,dy)dt \\
&= \lim_{n\to\infty}\frac{1}{T_n}\left\{\int_0^{T_n}\int_{\mathbb{R}^d} f(y)P(t,x,dy)dt - \int_0^s\int_{\mathbb{R}^d} f(y)P(t,x,dy)dt\right. \\
&\qquad \left. + \int_{T_n}^{T_n+s}\int_{\mathbb{R}^d} f(y)P(t,x,dy)dt\right\} \\
&= \int_{\mathbb{R}^d} f(x)\mu(dx) = \mu(f),
\end{aligned}$$

as the last two integrals are bounded by $2s\|f\|/T_n \to 0$ as $n \to \infty$. Thus μ is an invariant probability measure for X_t. In fact by Theorem 2.5,

$$\begin{aligned}\int_{\mathbb{R}^d} |y|^p \mu(dy) &= \lim_{k\to\infty} \frac{1}{T_k}\int_0^{T_k}\int_{\mathbb{R}^d} |y|^p P(t,x,dy)dt \\ &\le \lim_{k\to\infty} \frac{1}{T_k}\int_0^{T_k} \left(K_p + |x|^{\alpha_p} e^{-r_p t}\right) dt \\ &\le \lim_{k\to\infty} \frac{1}{T_k}\left[K_p T_k + |x|^{\alpha_p}(1-e^{-r_p T_k})/r_p\right] = K_p,\end{aligned}$$

and so μ has all finite moments up to $n \wedge (n-2+\alpha)$. Similarly if we define $\mu_T^0 = \int_0^T P^0(t,x,\cdot)dt/T$, then $\{\mu_{T_n}^0\}_{T_n>0}$ is tight in $P(\overline{\mathbb{R}}^d)$ and so there exists a subsequence, which we denote it by the same sequence T_n without loss of generality such that $\mu_{T_n}^0 \Rightarrow \mu^0$ weakly in $P(\overline{\mathbb{R}}^d)$ as $T_n \to \infty$. Indeed $\mu^0 \in P(\mathbb{R}^d)$ is an invariant probability measure of X_t^0 and has all finite moments up to n. It remains to prove the relations (2.21) and (2.22). By (2.19),

$$\begin{aligned}&\frac{1}{T_n}\int_0^{T_n} P(t,x,A)dt - \frac{1}{T_n}\int_0^{T_n} P^0(t,x,A)dt \qquad (2.24)\\ &= \frac{1}{T_n}\int_0^{T_n}\int_0^t\int_Z\int_{\mathbb{R}^d} P^0(t-s,x,du)[P(s,\tilde{c}(u,z),A) - P(s,u,A)]\lambda(dz)dsdt \\ &= \frac{1}{T_n}\int_0^{T_n}\int_s^{T_n}\int_Z\int_{\mathbb{R}^d} P^0(t-s,x,du)[P(s,\tilde{c}(u,z),A) - P(s,u,A)]\lambda(dz)dtds \\ &= \int_0^{T_n}\int_Z\int_{\mathbb{R}^d} \frac{1}{T_n}\int_0^{T_n-s} P^0(t,x,du)dt[P(s,\tilde{c}(u,z),A) - P(s,u,A)]\lambda(dz)ds.\end{aligned}$$

As $n \to \infty$, the left hand side of (2.24) converges to $\mu(A) - \mu^0(A)$. Then the right hand side of (2.24) converges and the limit is

$$\int_0^\infty \int_Z\int_{\mathbb{R}^d} \mu^0(du)[P(s,\tilde{c}(u,z),A) - P(s,u,A)]\lambda(dz)ds$$

as asserted in (2.21). The result (2.22) follows by a similar argument using (2.19).

Suppose the nondegenerate condition (2.23) holds. Then the transition probability for (2.1) has a density $p^0(t,x,y) \in C^\infty(\mathbb{R}_+ \times \mathbb{R}^d \times \mathbb{R}^d)$ as assured by [10, Theorem 3] and so $p(t,x,y) \in C(\mathbb{R}_+ \times \mathbb{R}^d \times \mathbb{R}^d)$ by Lemma 2.1. Then the diffusion (2.1) and the jump-diffusion (1.1) are strong Feller processes on $\mathbb{R}^d$. By [10, Theorem 6.1] μ^0 has a strictly positive density ψ^0 on $\mathbb{R}^d$. By [10, Proposition 5.1] $p^0(t,x,y) > 0$ for sufficiently large t, *a.e.* $y \in \mathbb{R}^d$, all $x \in \mathbb{R}^d$ and so do $p(t,x,y)$ by Lemma 2.1. By Lemma 2.2 μ has a continuous and strictly positive density

$$\psi(x) = \int_0^\infty \int_Z\int_{\mathbb{R}^d} e^{-\tau\lambda(Z)} p^0(\tau,\tilde{c}(u,z),x)\mu(du)\lambda(dz)d\tau \qquad (2.25)$$

on $\mathbb{R}^d$. Thus μ and μ^0 are unique invariant measures for (1.1) and (2.1), respectively. □

Example 2.1. Consider a nonlinear stochastic differential equation in $\mathbb{R}$ as follows:

$$dX_t = -a|X_t|^\alpha \mathrm{sgn}(X_t)dt + b|X_t|^\beta dW_t + cX_t dN_t \tag{2.26}$$

where N_t is a Poisson point process with intensity λ, and $\alpha > 0$, $\beta < (\alpha+1)/2$. The coefficients are at most linear growth and are only locally Lipschitz continuous on $\mathbb{R} \setminus (-\varepsilon, \varepsilon)$ for each $\varepsilon > 0$ if $\alpha,\ \beta < 1$, so that we consider the initial condition $X_0 \neq 0$. The infinitesimal generator is

$$Lf(x) = \frac{1}{2}b^2|x|^{2\beta}f''(x) - a|x|^\alpha \mathrm{sgn}(x)f'(x) + \lambda[f((1+c)x) - f(x)].$$

Then the corresponding adjoint operator is

$$L^*g(x) = \frac{1}{2}b^2[|x|^{2\beta}g(x)]'' + a[|x|^\alpha \mathrm{sgn}(x)g(x)]' + \lambda\left[\frac{1}{|1+c|}g\left(\frac{x}{1+c}\right) - g(x)\right].$$

The solution of the adjoint equation $L^*g = 0$ is δ_0 and

$$\begin{aligned} g(x) \quad &= \quad \frac{2K_1}{b^2|x|^{2\beta}} \exp\left\{-\frac{2a|x|^{\alpha-2\beta+1}}{b^2(\alpha-2\beta+1)}\right\} \int_0^{|x|} \exp\left\{\frac{2au^{\alpha-2\beta+1}}{b^2(\alpha-2\beta+1)}\right\} du \\ &\quad + \frac{K_2}{|x|^{2\beta}} \exp\left\{-\frac{2a|x|^{\alpha-2\beta+1}}{b^2(\alpha-2\beta+1)}\right\} \end{aligned}$$

if $c = 0$ or $c = -2$. In particular for $\alpha = 2\beta < 1$, an invariant probability measure has C^∞ strictly positive density

$$\psi(x) = \frac{K}{|x|^{2\beta}} e^{-2a|x|/b^2}, \quad x \in \mathbb{R} \setminus \{0\},$$

where

$$K = \frac{1}{2}\left[\int_0^\infty \frac{1}{|x|^{2\beta}} \exp(-ax/b^2)dx\right]^{-1} < \infty.$$

In general for $\beta \geq 1/2$, the point measure $\delta_0(x)dx$ is the only invariant probability measure while $g(x)$ is an invariant density but not an invariant probability density for (2.26). The non-uniqueness of invariant measures is due to the fact that the nondegenerate condition (2.23) does not hold at $x = 0$.

3. Lyapunov exponents

Before investigating Lyapunov exponents of the jump-diffusion (1.1), we give the following asymptotic results for the jump-diffusion martingale.

Lemma 3.1. *Let the zero-mean right-continuous martingale be*

$$M_t = \int_0^t g'(X_s)dW_s + \int_0^t \int_Z h(X_{s-}, z)\widetilde{N}(ds, dz), \tag{3.1}$$

where $|g(x)|^2 \leq K_1|x|^{\delta_1}$ *and* $\int_Z |h(x,z)|^2\lambda(dz) \leq K_2|x|^{\delta_2}$ *for some* $\delta_1 \geq 0$, $\delta_2 \geq 0$. *Then* $\lim_{t\to\infty} M(t)/t = 0$ *a.s.*

Proof. Note that the conditional quadratic variation of M_t is

$$\langle M\rangle_t = \int_0^t |g(X_s)|^2 ds + \int_0^t \int_Z h^2(X_s, z)\lambda(dz)ds.$$

Then

$$\begin{aligned} \sup_{t\geq 0} \mathbb{E}|g(X_t)|^2 &\leq \sup_{t\geq 0} K_1 \mathbb{E}|X_t|^{\delta_1} \\ &\leq \sup_{t\geq 0} K_1(K_{\delta_1} + |x|^{\delta_1} e^{-r_{\delta_1} t}) \\ &= K_1(K_{\delta_1} + |x|^{\delta_1}) < \infty, \end{aligned}$$

$$\begin{aligned} \sup_{t\geq 0} \mathbb{E} \int_Z h^2(X_t, z)\lambda(dz) &\leq \sup_{t\geq 0} K_2 \mathbb{E}|X_t|^{\delta_2} \\ &\leq \sup_{t\geq 0} K_2(K_{\delta_2} + |x|^{\delta_2} e^{-r_{\delta_2} t}) \\ &= K_2(K_{\delta_2} + |x|^{\delta_2}) < \infty. \end{aligned}$$

By a lemma in [4, p. 459], the result follows. □

Now we state a result concerning about the Lyapunov exponents of nonlinear stochastic differential equations with jumps.

Theorem 3.2. *Suppose* (2.7) *holds for* $\alpha > 2$ *and*

$$\|b(x)\|^2 \leq K_b|x|^\beta, \quad 2 \leq \beta < \alpha, \tag{3.2}$$

$$\int_Z |c(x, z)|^2 \lambda(dz) \leq K_c|x|^\gamma, \quad 2 \leq \gamma < \alpha, \tag{3.3}$$

for some constants $K_b > 0$ *and* $K_c > 0$. *Then*

$$\lim_{t\to\infty} \frac{\ln(X_t(x))}{t} = q_\mu \equiv \int_{\mathbb{R}^d} Q(x)\mu(dx), \tag{3.4}$$

for all $x \in \operatorname{supp} \mu$, *where* μ *is an extremal invariant measure of the jump-diffusion* (1.1) *and*

$$Q(x) = Q^0(x) + \int_Z \ln\left(\frac{|\tilde{c}(x, z)|}{|x|}\right)\lambda(dz), \tag{3.5}$$

$$Q^0(x) = \frac{\langle x, a(x)\rangle}{|x|^2} + \frac{1}{2}\sum_{i=1}^d \sum_{k=1}^m \frac{b_{ik}^2(x)}{|x|^2} - \frac{|b'x|^2}{|x|^4}. \tag{3.6}$$

Proof. Let $\rho_t = \ln|X_t|$. Then $d\rho_t = Q(X_t)dt + M_t$ with the martingale

$$M_t = \frac{X_t' b(X_t)}{|X_t|^2} dW_t + \int_Z \ln\left(\frac{|\tilde{c}(X_{t-}, z)|}{|X_{t-}|}\right)\widetilde{N}(dt, dz),$$

and so

$$|g(x)|^2 = \frac{x'b(x)b'(x)x}{|x|^4} \leq \frac{\|b(x)\|^2}{|x|^2} \leq K_b|x|^{\beta-2}, \quad \beta \geq 2.$$

If $\tilde{c}(X_{t_1}, z) = 0$, then $X_t = 0$ for all $t \geq t_1$ and so the Lyapunov exponent $q = \lim_{t\to\infty} \rho_t/t = -\infty$. For the non-trivial case, we may assume $|\tilde{c}(x,z)| \geq \varepsilon(z)|x|$ for some $\varepsilon(z) > 0$ and $\int_Z \ln^2 \varepsilon(z)\lambda(dz) < \infty$. Using the fact $\ln(1+x) \leq x$ for $x \geq 0$, we have

$$\begin{aligned}
\int_Z h^2(x,z)\lambda(dz) &= \int_Z \ln^2\left(\frac{|\tilde{c}(x,z)|}{|x|}\right)\lambda(dz) \\
&\leq \int_Z \ln^2\left(1+\frac{|c(x,z)|}{|x|}\right)\lambda(dz) \bigvee \int_Z \ln^2\varepsilon(z)\lambda(dz) \\
&\leq \int_Z \frac{|c(x,z)|^2}{|x|^2}\lambda(dz) \bigvee \int_Z \ln^2\varepsilon(z)\lambda(dz) \\
&\leq K_c|x|^{\gamma-2} \bigvee \int_Z \ln^2\varepsilon(z)\lambda(dz), \quad \gamma \geq 2,
\end{aligned}$$

By Lemma 3.1 and the ergodic property of X_t,

$$\lim_{t\to\infty}\frac{\rho_t}{t} = \lim_{t\to\infty}\frac{\rho_0}{t} + \lim_{t\to\infty}\frac{1}{t}\int_0^t Q(X_s)ds + \lim_{t\to\infty}\frac{M_t}{t} = \int_{\mathbb{R}^d} Q(x)\mu(dx)$$

for all $x \in \text{supp}\ \mu$. □

Corollary 3.3. *Suppose* (2.7) *holds for* $\alpha > 1$ *and*

$$\|b(x)\|^2 \leq K_b(|x|^{\beta_1} \wedge |x|^{\beta_2}), \quad 0 \leq \beta_1 < \beta_2,\ \beta_2 \geq 2,\ \beta_1 < \alpha, \quad (3.7)$$

$$\int_Z |c(x,z)|^2\lambda(dz) \leq K_c(|x|^{\gamma_1} \wedge |x|^{\gamma_2}), \quad 0 \leq \gamma_1 < \gamma_2,\ \gamma_2 \geq 2,\ \gamma_1 < \alpha, \quad (3.8)$$

for some constants $K_b > 0$ *and* $K_c > 0$. *Then the result* (3.4) *still holds for all* $x \in \text{supp}\ \mu$.

References

[1] L. Arnold, Lyapunov exponents of nonlinear stochastic systems, in *Nonlinear Stochastic Dynamic Engineering Systems*, Springer-Verlag, Berlin, 1988, pp. 181–201

[2] L. Arnold, *Random Dynamical Systems*, Springer-Verlag, Berlin, 1998.

[3] P. H. Baxendale, Invariant measures for nonlinear stochastic differential equations, in *Lyapunov Exponents*, Lecture Notes in Mathematics, **1486**, Springer-Verlag, Berlin, 1991, pp. 123–140.

[4] G. L. Blankenship and G. C. Papanicolaou, Stability and control of stochastic systems with wide-band noise disturbances I, SIAM J. Appl. Math., **34** (1978), 437–476.

[5] K. J. Engel and R. Nagel, *One-parameter Semigroups for Linear Evolution Equations*, Springer, New York, 2000.

[6] M. G. Garroni and J. L. Menaldi, *Second Order Elliptic Integro-differential Problems*, Chapman & Hall/CRC, Boca Raton, 2002.

[7] M. G. Garroni and J. L. Menaldi, *Green Functions for Second Order Parabolic Integro-differential Problems*, Longman Scientific & Technical, Essex, 1992.

[8] A. Grorud and D. Talay, Approximation of Lyapunov exponents of nonlinear stochastic differential equations, SIAM J. Appl. Math., **56** (2) (1996), 627–650.

[9] R. Z. Has'minskii, *Stochastic Stability of Differential Equations*, Sijthoff & Noordhoff, 1980.

[10] K. Ichihara and H. Kunita, A classification of the second order degenerate elliptic operators and its probabilistic characterization, Z.Wahrscheinlichkeitstheorie. verw. Gebiete, **30** (1974), 235–254; **39** (1977), 81–84.

[11] C. W. Li and G. L. Blankenship, Almost sure stability of linear stochastic systems with Poisson process coefficients, SIAM J. Appl. Math., **46** (1986), 875–911.

[12] C. W. Li and Z. Dong, On kernel estimations and invariant measures of stochastic jump-diffusions, Dynamics of Continuous, Discrete and Impulsive Systems, Series A, **10** (3) (2003), 373–387.

[13] C. W. Li, Z. Dong and R. Situ, Almost sure stability of linear stochastic differential equations with jumps, Probab. Theory Relat. Fields, **123** (2002), 121–155.

[14] J. L. Menaldi and M. Robin, Invariant measure for diffusions with jumps, Appl. Math. Optim., **40** (1999), 105–140.

[15] V. I. Oseledec, A multiplicative ergodic theorem, Lyapunov characteristic numbers for dynamical systems, Trans. Moscow Math. Soc., **19** (1968), 197–231.

[16] M. A. Pinsky, Stochastic stability and its Dirichlet problem, Comm. Pure & Appl. Math., **27** (1974), 311–350.

[17] A. V. Skorokhod, *Asymptotic Methods in the Theory of Stochastic Differential Equations*, American Mathematical Society, Providence, 1989.

[18] I. S. Wee, Stability for multidimensional jump-diffusion processes, Stochastic Processes Appl., **80** (1999), 193–209.

Department of Mathematics, City University of Hong Kong, 83 Tat Chee Avenue, Kowloon, Hong Kong
E-mail address: `macwli@cityu.edu.hk`

Progress in Probability, Vol. 56, 353–365

Stochastic Differential Equations with Additive Fractional Noise and Locally Unbounded Drift

David Nualart and Youssef Ouknine

Abstract. Let $\{B_t^H, t \in [0,T]\}$ be a fractional Brownian motion with Hurst parameter $H < \frac{1}{2}$. We prove the existence and uniqueness of a strong solution for a stochastic differential equation of the form $X_t = x_0 + B_t^H + \int_0^t b(s, X_s)ds$, where $b(s,x)$ is not locally bounded and satisfies a suitable integrability condition.

1. Introduction

Let $B^H = \{B_t^H, t \in [0,T]\}$ be a fractional Brownian motion with Hurst parameter $H \in (0,1)$. That is, B^H is a centered Gaussian process with covariance

$$R_H(t,s) = \mathbb{E}(B_t^H B_s^H) = \frac{1}{2}\left(t^{2H} + s^{2H} - |t-s|^{2H}\right).$$

If $H = \frac{1}{2}$ the process B^H is a standard Brownian motion. We have

$$\mathbb{E}\left(B_t^H - B_s^H\right)^2 = |t-s|^{2H},$$

and this implies that the process B^H has $(H-\varepsilon)$-Hölder continuous trajectories for all $\varepsilon > 0$.

Consider the following stochastic differential equation

$$X_t = x_0 + B_t^H + \int_0^t b(s, X_s)ds, \; 0 \le t \le T, \tag{1.1}$$

where $b : [0,T] \times \mathbb{R}$ is a Borel function. The purpose of this paper is to establish the existence and uniqueness of a strong solution to this equation in the case $H \in (0, \frac{1}{2})$ under the following integrability assumption on the coefficient $b(t,x)$:

$$b(t,x)^2 \le C + F(t,x), \tag{1.2}$$

2000 *Mathematics Subject Classification.* Primary 60H10; Secondary 60G18.
Key words and phrases. Fractional Brownian motion. Stochastic differential equations. Girsanov theorem.
Supported by the Spanish DGES grant BFM2000-0598 and the Moroccan Program PARS MI 37.

where F is a nonnegative Borel function satisfying

$$\int_0^T \left(\int_{\mathbb{R}} |F(t,x)|^p \, dx \right)^{\frac{\beta}{p}} dt < \infty,$$

for some $p > 1$ and $\beta > \frac{p}{p-H}$.

Note that the drift can be locally unbounded. In a previous paper [9] we have proved the existence of a unique solution to Equation (1.1) when the coefficient b satisfies the linear growth condition

$$|b(t,x)| \leq C(1 + |x|). \tag{1.3}$$

On the other hand, if $H > \frac{1}{2}$, the process B^H has a smaller regularization effect, and to establish the existence of a unique strong solution to Equation (1.1) we need the Hölder continuity assumption

$$|b(t,x) - b(s,y)| \leq C\left(|x-y|^\alpha + |t-s|^\gamma\right),$$

where $1 > \alpha > 1 - \frac{1}{2H}$ and $\gamma > H - \frac{1}{2}$ (see [9]).

In the case $H = \frac{1}{2}$, the existence and uniqueness of a strong solution when b satisfies the linear growth condition (1.3) has been first proved in the works of Zvonkin [14] and Veretennikov [13]. See also the work by Nakao [8] and its generalization by Ouknine [11]. In these papers the equation may contain a nonconstant and nondegenerate diffusion coefficient which is of bounded variation. The existence of strong solutions via approximations in the case where the drift does not satisfy the linear growth condition has been proved by Gyöngy and Krylov in [4].

In order to prove our result we first apply a suitable version of Girsanov theorem for the fractional Brownian motion and we obtain the weak existence and uniqueness of solution. This is the contents of Section 2. The existence of a strong solution is established in Section 3 using Krylov-type estimates and the comparison theorem. This method has also been used to handle one-dimensional heat equations with additive space-time white noise in [5].

Although we carry out the proofs only in the case $H < \frac{1}{2}$, our arguments could easily be extended to the case $H = \frac{1}{2}$, providing a new existence and uniqueness result for ordinary one-dimensional stochastic differential equations with locally unbounded drift.

2. Preliminaries

2.1. Fractional calculus

An exhaustive survey on classical fractional calculus can be found in [12]. We recall some basic definitions and results.

For $f \in L^1([a,b])$ and $\alpha > 0$ the *left-sided fractional Riemann-Liouville integral* of f *of order* α on (a,b) is given at almost all x by

$$I_{a+}^\alpha f(x) = \frac{1}{\Gamma(\alpha)} \int_a^x (x-y)^{\alpha-1} f(y) dy,$$

where Γ denotes the Euler function.

This integral extends the usual n-order iterated integrals of f for $\alpha = n \in \mathbb{N}$. We have the first composition formula

$$I_{a+}^{\alpha}(I_{a+}^{\beta} f) = I_{a+}^{\alpha+\beta} f.$$

The fractional derivative can be introduced as the inverse operation. We assume $0 < \alpha < 1$ and $p > 1$. We denote by $I_{a+}^{\alpha}(L^p)$ the image of $L^p([a,b])$ by the operator I_{a+}^{α}. If $f \in I_{a+}^{\alpha}(L^p)$, the function ϕ such that $f = I_{a+}^{\alpha}\phi$ is unique in L^p and it agrees with the *left-sided Riemann-Liouville derivative of f of order α* defined by

$$D_{a+}^{\alpha} f(x) = \frac{1}{\Gamma(1-\alpha)} \frac{d}{dx} \int_a^x \frac{f(y)}{(x-y)^{\alpha}} dy.$$

When $\alpha p > 1$ any function in $I_{a+}^{\alpha}(L^p)$ is $\left(\alpha - \frac{1}{p}\right)$- Hölder continuous. On the other hand, any Hölder continuous function of order $\beta > \alpha$ has fractional derivative of order α. The derivative of f has the following Weil representation:

$$D_{a+}^{\alpha} f(x) = \frac{1}{\Gamma(1-\alpha)} \left(\frac{f(x)}{(x-a)^{\alpha}} + \alpha \int_a^x \frac{f(x)-f(y)}{(x-y)^{\alpha+1}} dy \right) \mathbf{1}_{(a,b)}(x), \tag{2.1}$$

where the convergence of the integrals at the singularity $x = y$ holds in L^p-sense.

Recall that by construction for $f \in I_{a+}^{\alpha}(L^p)$,

$$I_{a+}^{\alpha}(D_{a+}^{\alpha} f) = f$$

and for general $f \in L^1([a,b])$ we have

$$D_{a+}^{\alpha}(I_{a+}^{\alpha} f) = f.$$

If $f \in I_{a+}^{\alpha-\beta}(L^1)$, $\alpha \geq 0$, $\beta \geq 0$, $\alpha + \beta \leq 1$ we have the second composition formula

$$D_{a+}^{\alpha}(D_{a+}^{\beta} f) = D_{a+}^{\alpha+\beta} f.$$

3. Existence of a weak solution, and pathwise uniqueness property

3.1. Girsanov transform

Let $B^H = \{B_t^H, t \in [0,T]\}$ be a fractional Brownian motion with Hurst parameter $0 < H < \frac{1}{2}$ defined on the probability space $(\Omega, \mathcal{F}, \mathbb{P})$. For each $t \in [0,T]$ we denote by $\mathcal{F}_t^{B^H}$ the σ-field generated by the random variables $\{B_s^H, s \in [0,t]\}$ and the sets of probability zero.

We denote by $\mathcal{E}$ the set of step functions on $[0,T]$. Let $\mathcal{H}$ be the Hilbert space defined as the closure of $\mathcal{E}$ with respect to the scalar product

$$\langle \mathbf{1}_{[0,t]}, \mathbf{1}_{[0,s]} \rangle_{\mathcal{H}} = R_H(t,s).$$

The mapping $\mathbf{1}_{[0,t]} \longrightarrow B_t$ can be extended to an isometry between $\mathcal{H}$ and the Gaussian space $H_1(B)$ associated with B. We will denote this isometry by $\varphi \longrightarrow B(\varphi)$.

The covariance kernel $R_H(t,s)$ can be written as

$$R_H(t,s) = \int_0^{t\wedge s} K_H(t,r)K_H(s,r)dr,$$

where K_H is a square integrable kernel given by (see [2]):

$$K_H(t,s) = \Gamma(H+\frac{1}{2})^{-1}(t-s)^{H-\frac{1}{2}}F(H-\frac{1}{2},\frac{1}{2}-H,H+\frac{1}{2},1-\frac{t}{s}),$$

$F(a,b,c,z)$ being the Gauss hypergeometric function. Consider the linear operator K_H^* from $\mathcal{E}$ to $L^2([0,T])$ defined by

$$(K_H^*\varphi)(s) = K_H(T,s)\varphi(s) + \int_s^T (\varphi(r)-\varphi(s))\,\frac{\partial K_H}{\partial r}(r,s)dr.$$

We have $\left(K_H^*\mathbf{1}_{[0,t]}\right)(s) = K_H(t,s)$. As a consequence, for any pair of step functions φ and ψ in $\mathcal{E}$ we have

$$\langle K_H^*\varphi, K_H^*\psi\rangle_{L^2([0,T])} = \langle\varphi,\psi\rangle_{\mathcal{H}}.$$

Thus, the operator K_H^* provides an isometry between the Hilbert spaces $\mathcal{H}$ and $L^2([0,T])$. Then, the process $W = \{W_t, t\in[0,T]\}$ defined by

$$W_t = B((K_H^*)^{-1}(\mathbf{1}_{[0,t]})) \tag{3.1}$$

is a Wiener process, and the process B has an integral representation of the form (see [1])

$$B_t^H = \int_0^t K_H(t,s)dW_s. \tag{3.2}$$

On the other hand, the operator K_H on $L^2([0,T])$ associated with the kernel K_H is an isomorphism from $L^2([0,T])$ onto $I_{0+}^{H+1/2}(L^2([0,T]))$ and it can be expressed in terms of fractional integrals as follows (see [2]):

$$(K_Hh)(s) = I_{0+}^{2H}s^{\frac{1}{2}-H}I_{0+}^{\frac{1}{2}-H}s^{H-\frac{1}{2}}h, \tag{3.3}$$

where $h\in H$.

Henceforth we consider a right-continuous increasing family of σ-fields

$$\{\mathcal{F}_t, t\in[0,T]\}$$

on $(\Omega,\mathcal{F},\mathbb{P})$ such that $\mathcal{F}_0$ contains the sets of probability zero.

Definition 3.1. A fractional Brownian motion $B^H = \{B^H, t\in[0,T]\}$ is called an $\mathcal{F}_t$-fractional Brownian motion if the process W defined in (3.1) is an $\mathcal{F}_t$-Wiener process.

Given an $\mathcal{F}_t$-adapted process with integrable trajectories $u = \{u_t, t\in[0,T]\}$ and an $\mathcal{F}_t$-fractional Brownian motion B^H, we introduce the transformation

$$\widetilde{B}_t^H = B_t^H + \int_0^t u_s ds. \tag{3.4}$$

We can write

$$\widetilde{B}_t^H = \int_0^t K_H(t,s)dW_s + \int_0^t u_s ds = \int_0^t K_H(t,s)d\widetilde{W}_s,$$

where

$$\widetilde{W}_t = W_t + \int_0^t K_H^{-1}\left(\int_0^\cdot u_s ds\right)(r)dr. \tag{3.5}$$

Notice that $K_H^{-1}\left(\int_0^\cdot u_s ds\right)$ belongs to $L^2([0,T])$ a.s. if and only if $\int_0^\cdot u_s ds \in I_{0+}^{H+1/2}(L^2([0,T]))$ a.s. As a consequence we deduce the following version of Girsanov theorem for the fractional Brownian motion:

Theorem 3.2. *Consider the shifted process* (3.4) *defined by an* $\mathcal{F}_t$*-adapted process* $u = \{u_t, t \in [0,T]\}$ *with integrable trajectories. Assume that:*

i) $\int_0^\cdot u_s ds \in I_{0+}^{H+1/2}(L^2([0,T]))$*, almost surely.*
ii) $\mathbb{E}(\xi_T) = 1$*, where*

$$\xi_T = \exp\left(-\int_0^T K_H^{-1}\left(\int_0^\cdot u_s ds\right)(s)dW_s - \frac{1}{2}\int_0^T K_H^{-1}\left(\int_0^\cdot u_s ds\right)^2(s)ds\right).$$

Then the shifted process $\widetilde{B}^H$ *is an* $\mathcal{F}_t$*- fractional Brownian motion with Hurst parameter* H *under the new probability* $\widetilde{\mathbb{P}}$ *defined by* $\frac{d\widetilde{\mathbb{P}}}{d\mathbb{P}} = \xi_T$.

Proof. By the standard Girsanov theorem applied to the adapted and square integrable process $K_H^{-1}\left(\int_0^\cdot u_s ds\right)$ we obtain that the process $\widetilde{W}$ defined in (3.5) is an $\mathcal{F}_t$- Brownian motion under the probability $\widetilde{\mathbb{P}}$. Hence, the result follows. □

From (3.3) the inverse operator K_H^{-1} is given by

$$K_H^{-1}h = s^{\frac{1}{2}-H} D_{0+}^{\frac{1}{2}-H} s^{H-\frac{1}{2}} D_{0+}^{2H} h, \tag{3.6}$$

for all $h \in I_{0+}^{H+\frac{1}{2}}(L^2([0,T]))$. If h is absolutely continuous it can be proved that

$$K_H^{-1}h = s^{H-\frac{1}{2}} I_{0+}^{\frac{1}{2}-H} s^{\frac{1}{2}-H} h'. \tag{3.7}$$

From (3.7) it follows that a sufficient condition for i) is $\int_0^T u_s^2 ds < \infty$.

3.2. A priori estimates

Let $p, \beta > 1$. We say that a Borel function

$$F : [0,T] \times \mathbb{R} \to \mathbb{R}$$

belongs to $L_{p,\beta}$ if

$$\|F\|_{p,\beta} = \left(\int_0^T \left(\int_{\mathbb{R}} |F(t,x)|^p \, dx\right)^{\frac{\beta}{p}} dt\right)^{\frac{1}{\beta}} < \infty.$$

Consider the following assumption:

(H) The Borel function $b : [0,T] \times \mathbb{R} \to \mathbb{R}$ satisfies

$$b(t,x)^2 \leq K + F(t,x),$$

for almost all $(t,x) \in [0,T] \times \mathbb{R}$, where $K > 0$ is a constant and F is a nonnegative function in $L_{p,\beta}$ for some $p > 1$ and $\beta > \frac{p}{p-H}$.

By a *weak solution* to equation (1.1) we mean a couple of adapted processes (B^H, X) on a filtered probability space $(\Omega, \mathcal{F}, \mathbb{P}, \{\mathcal{F}_t, t \in [0,T]\})$, such that:

i): B^H is an $\mathcal{F}_t$ -fractional Brownian motion in the sense of Definition 3.1.
ii): X and B^H satisfy (1.1).

Theorem 3.3. *Assume (H). Let X be a weak solution of Equation (1.1). Then the following estimates hold:*

i) *For every Borel function $h : [0,T] \times \mathbb{R} \to \mathbb{R}_+$*

$$\mathbb{E} \int_0^T h(t, X_t) dt \leq C \|h\|_{q,\alpha}, \tag{3.8}$$

for all $q > 1$ and $\alpha > \frac{q}{q-H}$, where C is a constant depending on K, p, β, q, α and $\|F\|_{p,\beta}$.

ii) *There exists a real analytic function A such that for every Borel function $h : [0,T] \times \mathbb{R} \to \mathbb{R}_+$*

$$\mathbb{E} \exp \left(\int_0^T h(t, X_t) dt \right) \leq A \left(\|h\|_{q,\alpha} \right),$$

for every $q > 1$ and $\alpha > \frac{q}{q-H}$.

In order to prove this theorem we will show some technical lemmas.

Lemma 3.4. *Let $B^H = \{B^H, t \in [0,T]\}$ be an $\mathcal{F}_t$-fractional Brownian motion. For every Borel function $h : [0,T] \times \mathbb{R} \to \mathbb{R}_+$ we have*

$$\mathbb{E} \left(\int_{t_0}^T h(t, B_t^H) dt | \mathcal{F}_{t_0} \right) \leq C_1 \left(\int_{t_0}^T \left(\int_{\mathbb{R}} h(t,x)^q dx \right)^{\frac{\alpha}{q}} dt \right)^{\frac{1}{\alpha}},$$

for all $q > 1$, $\alpha > \frac{q}{q-H}$, $t_0 \in [0,T]$ and with a constant $C_1 > 0$ depending only on q, α, H and T.

Proof. Write

$$B_t^H = \int_{t_0}^t K_H(t,s) dW_s + \int_0^{t_0} K_H(t,s) dW_s.$$

Note that $\int_{t_0}^t K_H(t,s) dW_s$ is a Gaussian random variable independent of $\mathcal{F}_{t_0}$ and with variance

$$\int_{t_0}^t K_H(t,s)^2 ds \geq C_H (t - t_0)^{2H},$$

where $C_H > 0$ is a constant depending on H. On the other hand, $\int_0^{t_0} K_H(t,s)dW_s$ is $\mathcal{F}_{t_0}$-measurable. By easy computations we obtain

$$\mathbb{E}\left(\int_{t_0}^{T} h(t, B_t^H)dt|\mathcal{F}_{t_0}\right) \leq C\int_{t_0}^{T}(t-t_0)^{-\frac{H}{q}}\left(\int_{\mathbb{R}} h(t,x)^q dx\right)^{\frac{1}{q}} dt,$$

where C depends on H and q. The result follows from Hölder inequality. □

Lemma 3.5. *The estimates i) and ii) of Theorem 3.3 are true for $X = B^H$.*

Proof. First notice that i) follows from Lemma 3.4 with $t_0 = 0$. To show ii) we use a method similar to that of Khasminski (see also [6]).

$$\mathbb{E}\exp\left(\int_0^T h(t, B_t^H)dt\right) = 1 + \sum_{n=1}^{\infty} I_n,$$

where

$$I_n = \mathbb{E}\left(\int_0^T \int_{t_1}^T \cdots \int_{t_{n-1}}^T h(t_1, B_{t_1}^H)\cdots h(t_n, B_{t_n}^H)dt_n\cdots dt_1\right).$$

Using Lemma 3.4 we get

$$I_n \leq C_1^n\,(n!)^{-\frac{1}{\alpha}}\left(\left(\int_0^T \left(\int_{\mathbb{R}} |h(t,x)|^q\,dx\right)^{\frac{\alpha}{q}} dt\right)^{\frac{n}{\alpha}}\right),$$

hence the result is true with the analytic function

$$A(z) = 1 + \sum_{n=1}^{\infty} C_1^n\,(n!)^{-\frac{1}{\alpha}}\,z^n. \tag{3.9}$$

□

Lemma 3.6. *Assume (H) and let X be a weak solution of Equation* (1.1)*. Set*

$$v_s = K_H^{-1}\left(\int_0^{\cdot} b(r, X_r)dr\right)(s). \tag{3.10}$$

Then

$$\xi_T = \exp\left(-\int_0^T v_s dW_s - \frac{1}{2}\int_0^T v_s^2 ds\right) \tag{3.11}$$

defines a random variable such that the measure $\widetilde{\mathbb{P}}$ given by $d\widetilde{\mathbb{P}} = \xi_T d\mathbb{P}$ is a probability measure equivalent to $\mathbb{P}$. Moreover, for every $\theta \in \mathbb{R}$ there exists a constant K_θ depending on θ, T, K, H, p, β and on $\|F\|_{p,\beta}$ such that

$$\mathbb{E}^{\widetilde{\mathbb{P}}}\xi_T^\theta + \mathbb{E}^{\mathbb{P}}\xi_T^\theta \leq K_\theta. \tag{3.12}$$

Proof. Assume first that b is bounded. From the results of [9] we know that for all $\theta \in \mathbb{R}$,

$$\mathbb{E}^{\widetilde{\mathbb{P}}} \exp\left(-2\theta \int_0^T v_s dW_s - 2\theta^2 \int_0^T v_s^2 ds\right) = 1$$

and $X - x_0$ is a fractional Brownian motion with the Hurst parameter H under $\widetilde{\mathbb{P}}$. Hence,

$$\begin{aligned}
\mathbb{E}^{\widetilde{\mathbb{P}}} \xi_T^\theta &= \mathbb{E}^{\widetilde{\mathbb{P}}} \exp\left(-\theta \int_0^T v_s dW_s - \frac{\theta}{2} \int_0^T v_s^2 ds\right) \\
&= \mathbb{E}^{\widetilde{\mathbb{P}}} \left(\exp\left(-\theta \int_0^T v_s dW_s - \theta^2 \int_0^T v_s^2 ds\right) \exp\left(\left(\theta^2 + \frac{\theta}{2}\right) \int_0^T v_s^2 ds\right)\right) \\
&\le \left(\mathbb{E}^{\widetilde{\mathbb{P}}} \exp 2\left|\theta^2 + \frac{\theta}{2}\right| \int_0^T v_s^2 ds\right)^{\frac{1}{2}}.
\end{aligned}$$

On the other hand, using (3.7) we get

$$\begin{aligned}
|v_s| &= \left| s^{H-\frac{1}{2}} I_{0+}^{\frac{1}{2}-H} s^{\frac{1}{2}-H} b(s, X_s)\right| \\
&\le \frac{1}{\Gamma(\frac{1}{2}-H)} s^{H-\frac{1}{2}} \int_0^s (s-r)^{-\frac{1}{2}-H} r^{\frac{1}{2}-H} |b(r, X_r)| dr \\
&\le C_2 + C_3 s^{H-\frac{1}{2}} \int_0^s (s-r)^{-\frac{1}{2}-H} r^{\frac{1}{2}-H} F(r, X_r)^{\frac{1}{2}} dr,
\end{aligned}$$

where C_2 and C_3 depend on K, H and T. Then, we have

$$\begin{aligned}
\int_0^T v_s^2 ds &\le 2C_2^2 T \\
&\quad + 2C_3^2 \int_0^T s^{2H-1} \left(\int_0^s (s-r)^{-\frac{1}{2}-H} r^{\frac{1}{2}-H} F(r, X_r)^{\frac{1}{2}} dr\right)^2 ds \\
&\le C_4 + C_5 \int_0^T F(r, X_r) dr, \qquad (3.13)
\end{aligned}$$

for some constants C_4 and C_5 depending on K, H and T. Therefore, applying Lemma 3.5, we obtain

$$\mathbb{E}^{\widetilde{\mathbb{P}}} \xi_T^\theta \le \exp\left(2C_4 \left|\theta^2 + \frac{\theta}{2}\right|\right) \left(A\left(2C_5 \left|\theta^2 + \frac{\theta}{2}\right| \|F\|_{p,\beta}\right)\right)^{\frac{1}{2}},$$

where A is the analytic function defined in (3.9). Notice also that

$$\mathbb{E}^{\mathbb{P}} \xi_T^\theta = \mathbb{E}^{\widetilde{\mathbb{P}}} \xi_T^{\theta - 1}.$$

So, the proof of (3.12) is finished in the case of a bounded drift. In order to deduce the estimate (3.12) in the general case it suffices to set $b_n = b 1_{|f| \le n}$ and let n tend to infinity. □

Proof of Theorem 3.3. Applying Hölder's inequality like in page 222 of [6] we can write for any $\delta > 1$

$$\mathbb{E}^{\mathbb{P}} \int_0^T h(t, X_t)dt \leq T^{\delta-1} \left(\mathbb{E}^{\widetilde{\mathbb{P}}} (\xi_T)^{-\frac{\delta}{\delta-1}}\right)^{\frac{\delta-1}{\delta}} \left(\mathbb{E}^{\widetilde{\mathbb{P}}} \int_0^T h(t, X_t)^{\delta} dt\right)^{\frac{1}{\delta}},$$

$$\mathbb{E}^{\mathbb{P}} \exp \int_0^T h(t, X_t)dt \leq \left(\mathbb{E}^{\widetilde{\mathbb{P}}} (\xi_T)^{-\frac{\delta}{\delta-1}}\right)^{\frac{\delta-1}{\delta}} \left(\mathbb{E}^{\widetilde{\mathbb{P}}} \exp \delta \int_0^T h(t, X_t)^{\delta} dt\right)^{\frac{1}{\delta}}.$$

We know that $X - x_0$ is a fractional Brownian motion with the Hurst parameter H under $\widetilde{\mathbb{P}}$. Taking δ close to 1 and applying Lemma 3.5 the result follows. □

3.3. Existence of a weak solution

Theorem 3.7. *Suppose that* $b(t, x)$ *satisfies Hypothesis (H). Then Equation* (1.1) *has a weak solution.*

Proof. Set $\widetilde{B}_t^H = B_t^H - \int_0^t b(s, B_s^H + x_0)ds$. We claim that the process $u_s = -b(B_s^H + x_0)$ satisfies conditions i) and ii) of Theorem 3.2. If this claim is true, under the probability measure $\widetilde{\mathbb{P}}$, $\widetilde{B}^H$ is an $\mathcal{F}_t^{B^H}$-fractional Brownian motion, and $\left(B^H, \widetilde{B}^H\right)$ is a weak solution of (1.1) on the filtered probability space

$$(\Omega, \mathcal{F}, \widetilde{\mathbb{P}}, \left\{\mathcal{F}_t^{B^H}, t \in [0, T]\right\}).$$

Set

$$v_t = K_H^{-1} \left(\int_0^{\cdot} b(s, B_s^H + x_0)ds\right)(t).$$

From (3.13) we obtain

$$\int_0^T v_s^2 ds \leq C_4 + C_5 \int_0^T F(s, B_s^H + x_0)ds. \tag{3.14}$$

Notice that from (3.7) it follows that the operator K_H^{-1} preserves the adaptability property. Hence, the process v_r is adapted and condition ii) can be proved using Novikov criterion. Indeed, it suffices to show that for some $\lambda > 1$

$$E\left(\exp\left(\lambda \int_0^T v_s^2 ds\right)\right) < \infty, \tag{3.15}$$

which is an immediate consequence of (3.14) and Lemma 3.5. □

3.4. Uniqueness in law and pathwise uniqueness

Let (X, B^H) be a weak solution of the stochastic differential equation (1.1) defined in the filtered probability space $(\Omega, \mathcal{F}, \mathbb{P}, \{\mathcal{F}_t, t \in [0, T]\})$. Define v_s and $\widetilde{\mathbb{P}}$ by (3.10) and (3.11), respectively. By Theorem 3.3 the process v_s satisfies conditions i) and ii) of Theorem 3.2. By the classical Girsanov theorem, the process

$$\widetilde{W}_t = W_t + \int_0^t v_s dr$$

is an $\mathcal{F}_t$-Brownian motion. In terms of the process $\widetilde{W}_t$ we can write

$$X_t = x_0 + \int_0^t K_H(t,s)d\widetilde{W}_s.$$

Hence, then $X - x_0$ is an $\mathcal{F}_t$-fractional Brownian motion with respect to the probability $\widetilde{\mathbb{P}}$ with Hurst parameter equal to H. As a consequence, the processes $X - x_0$ and $\widetilde{B}_t^H$ have the same distribution under the probability $\mathbb{P}$. In conclusion we have proved the following result:

Theorem 3.8. *Suppose that $b(t,x)$ satisfies the assumptions of Theorem* 3.7. *Then any weak solution has the same distribution.*

As a corollary we deduce the pathwise uniqueness of the solution to Equation (1.1):

Theorem 3.9. *Suppose that $b(t,x)$ satisfies the assumptions of Theorem* 3.7. *Then two weak solutions defined on the space filtered probability space must coincide almost surely.*

Proof. Let X^1 and X^2 two weak solutions defined on the same filtered probability space $(\Omega, \mathcal{F}, P, \{\mathcal{F}_t, t \in [0,T]\})$ with respect to the same fractional Brownian motion. It is easy to see that $\sup(X^1, X^2)$ and $\inf(X^1, X^2)$ are also solutions, then they have the same laws which implies that $X^1 = X^2$. □

4. Existence of strong solutions

We first show the following convergence result.

Proposition 4.1. *Consider a sequence $b_n(t,x)$ of measurable functions satisfying*

$$b_n(t,x)^2 \leq C + F(t,x), \tag{4.1}$$

where $F \in L_{p,\beta}$ for some $p > 1$ and $\beta > \frac{p}{p-H}$, and

$$\lim_{n\to\infty} b_n(t,x) = b(t,x)$$

for almost all $(t,x) \in [0,T] \times \mathbb{R}$. Suppose also that the processes $X_t^{(n)}$ satisfy

$$X_t^{(n)} = x_0 + B_t^H + \int_0^t b_n(s, X_s^{(n)})ds, \; 0 \leq t \leq T,$$

and converge a.s. to some process X_t for all $t \in [0,T]$. Then the process X_t is a solution of Equation (1.1).

Proof. It suffices to show that

$$\lim_n \mathbb{E} \int_0^T \left| b_n(s, X_s^{(n)}) - b(s, X_s) \right| ds = 0.$$

We can write

$$J(n) := \mathbb{E}\int_0^T \left| b_n(s, X_s^{(n)}) - b(s, X_s)\right| ds \leq J_1(n) + J_2(n),$$

where

$$J_1(n) := \sup_k \mathbb{E}\int_0^T \left| b_k(s, X_s^{(n)}) - b_k(s, X_s)\right| ds$$

and

$$J_2(n) := \mathbb{E}\int_0^T |b_n(s, X_s) - b(s, X_s)|\, ds.$$

Let $\kappa : \mathbb{R} \to \mathbb{R}$ be a smooth function such that $0 \leq \kappa(z) \leq 1$ for every z, $\kappa(z) = 0$ for $|z| \geq 1$ and $\kappa(0) = 1$. Fix $\varepsilon > 0$ and choose $R > 0$ such that

$$\left(\mathbb{E}\int_0^T [1 - \kappa(X_t/R)]^\delta \, dt\right)^{\frac{1}{\delta}} < \varepsilon, \tag{4.2}$$

where $\delta > 1$ will be chosen later. The sequence of functions b_k is relatively compact in $L_{p,\beta}([0,T] \times [-R, R])$. Hence, we can find finitely many bounded smooth functions $H_1, \ldots, H_N$ such that for every k

$$\left(\int_0^T \left(\int_{-R}^R |b_k(t,x) - H_i(t,x)|^p \, dr\right)^{\frac{\beta}{p}} dt\right)^{\frac{1}{\beta}} < \varepsilon \tag{4.3}$$

for some H_i. We have

$$\begin{aligned}
\mathbb{E}\int_0^T \left| b_k(t, X_t^{(n)}) - b_k(t, X_t)\right| dt \;&\leq\; \mathbb{E}\int_0^T \left| b_k(t, X_t^{(n)}) - H_i(t, X_t^{(n)})\right| dt \\
&\quad + \sum_{j=1}^N \mathbb{E}\int_0^T \left| H_j(t, X_t^{(n)}) - H_j(t, X_t)\right| dt \\
&\quad + \mathbb{E}\int_0^T |b_k(t, X_t) - H_i(t, X_t)|\, dt \\
:&= I_1(n,k) + I_2(n) + I_3(k).
\end{aligned}$$

By Theorem 3.3 and (4.3)

$$\begin{aligned}
I_1(n,k) &= \mathbb{E}\int_0^T \kappa(X_t^{(n)}/R) \left| b_k(t, X_t^{(n)}) - H_i(t, X_t^{(n)})\right| dt \\
&\quad + \mathbb{E}\int_0^T \left[1 - \kappa(X_t^{(n)}/R)\right] \left| b_k(t, X_t^{(n)}) - H_i(t, X_t^{(n)})\right| dt \\
&\leq C\varepsilon + L\mathbb{E}\int_0^T \left[1 - \kappa(X_t^{(n)}/R)\right] \\
&\quad + \left(\mathbb{E}\int_0^T \left[1 - \kappa(X_t^{(n)}/R)\right]^\delta dt\right)^{\frac{1}{\delta}} \left(\mathbb{E}\int_0^T \left| F(t, X_t^{(n)})\right|^\gamma dt\right)^{\frac{1}{\gamma}},
\end{aligned}$$

where L is a positive constants, C is the constant from (3.8) and $\frac{1}{\delta}+\frac{1}{\gamma}=1$. Hence, choosing γ sufficiently close to one we get

$$I_1(n,k) \le C\varepsilon + L\mathbb{E}\int_0^T \left[1-\kappa(X_t^{(n)}/R)\right] + M\left(\mathbb{E}\int_0^T \left[1-\kappa(X_t^{(n)}/R)\right]^\delta dt\right)^{\frac{1}{\delta}}$$

for some positive constant M. Hence, using (4.2) we get

$$\overline{\lim}_{n\to\infty} \sup_k I_1(n,k) \le (C+L+M)\varepsilon.$$

Similarly,

$$\sup_k I_3(k) \le (C+L+M)\varepsilon.$$

Consequently,

$$\overline{\lim}_{n\to\infty} \sup_k I(n,k) \le 2(C+L+M)\varepsilon,$$

and this implies that $\lim_{n\to\infty} J_1(n) = 0$. For the term $J_2(n)$ we can use the same arguments as before. □

Theorem 4.2. *Assume that $b(t,x)$ satisfies the growth condition (H). Then, there exists a unique strong solution to equation* (1.1).

Proof. We already know that pathwise uniqueness holds. For any $n > 0$ define $b_n(t,x) = (b(t,x) \vee (-n)) \wedge n$. It is clear that

$$b_n(t,x)^2 \le C + F(t,x).$$

where $F \in L_{p,\beta}$ for some $p > 1$ and $\beta > \frac{p}{p-H}$, and

$$\lim_{n\to\infty} b_n(t,x) = b(t,x).$$

Note that b_n is a bounded measurable function.

By the result of Nualart and Ouknine [9], the equation

$$X_t^{(n)} = x + B_t^H + \int_0^t b_n(s, X_s^{(n)})ds, \ 0 \le t \le T,$$

has a unique strong solution. But b_n increases and the comparison theorem implies that $X_s^{(n)}$ increases a.s. to some process X_t for all $t \in [0,T]$.

We deduce from proposition 4.1 that the process X_t is a solution of Equation (1.1). □

Acknowledgment

The work was carried out during a stay of Youssef Ouknine at the IMUB (Institut de Matemàtica de la Universitat de Barcelona). He would like to thank the IMUB for hospitality and support.

References

[1] E. Alòs, O. Mazet and D. Nualart: Stochastic calculus with respect to Gaussian processes. *Annals of Probability* **29** (2001), 766–801.

[2] L. Decreusefond and A. S. Üstunel: Stochastic Analysis of the fractional Brownian Motion. *Potential Analysis* **10** (1999), 177–214.

[3] X. M. Fernique: Regularité des trajectories de fonctions aléatoires gaussiennes. In École d'Été de Saint-Flour IV (1974), *Lecture Notes in Mathematics* **480**, 2–95.

[4] I. Gyöngy and N. Krylov: Existence of strong solutions for Itô's stochastic equations via approximations *Probab. Theory Rel. Fields* **105** (1996), 143–158.

[5] I. Gyöngy and E. Pardoux: On quasi-linear stochastic partial differential equations. *Probab. Theory Rel. Fields* **94** (1993) 413–425.

[6] I. Gyöngy and E. Pardoux: On the regularization effect of space-time white noise on quasi-linear parabolic partial differential equations. *Probab. Theory Rel. Fields* **97** (1993) 211–229.

[7] S. Moret and D. Nualart: Onsager-Machlup functional for the fractional Brownian motion. *Probab. Theory Rel. Fields* **124** (2002) 227–260.

[8] S. Nakao: On the pathwise uniqueness of solutions of one-dimensional stochastic differential equations. *Osaka J. Math.* **9** (1972) 513–518.

[9] D. Nualart and Y. Ouknine Regularization of differential equations by fractional noise. *Stoch Proc. Appl.* **102** (2002) 103–116.

[10] I. Norros, E. Valkeila and J. Virtamo: An elementary approach to a Girsanov formula and other analytical results on fractional Brownian motion. *Bernoulli*

[11] Y. Ouknine: Généralisation d'un Lemme de S. Nakao et Applications. *Stochastics* **23** (1988) 149–157.

[12] S. G. Samko, A. A. Kilbas and O. I. Mariachev: Fractional integrals and derivatives (Gordon and Breach Science 1993).

[13] A. Ju. Veretennikov: On strong solutions and explicit formulas for solutions of stochastic integral equations. *Math. USSR Sb.* **39** (1981) 387–403.

[14] A. K. Zvonkin: A transformation of the phase space of a diffusion process that removes the drift. *Math. USSR Sb.* **22** (1974) 129–149.

Facultat de Matemàtiques, Universitat de Barcelona, Gran Via, 585,
08007, Barcelona, Spain
E-mail address: `nualart@mat.ub.es`

Faculté des Sciences Semlalia, Département de Mathématiques, Université Cadi Ayyad, BP 2390, Marrakech, Maroc
E-mail address: `ouknine@ucam.ac.ma`

MIX
Papier aus verantwortungsvollen Quellen
Paper from responsible sources
FSC® C105338

If you have any concerns about our products,
you can contact us on
ProductSafety@springernature.com

In case Publisher is established outside the EU,
the EU authorized representative is:
Springer Nature Customer Service Center GmbH
Europaplatz 3, 69115 Heidelberg, Germany

Printed by Libri Plureos GmbH
in Hamburg, Germany